Parasitoids

MONOGRAPHS IN
BEHAVIOR AND ECOLOGY

Edited by John R. Krebs and
Tim Clutton-Brock

Parasitoids

Behavioral and Evolutionary Ecology

H.C.J. GODFRAY

Princeton University Press
Princeton, New Jersey

Copyright © 1994 by Princeton University Press
Published by Princeton University Press,
41 William Street, Princeton, New Jersey 08540
In the United Kingdom: Princeton University Press,
Chichester, West Sussex

Library of Congress Cataloging-in-Publication Data
Godfray, H. C. J., 1958–
Parasitoids : behavioral and evolutionary ecology /
H.C.J. Godfray
    p.   cm. — (Monographs in behavior and ecology)
Includes bibliographical references and indexes.
ISBN 0-691-03325-0 (cloth) — ISBN 0-691-00047-6 (pbk.)
1. Parasitic insects—Behavior.   2. Insects—Parasites.
3. Behavior evolution.   I. Title.   II. Series.
QL496.G59 1993
595.7′053—dc20   93-13158

This book has been composed in Times Roman

Princeton University Press books
are printed on acid-free paper, and meet the guidelines
for permanence and durability of the Committee on
Production Guidelines for Book Longevity of the
Council on Library Resources

Printed in the United States of America

10 9 8 7 6 5 4 3 2 1
10 9 8 7 6 5 4 3 2 1 (pbk.)

This book is dedicated to Caroline Elmslie

Contents

Acknowledgments

I am particularly grateful to Jeff Waage and Mike Hassell, who first introduced me to the evolutionary biology and population dynamics of parasitoids. I have also learned an enormous amount about parasitoids from Jacques van Alphen, Dick Askew, Ian Gauld, Mark Shaw, Mike Strand, and Jack Werren. Many people have read and commented critically on one or more chapters, or have shown me their unpublished research. They include Jacques van Alphen, Robert Belshaw, Carlos Bernstein, Man Suen Chan, James Cook, Ian Gauld, Ian Hardy, Mike Hassell, Brad Hawkins, Lex Kraaijeveld, John Lawton, Jane Memmott, Mark Ridley, Richard Stouthamer, Mike Strand, Clare Towner, Ted Turlings, Louise Vet, Jeff Waage, Jack Werren, and Bob Wharton. Kate Lessells, Molly Hunter, and Marcel Visser read the whole book in draft. I would also like to thank Jacques van Alphen, Ric Charnov, Alan Grafen, Paul Harvey, Bill Hamilton, Mike Hassell, Alisdair Houston, Molly Hunter, Tony Ives, Kate Lessells, Alex Kacelnik, Geoff Parker, Marcel Visser, Jeff Waage, and Jack Werren for valuable discussion and advice on the application of population biology theory to parasitoids. I am grateful to David Agassiz, Paul Desert, Nigel Ferguson, John LaSalle, John Noyes, Mark Shaw, Henk Vlug, Bob Wharton, and Mike Wilson for advice on taxonomic problems. The editors of the series, John Krebs and Tim Clutton-Brock, provided constant encouragement. Finally, I am very grateful to David Nash for the line drawings, and to Alice Calaprice for the careful editing.

I AM ALSO very grateful to the following publishers for permission to reproduce figures from published papers: Plenum Publishing Corporation (figs. 2.4, 2.5, 8.4); Blackwell Scientific Publications (figs. 2.8, 2.10–2.15, 2.17, 3.1, 3.4, 3.6, 4.2–4.5, 4.11, 7.1, 7.3); *Oecologia* (figs. 3.7, 4.12); E. J. Brill (fig. 3.11); Academic Press Ltd. (figs. 3.13, 6.1); The Society for the Study of Evolution (figs. 4.4, 4.8, 4.10, 8.1, 8.3); Macmillan Science Magazines Ltd. (figs. 4.6, 4.14, 4.15, 6.1); The American Association for the Advancement of Science (figs. 4.7, 4.9, 8.8); *Entomophaga* (fig. 5.1); University of Chicago Press (fig. 8.2).

Parasitoids

I cannot persuade myself that a beneficent and omnipotent God would have designedly created the Ichneumonidæ with the express intention of their feeding within the living bodies of Caterpillars, or that a cat should play with mice. Not believing this, I see no necessity in the belief that the eye was expressedly designed. On the other hand, I cannot anyhow be contented to view this wonderful universe, and especially the nature of man, and to conclude that everything is the result of brute force. I am inclined to look at everything as resulting from designed laws, with the details, whether good or bad, left to the working out of what we may call chance. Not that this notion at all satisfies me. I feel most deeply that the whole subject is too profound for the human intellect. A dog might as well speculate on the mind of Newton. Let each man hope and believe what he can.

—Charles Darwin
22 May 1860, Letter to Asa Gray

1   Introduction

Alien, a very popular motion picture of 1979, described the fate of the occupants of a spaceship infiltrated by a deadly alien life form. Larvae of these creatures entered the bodies of the crew where they developed and grew, their eventual emergence resulting in the (spectacular) death of the infected human. Why did this film achieve such widespread popularity (as I write, *Aliens 3* is about to be released)? An imaginative script, an attractive star, and the survival of the ship's cat certainly helped, but possibly also the assurance that the events depicted could not happen on earth. This is certainly true for humans: many of us are plagued by parasites, but by nothing quite as gruesome as this. Yet the creature in *Alien* is immediately recognizable as a parasitoid—specifically a primary, solitary, endoparasitoid with a planidial larva—differing in minor detail (no DNA, silicon-based biochemistry) from thousands, possibly millions, of species of insects that attack other arthropods in nearly every terrestrial ecosystem. This book is about the behavioral and evolutionary ecology of earth-based parasitoids.[1]

Parasitoids are insects whose larvae develop by feeding on the bodies of other arthropods, usually insects. Larval feeding results in the death of the parasitoid's host. Although the natural history and identity of parasitoids are little known among nonbiologists, they are of immense importance in natural and agricultural ecosystems where they influence or regulate the population density of many of their hosts. Much research on parasitoids has been stimulated by their frequent success in biological control programs; many species have been released to combat agricultural pests, and while effective control is by no means assured, huge savings, both in financial and human terms, have resulted from successful programs.

Research on the parasitoids of agricultural and other pests has generated a huge amount of information on the behavior and ecology of many different species. In addition there is a large and increasing number of studies of the fundamental biology of parasitoids. In this book, I attempt to review recent research on parasitoid behavioral and evolutionary ecology. I aim to show that parasitoids provide marvelous systems for investigating outstanding problems in behavioral and evolutionary ecology, and that fundamental research can

---

[1] Thus I exclude discussion of the mutational response of the braconid wasp *Bracon hebetor* (=*Habrobracon juglandis*) in the Biosatellite II experiment (von Borstel et al. 1968).

illuminate many aspects of parasitoid biology that are important to applied entomologists.

My approach to many of the evolutionary questions discussed in this book is that of modern behavioral ecology. This research program is often the subject of criticism, both fair and unfair, and a brief word is needed to explain my use of the method. Natural selection is an optimization process that tends to maximize the efficiency with which genes are transmitted to future generations. This property of natural selection can be used to make predictions about the morphology, physiology, or behavior of animals or plants. The behavioral ecologist attempts to understand how animal behavior interacts with the environment to determine fitness. He or she tries to distinguish the behavioral options available to the animal (the strategy set), the consequences for the animal of adopting different strategies, and how the consequences translate into Darwinian fitness. These hypotheses constitute a model of animal behavior, and the optimizing property of natural selection is used to make the prediction that the behavior observed in the field is that which maximizes fitness. What is at test is not the assumption that the animal is behaving optimally—that is axiomatic—but the model of animal behavior.

Use of the behavioral ecological method has led to great insights into animal behavior, and I hope to convince the reader that some of the best examples of this technique are provided by studies of parasitoids. Nevertheless, there are potential pitfalls. Problems begin when animals fail to conform to predicted behavior. There are at least two potential explanations for this failure. First, the model of animal behavior may be incorrect. The manner in which the test failed often provides useful information about important aspects of the animal's biology that have been omitted from the model. Typically, the model is revised, and a new prediction made; ideally, the revised prediction is tested with a new experiment. Successive iteration can be a valuable way of dissecting the functional significance of animal behavior; however, there is also a danger of overinterpretation—of making a posteriori modifications to the model to force it to fit the facts. It is important to distinguish between new ideas stimulated by a failed test, and hypotheses that have been subject to independent experiments. The second explanation for a failed test is that the underlying assumption that the animal is behaving optimally is untrue. Perhaps the animal has not yet had time to adapt to the environment, or perhaps the mechanics of the genetic process prevent sufficiently good adaptation. There are no general solutions to these problems, and biologists working within the behavioral ecological research program must be constantly aware of these potential difficulties.

A recurring problem in behavioral ecology is understanding how different behaviors affect Darwinian fitness. Typically, a surrogate measure of fitness is used; for example, in studies of foraging behavior, it is often assumed that the foraging strategy that maximizes the rate of energy intake also maximizes fitness. The study of many aspects of parasitoid reproductive strategy is simplified by a very direct link between behavior and fitness: the consequences of failing to find a host, or making an incorrect oviposition decision after locating a host, are obvious and relatively easy to measure. This simplicity makes parasitoids an important model system in the development of behavioral ecological methods.

In my discussion of behavioral ecological hypotheses, I sometimes use an informal shorthand and write, for example, that a parasitoid seeks to locate as many hosts as possible, or to lay a clutch size that maximizes the number of offspring that can develop on a host. Such phrases do not of course imply any conscious motivation or calculation on the part of the parasitoid, but just avoid the constant repetition of long sentences detailing precisely how natural selection is assumed to maximize fitness.

Many aspects of parasitoid behavior have both an evolutionary and a mechanistic explanation. Consider the question of the relationship between clutch size and host size in gregarious parasitoids. In chapter 3 I describe behavioral ecological models designed to predict the optimal clutch size on hosts of different size. These models attempt to provide an ultimate or evolutionary explanation for clutch size behavior. At a different level, questions can be asked about the behavioral mechanisms responsible for clutch size. For example, what properties of the host cause the parasitoid to lay a particular number of eggs? It is important not to fall into the trap of treating answers to evolutionary "why" questions and mechanistic "how" questions as alternatives: they are instead complementary. Indeed, one of the most interesting challenges in modern behavioral ecology is to dissect the behavioral rules that allow an animal to pursue behavioral strategies favored by natural selection. Work with parasitoids is likely to be important in exploring this problem, both because of the straightforward link between behavior and fitness, and also because many oviposition decisions faced by parasitoids are relatively simple and amenable to experimental manipulation. There are of course other levels of questions that might be asked about behavior in addition to the evolutionary and mechanistic—for example, questions about the neurological or hormonal basis of behavior.

The rest of the chapter is organized as follows. The next section deals with different definitions of the term "parasitoid" while the following reviews the main features and variants of parasitoid life histories. The study of parasitoids has generated its fair share of specialist terminology, and although an attempt has been made to use as little jargon as possible, a minimal set is introduced in this section. The third section describes the natural history of a few more specialized forms of parasitoids that will be referred to later in the book. The fourth section comprises a brief overview of parasitoid taxonomy, while the fifth section discusses the origins of the parasitoid habit and describes the natural history of groups that have evolved from parasitoids.

1.1 Parasitoid Definitions

A parasitoid is defined by the feeding habit of its larva. The larva feeds exclusively on the body of another arthropod, its host, eventually killing it. Only a single host is required for the parasitoid to complete development, and often a number of parasitoids develop gregariously on the same host. In many ways parasitoids are intermediate between predators and parasites: like predators, they always kill the host they attack; like many parasites, they require just a single host on which to mature. After attacking a host, the female parasitoid does not attempt to move the host to a prepared cache or nest. This distinguishes parasitoids from some solitary wasps which in other respects they closely resemble. The life cycle of all parasitoids can be divided into four stages—egg, larva, pupa, and adult—in other words, they belong to the holometabolous insect orders.

Although the term "parasitoid" was introduced by Reuter in 1913, it is only in the last twenty years that it has become universally accepted. Before that, parasitoids were most commonly referred to as insect parasites. Understandably perhaps, the clumsy words "parasitoidize" and "parasitoidism" have never found favor, and I shall follow normal practice and use "parasitize" and "parasitism." The term "protelean parasite" is sometimes used both to refer to parasitoids and to insects that are true parasites in their larval stage (e.g., Askew 1971). Flanders (1973) attempted, without success, to replace "parasitoid" with "carniveroid."

Several authors have attempted to expand parasitoid to include other organisms with related life histories. Eggleton and Gaston (1990) argued that the term should be used for all organisms that complete their development on, and then kill, a single animal host. This definition includes solitary wasps, and organisms as diverse as fungi and nematodes. Price (1975), on the other hand, accepts that parasitoids are insects but allows their hosts to be plants; seed weevils (Bruchidae) whose larvae develop on and kill a single seed are thus included in this definition. These authors are making important points about organisms with similar trophic functions and adaptations. Nevertheless, I prefer to retain the term "parasitoid" in its more restricted sense, both because this is the sense in which it is used by most biologists, and because parasitoids, defined in this way, are faced with a multitude of similar biological problems that they have to solve in order to survive.

According to Silvestri (1909), the first published observation of a parasitoid emerging from a host was by U. Aldrovandi in 1602, and the first illustration was that of Johannes Goedaert in 1662. Apparently neither author fully understood his observations. Silvestri credits Antonio Vallisnieri for the first correct interpretation of insect parasitism in 1706, but DeBach (1974) says that van Leeuwenhoek in 1701 correctly described the parasitism of a willow sawfly. John Ray, however, in 1710 published an account of parasitism of white but-

terflies by what is now called *Cotesia (=Apanteles) glomeratus* based on observations made in 1658 (Mickel 1973; Shaw 1981b). Whoever may have first
understood the parasitoid life cycle, its true nature became widely known in
entomological circles in the first half of the eighteenth century.

1.2 Parasitoid Natural History

The free-living adults of parasitoids generally look much like their closest
nonparasitoid relatives (fig. 1.1). Hosts are usually located by the adult female
who lays her eggs either directly on the host or in its immediate vicinity.
Hymenopteran parasitoids have highly specialized ovipositors which are used
both to manipulate eggs and to sting the host. The sting causes paralysis that
may be permanent, or the host may recover and continue feeding. Parasitoids
that attack concealed hosts often have long ovipositors which, when not in use,
either extend beyond the end of the abdomen enclosed between protective
valves, or are coiled inside the abdomen of the female. Cutting ridges at the
end of the ovipositor allow wasps to drill through plant tissue and even wood
to locate hidden hosts. In exceptional cases, the ovipositor can be eight times
the length of the rest of the body (Askew 1971). Special adaptations are needed
to allow the egg to pass down long and thin ovipositors; often the egg is very
small and expands enormously within the host's body. In a number of hymenopteran and dipteran parasitoids the whole abdomen is laterally or dorso-
ventrally compressed so that it can be slid into narrow openings, for example
to locate hosts between the gill slits of fungi. Adult parasitoids may feed from
flowers, sap fluxes, and other energy sources, and many also feed on potential
hosts (*host feeding*).

In some cases, the adult female does not lay her eggs on the host but on the
host's foodplant. Parasitism occurs if the host eats the eggs. There are also
some parasitoids that lay their eggs away from the host but which have active
free-living first instar larvae that are responsible for host location (sec. 2.2.5).

The hosts of parasitoids are almost exclusively insects themselves, although
spiders and even centipedes are occasionally parasitized. The juvenile stages
of insects are most frequently attacked, although a few groups attack adult
insects. The parasitoids of holometabolous insects such as Lepidoptera, beetles, and flies can be classified by the stage they attack. Thus a host may suffer
attack from *egg parasitoids, larval parasitoids, pupal parasitoids*, or *adult
parasitoids.* Some parasitoids lay eggs in one host stage but their progeny do
not kill the host until it has entered a later stage, for example *egg-larval* and
larval-pupal parasitoids. Hemimetabolous insects (with no pupal stage) are
also attacked by egg parasitoids, but there is less of a distinction between
parasitoids attacking nymphal and adult stages.

Parasitoids can be divided into two classes by the feeding behavior of their
larvae. Some parasitoids develop within the body of their host, feeding from

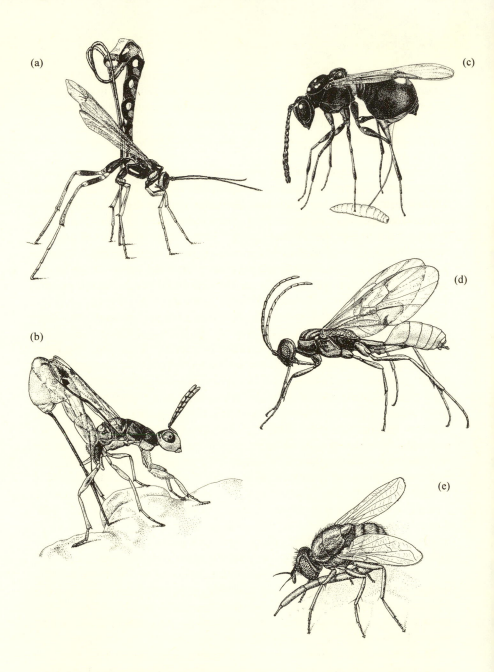

Figure 1.1 The adults of some of the main taxonomic groups of parasitoids. Insects (a)–(d) are Hymenoptera, and (e) is Diptera: (a) *Megarhyssa* sp. (Ichneumonoidea: Ichneumonidae) ovipositing into a concealed larva in wood; after a photograph in Gauld and Bolton 1988; (b) *Megastigmus stigmatizans* (Chalcidoidea: Torymidae) ovipositing into a cynipid gall; from a photograph taken by Graham Stone; (c) *Leptopilina clavipes* (Cynipoidea: Eucoilidae) attacking a *Drosophila* larva; from a photograph taken by Marcel Visser; (d) *Xiphyropronia tianmushanensis* (Proctotrupoidea: Roproniidae) after He and Chen 1991; (e) *Nemorilla floralis* (Tachnidae) about to oviposit; after Herting 1960.

the inside, and are known as *endoparasitoids*. *Ectoparasitoids*, on the other hand, live externally, normally with their mouthparts buried in the body of their host. The majority of parasitoids fall naturally into one of these groups though there are a minority of species that spend their first few instars as ectoparasitoids before burrowing into their hosts to become endoparasitoids, or vice versa.

Parasitoids that feed alone on a host are known as *solitary* parasitoids as opposed to *gregarious* parasitoids, where from two to several thousand individuals feed together on a single host. If further eggs are deposited on the host by the same species of parasitoid, *superparasitism* is said to occur. If a second female of a different species lays her eggs on the host, one of two things may happen. If the larvae of the second species compete with the resident larvae for host resources, *multiparasitism* occurs. However, if the larvae of the second species feed, not on the host, but on the parasitoid larvae already present, *hyperparasitism* occurs. Hyperparasitism is generally of two kinds: *facultative hyperparasitoids* are able to attack unparasitized host individuals and only develop as hyperparasitoids when eggs are laid on a previously parasitized host; in contrast *obligate hyperparasitoids* are only able to develop as parasitoids of parasitoids. Hyperparasitoids are often referred to as *secondary parasitoids* and cases are even known of *tertiary parasitoids*. A final, rather rare type of parasitism is *cleptoparasitism*. A cleptoparasitoid obligatorily requires the presence of another species of parasitoid, though, unlike a hyperparasitoid, does not feed on it. Good examples of cleptoparasitism are certain parasitoids of insects living inside dead wood that lack a boring ovipositor and which can only attack previously parasitized insects after the first parasitoid has drilled a hole for oviposition (Spradbery 1969).

Parasitoids that allow hosts to continue to grow in size after parasitism are called *koinobionts* as opposed to *idiobionts*, where the parasitoid larvae must make do with the host resources present at oviposition (Haeselbarth 1979; Askew and Shaw 1986). Egg, pupal, and adult parasitoids are usually idiobionts, as are those larval parasitoids whose sting causes permanent paralysis. The most important types of koinobionts are egg-larval and larval-pupal parasitoids, and those larval parasitoids which do not permanently paralyze their host at oviposition. Koinobiont parasitoids normally suspend their development as first instar larvae while the host continues to feed and grow, or they begin to grow but refrain from feeding on the vital organs of their host.

There is much variation in pupation site among parasitoids. Egg, pupal, or larval-pupal parasitoids usually pupate within the egg or pupa of the host, taking advantage of the relative security of one of the host's more protective stages. Species attacking hosts such as moth larvae or aphids frequently cement the eaten-out husk of the host to the substrate and pupate within this shelter, usually referred to as a *mummy*. Parasitoids of hosts living within galls, mines, or in other concealed habitats often form naked pupae near the remains of the host. Parasitoids of hosts that feed in exposed situations usually pupate within protective cocoons of silk produced by the larvae themselves. The pupae of some wasps hang suspended on a silken thread from a leaf or other substrate. Where parasitoids have several generations a year and overwinter in the pupal stage, the winter cocoon is often considerably thicker and tougher than cocoons made during the summer.

There are several books that provide very good introductions to parasitoid biology. Askew's (1971) excellent book, *Parasitic Insects*, provides a wealth of interesting biological and natural history detail and also deals with true parasites such as fleas, lice, and biting flies. Gauld and Bolton's (1988) *Hymenoptera* surveys the order from a taxonomic viewpoint but includes much biology and natural history. Oldroyd's (1964) *Natural History of Flies* includes a good introduction to fly parasitoids. Though now rather dated, Clausen's (1940a) classic, *Entomophagous Insects*, provides an encyclopedic compendium of parasitoid biology. Finally, the edited volume by Waage and Greathead (1986) contains an important series of reviews on many aspects of parasitoid biology.

1.3 Unusual Life Histories

Several groups of parasitoids with unusual or bizarre life histories provide interesting tests of evolutionary hypotheses. As they will be referred to in several places in the book, their natural history and biology are described here to avoid unnecessary repetition.

1.3.1 POLYEMBRYONIC PARASITOIDS

One of the most spectacular sights in parasitoid biology is the emergence of 2000 small wasps from the eaten-out husk of a moth caterpillar. These wasps have arisen from the asexual division of one or two eggs laid by an adult female into the egg of the moth. This form of asexual division is called *polyembryony* and is known from four families of parasitoid wasps (Ivanova-Kasas 1972), each representing an independent evolutionary event. Outside parasitoids, polyembryony is found rarely and sporadically throughout the animal kingdom, from planarians to armadillos.

Polyembryony in parasitoids was first described by Marchal (1898) in the chalcidoid family Encyrtidae, where it is found in a series of closely related genera (tribe Copidosomatini) of egg-larval parasitoids of Lepidoptera. Recently, Strand and his colleagues have used a variety of modern techniques from physiology and molecular biology to investigate polyembryony in the encyrtid wasp *Copidosoma floridanum*. Here I provide a brief description of polyembryony in *C. floridanum* based on Strand's findings (1989a, 1989b, 1989c, 1992, pers. comm.).

A female *C. floridanum* lays either a single egg (male or female) or two eggs (always one male plus one female) into its host, the egg of a noctuid moth. The moth egg hatches into a larva which develops until its final instar. During this period, a single parasitoid egg can divide to produce up to 1500 separate individuals (3000 in related species). However, the host suffers no major ill effects of parasitism until its final instar, when the parasitoids develop quickly, consume the host, and then pupate within the skin of the exhausted caterpillar.

There are some subtle differences in the development of male and female eggs. Consider female eggs first. The egg initially appears to divide normally to produce a mass of similar-looking cells enclosed within a serosal membrane, derived from the polar bodies. In a normal parasitoid, the egg would then develop through a blastula and then a gastrula stage. In *C. floridanum*, the cell mass divides to produce a number of "daughter" masses, each contained within a membrane, the assemblage bounded by the original serosal membrane. The term "polygerm" is often used to describe the collection of individual cell masses and their associated membranes, while the maturing cell masses are called "morulae." As the host larva ages, the morula-stage embryos proceed through a number of rounds of division, perhaps synchronized by cycles of endocrine hormones associated with the progression of host instars. In the last instar, again triggered by changes in host hormone titres (Strand et al. 1989, 1990, 1991a, 1991b), the now numerous embryos embark on a normal path of development.

Early embryologists noted that a few embryos began development very much earlier than the majority, and that they developed into atypical larvae with relatively large mandibles that eventually died (Silvestri 1906). Cruz (1981) proposed that the function of these larvae was to protect their (genetically identical) siblings from competition from other parasitoids. Female eggs typically produce five to eight defensive larvae during the early stages of morula division. The mechanism that allows only a few embryos to develop precociously is not yet known. Interestingly, artificial elevation of host juvenile hormone titer or starvation of the host increases the number of defensive larvae produced.

The development of male eggs is similar except that instead of developing in the anterior of the host larva, they develop more posteriorly, associated with the host fat body. Male eggs also produce defensive larvae, but far fewer, and only when the host larva is quite large.

Polyembryonic species of other families of wasps do not produce the very large broods found in some encyrtids. The Platygastridae are a rather poorly known family of parasitoid wasps that attack a variety of hosts, especially cecidomyiid gall midges. It is thought that a number of species in the genus *Platygaster* are polyembryonic, though it appears that the female deposits several eggs in each host, not all of which need divide polyembryonically. The maximum number of individuals emerging from a single host is about twenty. A rather large, and sometimes contradictory, early literature on platygasterid development is reviewed by Clausen (1940a); there seem to be no modern investigations. Polyembryony is known from some members of the braconid genus, *Macrocentrus*, egg-larval parasitoids of lepidopterans. Again, an undifferentiated mass of cells divides to form up to about forty embryos (Voukasso-vitch 1927; Parker 1931). In one species, *M. ancylivorus*, development initially proceeds as in polyembryonic species, but when one individual reaches the larval stage, it appears to inhibit further development by its siblings (Daniel 1932). If two individuals develop simultaneously, they fight until just one survives. There is some evidence for a similar form of incipient polyembryony in *Platygaster* (Clausen 1940a). The Dryinidae are parasitoids of planthoppers (Auchenorrhyncha) and one species is known to be polyembryonic: the aberrant *Crovettia (=Aphelopus) theliae* is an endoparasitoid of membracid nymphs and one egg divides to produce up to seventy larvae (Kornhauser 1919).

1.3.2 HETERONOMOUS APHELINIDS

In most parasitoids, male and female progeny develop on the same type of host. Quite often, male eggs tend to be laid on smaller hosts and female eggs on larger hosts (see Section 4.4), but at least potentially either sex can develop on large and small hosts. The chalcidoid family Aphelinidae is unique in containing many species where the two sexes are obligatorily restricted to developing on different types of host (Viggiani 1984), a feature first described by Flanders (1936). The female wasp always develops as an endoparasitoid of a homopteran, for example a mealy bug, scale insect, or whitefly, while the site of development of the male is variable. Walter (1983a, 1983b) coined the term *heteronomous aphelinids* for species with sexually dimorphic development, and distinguished three main types of life history (this classification has largely superseded earlier systems by Flanders 1959, 1967; Zinna 1961, 1962; and Ferrière 1965).

Diphagous parasitoids. Males develop, like females, as parasitoids of homopterans but as ectoparasitoids, rather than endoparasitoids. Here the mode of development differs between the sexes, but not the type of host.

Heteronomous hyperparasitoids (sometimes called autoparasitoids or adelphoparasitoids). Males develop as hyperparasitoids of homopterans, attacking females of their own or another species of parasitoid. Heteronomous hyperparasitoids vary in the range of parasitoid species suitable for male development; Walter (1983a) suggests some wasps always avoid, and others are restricted to, females of their own species. Some species only place male eggs in homopterans already containing a suitable female host while others lay male eggs in unparasitized homopterans even though they will only survive if a second wasp places a female egg in the same insect. Finally, the male egg may develop either as an endoparasitoid or an ectoparasitoid of a female larva.

Heterotrophic parasitoids. Males develop on completely different hosts: the eggs of Lepidoptera. However, in a review of egg parasitism by Aphelinidae, Polaszek (1991) has recently questioned whether males of these species are in fact obligate parasitoids of lepidopteran eggs and concludes that heterotrophic parasitism is "a phenomenon which, in all probability, does not exist."

How might heteronomy have evolved? Walter (1983a) points out that in some species of nonheteronomous Aphelinidae, where both males and females develop as primary endoparasitoids of homopterans, the oviposition site of male and female eggs is subtly different. For example, *Aphytis melinus*, a parasitoid of red scale (*Aonidiella aurantii*), lays female eggs on the dorsum of the host, beneath the scale cover, while male eggs are laid under the scale-insect's body (Abdelrahman 1974; Luck et al. 1982; curiously, the related *A. lignanensis* lays male and female eggs in both positions). The explanation of this behavior is not known, but once male and female eggs are deposited in different sites, genes with sex-limited expression might allow the development of the two sexes to evolve independently. Walter (1983a) suggests that diphagous parasitoids evolved first as this life history involves little change in the behavior of the adult wasp. Once diphagy was established, the adult female wasp might be selected to adjust her behavior leading to the other forms of heteronomous parasitism.

1.3.3 FIG WASPS

There are about eight hundred species of fig (Moraceae, *Ficus*), the majority trees but also climbers and shrubs, distributed throughout the tropics with a few temperate species (including the edible fig). All sexual fig species are pollinated by chalcidoid wasps in the family Agaonidae which develop as mutualists in galls within the fig. With a few possible exceptions, every species of fig has its own species of pollinating wasp. Pollinating fig wasps are not parasitoids although they have certainly evolved from parasitoids. However, the larvae of pollinating fig wasps are attacked by other species of fig wasps that are true parasitoids. The bizarre life history of many fig wasps has attracted much attention by evolutionary biologists. Because the biology of fig wasp pollinators and parasitoids is so interwoven, both types of insects are considered here. For recent studies of fig wasp biology and an entry into the earlier literature, see Hamilton (1979), Janzen (1979); Boucek et al. (1981), Frank (1984); Kjellberg and Valdeyron (1984); Herre (1985, 1987, 1989); Kjellberg et al. (1987); Murray (1985, 1987, 1989, 1990); Godfray (1988); Bronstein (1988a, 1988b); and Grafen and Godfray (1991).

The flowers of the fig plant line the inside walls of the hollow fig "fruit," or syconium. When the fig is young, female pollinators crawl into the central cavity through a small pore or ostium. The pore is so narrow that wasps frequently lose wings, legs, or antennae and they never again leave the fig. Fig trees may be either monoecious or dioecious. The figs of monoecious species contain three types of flowers: male flowers, female flowers that if fertilized develop into seeds, and female flowers that support the development of a fig wasp. Dioecious fig trees may be either male or female: the figs on male trees contain male flowers and female flowers capable of supporting fig wasps, while the figs on female trees only contain female flowers from which seeds will develop (as "male" figs contain female flowers, albeit not flowers from which seeds will develop, dioecious species should strictly be called gynodioecious). Once inside the fig, the female pollinating wasp seeks to lay her eggs in flowers that support the development of her offspring. While ovipositing, she pollinates the female flowers using pollen that has either adhered to her body, or that she collected and placed in special receptacles on her body before leaving the fig in which she was born. Note that in dioecious fig trees, wasps that enter female figs fail to produce any offspring.

The young fig wasp develops on the fleshy endosperm within the female flower. As the fig grows, the pore closes and the insects are completely isolated from the outside world. The males and females of the pollinating fig wasps are extraordinarily sexually dimorphic. The adult female has a fairly typical chalcidoid appearance, though the face is characteristically projected forward. The male is a very curious insect. All traces of wings and pigmentation have disappeared and the compound eye is reduced to a few ommatidia; the abdomen is long and tubular and often reflexed underneath the body. The male never leaves the fig and its strange morphology can be viewed as an adaptation to its sole aim in life, that of finding and inseminating females in the dark world of the interior of the fig. Typically, males emerge from the pupae first and roam through the fig until they find a gall containing a female. They chew through the wall of the gall and then insert their tubular abdomen, inseminating the female prior to her emergence. When the females hatch, they collect pollen from the male flowers and leave the fig. Sometimes the pore reopens, allowing the females to escape, while in other species the male wasps dig a hole through the side of the fig.

A variety of species parasitizes the mutualism between fig and fig wasp. Some agaonids have identical life histories to the pollinating wasps except that they carry no pollen. As the endosperm required for the developing larva develops after pollination, individuals of these wasps can produce offspring only in figs that have also been entered by the legitimate pollinator, with which they compete for resources. Other species of wasps lay their eggs in the gall flowers without ever entering the fig. A large and diverse group of wasps traditionally placed in the family Torymidae[2] are characterized by extremely long ovipositors which they use to pierce the wall of the fig and to lay an egg in a gall containing a developing fig wasp. The larva of the wasp develops as a parasitoid of the fig wasp though also probably feeds on plant tissue. While Torymidae are the most common parasitoids of the pollinators, a few other chalcidoid groups have also evolved to parasitize the mutualism. Some pteromalids lay their eggs into developing gall flowers although, unlike the Torymidae, these insects often enter the fig to oviposit.

The females of nonpollinating fig wasps tend to resemble typical chalcids

---

[2] The taxonomic status of these species is controversial. Traditionally placed in the Torymidae (subfamily Idarninae), Boucek (1988) has recently argued that they are more properly placed with the true fig wasps in the family Agaonidae.

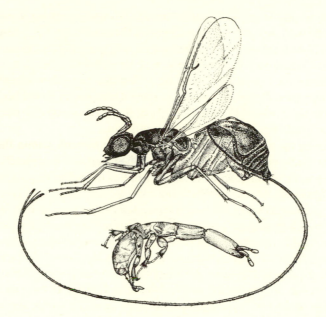

Figure 1.2 Male (lower) and female (upper) of the parasitoid figwasp *Apocrypta perplexa* (Chalcidoidea: Torymidae). After Ulenberg 1985.

except that species which lay eggs from the outside of the fig have extremely long ovipositors. The males of some species never leave the fig and have developed extremely specialized morphologies, often paralleling the adaptations of the pollinating wasps (fig. 1.2). As discussed further in sec. 7.2.1, some species have evolved fearsome mandibles and fight among themselves for the privilege of mating the conspecific females in the fig. The males of other species are similar to females, while yet other species have two types of male: normal males that leave the fig and highly modified males that compete for mates within the fig (see sec. 7.2.1).

1.3.4 AQUATIC PARASITOIDS

A few parasitoids have become adapted to attack aquatic and semiaquatic hosts. The eulophid *Mestocharis bimaculatus* parasitizes water beetle eggs, but only when exposed by fluctuating water levels. It can walk on the surface film but drowns if fully submerged (Jackson 1964). The mymarid *Caraphractus cinctus* and the trichogrammatid *Prestwichia aquatica* also attack the eggs of water beetles and other freshwater insects. These species are fully aquatic, using their wings and legs respectively as oars (Lubbock 1862; Jackson 1958; Askew 1971). They are minute insects, which probably allows them to obtain sufficient oxygen by diffusion across the surface of their bodies. The much larger ichneumonid *Agriotypus armatus* searches for caddis larvae under water: it is able to stay submerged for thirty minutes using oxygen in air trapped in the thick pubescence that clothes its body. Before pupating, the wasp's larva constructs a rigid ribbon of silk containing a bubble of air which acts as a plastron supplying the pupa's oxygen needs (Askew 1971).

1.4 Parasitoid Taxonomy

The majority of parasitoids are either members of the order Hymenoptera (the sawflies, ants, bees, and wasps) or the order Diptera (true flies). There are probably about 50,000 described species of hymenopteran parasitoids (Gaston 1991; LaSalle and Gauld 1991), 15,000 described species of dipteran parasitoids, plus about 3,000 species in other orders (Eggleton and Belshaw 1992), giving a grand total of about 68,000 described species. In all there are between 750,000 and 850,000 described species of insects (Gaston 1991) so parasitoids constitute about 8.5% of all insects (and a little over 4% of all metazoans). There is considerable debate about the total number of described and undescribed species of insects although there is most support for a figure of around 8 million (Gaston 1991). If parasitoids make up the same proportion of described and undescribed species, then there are around 800,000 species of parasitoid. However, many workers argue that parasitoids are relatively poorly known. LaSalle and Gauld (1991) suggest that parasitic Hymenoptera alone might constitute up to 20% of all insects, and Crosskey (1980) states that the Tachinidae, the large family of dipteran parasitoids, may be the most species-rich family of flies. Assuming that parasitoids constitute 20%–25% of the 8 million species of insects, these estimates put an upper bound of around 1.6–2 million on the number of species of parasitoid on earth.

1.4.1 HYMENOPTERA

The great insect order Hymenoptera is divided into two suborders, the Symphyta and the Apocrita. The Symphyta contains the sawflies, a primarily phytophagous group of insects though it contains a few species of parasitoids. The Aprocrita comprises the ants, bees, and wasps and is itself split into two major divisions, the Parasitica and the Aculeata. As their name suggests, the Parasitica are almost exclusively parasitoids. Most species in the Aculeata are predatory or collect pollen, but a few species are parasitoids. The eusocial Hymenoptera all belong to the Aculeata.[3]

Table 1.1 summarizes the taxonomy of hymenopterous parasitoids and lists the families mentioned in this book with some estimates of their relative abun-

---

[3] Unfortunately, the divisions of the Hymenoptera outlined here cannot be justified as part of a phylogenetic (cladistic) classification. A valid taxon must be holophyletic, that is, have a single evolutionary origin (monophyly) and contain all descendant species. The Symphyta is mono-

Table 1.1

Families of Hymenoptera containing parasitoids and mentioned in the text, and the number of representatives in the British fauna, in a large collection from Sulawesi (Indonesia), and in the estimated number of described species in the world fauna.

Suborder Division Superfamily Family	British Fauna[a]	Sulawesi Fauna[b]	World Fauna[c]
Symphyta			
Orussoidea			
Orussidae	1	†	75
Apocrita			
Parasitica			
Trigonalyoidae			
Trigonalyidae	1	†	70
Evanoidea			
Evaniidae*	2	15	400
Aulacidae	1	†	150
Gasteruptiidae	5	†	500
Cynipoidea			
Ibaliidae	2	†	9
Figitidae	34	3	125
Eucoilidae	55	365	1000
Charipidae	39	†	1200
Chalcidoidea			
Leucospidae	0	†	139
Chalcididae	7	†	1500
Eurytomidae	89	26	1100
Torymidae	74	14	1500
Agaonidae [Fig Wasps]	0	50	800
Eucharitidae	0	†	350
Perilampidae	9	†	200
Pteromalidae	528	115	3100

Table continues on following page

dance. Typical examples of adults of the major superfamilies are illustrated in figure 1.1.

The taxonomy of parasitoid Hymenoptera presents some of the greatest challenges facing systematic entomologists today. Even in regions with ex-

---

phyletic but gave rise to the Apocrita and is thus paraphyletic. The Apocrita is generally considered holophyletic but the Parasitica is paraphyletic, since the Aculeata evolved from members of this group. Finally, the Aculeata are probably holophyletic. In the absence of any serviceable phylogenetic classification, nearly all authors are content to stick with the traditional classification. For a full discussion of these taxonomic problems, see Gauld and Bolton (1988).

Table 1.1 (*continued*)

Suborder *Division* Superfamily Family	British Fauna[a]	Sulawesi Fauna[b]	World Fauna[c]
Chalcidoidea (cont.)			
Signiphoridae	2	†	75
Encyrtidae	191	254	>3000
Aphelinidae	37	229	900
Eulophidae	382	484	>3000
Trichogrammatidae	29	51	532
Mymaridae	84	165	1300
Proctotrupoidae			
Proctotrupidae	36	12	334
Diapriidae	298	181	2028
Scelionidae	102	250	2768
Platygastridae	157	120	987
Roproniidae	0	0	17
Ceraphronoidea			
Megaspilidae	67	9	250[d]
Ceraphronidae	26	215	
Ichneumonoidea			
Ichneumonidae	2029	420	15000
Braconidae	1163	431	10000
Aculeata			
Chrysidoidea			
Dryinidae	44	16	850
Bethylidae*	20	55	2000
Chrysididae*	31	20	3000
Vespoidea			
Tiphiidae	4	7	1500
Pompilidae*	41	46	4000

* Family contains many species that are not true parasitoids. Numbers given refer to the total species in the family.

† Data not given: absent or very rare.

[a] Data from Noyes 1989, Gauld and Bolton 1988, and Fitton et al. 1978.

[b] Data from Noyes 1989.

[c] Estimated numbers of described species; data from Gauld and Bolton 1988, Gaston 1991, Lasalle and Gauld 1991, Vlug 1993, Johnson 1992.

[d] P. Desert (pers. comm.) estimates there are approximately 250 valid descriptions of Ceraphronidae but a very large number of both undescribed species and invalid names.

tremely well-known faunas, such as northern Europe, there are some genera and subfamilies in which it is not possible to identify reliably individual specimens at the species level. The chief reasons for this taxonomic intractability are the amount of variation commonly observed within species, the paucity of character states of use to the taxonomist, and the frequent occurrence of convergent evolution, parallel evolution, and character reversal (Gauld 1986a). Compounding the inherent difficulties with the group, the activities of the first generation of Hymenoptera systematists tended to increase the taxonomic confusion. Frequently unaware of intraspecific variation, nineteenth-century taxonomists, such as the legendary Francis Walker, described huge numbers of species. Untangling the nomenclatural chaos so created has been the life work of several of this century's most eminent taxonomists (e.g., Graham 1969). Gauld and Bolton (1988) provide an entry into the taxonomic literature.

1.4.2 DIPTERA

Traditionally, the true flies are divided into three suborders, the Nematocera (crane flies, midges, mosquitoes, etc.), the Brachycera (horseflies, robber flies, bee, flies, etc.), and the Cyclorrhapha (higher flies). The taxonomic subdivision of the Cyclorrhapha is quite complex but includes two major assemblages that contain parasitoids, the Acalypterae (a group containing many families with varied life histories) and the Calypterae (houseflies and relatives). Parasitoids are very rare in the Nematocera, but several important families of Brachycera, Acalypterae, and Calypterae are exclusively parasitoids. Table 1.2 lists

Table 1.2

Families of Diptera containing parasitoids and mentioned in the text, and the number of described species in the British and world faunas.

Suborder *Division* Family	British Fauna[a]	World Fauna[b]
Nematocera		
Cecidomyiidae*	0/630	6/4500
Brachycera		
Acroceridae	3	475
Bombylidae	10	3000
Nemestrinidae	0	300
Cyclorrhapha		
Acalypterae		
Phoridae*	0/250	300/3000
Pipunculidae	75	600
Conopidae	24	800
Calypterae		
Sarcophagidae*	53	1250/2500
Tachinidae	234	8200

* Family contains many species that are not true parasitoids. Where one figure is given, it refers to the size of the whole family; where two figures are given, e.g., 10/20, 10 species out of 20 are parasitoids. In Phoridae and Sarcophagidae, the figure for the total number of parasitoid species is obtained by extrapolation from the proportions with known biology.
[a] Data from K.G.V. Smith 1976.
[b] Data from Eggleton and Belshaw 1992, and Belshaw, pers. comm.

Table 1.3

Families of parasitoids, other than those in the Hymenoptera and Diptera, mentioned in the text, and the number of described species in the British and world faunas.

Order _Family_	_British Fauna_[a]	_World Fauna_[b]
Coleoptera		
Carabidae*	362	470/30000
Staphylinidae*	990	500/30000
Rhipiphoridae	1	400
Meloidae*	9	2000/3000
Stylopoidea (=Strepsiptera)[c]*	15	10/400[d]
Lepidoptera		
Pyralidae	0/208	1/20000
Epipyropidae	0	10/20
Neuroptera		
Mantispidae	0	50/250

* Family contains many species that are not true parasitoids. Where one figure is given, this refers to the size of the whole family, where two figures are given, e.g., 10/20, 10 species out of 20 are parasitoids.
[a] Data for Coleoptera from Pope 1977.
[b] Data from Eggleton and Belshaw 1992, and Belshaw, pers. comm.
[c] The Stylopoidea is a superfamily.
[d] The biology of the Stylopoidea is difficult to categorize; see text.

the families of dipterous parasitoids mentioned in the text with some information on their numerical importance; figure 1.1 illustrates the adults of two important families. Eggleton and Belshaw (1992) provide a catalog of the distribution of parasitoids among dipteran families.

1.4.3 COLEOPTERA AND OTHER ORDERS

A small number of beetle families contain parasitoids (table 1.3). The large families Carabidae (ground beetles) and Staphylinidae (rove beetles) include a few species that are parasitoids of soil arthropods. The Rhipiphoridae and Meloidea contain species that are parasitoids of larval bees and wasps. Eggleton and Belshaw (1992) catalog the occurrence of parasitoids among beetle families.

The superfamily Stylopoidea was until recently accorded ordinal status as the Strepsiptera. One primitive family (the Mengeidae) contains parasitoids of Thysanura (apterygote insects including the silverfish) while other families attack ants, bees, wasps, and hemipterans. This latter group are perhaps best regarded as true parasites: male and female larvae develop in the host hemocoel until the final instar, when they force their heads through an intersegmental membrane and pupate. The adult female remains in situ throughout her adult life and produces a very large number of eggs while the adult male leaves the host and searches for females. Insects that have been attacked by female stylopoids normally do not die, although very frequently they are castrated. Insects attacked by some species of males do die, and in these cases the male stylopoid can be considered a parasitoid. The stylopoid life history is thus in some ways intermediate between parasitoids and parasites. Further details of their biology can be found in Askew (1971) and Waloff and Jervis (1987).

There are a very few species of parasitoid among the moths (Lepidoptera) and the lacewings (Neuroptera) (Askew 1971). Very recently, a caddisfly (Trichoptera) has been found to develop as a parasitoid of other caddisflies (Wells 1992).

1.5 Evolutionary Transitions

In this section I briefly discuss ideas about the evolution of the parasitoid habit and the biology of species that have evolved from parasitoids. This subject has been recently reviewed by Eggleton and Belshaw (1992) in a careful cladistic analysis of the available evidence.

1.5.1 THE EVOLUTION OF THE PARASITOID HABIT

The parasitoid habit has probably evolved just once within the order Hymenoptera, although many times in the Diptera and Coleoptera. Within the Hymenoptera, the Apocrita plus the single symphytan family Orussidae form a holophyletic assemblage containing all the parasitoids and groups that evolved from parasitoids. Where the biology is known, the Orussidae are parasitoids of insects living in dead wood, including other sawflies. Some idea of how the parasitoid habit may have evolved can be obtained by examining the biology of what is possibly the sister group to the Aprocrita+Orussidae clade, the sawfly superfamily Siricoidea (*sensu lato*). The sawflies of the Siricoidea (the best known species is probably the Wood Wasp *Urocerus gigas*) generally feed on dead wood that has been digested by symbiotic fungi. Some siricoids lack symbiotic fungi and make use of fungi associated with other species. A plausible hypothesis for the evolution of the parasitoid habit is that a species of siricoid which did not possess a fungus evolved to kill other species which did, and then progressed from only killing the donor to eating it as well (Eggleton and Belshaw 1992; see Malyshev (1968) for a related idea based upon a different phylogenetic hypothesis). If this hypothesis is correct, the parasitoid habit evolved from mycophagy in the Hymenoptera. Several groups of coleopterous parasitoids have evolved from mycophagous ancestors in dead wood (for example, the Rhipiphoridae). Crowson (1981) has proposed an essentially similar hypothesis to explain the evolution of the parasitoid habit in these groups.

Another important evolutionary pathway to the parasitoid habit is through feeding on dead and decaying insects (saprophagy). The dipteran family Sarcophagidae contains many species that feed on dead and decaying insect larvae and some species have evolved into true parasitoids that attack and cause the death of their hosts. The Sarcophagidae are closely related to the large and important family of dipteran parasitoids, the Tachinidae, and it is likely that they also evolved from saprophages.

The two beetle families Staphylinidae and Carabidae are almost exclusively predatory in habit but contain a few small-sized species that have become parasitoids. The transition from a predator to a parasitoid is particularly straightforward since a predator that requires only a single prey item as a larva is by definition a parasitoid. Presumably, the parasitoid habit might evolve by a reduction in the size of the predator, or by an increase in the size of the prey. Eggleton and Belshaw (1992) suggest that a number of dipteran parasitoid groups also evolved from predators.

1.5.2 Life Histories Derived from Parasitoids

The parasitoid habit has been secondarily lost in a variety of different groups. In some cases a single species or genus has adopted a variant life history, but there are a number of examples of major nonparasitoid lineages whose origins can be traced to a parasitoid ancestor.

PROVISIONING PREDATORS

The indistinct boundary between parasitoids and insects that paralyze and move a host to a concealed site prior to oviposition was noted at the beginning of this chapter. In some families of wasps such as the Bethylidae, both types of life history are common. Movement to a concealed site probably evolved as a means to reduce predation, superparasitism, or hyperparasitism of the developing young. A significant consequence of movement is that it allows the parent to add further prey items to the first and so create a cache of food; insects with this life history are termed "provisioning predators." Within the aculeate Hymenoptera, provisioning predation has evolved a number of times from the parasitoid habit and has, in its turn, given rise to groups with other life histories. For example, many bees (Apidae) feed solely on pollen or nectar, while ants (Formicidae) are frequently omnivorous but may also be highly specialized mycophages or seed eaters. Another consequence of the movement of hosts to a concealed site is the evolution of nest-making behavior, an activity that achieves enormous sophistication in the social Hymenoptera.

PREDATION

A number of otherwise typical parasitoids attack more than one host and so must be classed as predators. For example, Taylor (1937) studied the parasitoid community associated with beetles of the genus *Promecotheca*, leaf miners in palms such as coconut and oil palm. Several chalcidoid species such as the eulophid *Hispinocharis (=Achrysocharella) orientalis* and a eupelmid *Eupelmus* sp. were chiefly reared as hyperparasitoids of other parasitoids attacking the beetle. However, a single host was frequently insufficient to complete development, and Taylor observed larvae of both species marauding through the mine consuming any chalcid larva they chanced upon. Similar examples occur among eurytomid, pteromalid, and eulophid parasitoids of gregarious gall formers (Gauld and Bolton 1988). Another eulophid, *Aprostocetus (=Tetrastichus) mandanis*, begins life as a typical egg parasitoid of delphacid homopterans (planthoppers) but then emerges from the egg and searches for other eggs as a predator (Rothschild 1966); several eurytomids have a similar life history (Clausen 1940a). Some phygadeuontine ichneumonids attack a number of social bee larvae in adjacent cells (Daly 1983).

A number of wasps have evolved to oviposit into the egg masses of their hosts; each larva feeds on many host eggs and is thus an egg predator. For example, wasps in the family Evaniidae feed in the egg capsules (oothecae) of cockroaches while the Podagrioninae, a subfamily of the Torymidae, are specialist predators in mantid oothecae. Spider egg masses are also attacked by a number of ichneumonid wasps (Austin 1984, 1985; Fitton et al. 1987).

PHYTOPHAGY (NON-GALL FORMERS)

Several lineages derived from parasitoids have become phytophagous. Most species have close relatives that are parasitoids of insects feeding internally in plant tissue, and it appears that these species have switched from feeding on the host to feeding on the host's food. Possibly, phytophagy evolved originally as a means of supplementing the food resources provided by the host. Phytophagy is particularly common in the chalcidoid family Eurytomidae. Wasps in the genus *Tetramesa* mine grass and cereal stems, and those in the genera *Systole* and *Bruchophagus* (possibly a subgenus of *Eurytoma*) feed on the seeds of Umbelliferae and Leguminosae respectively (Claridge 1959, 1961). Some eurytomids begin their larval development as parasitoids but consume plant tissue as they grow older (Varley 1937). A few species of *Gasteruption* (Gasteruptiidae) and *Grotea* (Ichneumonidae) parasitize social insects, eating an egg or larvae, but obtaining most of their nourishment from stored pollen and nectar (Gauld and Bolton 1988).

PHYTOPHAGY (GALL FORMERS)

A number of groups of gall-forming insects are derived from parasitoids. A few *Tetramesa* species (see last paragraph) form modest galls, while extremely complex galls are found in the Cynipidae. The latter family is exclusively phytophagous with most species forming galls although a sizable minority live as inquilines within the galls of other species. Cynipidae are associated with many plant families, with a majority of species forming galls on oaks (Fagaceae). Many species show alternation of generations: a sexual generation followed by an asexual generation. Galls produced by the two generations are frequently different in morphology and may be produced on different plant tissues or even different plant species. Gall morphology can be extremely elaborate but is very consistent, and there has been some debate about how the insect causes the plant to produce such structures. One interesting hypothesis is that the insect injects DNA or RNA into the plant which is incorporated and expressed in the plant genome (Cornell 1983). This idea has yet to be confirmed but, if true, it will be interesting to see if there is any homology between the DNA injected into the plant by cynipids and the DNA injected by parasitoids (as polydnavirus, see sec. 6.3.2) to counteract host defences.

Finally, the natural history of pollinating fig wasps has already been discussed in this chapter (sec. 1.3.3). These wasps provide another example of the transition from the parasitoid habit to gall-forming phytophagy.

PARASITISM

There are a very few example of parasitoids of adult insects that allow their hosts to reproduce before killing them (Askew 1971). Strictly, such species should be called parasites rather than parasitoids. Some scale insects, mealy bugs, and aphids can survive long enough to reproduce after parasitism. A number of euphorine braconids (e.g., *Perilitus*) oviposit into adult beetles and may emerge without killing their host (Shaw and Huddleston 1991). The host continues to feed and reproduce and, exceptionally, acts as a host for a second generation of wasp (Timberlake 1916). Adult Hemipterans also sometimes recover from attack by tachinid flies (Worthley 1924). Recovery from parasitism by larval insects is much rarer; the examples I know all concern lepidopteran caterpillars attacked by tachinid flies (Richards and Waloff 1948; DeVries 1984; and English-Loeb et al. 1990). In the last case, English-Loeb et al. (1990) found that about 25% of caterpillars of the arctiid moth *Platyprepia virginalis* that were attacked by the gregarious tachinid *Thelairia bryanti* survived their ordeal. Although they took longer to develop, their fecundity as adults was not impaired.

The curious biology of the Stylopoidea has already been discussed in this chapter, where it was suggested they are best regarded as parasites. The most primitive members of the superfamily as parasitoids and it is possible that all members of the group are derived from parasitoid ancestors.

1.6 Conclusions

Parasitoids are abundant components of nearly all terrestrial ecosystems, both in terms of numbers of species and numbers of individuals. Although some unusual variants exist, the majority of parasitoids have broadly similar life histories and face many similar evolutionary challenges, which are explored in subsequent chapters. Parasitoids must locate hosts in a complicated and heterogeneous environment (chapter 2) and on finding a host make a series of reproductive decisions. The parasitoid must decide whether the host is suitable for oviposition and, in the case of gregarious species, decide how many eggs to lay (chapter 3). Hymenopterous parasitoids, whose females have proximate control of the sex ratio, must also decide whether to produce male or female eggs (chapter 4). The study of sex allocation in parasitoid wasps is complicated by the recent discovery of a variety of non-Mendelian factors that can influence observed sex ratios (chapter 5). The fate of the developing parasitoid is strongly influenced by host quality and, in the case of koinobiont species, by defenses against parasitism mounted by the host. Parasitoids, in their turn, have evolved counteradaptations to host defenses (chapter 6). The size and fitness of the adult parasitoids is also influenced by the quality of the host in which it developed. The adult parasitoid faces many of the same challenges as other insects, such as finding a mate and avoiding predation (chapter 7). Finally, host ecology and the presence of competing species of parasitoid combine together with phylogenetic considerations to determine the overall life history of the parasitoid: for example, the division of resources between reproductive and trophic functions, and the balance between egg size and number (chapter 8).

I hope to show in this book that parasitoids are not only fascinating organisms for study in their own right, but that their often unique biology allows valuable insights into many aspects of natural selection and adaptation.

2

Host Location

Considering the small size of both parasitoids and their hosts, and also the structural complexity of the environments inhabited by most parasitoids, finding a suitable host appears a formidable task. This chapter is concerned with the behavioral ecology of host location in parasitoids.

Research into host location by parasitoids falls into two main schools. One school has concentrated on trying to understand the behavioral mechanisms used by the parasitoid to locate their hosts. This research program, started in the 1930s, has been spectacularly successful in revealing the complex assemblages of cues used by parasitoids in host location. A major motivation behind this research has been the prospect of manipulating the stimuli perceived by the parasitoid to enhance biological control. Recent research in this field has emphasized the plasticity of response by parasitoids to different cues, and the importance of learning. The origins of the second school are more recent and lie in the explosion of interest in behavioral ecology in the 1970s. A cornerstone of the new field is optimal foraging theory, which seeks to predict the feeding behavior of animals on the assumption that behavior is optimized by natural selection. Searching for hosts has much in common with foraging for food, and the classical models of foraging theory were soon applied both to host location and host acceptance (chapter 3).

In the first section of this chapter I discuss a number of broad conceptual models that have been employed to organize discussion of host location in parasitoids. The second section is a brief review of the amazing variety of host location mechanisms that have been discovered in parasitoids, while the third section describes the evidence for plastic responses and learning. Comparative studies of host location in parasitoids are still in their infancy, but the fourth section describes pioneering work on a guild of parasitoids of Diptera. In the final section I discuss how parasitoids respond to the spatial distribution of their hosts and, in particular, the application of ideas from foraging theory to parasitoid searching.

2.1 Conceptual Models of Host Location

Host location and attack is traditionally discussed using a conceptual model first developed by Salt (1935) and Laing (1937). Salt divided host location and attack into "ecological" and "psychological" components, the former incorporating habitat and to a certain extent host location, while the latter referred chiefly to host acceptance. Successful parasitism also required host suitability. Laing developed this theme by arguing that host finding was a two-stage process involving host habitat location followed by host location.

The division of successful parasitism into the hierarchical process of host habitat location, host location, host acceptance, and host suitability has been immensely influential and has been adopted by nearly all authors reviewing the subject (Flanders 1953; Doutt 1959, 1964; Vinson 1976, 1984, 1985; Vinson and Iwantsch 1980a; Nordlund et al. 1981; van Alphen and Vet 1986; Wellings 1991). Some authors have inserted further divisions, for example dividing host acceptance into examination, probing, drilling, and oviposition (Vinson 1985). Although there has been general recognition that these "divisions are primarily for our convenience in thought and communication" (Vinson 1981), this conceptual model has tended to emphasize a static hierarchical view of parasitoid behavior.

A much more dynamic model has been proposed recently by Lewis et al. (1990) and Vet et al. (1990). They first point out that stimuli will vary in their information content and that the parasitoid should respond to the stimulus most closely associated with the host. Thus host habitat location is redundant if the parasitoid is able to locate the host directly. They envisage a naive parasitoid being born with an innate set of "response potentials" to different stimuli; a parasitoid presented with a number of stimuli will react to the one with the highest response potential (fig. 2.1). The ranking of different stimuli will be fine-tuned by natural selection to maximize the parasitoid's chance of successful host location.

Lewis, Vet, and colleagues stress that the ranking of different stimuli will change over the life of the parasitoid. In particular, if a parasitoid finds that a certain stimulus is associated with the presence of hosts, its ranking may increase. Thus a naive parasitoid might initially locate a host using a chemical stimulus emitted by the host or even by chance. After finding a host on a particular food plant, it might then use volatile chemicals associated with that food plant in future host location: a latent response to the food plant is promoted through experience (fig. 2.1). Within a species, not all individuals will rank stimuli in the same way. Genetic differences will arise due to local adaptation, and Lewis et al. speculate that there may be within-population genetic differences as well. Finally, response potentials will be modified by physiological state; a hungry individual may not respond to stimuli associated with host location, but prefer to forage for food instead.

This model is an advance on the strictly hierarchical view of host location. However, in stressing the behavioral responses to chemical stimuli, the model leaves little room for other strategies that explicitly involve movement in space, such as systematic search. It thus may also be useful to think of host location by parasitoids in terms of the model illustrated in figure 2.2. Superimposed on the real world is a surface, the height of which represents the parasitoid's esti-

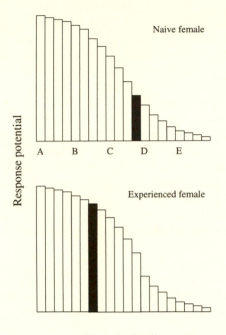

Figure 2.1 Lewis, Vet, and colleagues assume that a naive female wasp has an innate tendency to respond to an array of different stimuli (a response potential). They speculate that the distribution of response potentials is sigmoid, although this is not essential for their argument. The stimuli most closely associated with the host will have the highest response potential; thus A may be a volatile chemical produced by the host, B a chemical associated with the frass, and so on down to the lowest-ranking stimuli, which might be associated simply with the host habitat. Suppose that the filled bar in the top figure is the response potential associated with a potential food plant of the host. If a female finds hosts on this particular plant species, the response potential to the host plant stimulus may increase (bottom figure, filled bar).

mation of the presence of a host (in reality three-dimensional, though shown for simplicity in one dimension). The parasitoid will be selected to move toward a host; it will do this most efficiently by climbing the steepest slope of the likelihood surface. Directional stimuli with different information values represent different slopes of this surface. Of course, the parasitoid's estimation of the likelihood of discovering a host may be flawed, and one can imagine a parallel surface representing the true likelihood of host discovery. However, natural selection will act to make the two surfaces as congruent as possible.

The pictorial model can also be used to illustrate the parasitoid's actions on entering a region where the probability of locating a host is high but where there are no directional stimuli: the insect is now on the edge of a plateau, the plateau containing a further, invisible peak. The parasitoid should now search the plateau systematically, turning back on encountering the edge of the pla-

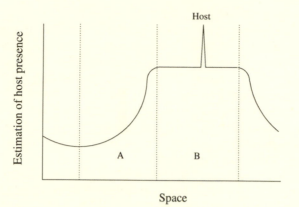

Figure 2.2 The relationship between directional search and patch use. The parasitoid's estimation of the presence of a host is plotted in a one-dimensional space. The insect will be selected to make use of stimuli that carry directional information; an individual in region A will thus move to the right, perhaps up the concentration gradient of a volatile chemical associated with the host. When it is in region B, the parasitoid obtains no directional information and makes use of other searching strategies to locate the host. The insect will, however, often turn back at the edge of the "plateau." As discussed in the text, the shape of the estimation surface may be influenced by previous experience and the presence of other searching parasitoids.

teau, and reacting to any further increase in height that may indicate the precise location of the host.

After the parasitoid has located and parasitized a host, the contours of the likelihood surface change in a manner that reflects the spatial distribution of the host. If the host tends to be solitary, what was a peak becomes either a hole or a point on a plain, and the parasitoid moves off in search of further hosts. If the host tends to be gregarious, the height of the plateau may be reduced but the insect may still remain in the area, attracted to a nearby peak, or systematically searching the plateau in the hope of further host encounters. The shape of the likelihood surface will also be influenced by past experience; as Lewis et al. (1990) and Vet et al. (1990) stress, a stimulus may offer much greater information about the location of a host if the parasitoid has already encountered a host associated with the source of the stimulus. Finally, the decisions made by the parasitoid may be affected not only by the spatial distribution of the host, but also by the distribution of searching competitors.

2.2 Mechanisms of Host Location

Parasitoid biologists have made enormous advances in recent years in understanding the cues and stimuli used by parasitoids to locate hosts (important reviews: Vinson 1976, 1981, 1984, 1985; Waage 1978; R. L. Jones 1981;

Weseloh 1981; Nordlund et al. 1988; Vet and Dicke 1992). Here, no attempt will be made to provide a comprehensive review of this field, though a range of examples of the mechanisms involved in host location will be described. The intense selection pressure that parasitoids experience in locating hosts is well illustrated by the variety of subtle cues used in host searching. I will distinguish three broad categories of information that are used in host location: stimuli from the host microhabitat or foodplant, stimuli indirectly associated with the presence of the host, and stimuli arising from the host itself. Although the categories blend into each other, the ranking roughly reflects increasing importance as indicators of host presence.

Another important distinction, particularly appropriate to chemical cues, is whether or not the stimulus imparts directional information. Dethier et al. (1960; see also Waage 1978) distinguished between *attractant chemicals* that insects use to locate hosts, and *arrestant chemicals* which, while not providing directional information, reveal the possible presence of the host in the near vicinity. In terms of the metaphor of figure 2.2, attractants determine the slope of the host likelihood surface while arrestants define the boundaries of the plateaus. Arrestant chemicals tend to have higher molecular weight and lower volatility in comparison with attractant chemicals. The study of chemical arrestants has been an extremely active area of research because of their possible economic significance: it has been suggested that the application of these chemicals to crops may result in parasitoids remaining longer in the vicinity of the crop, thus destroying more pests (Gross 1981).

Chemicals that convey information between two species are sometimes called *allelochemicals*. If both the receiver and signaler benefit from the exchange of information, the allelochemical is called a *synomone*; if the receiver alone benefits, the chemical is called a *kairomone*; and if the signaler alone benefits, an *allomone*. Chemicals that convey information between the members of one species are called *pheromones*. Allelochemicals and pheromones are the two classes of *infochemical* (Nordlund and Lewis 1976; Dicke and Sabelis 1988; Vet and Dicke 1992).

In this section, I first describe examples of host location by parasitoids using cues associated with the host habitat, indirectly with the host, and with the host itself. I also discuss host location by phoresy—hitching a ride on the adult host. The final part of this section describes host location in parasitoids which oviposit away from the host and which require either the ingestion of the parasitoid egg or active search by the parasitoid larva.

2.2.1 Cues from the Microhabitat and Host Plant

It is well established that chemical cues from the host's microhabitat can attract parasitoids in the absence of the host itself. As long ago as 1937, Laing demonstrated that the braconid *Alysia manducator* and the pteromalid *Nasonia* (=*Mormoniella*) *vitripennis*, both parasitoids of carrion flies (*Calliphora*), were attracted to uninfested meat. In the same year, Thorpe and Jones (1937) showed that the ichneumonid *Venturia (=Nemeritis) canescens*, which attacks stored product moths, was attracted to clean oatmeal.

More recently, Vet, van Alphen and their coworkers have conducted extensive investigations on microhabitat location in various parasitoids of Drosophilidae. Drosophilid parasitoids, chiefly braconids and eucoilids, tend to specialize on flies living in different microhabitats such as fungi, decaying leaves, fruit, and sap exuding from trees. Vet (1983, 1985a), Vet et al. (1983, 1984a), and van Alphen et al. (1991) found that most microhabitat specialists were attracted to odors produced in that microhabitat, often by yeasts (Dicke et al. 1984). In some cases, a more fine-tuned location mechanism was found. Thus the eucoilid *Leptopilina clavipes* is only attracted to mature fungi, just beginning to decay, the stage at which it is attacked by the wasp's host (Vet 1983). In another case, individuals of one species, the braconid *Asobara tabida*, appeared to be either attracted to fruit or to decaying leaves. Further investigation revealed that *A. tabida* was in fact composed of two closely related sibling species, each specialised on its own microhabitat (Vet *et al.* 1984a).

The importance of the host plant for parasitoid searching is suggested by several lines of indirect evidence. Taxonomically unrelated hosts feeding on the same species of plant frequently share the same parasitoids, indicating that host location is influenced by the host plant (Picard and Rabaud 1914). Similarly, the amount of parasitism suffered by a polyphagous host species often depends on the food plant it attacks (Vinson 1981, 1985; Nordlund et al. 1988).

Stronger evidence for the importance of host plant odors has been obtained from behavioral studies in the laboratory, first using Y-tube olfactometers, but more recently from multiple-choice olfactometer (Vet et al. 1983) and wind tunnel experiments (e.g., Elzen et al. 1986; Drost et al. 1986). Thorpe and Caudle (1938) observed that the ichneumonid *Coccophagus turionellae* (= *Pimpla examinator*) was attracted to the odor of pine trees, the food plant of its host, the pine shoot moth *Rhyacionia buoliana*. Curiously, after emergence the wasp takes three or four weeks to mature its eggs and during this period it is repelled by the smell of pine. Arthur (1962) found that the ichneumonid *Itoplectis conquisitor* which also attacks *Rhyacionia buoliana* is attracted to the odor of Scots pine (*Pinus sylvestris*) to a far greater degree than to red pine (*Pinus resinosa*), a preference reflected in the distribution of parasitism in the field.

Read et al. (1970) studied host location by the braconid wasp *Diaeretiella rapae*, which attacks a variety of aphid species, especially those on crucifers. Using olfactometer experiments, they demonstrated that the wasp was attracted to volatile mustard oils released by the host plant. Although the wasp could develop on many aphid species, its response to crucifer volatiles resulted in a restricted host range in nature. Aphids feeding on sugar beet were more frequently parasitized if cabbage (collards) were growing nearby because parasitoids were attracted by the crucifer. In a series of studies, Elzen et al. (1983, 1984a, 1984b, 1986, 1987 and H. J. Williams et al. (1988) have dissected the behavioral response of the ichneumonid *Campoletis sonorensis* to volatile terpenoids produced by cotton, the food plant of its host, larvae of the moth *Heliothis virescens*. The relative attractiveness of different cotton cultivars depends on their production of these volatiles, which are associated with glands on the leaf of the plant. Odors emanating from plants need not always assist the parasitoid in host location. Monteith (1960) suggested that the low rates of parasitism of the larch sawfly (*Pristiphora erichsonii*) by the tachinids *Bessa harveyi* and *Drino bohemica* found in mixed forests as compared with pure stands of larch occur because the odor of the host plant is masked by the volatiles of many different plant species.

Visual and tactile microhabitat cues are also important in host location. Van Alphen (quoted in van Alphen and Vet 1986) observed that the ichneumonid *Diaparsis truncatis* was attracted to wooden models of asparagus berries, the feeding site of its host. Similarly, the braconid *Opius (=Diachasma) alloeum* is attracted to hawthorn berries where its host feeds (Glas and Vet 1983). Visual cues are also used by parasitoids in the final stages of approach and landing after they have been attracted to the host plant by large-range stimuli such as volatile chemicals (McAuslane et al. 1990a; Wäckers and Lewis 1992).

2.2.2 INDIRECT CUES FROM THE HOST

Parasitoids frequently orientate toward cues that are derived from the activity of the host though not actually from the host itself. Again, the majority of these stimuli are chemical in nature.

A number of parasitoids respond to odors released by the feeding activity of their hosts. Damaged pine trees release the terpene α-pinene which attracts the pteromalid wasp *Heydenia unica*, a parasitoid of the bark beetle *Dendroctonus frontalis* (Camors and Payne 1972). Bragg (1974) discovered that the ichneumonid *Phaeogenes cynarae* was attracted to damaged thistles and globe artichokes. When the plant was damaged, either accidentally or by the host, a plume moth (*Platyptilia carduidactyla*), parasitoids could be observed flying upwind to inspect the damaged tissue. The braconid *Cotesia (=Apanteles) rubecula* is also attracted to plants damaged by its host (the butterfly *Pieris rapae*), but in this case artificial damage fails to elicit the same response (Nealis 1986).

Recent studies are beginning to reveal the complexity of the tritrophic interaction between host plant, host, and parasitoid. Attack by the cassava mealy bug (*Phenacoccus manihoti*) causes extensive changes to the physiology of the host plant. The encyrtid, *Epidinocarsis lopezi*, is attracted to damaged cassava though not to either cassava alone or the mealy bug alone. Uninfested leaves from infested plants are also attractive (Nadel and van Alphen 1987). The braconid parasitoid *Cotesia (=Apanteles) marginiventris* reacts to a variety of stimuli emanating from the host and host plant. Turlings et al. (1991a) found that the weakest response was to host larvae, the next strongest was to host frass, but by far the most important response was to damaged leaves. The wasp responds much more strongly to leaves that have been damaged by the host than to artificially damaged leaves (Turlings et al. 1990b). However, if saliva from the host caterpillar (the fall army worm, *Spodoptera exigua*) is placed on artificially damaged leaves, the wasp responds as if to host feeding. It appears that a chemical in the caterpillar's saliva causes the plant to release heavy terpenoids and indole, which are attractive to the parasitoid (Turlings et al. 1990b, 1991a, 1991b). The chemicals are released not only from the site of attack, but systemicly by the rest of the plant (Turlings and Tumlinson 1992). Many of the responses of the parasitoid to host and plant-derived stimuli increase dramatically with experience (see sec. 2.3). These studies raise the intriguing possibility that the plant may be selected to produce volatile chemicals that attract natural enemies of its herbivores (see sec. 8.2.5).

Even when the host does not actively damage the plant, there may be interactions between odor cues derived from the host and the host plant. For example, Kaiser et al. (1989) found that naive females of the egg parasitoid *Trichogramma maidis* (Trichogrammatidae) did not respond to odor from host eggs (the European corn borer, *Ostrinia nubilalis*), to host sex pheromone or to an extract from the host plant (maize). However, they did respond to a combination of the three odors. These examples emphasize the artificiality of separating host habitat location from host location.

Volatiles released by other organisms, as well as by the host, may be used in host location; their usefulness naturally depends on the closeness of their association with the host. The braconid *Diachasmimorpha (=Biosteres) longicaudatus*, which attacks tephritid fruit flies, is attracted to acetaldehyde, ethanol, and acetic acid released by a fungus that grows on peaches (Greany et al. 1977). Many of the volatile chemicals used by *Drosophila* parasitoids in host location are produced by yeasts in the substrate (Dicke 1988). A much stronger association is that between the parasitoids of wood wasps (*Urocerus (=Sirex)* spp.) and their symbiotic fungus (*Amylostereum* sp.). The larvae of the wood wasp bore into timber but can feed only on wood attacked by the fungus; the parent wasp inoculates the tree with the fungus at oviposition. A range of ichneumonids in the genera *Rhyssa* and *Megarhyssa* and of ibaliids in the genus *Ibalia* are all attracted to volatile chemicals produced by the fungus (Madden 1968; Spradbery 1970a, 1970b). This story is further complicated by the activities of another ichneumonid, the cleptoparasitoid *Pseudorhyssa sternata*. This species is unable to drill its own oviposition shaft and can only oviposit using the shafts made by *Rhyssa*. Like *Rhyssa*, it is attracted to wood infested by the host fungi though, in addition, it responds to a glandular secretion of *Rhyssa* which it uses to locate the oviposition shaft (Spradbery 1969). A similar example is provided by the ichneumonid *Temelucha interruptor* which uses the odor of the braconid *Orgilus obscurator*, to locate its host, the pine shoot moth *Rhyacionia buoliana* (Arthur et al. 1964)—the ichneumonid tends to win in competition with the braconid. *Ichneumon eumerus* (Ichneumonidae) parasitizes the caterpillars of blue butterflies (*Maculinea rebeli*) which feed inside ant nests (Thomas and Elmes 1993). The wasp detects the entrance of nests using chemical cues from the ants and is able to distinguish ant nests of the correct species (*Myrmica schencki*) from those of several congeners in the same habitat. Wasps approach all nests of the right species but only enter nests containing caterpillars. How they detect the presence of the butterfly at close range is not known. Chemical cues may be involved although auditory cues are also a possibility; *Maculinea* caterpillars produce sounds that are similar to, though clearly distinguishable from, ant workers.

The activity of the adult host at oviposition is the source of a number of arrestant chemicals. One of the best studied short-range chemical stimuli is a substance, tricosane, found on the scales of moths that are dislodged during oviposition. Trichogrammatid egg parasitoids (*Trichogramma* spp.) are less likely to disperse from an area after detecting the arrestant stimulus (Laing 1937; Lewis et al. 1971a, 1972, 1975a, 1975b; R. L. Jones et al. 1973; Nordlund et al. 1977). Although the scales help parasitoids to locate hosts, they probably also protect the egg batch from predators. The braconid *Opius lectus* is able to detect an oviposition site marker deposited by its host, the tephritid fruit fly *Rhagoletis pomonella* (Prokopy and Webster 1978). The marker is placed by the fly to deter conspecific oviposition. Similarly, the pteromalid *Halticoptera rosae* detects a site marker deposited by *Rhagoletis basiola* (Roitberg and LaLonde 1991). Female *R. basiola* that do not mark their oviposition sites are found at low frequencies in the field, and Roitberg and LaLonde suggest parasitoid attack may lead to a polymorphism in the host population (see sec. 3.3.3).

Some parasitoids use stimuli produced by host adults to help in the location of the immature stage which they attack. The usefulness of such stimuli obviously depends on the closeness of the association between adult and juvenile. There is a close association in bark beetles (Scolytidae); the adults inhabit the same galleries in the bark as the larvae. Bark beetles emit an aggregation pheromone which in a number of cases has been synthesized and used in pest management. Kennedy (1979) discovered that one such preparation, "multilure," attracted a variety of pteromalid, eulophid, and braconid parasitoids of bark beetle larvae (*Scolytus multistriatus*). In addition, a pteromalid hyperparasitoid of the beetle also responded to the pheromone. The braconid parasitoid *Aphidius ervi* uses aphid alarm pheromone in host location (F. Pennacchio, quoted by Vinson 1990a). The juvenile stage most closely associated with the adult insect is the egg, and some egg parasitoids use adult sex pheromone in host location. Lewis et al. (1982) found that egg parasitoids (*Trichogramma* sp.) responded to the sex pheromone of their host, the moth *Heliothis virescens* (see also Noldus and van Lenteren 1985; Noldus 1989). The sex pheromone is adsorped and retained on the surface of the leaf and thus provides information about the past presence of a sexually active adult (Noldus et al. 1991). Three species of braconid wasp in the genus *Praon* were attracted to the synthetic sex pheromone of their aphid host (Hardie et al. 1991). Read et al. (1970) discovered that males of the braconid parasitoid of aphids, *Diaeretiella rapae*, were attracted to a volatile chemical produced by female wasps, and this chemical also attracted the cynipid *Alloxysta (Charips) brassicae*, a hyperparasitoid attacking *D. rapae* larvae. The hyperparasitoid showed no attraction to plants or aphids and thus only orientated toward parasitized aphid colonies.

Other important sources of arrestant or short-range attractant chemicals include frass and honeydew. For example, the braconid *Microplitis croceipes*, a parasitoid of the corn earworm (*Helicoverpa zea*), responds by antennation to 13-methylhentriacontane, a chemical in the host frass (Lewis 1970; Lewis and Jones 1971; R. L. Jones et al. 1971). Many homopterans produce large quantities of honeydew which both reveal their presence and provide food for parasitoids. The encyrtid *Microterys nietneri (=flavus)* responds to fructose and sucrose as well as to some other unidentified compounds in the honeydew secreted by its host *Coccus hesperidum*, the brown soft scale (Vinson et al. 1978). Aphid parasitoids frequently respond to honeydew (Bouchard and Cloutier 1984, 1985; Ayal 1987). The braconid *Diaeretiella rapae* searches crucifers for its host, the cabbage aphid *Brevicoryne brassicae*. It flies first to the base of the crucifer and only if it discovers honeydew, which tends to drip or get washed to the base, does it embark on a more careful search of the plant (Ayal 1987). Some scale insect parasitoids respond to chemicals present in the wax of their host (Takabayashi and Takahashi 1985).

Mandibular and labial gland secretions, chiefly of lepidopterous hosts, are an important source of short-range attractants and arrestants. Stored product moths (Pyralidae, Phycitinae) secrete chemicals from their mandibular gland that may act as a dispersal pheromone (Corbet 1971). The well-studied ichneumonid *Venturia canescens* uses these substances (2-acylcyclohexane-1,3-diones) as an arrestant stimulus (Mayer 1934; Thorpe and Jones 1937; Corbet 1971, 1973; Mudd and Corbet 1973, 1982; Mudd et al. 1984; Waage 1978). The braconid *Bracon hebetor*, which attacks the same host, also uses the mandibular secretions for the same purpose (Strand et al. 1989). *B. hebetor*, but not *V. canescens*, will follow trails made by the host containing traces of the kairomone. The reason for this difference in behavior is that *B. hebetor* attacks mature larvae and *V. canescens* young larvae, and that older larvae enter a wandering phase prior to pupation. Another very well studied behavior is the arrestant response of the braconid *Cardiochiles nigriceps* to mandibular secretions of its hosts, moths in the genus *Heliothis* (Vinson and Lewis 1965; Vinson 1968; Vinson et al. 1975). Substances associated with silk produced by the labial gland are also known to act as short-range cues; a good example of this is provided by the braconid parasitoid *Cotesia (=Apanteles) melanoscelus*, a parasitoid of the gypsy moth (*Lymantria dispar*) (Weseloh 1976a, 1977, 1981).

There are some examples of indirect visual cues that are used in host location. Arthur (1966) found that the ichneumonid *Itoplectis conquisitor* was attracted to leaf rolls made by its host, the moth *Rhyacionia buoliana*. Leaf-mining insects leave a visual trace of their feeding activities which is often visible from some distance. A number of their parasitoids are known to respond to this visual cue by alighting preferentially on mined leaves (Kato 1984; Sugimoto et al. 1986, 1988b, 1988c; Casas 1989).

Some hosts leave "trails" over their environment which can be followed by parasitoids to their source (Klomp 1981). For example, many leaf-mining insects produce sinuous linear mines. The braconid *Dapsilarthra rufiventris* and the eulophid *Chrysocharis pentheus* (=*Kratochviliana* sp.), which attack an agromyzid fly (*Phytomyza ranunculi*) in buttercup leaves (*Ranunculus glaber*), move over the surface of the leaf until a mine is discovered, which is then followed until a host is encountered. If the mine begins to narrow, the parasitoids realize they are going the wrong way (i.e., toward the egg instead of towards the host) and reverse direction (Sugimoto 1977; Sugimoto et al. 1986, 1988a, c). Kato (1984, 1985) has suggested that some insects produce mines that are shaped to confuse parasitoids.

2.2.3 DIRECT CUES FROM THE HOST

A famous anecdote about long-range parasitoid orientation to chemical cues emanating from the host concerns the response of the ichneumonid *Pimpla bicolor* to cocoons of its host, the lymantriid moth *Euproctis terminalia*. "If a cocoon of the moth be broken open in the forest, both pupa and the hands and arms of the observer are covered by a swarm of the parasite females within the matter of a few minutes, although few or no parasites may have been observed in the vicinity previously. The range over which this attraction becomes effective so rapidly must be comparatively extensive to produce this phenomenon. The normal attraction of the pupa within the cocoon is no doubt intensified by breaking open the latter" (Ullyett 1953).

Selection will normally act on hosts to reduce the emission of volatile chemicals if they are used by parasitoids for host location. However, there are cases where hosts deliberately emit volatiles for their own purposes, as sex pheromones or aggregative pheromones for example, and these chemical advertisements are exploited by the parasitoid. The green stink bug (*Nezara viridula*, Pentatomidae) emits a chemical that acts as an aggregative and possibly also a sex pheromone. The tachinid *Trichopoda pennipes* which lays its eggs on the adult insect uses the chemical to locate its host (Mitchell and Mau 1971; Harris and Todd 1980). Clausen (1940a) remarks that many tachinid flies which parasitize the adult stages of a variety of insects are largely reared from the female sex, and it is possible that host location by sex pheromones is responsible. Feeding by bark beetles on trees is facilitated by mass attack, which is accomplished by the emission of an aggregative pheromone. The pteromalid *Tomi-*

cobia tibialis attacks adult bark beetles in the genus *Ips* and is attracted to the aggregative pheromone of its hosts, but not to that of closely related species (Rice 1968, 1969). The response of the parasitoid to host kairomones derived from geographical strains of the same species also varies (Lanier et al. 1972). Sex pheromone produced by the California Red Scale (*Aonidiella aurantii*) serves as an attractant chemical for aphelinid parasitoids (*Aphytis* spp.) (Sternlicht 1973).

Though the detection of chemical cues seems to be the most frequent method of host location, some parasitoids make use of other senses. A few parasitoid flies are known to be attracted by the sound of their host: the tachinid *Euphasiopteryx ochracea* to crickets (Cade 1975, 1981, 1984; Mangold 1978) and the aptly named sarcophagid *Colcondamyia auditrix* to cicadas (Soper et al. 1976). Both species are attracted by tape recordings of their host. Richerson and Borden (1972) suggested that the braconid *Coeloides brunneri* used infrared radiation to detect its host, a bark beetle. They discovered that the wasp would investigate areas of bark heated by as little as 1°C. However, it is possible that convection or conduction rather than radiation were responsible for heat perception (Weseloh 1981).

Substrate vibration is often used by parasitoids, especially those attacking concealed hosts. Oviposition by the braconid *Coeloides brunneri*, a bark beetle parasitoid, can be induced by scratching the undersurface of the bark with a pin (Ryan and Rudinsky 1962; but see Richerson and Borden 1972). Similarly, the braconids *Opius melleus, Diachasmimorpha (=Biosteres) longicaudatus* and *Opius (=Diachasma) alloeum*, all of which parasitize larval tephritid fruit flies, locate hosts through their movement (Lathrop and Newton 1933; Lawrence 1981a; Glas and Vet 1983). Many eucoilid and braconid parasitoids of *Drosophila* use substrate vibration (Vet and van Alphen 1985; Vet and Bakker 1985), as do parasitoids of grain weevils (van den Assem and Kuenen 1958) and leaf-mining flies (Sugimoto et al. 1988a, 1988b).

Movement by the host, detected visually, frequently guides parasitoids in the final stage of host location. The tachinid *Drino bohemica* which attacks sawfly larvae (*Neodiprion lecontei*) detects nearby hosts by their movement (Monteith 1956, 1963). Adult insects tend to be more mobile than larvae and their parasitoids, in particular, use movement to locate hosts; two good examples are the tachinid *Trichopoda pennipes*, which attacks adult stink bugs (*Nezara viridula*) (Mitchell and Mau 1971), and the pteromalid *Tomicobia tibialis*, which attacks adult bark beetles (*Ips* sp.) (Rice 1968). The braconid subfamily Euphorinae contains many species that attack adult beetles, particularly weevils, and host movement is normally important in the final stages of host location (Shaw and Huddleston 1991).

A relatively small number of parasitoids attack swiftly moving adult insects which they intercept in flight. Hosts are detected visually and these parasitoids often have large eyes. Conopids are large robust flies that mostly parasitize bees and wasps. The fly may mimic the flight of its host before pouncing and laying an egg in midflight (Raw 1968). Female *Conops scutellatus* loiter near the entrance of wasp nests and pounce on insects as they exit or enter (Clausen 1940a). A tachinid fly, *Rondanioestrus apivorus*, deposits larvae (from eggs it has incubated in its reproductive tract) on worker bees in flight (Skaife 1921). The aberrant conopid genus *Stylogaster* follows columns of army ants and parasitizes a variety of adult insects that are flushed by the ants (Askew 1971). Pipunculids are flies with very large eyes that parasitize homopteran nymphs (or occasionally adults). Members of this family are famed aerial acrobats and are able to hover and even fly backwards. They locate their hosts visually, swooping down to carry them into the air where parasitism occurs (Clausen 1940a). Another famous example of aerial attack is provided by a phorid fly in the genus *Apocephalus*. It again locates its hosts, workers of leaf-cutter ants (*Atta*), by sight and flies down to lay an egg quickly on the ant's neck. The ant is able to defend itself with its mandibles, except when it is returning to the nest carrying a leaf fragment. However, in these circumstances a minute worker of a separate caste rides shotgun on the leaf and protects the larger worker (Eibl-Eibesfeldt and Eibl-Eibesfeldt 1968).

2.2.4 PHORESY BY ADULT PARASITOIDS

The majority of parasitoid hosts are immature insects. One way of locating immatures is to hitch a ride on the adult and wait until it oviposits or returns to a nest. Phoresy has been recorded in a number of parasitoids, and has been reviewed by Clausen (1976) and Vinson (1985).

Egg parasitoids in particular are likely to benefit from phoresy because of the physical contact between the adult and the egg. For example, the scelionid *Mantibaria (=Rielia) mantis* attaches itself to adult mantids (*Mantis religiosa*), where it loses its wings and waits until the mantid oviposits. If it finds itself attached to a male it transfers to the female during mating. There is some evidence that the female wasp may feed from the adult mantid as a true parasite. When the mantid oviposits, the wasp jumps off and parasitizes the egg. The now wingless wasp is not deterred by the frothy liquid used by the mantid to cover her eggs. After parasitism, the wasp attempts to remount the adult mantid (Rabaud 1922; Chopard 1923). The parasitoid is unable to attack host egg masses after the frothy liquid has hardened and the necessity of locating newly laid egg batches is likely to have been important in the evolution of phoresy. The members of the family Scelionidae are all egg parasitoids and a number are phoretic on grasshoppers, moths, planthoppers, and even dragonflies (Clausen 1976). The Trichogrammatidae are also exclusively egg parasitoids and a few phoretic species are known. A species of *Xenufens* attaches itself to the base of butterfly wings (*Caligo eurilochus*). In one population, Malo (1961) found an average of 75 parasitoids per host with a maximum of

250 parasitoids on a single butterfly. The main advantages of phoresy by egg parasitoids are that it facilitates host location and brings the parasitoid into contact with newly laid eggs where the chances of successful parasitism and development are usually greatest (Strand 1986).

A rather different type of phoresy has been noted by Askew (1971) and Takagi (1986). The pteromalid wasp *Pteromalus puparum* attacks butterfly pupae. Occasionally wasps can be found riding on the backs of full-grown caterpillars in their wandering phase before pupation.

2.2.5 OVIPOSITION AWAY FROM THE HOST

While the vast majority of parasitoids deposit their eggs directly on the host or in its very near vicinity, a substantial minority of species oviposit away from the host. There are three main mechanisms of host infection: the host may eat the parasitoid egg; the parasitoid larva may actively search for suitable hosts, or the parasitoid larva may be carried phoretically to a suitable host. A very thorough review of the morphological adaptations involved is provided by Hagen (1964).

EGGS INGESTED BY HOSTS

Many members of the fly family Tachinidae, and all members of the wasp family Trigonalyidae, rely on the host to ingest their eggs. Parasitoids with this attack strategy typically oviposit a large number of very small eggs, termed microtype eggs. Some trigonalyids lay over 10,000 eggs. An ingested egg is stimulated to hatch by a combination of salivary juice, mechanical rupture, and the high pH of the insect's gut (Hagen 1964). On hatching, the larva burrows through the gut wall and often moves to a specific host tissue to continue its development.

Oviposition by tachinids is frequently in response to some indication of the presence of hosts. *Cyzenis albicans*, the well-known parasitoid of the winter moth (*Operophtera brumata*), deposits its eggs on foliage that has been damaged by caterpillar feeding (Hassell 1968). The fly appears to respond to sugars in the sap and can be persuaded to lay eggs on leaves sprayed with a sugar solution. *C. albicans* also responds to a long-range attractant produced by the damaged leaves of some, but not all, of its host's food plants (Roland 1986; Roland et al. 1989).

The life cycle of trigonalyids can be extremely complicated. Again, the eggs must be ingested by a host, normally a lepidopteran or sawfly caterpillar. In some species, the wasp larva then develops as a typical endoparasitoid, but in other species the larva remains dormant unless one of two fates befalls the host. If the host is parasitized by an endoparasitic ichneumonid or tachinid, the trigonalyid enters the primary parasitoid, waits until it has nearly consumed the host, and then develops as a hyperparasitoid, initially feeding endoparasitically but finally finishing off its new host as an ectoparasitoid. However, if the original host is preyed upon by a social wasp (*Vespa*), and its dismembered body, including the small trigonalyid larva, is fed to a wasp grub, the trigonalyid develops as a parasitoid of the *Vespa* (Clausen 1940a; Cooper 1954).

LARVAE ACTIVELY LOCATE HOSTS

Active host location by the larvae is rather rare among hymenopteran parasitoids (restricted to the two chalcidoid families Perilampidae and Eucharitidae, and the ichneumonid subfamily Eucerotinae) but widespread among dipteran and coleopteran parasitoids. The very few lepidopteran and neuropteran parasitoids also locate their hosts in this manner. Two main types of searching larvae can be distinguished: the triungulin and the planidium (fig. 2.3). Triungulin larvae have legs and are characteristic of beetle parasitoids. Planidial larvae lack legs and move by flexing elongated setae. There is remarkable convergence in structure among planidial larvae in the Diptera and Hymenoptera. Many Tachinidae with searching larvae have abandoned an external egg stage: the adults incubate the eggs internally within a distended part of the reproductive tract and give birth to larvae directly (larviposition).

The parent often assists the larvae to locate hosts by depositing eggs or larvae where there is evidence of host activities. For example, the tachinid *Lixiophaga diatraea* is stimulated to larviposit by host frass, but the larvae themselves must find the host (Roth et al. 1978). Many tachinid flies with this type of larvae attack concealed hosts in stems, fruits, and seeds and lay their eggs near entrance holes made by the hosts (Clausen 1940a). Several bee flies (Bombyliidae) seem to fire eggs into burrows made by their hosts, solitary bees (Clausen 1940a). The larvae themselves frequently detect hosts by movement. The larvae of some tachinids anchor themselves in the remains of the egg and stand vertically, pivoting toward any moving object (Hagen 1964). Some perilampid wasps attack lacewing larvae (*Chrysopa*). Lacewings lay eggs on long stalks and perilampids can be observed attached to the bottom of the stalk waiting for the eggs to hatch (Clausen 1940a). Other perilampids enter caterpillars but only develop, as hyperparasitoids, if the caterpillars are subsequently attacked by another parasitoid. An anomalous ichneumonid (*Euceros frigidus*) has a similar life history (Tripp 1961).

In some cases, oviposition seems to be independent of host distribution. Larvae of the dipterous family Acroceridae are endoparasitoids of spiders. Females oviposit in flight, shooting the adhesive eggs at a suitable surface. Often several females oviposit together, and Clausen (1940a) has estimated that a single individual may lay up to 10,000 eggs. The planidial larvae move by looping and jumping and can crawl along the thread of a spider web.

LARVAE CARRIED TO HOSTS

A number of parasitoid groups with active larvae attack the immature stages of nest-building Hymenoptera. They do not locate the nests directly but are transported phoretically by the adult host.

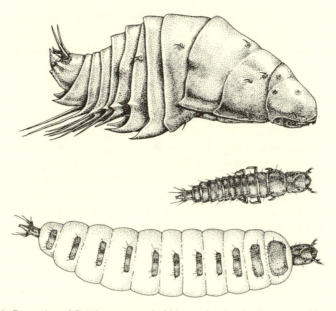

Figure 2.3 Examples of first instar parasitoid larvae involved in host searching. *Top*: The planidium larva of the chalcid wasp *Stilbula montana* (Eucharitidae) (after Heraty and Darling 1984). *Bottom*: The triungulin larva of the beetle *Aleochara curtula* (Staphylinidae) before (above) and after (below) feeding (after Kemner 1926).

Species of oil beetle (Meloidae) parasitize the larvae of solitary bees but also consume the food stored for the larva. Meloids have triungulin larvae and many species seize hold of passing bees which carry them to the nest. Adult females of some species lay their eggs near the entrance of the nest in which they developed; their larvae attach themselves to bees as they leave the nest and transfer from males to females during mating. Other species lay their eggs away from the nest, and the larvae climb flowers and wait for visiting bees (Fabre 1857; Askew 1971). Another beetle family, the Rhipiphoridae, also contains species that are parasitoids of nest-building Hymenoptera, for example social wasps (*Vespula*). Their triungulin larvae attach themselves to foraging workers; once in the nest they develop initially as endoparasitoids (unusual for a beetle) of wasp grubs but later feed ectoparasitically (Clausen 1940a; Askew 1971). Other species of both meloids and rhipiphorids attack nonhymenopteran hosts that are located directly by the triungulin larvae.

The Eucharitidae are a family of odd-looking chalcidoids that are exclusively parasitoids of ants. The adult females lay clutches of eggs in a very wide variety of sites, for example on or in leaves, hidden in fruit or plant buds, or in seed pods (Clausen 1940a, 1940b, 1940c, 1941; D. W. Johnson 1988; Heraty and Barber 1990). The larvae are planidia and capable of jumping, though they normally remain motionless waiting for an ant to pass. In at least one species, the egg contains copious fluid which is released when the larva hatches and appears to attract ants (Heraty and Barber 1990). Once attached to an ant, they are carried back to the nest where they parasitize ant larvae. Several species have the bizarre habit of ovipositing near the eggs of thrips. The larvae attach themselves to thrips as they hatch, and may even feed from them. What happens then is unknown, though presumably they eventually enter an ant nest (Clausen 1940a, 1940b; J. B. Johnson et al. 1986).

2.3 Learning

The spatial distribution of hosts is seldom fixed but varies, both between generations and possibly during the lifetime of a searching parasitoid. This is especially true for generalist parasitoids which attack a variety of host species whose relative abundances changes over time, and also for more specialist species that attack polyphagous hosts. Parasitoids will be strongly selected to gather information about current host distributions and to modify their search strategy accordingly. It has been known for many years that parasitoids have the capability of learning about the environment; however, it is only in the last ten years that the ubiquity of learning has been fully appreciated (Vet and Groenewold 1990; Turlings et al. 1992). The parasitoid can obtain useful information about the distribution of hosts both from its emergence site, and also from its experiences while searching for hosts.

2.3.1 Information from the Emergence Site

Consider a parasitoid that either attacks a variety of host species with different ecologies or one host species that lives in a number of different microhabitats. A newly emerged adult parasitoid is faced with the problem of which host or microhabitat to search for. In the absence of direct evidence of the current location of hosts, the parasitoid can obtain some information from its emergence site; at the very least, hosts were present in this type of microhabitat one generation ago. Naive parasitoids may thus be selected to orientate toward stimuli associated with the emergence site. In addition, traces of chemicals associated with the host may still be present at the site of emergence; for example, host frass is frequently entangled in parasitoid cocoons. There is thus an opportunity for a parasitoid to learn stimuli associated with the host before it begins to search.

The first evidence for the importance of the emergence site in adult searching behavior was obtained by Thorpe and Jones (1937). They studied the adult searching behavior of the ichneumonid *Venturia (=Nemeritis) canescens* reared on different host species and found that the adult tended to respond to the odor of the host on which it developed. They suggested that this searching preference was learned while the wasp was still a larva (see also Smith and

Cornell 1979; Powell and Zhang 1983) but today it is thought more likely that the adult wasp learns the host odor after emerging from the pupal stage (Corbet 1985; Vet and Groenewold 1990; Turlings et al. 1992). A number of other examples of the importance of larval habitat have been discovered. The eucoilid parasitoid of *Drosophila, Leptopilina clavipes,* is normally attracted to fungi. However, if reared on a fruit medium containing yeast, the adult wasp will also orientate toward decaying fruit (Vet 1983). The braconid parasitoid *Diaeretiella rapae* normally attacks aphids on crucifers but is more responsive to potato plants if it is reared on hosts feeding on potato (Sheehan and Shelton 1989). Of particular importance to biological control, the braconid *Microplitis demolitor* will respond to cowpea odor if reared on hosts (*Helicoverpa zea*) feeding on real cowpeas, but not if reared on artificial diet (Hérard et al. 1988). If the parasitoid pupa is removed from its cocoon, the ability to learn the odor of food plants disappears. Host frass attached to the cocoon appears to be the source of the learned stimulus. Although artificial diets can streamline and cheapen mass-rearing, they may result in parasitoids that perform poorly when released into the field.

The ability of parasitoids to learn the characteristics of their development site is a means by which information can be inherited nongenetically between generations—a form of cultural transmission (Godfray and Waage 1988). If this type of learning was very important it would serve to isolate populations on different microhabitats or hosts with significant consequences for the genetic subdivision of the population and speciation. However, the behavior of some parasitoids appears to be unaffected by their larval environment (e.g., Mueller 1983; McAuslane et al. 1990a) and, as discussed in the next section, parasitoids also obtain much information while searching. Although adult learning could act to reinforce a culturally transmitted preference, the great flexibility in response shown by many parasitoids will work against the establishment of subpopulations specialized on particular hosts or microhabitats.

2.3.2 Information Obtained While Searching

Although the emergence site offers some clues to the current distribution of hosts, the actual experience of host (or host-product) discovery as an adult is likely to be a far better guide. Many parasitoids exhibit associative learning. Associative learning occurs when a response to a new stimulus is added to a parasitoid's behavioral repertoire after it has been encountered together with a stimulus toward which the parasitoid already has an established response. The attraction to the new stimulus is sometimes called a conditioned response as opposed to the unconditioned response to the innately attractively stimulus. Care must be taken in interpreting a change in behavior as associative learning because the mere exposure to a stimulus may change behavior through habituation or sensitization (Papaj and Prokopy 1989). Sensitization (also called priming) is a general increase in responsiveness brought about by exposure to the host or a host product (Turlings et al. 1992).

Arthur (1966, 1967) demonstrated associative learning in the ichneumonid *Itoplectis conquisitor*, a polyphagous parasitoid of moth pupae concealed in leaf rolls and similar habitats. He created artificial shelters of various shapes, sizes, and colors and found that once a parasitoid had located a host in a particular type of shelter, it later preferentially examined similar shelters. The flexible behavior of this ichneumonid probably explains the ease with which it attacked *Thymelicus lineola*, a butterfly accidentally introduced to Canada which became a pest of meadow grass (Zwölfer 1971). Monteith (1963) kept tachinid flies (*Drino bohemica*) in a laboratory cage and provided them with hosts (the sawfly *Neodiprion lecontei*), which were slid into the cage on removable trays. The flies learned to associate the moving tray with the presence of hosts and would fly down and inspect empty trays. They also became habituated to the experimenter and his equipment, and could almost be described as becoming tame. Wardle and Borden (1985) also found that the ichneumonid *Exeristes roborator*, a polyphagous parasitoid of microlepidopterans, learned to recognize laboratory apparatus, in this case a device in which hosts were sandwiched between laboratory tissue for presentation to the wasp. More recently they have shown that *E. roboratur* uses both color and shape to distinguish different microhabitats containing hosts (Wardle 1990; Wardle and Borden 1990).

Drosophilid parasitoids studied in the laboratory by Vet (1983, 1985b), Vet and van Opzeeland (1984, 1985), Vet and Schoonman (1988), and Vet and Groenewold (1990) frequently showed flexibility in host-searching behavior. As mentioned above, these species tend to be microhabitat specialists. However, exposure to hosts in novel microhabitats leads them, to varying degrees, to be attracted to odors emanating from the new microhabitat and, on arrival, to spend more time searching. Species such as the eucoilid *Leptopilina heterotoma* which attack hosts in a variety of microhabitats seem to show the strongest associative learning. In an elegant field experiment, Papaj and Vet (1990) showed how associative learning could be important under natural conditions. Female *Leptopilina heterotoma* were assigned to three experimental treatments: naive females with no oviposition experience; females with a brief experience of oviposition on an apple-yeast medium; and females with a brief experience of oviposition on a mushroom medium. Marked wasps were released in a woodland at the center of a circular array of apple-yeast or mushroom baits. Wasps with prior oviposition experience were more likely to locate baits, and found them more quickly. Females were much more likely to be recovered on baits containing the medium they had previously experienced (fig. 2.4).

Two groups of USDA scientists at Gainesville and Tifton have carried out extensive investigations of the learning abilities of a generalist braconid para-

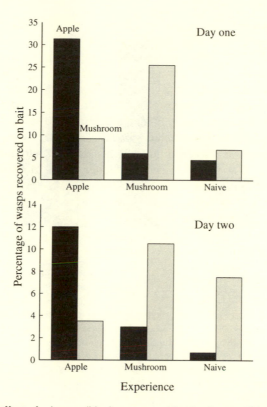

Figure 2.4 The effect of prior conditioning on searching by the eucoilid parasitoid *Lepto-pilina heterotoma*. Wasps were allowed to oviposit on hosts in fruit or mushroom media or were given no experience of oviposition (naive). They were released in a woodland containing apple and mushroom baits. The two figures show recaptures on the first and second days after release. Experienced wasps were recaptured more often than inexperienced wasps. Wasps tended to be recaptured on the bait with which they were familiar. The higher numbers of naive wasps which were caught at mushroom baits on the second day suggests that in the absence of prior experience the wasps prefer this bait. *Dark bars*: recaptures on apple bait; *light bars*: recaptures on mushroom bait. (From Papaj and Vet 1990.)

sitoid *Cotesia (=Apanteles) marginiventris*, which attacks a variety of lepidopteran caterpillars (Turlings et al. 1989, 1990a, 1990b, 1991a, 1991b), and of a specialist braconid, *Microplitis croceipes*, which attacks the larvae of *Heliothis* and *Helicoverpa*. Naive female *C. marginiventris* show little response to the odors of hosts and damaged plants in olfactometer experiments. After wasps were allowed to oviposit on either cabbage looper (*Trichoplusia ni*) feeding on cotton or fall army worm (*Spodoptera frugiperda*) feeding on maize, they showed a greater response to odors from either host (sensitization), though they responded most strongly to the host they had previously attacked

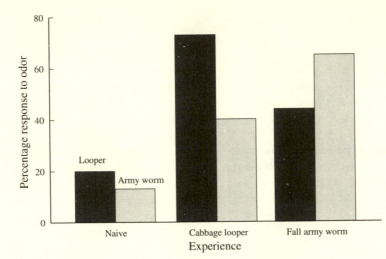

Figure 2.5 The effect of prior conditioning on searching by the braconid wasp *Cotesia marginiventris*. Wasps were allowed to (1) oviposit on cabbage loopers feeding on cotton, (2) oviposit on fall army worm feeding on maize, or (3) were not allowed to oviposit (naive). Response to odor from feeding cabbage loopers or fall army worms was then tested in a four-armed olfactometer. The odor source was placed in a single arm, the other three arms remaining empty; the bars in the figure show the percentage of wasps that responded to the odor. Experience increases the response to either odor but the strongest response is to the host the wasp had previously experienced. *Dark bars*: response to odor from cabbage looper on cotton; *light bars*: response to odor from fall army worm on maize. (From Turlings et al. 1989.)

(associative learning) (fig. 2.5). Other experiments showed that oviposition itself was not essential for learning, contact with frass and damaged leaves was sufficient, and that the learned response persisted for several hours. In choice experiments where the wasp was offered different odor combinations of the two hosts and the two food plants, an innate preference for corn over cotton and *S. frugiperda* over *T. ni* was modified by experience. The wasp is also able to learn to distinguish the same host species feeding on different host plants. A specific spectrum of volatile chemicals is produced by different food plants after attack by different species of caterpillar, which facilitates learning by the parasitoid. The possibility that the plant may be selected to emit volatiles to attract parasitoids is discussed in section 8.2.5.

The behavior of *Microplitis croceipes* shows many similarities (Drost et al. 1986, 1988; Lewis and Tumlinson 1988; Eller et al. 1992, Wäckers and Lewis 1993, McCall et al. 1993). Although a specialist on *Heliothis* and *Helicoverpa*, the wasp needs to search for hosts on a wide variety of food plants. An initial weak response to a host plant odor is markedly strengthened by experience with a plant-host complex (i.e., the host feeding on the plant). Again, oviposition itself is not essential, contact with a damaged plant (or to a lesser degree host frass) is sufficient for associative learning to occur. However, the response to host products without oviposition is often rather weak; for example, several exposures to host frass on a novel plant are required to change the wasp's preference. The attractiveness of an infested host plant is influenced by plant species and by growth phase, and experience on one host plant species can lead to a general increase in responsiveness to all host plant species. The wasp is able to learn to distinguish between odors produced by different parts of a cotton plant when fed on by the same species of host (*H. zea*). There appears to be an innate preference for old rather than freshly damaged plants. This may be adaptive as the volatiles released immediately after damage are not specific, while the volatiles produced later are a direct response to caterpillar saliva (see also sec. 2.2.2). *M. croceipes* displays particularly strong learned preferences when it encounters a host in a microhabitat containing both obvious olfactory and visual cues. Although a specialist, *M. croceipes* has only a poor ability to distinguish hosts from nonhosts before alighting on the plant, and decisive host recognition may require nonvolatile chemicals in the frass.

Often some of the sharpest increases in responsiveness occur when a parasitoid discovers hosts on an unusual food plant. Thus the braconid *Diaeretiella rapae*, which normally attacks aphids on crucifers searches potatoes more assiduously after successful oviposition on this host plant (Sheehan and Shelton 1989). The braconid *Macrocentris cingulum* (=*grandii*) attacks the European corn borer (*Ostrinia nubilalis*) on maize; naive females are attracted to this food plant but normally ignore sunflower, a rare host plant. However, females orientate to this host plant after oviposition on sunflower or after finding the frass of a host that has fed on sunflower (Ding et al. 1989).

A number of parasitoids can learn to associate novel chemicals with the presence of hosts. Thus Arthur (1971) found that *Venturia canescens* could be attracted to geraniol if it had previously encountered a host in an environment containing this chemical. Vinson et al. (1976, 1977) discovered, rather disconcertingly, that their laboratory strain of the braconid *Bracon mellitor* was using as an arrestant stimulus an antibiotic present in the artificial diet used to culture its host, the boll weevil (*Anthonomus grandis*). Lewis and Tumlinson (1988) showed that if *Microplitis croceipes* discovered host frass in association with a novel chemical (such as vanilla) they would subsequently orientate toward the new stimulus. In further experiments, Lewis and Takasu (1990) demonstrated that the same wasp could be trained to associate one volatile with food (a nectar source) and another with hosts (the chemicals used were chocolate and vanilla). In choice experiments in olfactometers, the wasp orientated to the chemical associated with food if unfed, but to the chemical associated with hosts if well fed. As the authors note, *M. croceipes* is displaying responses of a sophistication more normally associated with rats! These examples illustrate the extreme flexibility many parasitoids possess in the incorporation of new stimulus responses into their behavioral repertoire.

The ease with which a new stimulus elicits learning, and the speed with which a learned response decays, are likely to be related to its value in revealing the location of prey (Vet et al. 1990; Lewis et al. 1990; sec. 2.1). Oviposition tends to be most efficient at changing parasitoid behavior and the changes that result are slowest to decay (e.g., Vet and Groenewold 1990; Turlings et al. 1992). In the limit, stimuli that are invariably associated with the host should always elicit a response (unconditioned stimuli). The value of a stimulus is likely to be greater if encountered several times. Several contacts with a host product can lead to associative learning where one is ineffective (Eller et al. 1992). It is possible that contact with host products such as frass may, in the absence of an actual host contact, indicate that the site has been abandoned by the host. McAuslane et al. (1990b) suggest this may explain why the ichneumonid *Campoletis sonorensis* experiences reduced responsiveness after contacting damaged host plant or feces without oviposition. Learning of the visual and olfactory properties of a site may also be important in the avoidance of superparasitism (Waage 1979; Turlings et al. 1992).

Learning by parasitoids has also been discussed in the context of switching behavior. In population dynamics, switching occurs when exposure to one prey or host species leads to an increase in a predator or parasitoid's rate of attack on that species, and a decrease in the rate of attack on alternative prey or host species (Murdoch 1969; Murdoch and Oaten 1975). Typically, switching results in the natural enemy concentrating on the most common victim species. Switching may arise if a predator encountering many common prey items becomes more efficient at locating that type of prey. If hunting is visual, it is frequently suggested that the predator develops a search image for the prey. Cornell (1976) and Cornell and Pimentel (1978) have used switching to describe the behavior of parasitoids. In particular, they found that the pteromalid *Nasonia vitripennis*, a parasitoid of dipteran pupae, tended to switch to the most abundant host type when presented with a range of different species. This behavior can probably be explained as a type of learning. In other studies of switching, parasitoids have consistently preferred one host species over another irrespective of their relative densities (Dransfield 1979; Heong 1981; Gardner and Dixon 1985; Chow and Mackauer 1991, 1992).

2.4 Comparative Studies of Host Location

The raw material of natural selection is the heritable variation in a trait present in a population. The nature of such variation, and hence the response of a species to a new environmental challenge, is strongly influenced by phylogenetic history. Insights into the interaction between phylogenetic and ecological factors can be obtained by comparing the solutions adopted by different spe-

cies to similar selection pressures. There have been few such studies of parasitoids, one notable exception being a comparative study of host location by
parasitoids of higher Diptera.

Vet and van Alphen (1985) and Vet and Bakker (1985) studied host location in thirty-two braconid wasps of the subfamily Alysiinae and twenty-five
eucoilid wasps. All species are endoparasitoids of dipteran larvae, the adult
parasitoids emerging from the pupal hosts. The majority of species attack
drosophilids. Previous studies (sec. 2.2.1) had demonstrated that most species
located host patches by attraction to odors produced in the microhabitat. Vet
and her coworkers were interested in the host-finding behavior of parasitoids
after they had located the patch.

The wasps displayed three main types of searching behavior: (1) probing
with the ovipositor while walking across the substrate, (2) location of hosts by
detection of their movement (vibrotaxis), and (3) searching the substrate by
drumming or feeling with the antennae. Some species showed combinations of
searching strategy although there was little substantial variation within each
species (with the exception of two strains of one species from different conti-
nents). Species that detect host movement spend much time motionless, "lis-
tening." Ovipositor probing and vibrotaxis are thus to some extent mutually
exclusive.

Closely related species tend to show similar oviposition behavior, even
when they search for hosts on different substrates. Thus most species of the
eucoilid genus *Leptopilina* probe with their ovipositor and are unable to detect host movement, while species in the braconid genera *Asobara* and *Aphae-
reta*, and in the eucoilid genera *Ganaspis*, exclusively use vibrotaxis. Species
of *Kleidotoma* (Eucoilidae) and *Dinotrema* (Braconidae) are able to employ
both strategies. Only a few species use their antennae to feel for hosts: an
example is the braconid *Tanycarpa punctata* whose close relatives (including
the congeneric *T. bicolor*) search primarily by vibrotaxis. In this case, the
change in search behavior is probably related to the biology of *T. punctata*;
this species attacks hosts older than its relatives, and the spiracles of these
hosts protrude above the surface of the medium where they can be detected by
antennation.

The presence of both vibrotaxis and ovipositor probing in eucoilids and
braconids, members of different superfamilies, suggests possible convergent
evolution. On the other hand, the relatively uniform nature of host location
within genera, and the fact that different species of wasps use different techniques to locate the same species of host on the same substrate, suggest the
importance of phylogenetic history. Comparative studies such as these offer an
important insight into the evolutionary dynamics of host location. Further progress will be made by analyzing species behavior and ecology within an explicitly phylogenetic framework.

2.5 Patch Use

The hosts of few if any parasitoids are distributed randomly in the environment and often they occur in discrete patches. The behavior of parasitoids searching for patchily distributed hosts has attracted considerable attention from parasitoid biologists, and also from theoretical behavioral ecologists. In the first part of this section I discuss some of the classical behavioral studies of parasitoids searching in patchy environments. I then describe the application of patch-use theory from behavioral ecology and finish by reviewing experimental work in this area.

2.5.1 BEHAVIOR ON HOST PATCHES

AREA-RESTRICTED SEARCHING

Suppose that hosts are distributed in a clumped manner in the environment but that the parasitoid has no information about their position. If a parasitoid locates and parasitizes a host, it would make sense to remain in the immediate vicinity and search for other hosts. Such behavior is known as area-restricted searching (or success-motivated searching) and often involves both a decrease in the speed of searching (orthokinesis) and an increase in the rate of turning (klinokinesis). Similar behavior is observed when a parasitoid enters an area contaminated by a host-associated chemical (e.g., Waage 1978; Loke and Ashley 1984; Gardner and van Lenteren 1986).

If hosts are not clumped but isolated in the environment, dispersal immediately after oviposition may be the optimal strategy. Strand and Vinson (1982a) found that the braconid *Cardiochiles nigriceps* which attacks a solitary host (the tobacco budworm *Heliothis virescens*), searches a patch containing host-associated chemicals for some time but disperses immediately after locating and attacking the caterpillar. In laboratory experiments, Nealis (1986) observed similar behavior after the braconid *Cotesia (=Apanteles) rubecula* encountered its host, larvae of the butterfly *Pieris rapae*. However, this was misleading because field observations showed that the wasp subsequently returned to the site and groomed extensively. This species attacks a very patchily distributed host and shows great site fidelity after locating a suitable host.

RECOGNITION OF PATCH BOUNDARIES

Where hosts are found in circumscribed areas of the environment, it is obviously in the parasitoid's interest to recognize the boundary of the host patch and to turn back when the boundary is crossed. A number of studies have looked at the behaviors associated with detecting the edge of a patch.

Waage (1978) studied the ichneumonid *Venturia canescens* which attacks the larvae of stored product moths. As discussed in section 2.2.2, the parasitoid uses a chemical excreted by the mandibular gland of its host as an arrestant stimulus. The chemical comes to contaminate the host's foodstuff and serves to delimit a patch for searching parasitoids. Waage observed that when the parasitoid encountered an area contaminated by the chemical, it reduced speed and increased its rate of turning. When the parasitoid crossed the edge of the patch, it turned back and reentered the patch. The braconid *Cardiochiles nigriceps* shows essentially the same behavior when encountering leaves containing traces of the mandibular gland secretions of the tobacco budworm (Strand and Vinson 1982a), as does the trichogrammatid egg parasitoid *Trichogramma evanescens* when it encounters a leaf surface contaminated by moth wing scales (Gardner and van Lenteren 1986). Vet and van der Hoeven (1984) compared the responses of two species of the eucoilid genus *Leptopilina* after encountering host-associated chemicals. Although both species slowed down and began probing, *L. fimbriata* almost came to a stop while *L. heterotoma* moved more quickly across the substrate. The differences probably reflect the distributions of their respective hosts. *L. fimbriata* attacks the drosophilid *Scaptomyza pallida* in decaying plants; hosts are solitary and the area of contamination very small. *L. heterotoma* attacks *Drosophila* species in substrates such as fruit where many hosts feed together, contaminating a large area.

Parasitoids of leaf-mining insects recognize the mine as a patch within which, depending on the species of miner, one or more potential hosts are likely to be found. The boundaries of the mine are normally clearly visible, at least to the human eye, although the parasitoid may detect more subtle cues associated with the texture of the leaf surface. Casas (1988) studied two eulophid parasitoids (*Pnigalio soemius* and *Sympiesis sericeicornis*), which attack very small moths (*Phyllonorycter*) whose larvae make blisterlike mines in the leaves of many trees. When the parasitoid moves from an unmined part of the leaf to an area containing a mine, the mean and variance of its rate of turning increase markedly. On encountering the margin of the mine, the wasp turns back sharply.

PATCH MARKING

A number of parasitoids are known to leave chemical marks in the environment that are recognized by other parasitoids. Chemicals may be deposited as the parasitoid moves through the habitat (trail odors) or when a parasitoid leaves a patch. Price (1970a, 1972a) studied a guild of ichneumonid parasitoids of sawfly pupae which search for hosts on the forest floor. He found that female *Pleolophus basizonus* recognized and tended to avoid areas they had already searched. The wasp deposited a mark as it searched the leaf litter, and the trail odor was recognized not only by conspecifics but also by different species of ichneumonids attacking the same host. Price suggested that the pungent odor associated with many ichneumonids, and which previously had been assumed to be involved in predator defence (sec. 7.6), might be part of the trail-laying mechanism. Vinson (1972a) found that the braconid *Cardiochiles nigriceps* which attacks the tobacco budworm *Heliothis virescens* deposits a chemical on the tobacco leaf that discourages subsequent females from searching on that leaf. Patch marking has also been recorded in the braconid *Orgilus lepidus*, a parasitoid of the potato tuberworm *Phthorimaea operculella* (Greany and Oatman 1972), the braconid *Asobara tabida* which attacks drosophilids (Galis and van Alphen 1981), the encyrtid *Epidinocarsis lopezi* which parasitizes mango mealy bug (*Phenacoccus manihoti*) (van Dijken et al. 1992), and in the ichneumonid parasitoid of stored product moths *Venturia canescens* (Harrison 1985).

It is more common for a parasitoid to mark the host than the patch, although the two behaviors probably have evolved for similar reasons. A discussion of the evolution of patch and host marking is postponed until section 3.3.3.

INTERFERENCE

It is quite frequent for several parasitoids to be found searching a patch simultaneously. When two adult parasitoids meet in a patch, they may ignore each other and continue searching, suspend searching to interact with each other, or one or both parasitoids may leave the patch. Population dynamicists have long been interested in cases where interactions between parasitoids lead to a reduction in searching efficiency and have called this process "interference." Interference is density dependent and may be important in stabilizing the population dynamics of hosts and parasitoids (Hassell and Varley 1969).

Hassell (1978) reviews a large literature on interference, chiefly concerned with population dynamic issues. In laboratory studies, the probability of emigration from a patch is frequently a function of parasitoid density although there is seldom evidence of the proximate cause for dispersal. Some doubt has been cast about the relevance of laboratory observations to the field; the densities of parasitoids used in laboratory experiments are typically an order of magnitude higher than those encountered in the wild (Griffiths and Holling 1969). In the rest of this chapter I consider the evolution of patch-use strategies which often result in density-dependent reductions in searching efficiency that population dynamicists would describe as interference or pseudo-interference (see Visser and Driessen 1991 for further discussion).

2.5.2 PATCH-USE THEORY

CLASSICAL FORAGING MODELS

Two areas of foraging theory are particularly relevant to the study of parasitoids searching in patchy environments: the classical theory of patch use, and the theory of ideal free distributions. Here I briefly review the relevant theory; introductions to foraging theory are provided by Stephens and Krebs (1986) and Krebs and Kacelnik (1991).

Classical patch-use models predict when an animal, foraging alone, should leave one patch and begin searching for another. It is assumed that the animal depletes the patch as it forages, and that it is selected to maximize its rate of gain of fitness. In practice, some more convenient currency, such as the rate of gain of energy, is used to substitute for fitness. When the animal enters the patch, its rate of gain of fitness is initially high but then drops as the patch is depleted. Charnov's (1976) marginal value theorem shows that the optimum time to leave a patch is when the instantaneous rate of fitness gain (i.e., the marginal fitness) drops to the maximum average rate that can be achieved in that environment. This argument is much easier to explain graphically (fig. 2.6). Factors that reduce the maximum rate of fitness gain in that environment lead to greater patch residence times. For example, animals will spend more time per patch when travel time between patches is high. If the environment consists of patches of different quality, animals are predicted to spend a relatively longer amount of time in high-quality patches.

Classical patch-use models predict that the animal leaves the patch when its rate of gain of fitness drops to a threshold value. A frequent criticism of this model is that the animal is assumed not to gather information about the quality of the patch as it forages (Stewart-Oaten 1977; Green 1980, 1984; McNamara 1982). Where patches vary greatly in quality, and where this variation cannot be recognized prior to exploitation, alternative patch-leaving rules perform better. Stephens and Krebs (1986) discuss these criticisms and point out that, in general, large variances in patch quality will be recognizable prior to exploitation and that when variances are small, the marginal value theorem performs quite well. When patch assessment occurs, foragers will tend to stay longer on good patches and spend a shorter time on poor patches than would be predicted by a simple application of the marginal value theorem.

A second assumption of simple patch-use models, violated by some parasitoids, is that the animal sequentially visits a relatively large number of patches. (The mathematics underlying the marginal value theorem is based on renewal theory which requires this assumption.) Some parasitoids will visit only relatively few patches over their lifetime, and in these cases a different approach, based on dynamic programming, must be adopted (see below).

The second area of foraging theory that is important for the analysis of parasitoid searching addresses the problem of how groups of foraging individuals should distribute themselves across a patchy environment. Consider a group of individuals foraging in an environment which contains nondepletable patches of different quality, and suppose that travel costs between patches are minimal (note the contrast with the depletable patches and important travel costs assumed by classical patch-use theory). Further, suppose that individuals compete together on a patch for the resources it contains, that all animals are identical in competitive abilities, and that animals are able to identify the quality of a patch before entering. Fretwell and Lucas (1970; see also Parker 1970)

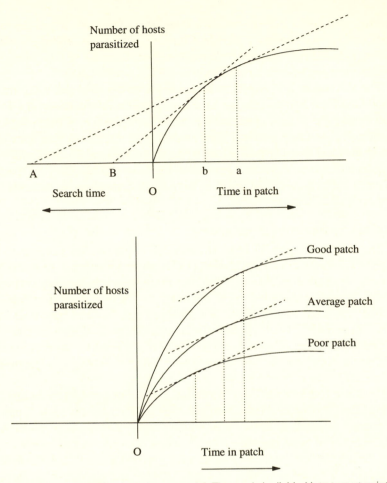

Figure 2.6 Charnov's optimal patch-use model. The *x* axis is divided into two at point *O*. The time spent in the patch is represented by the distance to the right of *O* and the travel time between patches by the distance to the left of *O*. *Top*: Identical patches. The curve represents the cumulative number of hosts attacked by a searching parasitoid. Suppose that the travel time between patches is *A*; the optimal patch residence time *a* is the point where a line rooted at *A* is just tangent to the cumulative parasitism curve. Shorter travel times (e.g., *B*) lead to shorter optimal patch residence times. *Bottom*: Patches vary in quality. The parasitoid should leave a patch when its marginal rate of gain of fitness (i.e., the slope of the cumulative parasitism curve) falls to the maximum average gain rate that can be achieved in that environment (represented by the slope of the three dashed lines). Parasitoids should remain longer in good patches but spend less time in poor patches.

argued that individuals will be selected to distribute themselves among patches such that all animals have the same fitness (this is known as the "ideal free distribution"). Thus better patches will contain more individuals than poorer patches. Any other distribution is evolutionarily unstable as individuals in patches with below average fitness will be selected to move to patches where their fitness would be above average. Competition between individuals may be purely exploitative, though interference competition can also be considered within the same framework (Sutherland 1983; Lessells 1985).

The theoretical and empirical investigation of classical patch-use models and of the ideal free distribution have largely advanced independently. This is unfortunate for studies of parasitoids as the combination of patch depletion and travel costs (as in marginal value models) with many searching individuals (as in the ideal free distribution) are both normally essential ingredients of the problem. However, there are a few studies that have addressed optimal patterns of patch use by parasitoids, and these have incorporated various elements from both strands of theory.

MODELS OF PARASITOID SEARCHING

The first two studies of optimal patch use in parasitoids, Cook and Hubbard (1977) and Comins and Hassell (1979), share a number of features. Both assume depletable patches and consider more than one foraging parasitoid. Within a patch, parasitoids search randomly so that the rate of host depletion can be calculated by a modified disk equation (the "random parasitoid equation," Rogers 1972). Both models also assume a high level of omniscience on the part of the parasitoid and that the costs of traveling between patches are negligible. However, the two studies make different assumptions about what natural selection maximizes.

Cook and Hubbard (1977) consider how a group of parasitoids, foraging for a fixed period of time, should distribute their searching effort across a number of patches of different quality. The quantity maximized in their model is the total number of hosts parasitized by all wasps. This model framework can be criticized on several grounds: first, natural selection maximizes individual fitness, not the fitness of groups of wasps; and second, parasitoids do not normally have a fixed time interval in which to search (Stewart-Oaten 1977). However, if travel time between patches is zero, and in the absence of interference competition, the group and individual optima coincide. This explains why Comins and Hassell (1979), who considered a group of parasitoids, all individually optimizing their rates of gain of fitness and foraging without a fixed time horizon, obtained almost identical results. Lessells (1985) provides a simple verbal model that also gives the same results.

The predictions of both the Cook and Hubbard (1977) and Comins and Hassell (1979) models are illustrated in figure 2.7. Suppose a group of parasitoids enters an environment containing four patches that differ in quality. I shall begin by assuming that parasitoids can recognize instantly previously parasitized hosts. Initially, all parasitoids search in the best-quality patch where they achieve the highest rate of gain of fitness. The best patch will be depleted until it contains the same number of unparasitized hosts as the second-best patch.

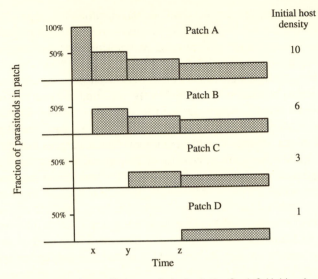

Figure 2.7 An example of a prediction made by both the Cook & Hubbard and Comins & Hassell models. Suppose a group of parasitoids forage in an environment composed of four patch types *A–D* with initial host densities 10, 6, 3, and 1, respectively. Initially all parasitoids should congregate in the best patch type where the rate of host attack is highest. As depletion proceeds, the rate of host attack drops until a point is reached (*x*) when searching in the second-best patch type becomes profitable. The two best patches are then depleted together until time *y*, when the third-best patch type is colonized, and so on until all patches are exploited.

At this point the parasitoids redistribute themselves so that half remain on the original top-quality patch and half search on the second-best patch. Depletion proceeds in parallel on both patches until the number of unparasitized hosts remaining in each patch equals those in the third patch. Another episode of redistribution occurs leading to equal numbers of parasitoids on the first, second, and third patches. After further depletion some parasitoids move to the fourth patch. Throughout the period of patch exploitation, no parasitoid can increase its instantaneous rate of gain of fitness by moving to a different patch. These models thus predict a time-varying ideal free distribution. If parasitoids are unable instantaneously to recognize hosts that have been previously parasitized, then the rate of gain of fitness in a patch will be influenced not only by the number of unparasitized hosts in the patch, but also by the number that have been previously attacked. This complicates the mathematics though the principle of the time-varying ideal free distribution remains the same. Comins and Hassell showed that with finite recognition time, the first parasitoid redistribution occurs before the number of hosts in the top-quality patch drops to the number in the second-best patch, and a little under half the parasitoids transfer from the best patch.

Yamada (1988) has discussed a modified model which he suggests is applicable to parasitoids whose reproductive success is limited by egg supply rather than by time. He assumes that parasitoids cannot detect previous parasitism so that as the exploitation of a patch proceeds, there is an increased risk of wasting an egg by attacking a previously parasitized host (such superparasitism is assumed never to be successful). Whereas the rate-limited Cook and Hubbard and Comins and Hassell models predict that the rate of gain of fitness should be the same in all exploited patches, the egg-limited Yamada model predicts that the risk of egg wastage should be constant. Yamada argues that the risk of wasting an egg is equal to the percentage of parasitized hosts in the patch, and that field observations of constant percentage parasitism across patches is evidence for his model. To obtain his results, Yamada makes the same assumptions as Cook and Hubbard about the action of natural selection on groups of foraging parasitoids and is open to the same criticisms. He also does not allow the strategy of the parasitoid to change as a function of egg load. Host acceptance in egg-limited parasitoids has been modeled using dynamic programming techniques (sec. 3.1.1), which is probably a better way to model patch-use strategies under the same assumptions. More seriously, the assumption of no host discrimination (identification of previously parasitized hosts) is unlikely to be correct for many species of parasitoids (see sec. 3.3.2).

MODELS INCORPORATING MORE REALISTIC BEHAVIOR

While the results of the Cook & Hubbard and Comins & Hassell models are intuitively understandable and mathematically tractable, this simplicity has been attained by making what the authors acknowledge are unrealistic biological assumptions. In particular, the lack of travel costs between patches and the assumption that all parasitoids know the exact distribution of parasitized and unparasitized hosts, as well as the distribution of other parasitoids, will normally be incorrect. To assess the importance of these simplifications, Bernstein et al. (1988) have taken a different approach, abandoning analytical simplicity for the greater realism allowed by simulation modeling.

Like the earlier models, Bernstein et al. (1988) consider a group of animals foraging in a patchy environment, each attempting to maximize its rate of gain of fitness. The particular model they analyze is of a predator (or a parasitoid that can recognize instantaneously previously parasitized hosts) though their qualitative conclusions are likely to be true for most types of parasitoids. The success of an individual predator is affected in two ways by the number of other predators in the patch. First, more predators result in greater competition for prey (exploitation competition). Second, following Sutherland (1983), Bernstein et al. assumed that predator efficiency declines when there are frequent encounters with conspecifics, perhaps because they fight or in other ways interfere with each other (interference competition). If the predators were completely omniscient, they would at all times distribute themselves across patches so that no individual could improve its fitness by changing patches. This would lead to a time-varying ideal free distribution, as predicted by the Cook & Hubbard and Comins & Hassell models, which could be achieved by the predator leaving a patch whenever its rate of gain fitness falls below a critical threshold. Bernstein et al. assumed that the predator had a rule of this sort, but one that had to be learned. In particular, they assumed that the leaving threshold was a linear combination of the current rate of food intake and of past experience; the relative importance of past experience was determined by a parameter called the "memory factor." After leaving a patch, the predator travels (instantaneously) to another, randomly chosen patch. Bernstein et al. compared the performance of their model animals against the ideal free distribution in environments with no depletion, and when depletion was rapid or slow.

In a nondepleting environment, the distribution of predators approaches the ideal free distribution quite quickly (fig. 2.8). Thus predator omniscience is not essential for a predator (or parasitoid) population to achieve an ideal free distribution in a biologically reasonable period of time. In a depleting environment, an ideal free distribution is established and maintained as long as depletion occurs slowly relative to learning (fig. 2.8). If depletion occurs quickly, the predator's estimate of the quality of the environment lags behind the true value and the ideal free distribution is never attained. The fit to the ideal free distribution was always best when the variance in patch quality was greatest—it is hard for the learning process to respond to small differences in patch quality. Bernstein et al. (1988) carried out a sensitivity analysis and concluded that these predictions were largely independent of the details of their model construction.

In a second study, Bernstein et al. (1991) examined the influence of the time taken to travel between patches. As the time costs of travel increased, the model predicted a greater departure from the ideal free distribution. The reason for this is that parasitoids in reasonably good patches will be selected not to leave, and suffer the costs of movement, because of the uncertainty of arriving in a better patch. Bernstein et al. also extended their analysis from purely patchy environments to environments where prey or host density varied continuously. The ability of a predator to respond to prey density is now influenced by the scale over which prey density changes relative to the mobility of the predator. Where prey density changes relatively slowly, the predator fails to obtain a good estimation of the distribution of the prey and a poor fit to the ideal free distribution results.

To summarize, these simulation results suggest that unless host depletion occurs very quickly, searching parasitoids will be able to obtain sufficient information about the distribution of hosts in the environment to make adaptive decisions about patch use. The conclusion is based on defensible assump-

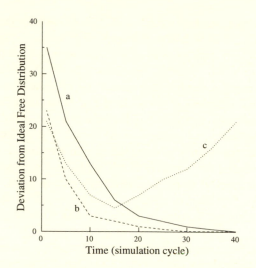

Figure 2.8 The attainment of the Ideal Free Distribution by parasitoids searching with incomplete information; typical results of simulation models developed by Bernstein et al. 1988. The model insect is allowed to search in (a) patchy environments where the supply of hosts in each patch is constant (nondepleting environments); (b) environments where host depletion occurs slowly; and (c) environments where host depletion occurs quickly. The figure shows how closely the behavior of the host matches the Ideal Free Distribution (see text). A good match is quickly attained except in environments with fast depletion, where the match deteriorates with time. (Adapted from Bernstein et al. 1988.)

tions about the nature of parasitoid learning. At the level of the population, an ideal free distribution occurs which changes over time as patch depletion proceeds.

BEHAVIORAL MECHANISMS: RULES OF THUMB

Within foraging theory there has been much interest in simple behavioral rules, so-called rules of thumb, which, when adopted by a forager, lead to behavior that closely approximates the evolutionary optimum. For example, a number of simple rules might be used to decide when to leave a patch: (a) leave after finding a certain number of prey; (b) leave after a fixed time; or (c) leave if no prey is found for a certain period of time (fixed giving-up-time). The relative performance of each rule depends quite critically on the details of prey distribution within and between patches; no one rule is always the best (McNair 1982; Green 1984; Iwasa et al. 1981; Stephens and Krebs 1986). A number of empirical studies have explored whether any of the three simple patch-leaving rules can explain patch residence times in parasitoids (see below).

The three simple rules take only a limited account of experience within a patch. The performance of rules of thumb that involve more sophisticated

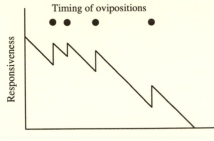

Figure 2.9 Waage's patch-leaving rule. Waage assumes that a parasitoid responds to chemicals secreted into the patch by hosts, and leaves the patch when its responsiveness drops to zero. On entering the patch, the responsiveness is set at a certain value that decays linearly over time unless a host is encountered. The effect of an oviposition is to increase responsiveness by an amount determined by the time since the last oviposition; if two hosts are attacked in quick succession, the second attack leads to a smaller increase in responsiveness than the first attack.

assessment have also been investigated. Green (1984) studied an increment-decay model in which a forager has a giving-up-time that increases by a fixed increment every time a prey item is encountered. In comparison with fixed giving-up-times, this rule performs better when there is variation in patch quality and leads to the forager remaining longer in good-quality patches. Increment-decay models have a long tradition in ethology and, as Stephens and Krebs (1986) note, have obvious neurological analogues.

The patch-leaving rule that has had the most influence in studies of parasitoid searching is due to Waage (1979) and was prompted by observations of searching by the ichneumonid wasp *Venturia canescens*. Waage suggested a parasitoid would leave a patch when its "responsiveness" to the arrestant stimuli on the patch dropped below a critical level (fig. 2.9). When a parasitoid enters a patch, the responsiveness is set at a level determined by the concentration of host-associated chemicals in the patch. More generally, the responsiveness is set by the animal's initial estimate of patch quality. If the parasitoid encounters no hosts, the responsiveness decays linearly so that the wasp leaves after a fixed giving-up-time. When a host is found, the responsiveness increases. However, in contrast to a fixed giving-up rule, the responsiveness does not always return to the same level after host discovery. Waage's rule resembles Green's increment-decay model except that the size of an increment is not constant; if two attacks occur in quick succession, the total increase in responsiveness is less than if the two attacks are widely separated in time.

Although prompted by experimental observations, Waage's model is really an a priori hypothesis with its antecedents in classical ethology rather than behavioral ecology. The main strength of the model is the explicit incorporation of many details of parasitoid biology. The rule undoubtedly performs

better than constant giving-up-times and the other simple rules, but its performance against possible alternatives of the same degree of complexity has not
been studied.

In recent years there has been a shift away from the investigation of specific
hypotheses that are meant to describe the patch-leaving behavior of parasitoids
to a more inductive approach where patch-leaving rules are deduced from the
statistical analysis of real behavior. This approach, described in the next section, leads to complicated, probabilistic rules, but is more likely to provide a
good description of parasitoid behavior.

2.5.3 EXPERIMENTAL STUDIES

TESTS OF PARASITOID SEARCHING MODELS

Hubbard and Cook (1978) performed the first experiments with parasitoids
explicitly designed to test patch-use theory. They worked with the ichneumonid *Venturia canescens* which attacks a variety of stored product moths feeding on bran and similar foodstuffs. Discrete host patches were constructed by
placing different numbers of host larvae in petri dishes filled with bran and
covered by terylene gauze; the gauze stops the host escaping but allows the
wasp to oviposit through the holes. The petri dishes were placed on the floor
of an arena and the whole area was covered with bran. The parasitoid thus
searched a uniform layer of bran with the host patches hidden below the surface. As described in section 2.2.2, *V. canescens* uses a chemical produced in
the mandibular glands of its host as an arrestant stimulus. It is essential that the
host patches are created several days before the experiment so that enough of
the arrestant is produced that the patches are recognizable by the wasp.

Hubbard and Cook's main experiments consisted of observing either one or
two parasitoids foraging for six hours in an arena containing five patches. The
patches contained 64, 32, 16, 8, or 4 hosts. During this period substantial
depletion of the host patches occurred and the parasitoids visited each patch a
number of times. A strong prediction of theory is that all patches should be
depleted to a level where the rate of discovery of new hosts is the same. Hubbard and Cook compared the rate of oviposition in different patches at the
beginning and end of the experiment—oviposition is obvious in *V. canescens*
which has a characteristic "cocking" action after an egg is laid when a new egg
travels down the ovipositor. Initially, oviposition rates were much higher in
the good patches but by the end of the experiment they were roughly equal
across patches (fig. 2.10). Theory also predicts that the wasps should initially
concentrate their search in the best patches and then spread to other patches as
depletion proceeds (e.g., fig. 2.7). The time spent in each patch over the course
of the experiment is shown in figure 2.11. The results are, at best, only weakly
consistent with theory and the match is particularly poor for the experiments
with two searching parasitoids. Wasps tend to spend too much time in the

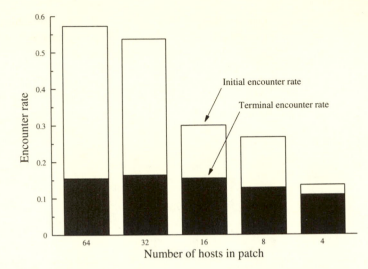

Figure 2.10 The oviposition rate of *V. canescens* in Hubbard and Cook's (1978) patch-use experiments. The open bars represent oviposition rates in patches of different densities at the start of the experiment, and the solid bars are rates at the end of the experiment. These results are corrected to exclude occasional superparasitism and are from experiments in which a single parasitoid searched in an arena. The results when two parasitoids searched are similar although the terminal oviposition rates are slightly more variable.

worst patches and too little time in the best patches, especially near the beginning of the experiments—behavior that perhaps reflects prior expectation. An important assumption of theory is that the wasps know the distribution of hosts across the environment, and clearly this is unlikely to be true for insects searching in a novel experimental arena. Part of the mismatch may thus be explained by the wasp sampling all patches to assess the quality of the environment. Alternatively, the wasp may be unable to estimate host density accurately.

Hubbard and Cook (1978) suggested that the wasps may be using constant giving-up-times as a behavioral rule to decide when to leave the patch. The time between the last oviposition and finally leaving the patch was roughly constant, though rather variable, across patches. They also showed that the time spent in a patch containing a particular number of hosts was influenced by the overall quality of the environment. In their main experiments, the highest-density patch contained sixty-four hosts but in preliminary experiments all patch densities were twice as high so that the second-ranking patch contained sixty-four hosts. Wasps spent more time in patches of this density when there were no higher-density patches in the environment. This suggests that the relationship between host density and searching effort is relative rather than absolute, as theory predicts.

Testing optimal search models in the field is difficult because of the small

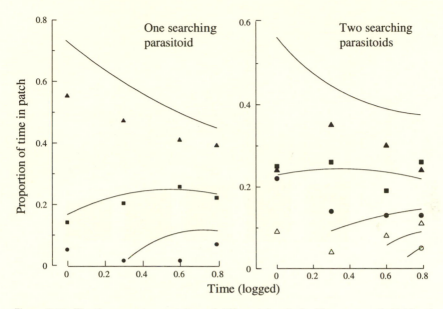

Figure 2.11 The time spent on patches of different density by *V. canescens* in Hubbard and Cook's (1978) patch-use experiments. *Left:* Single foraging parasitoid. *Right:* Two foraging parasitoids. The solid lines are predictions from Cook and Hubbard's (1977) foraging model. ▲, 64 hosts; ■, 32 hosts; ●, 16 hosts; 4, 8 hosts; ○, 4 hosts. In the figure on the left, the single parasitoid spent very little time in the two lowest density patches.

size of most parasitoids. The few studies that have manipulated host density and observed parasitoid behavior illustrate some of the problems likely to be encountered.

Waage (1983) placed variable numbers of larvae of the diamond back moth (*Plutella xylostella*) on small cabbage plants arranged in a 5 × 5 grid in a field. The number of larvae per plant varied from 0 to 16. Individuals of the ichneumonid *Diadegma eucerophaga* moved naturally into the grid from the surrounding vegetation and were observed through binoculars. Waage observed more wasps on higher density patches, as theory would predict (fig. 2.12a). However, percentage parasitism was roughly constant across patches (fig. 2.12b). One possible explanation is that all susceptible hosts were attacked during the experiment; the 30% of larvae that were left unattacked may have been unsuitable for parasitism, or in some way protected from wasp attack. Waage following Hassell (1982), suggested that long handling times might be responsible for the lack of observed density dependence. The host sometimes drops off the leaf on a thread and the parasitoid waits until it climbs back onto the leaf. However, Lessells (1985) has shown that an optimally searching, rate-maximizing parasitoid will always cause density-dependent parasitism, irrespective of handling time. Nealis (1990) also studied patch use by a parasitoid of a cabbage pest, in this case the braconid *Cotesia (=Apanteles) rubecula*

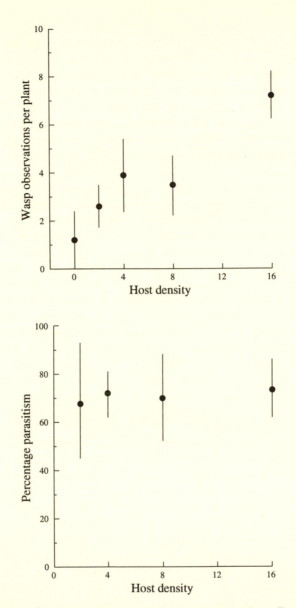

Figure 2.12 Aggregation by the ichneumonid *Diadegma eucerophaga*. *Top*: The number of wasps observed on plants with different densities of hosts. *Bottom*: The resulting percentage parasitism. Means and 95% confidence limits are shown. (From Waage 1983.)

which attacks the cabbage butterfly *Pieris rapae*. In large field cages, he too observed that parasitoids spent more time on high-density patches yet percentage parasitism was roughly constant across patches. *C. rubecula* is attracted by damaged cabbage leaves (Nealis 1986, 1990) and so tends to discover plants that have been severely chewed by many caterpillars. However, once on the patch, heavy feeding damage may actually reduce parasitoid searching efficiency because the wasps waste much time examining the feeding damage. In addition, high-density patches often contain mature larvae, and the parasitoids require to rest and groom for some time after attempting to attack a large caterpillar.

The larvae of *Euphydryas phaeton*, a checkerspot butterfly, live gregariously in large silken webs where they are attacked by the braconid parasitoid *Cotesia (=Apanteles) euphydryidis*. Wasps can be observed crawling over the surface of the web looking for hosts. Stamp (1982) manipulated the number of larvae per web and the number of webs at a site. Wasps tended to congregate on webs containing many larvae leading to greater percentage parasitism. Stamp predicted that as the season progressed, the distribution of wasps across webs would become more even as all webs were reduced to the same marginal profitability; this redistribution was predicted to occur most swiftly where the ratio of parasitoids to webs was high. However, parasitoid attack remained aggregated throughout the season. A number of factors may have frustrated this attempt to test Cook and Hubbard's (1977) model. The encounter rate with host patches (webs) may not have been uniform across the experiment, leading to the unexpected aggregation. Perhaps more importantly, it is unclear whether the patches were heavily exploited during the experiment as the overall percentage parasitism was low.

A few other studies have observed the behavior of adult parasitoids in the field as they searched for hosts in patches of different densities. A.D.M. Smith and Maelzer (1986) counted the numbers of aphelinids (*Aphytis melinus*) attacking oranges infested by red scale (*Aonidiella aurantii*). Fruit with high densities of scale tended to have more searching parasitoids though the relationship was weak; parasitism was density independent. Two other studies manipulated host densities. Summy et al. (1985) found that the aphelinid *Encarsia opulenta* aggregated in high-density patches of citrus blackfly (*Aleurocanthus woglumi*), while Hammond et al. (1993) found similar behavior among encyrtids (*Epidinocarsis lopezi*) attacking cassava mealy bug (*Phenacoccus manihoti*). Thus, field studies suggest that parasitoids do tend to aggregate in patches of high density.

STUDIES OF BEHAVIORAL MECHANISMS: A PRIORI MODELS

At about the same time as Cook and Hubbard, Waage (1978, 1979) was also studying foraging in *Venturia canescens*. The chief aims of his experiments were not explicitly to test particular foraging models, but to explore the mech-

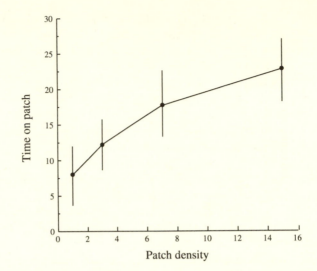

Figure 2.13 Searching by the ichneumonid wasp *Venturia canescens* on artificial laboratory patches. The time spent by naive wasps (means and 95% confidence limits) on their first visit to patches containing different number of hosts. (From Waage 1979.)

anistic behavioral rules employed by foraging parasitoids. The particular behavioral model he developed has already been described (fig. 2.9). Waage's experimental methods were broadly similar to Hubbard and Cook's although he tended to use lower host densities. In an ingenious first experiment (1978), he constructed an experimental chamber where parasitoids searched for hosts over a uniform surface of bran, beneath which were placed patches of hosts. The experimenter was able to swap patches without disturbing the searching wasps. Waage compared the patch residence times of wasps that had been allowed to attack five hosts in quick succession or five hosts spaced out over a period of fifteen minutes. Parasitoids remained longer on patches in the latter treatment, suggesting that the rate of oviposition is important in determining patch residence times.

In a second series of experiments, Waage (1979) explored the length of time that naive wasps remained on patches of different densities and whether they reduced all patches to the same level of profitability before leaving. In these experiments, patches were presented without choice to individual wasps. Waage demonstrated that wasps had both an innate tendency to remain longer on high-density patches (fig. 2.13) and that the marginal rate of oviposition prior to leaving the patch was independent of patch density (fig. 2.14). He found giving-up-times to be highly variable and often to be shorter than the time intervals between previous ovipositions. In a final set of experiments, Waage observed wasps searching in an arena containing six patches of three densities (1, 4, 8) in order to explore the effect of previous experience on the

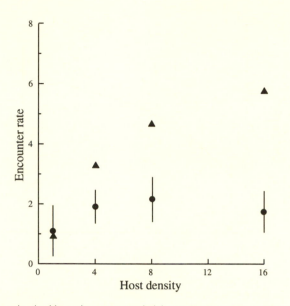

Figure 2.14 Foraging by *Venturia canescens* in laboratory arenas containing host patches of different densities; initial and terminal encounter rates. The solid circles indicate the average number of hosts (means and 95% confidence limits) encountered during the last three minutes of search (Waage 1979). The solid triangles are the average number of hosts encountered during the first three minutes of search (the confidence limits are omitted for clarity, but are of similar magnitude to those of the terminal encounter rates).

time spent visiting a patch. The results were complex and difficult to interpret. For example, the time spent on a patch dropped quickly with successive revisits, possibly because the parasitoid marks a patch before leaving. A curious result was that the wasp spent less time on a patch if the last patch it had visited was of the same density. Waage speculated that if a wasp enters a patch where the concentration of host-associated chemicals is identical to the patch it has just left, it may think that it has just reencountered the same patch. This may be a good assumption in nature where the chances of encountering two successive patches of the same density are remote.

The parasitoids of *Drosophila* are a second popular experimental system for studying patch use. Galis and van Alphen (1981; van Alphen and Galis 1983), building on earlier work by van Lenteren and Bakker (1978), conducted a number of experiments designed to investigate the determinants of patch residence time by the braconid wasp *Asobara tabida* attacking *Drosophila melanogaster* larvae. The wasp detects chemicals secreted by the host (kairomones) into its food medium and can also respond to yeast associated with the feeding sites of its host (see secs. 2.2.1 and 2.3.1). Single patches of hosts were presented to individual parasitoids and the behavior of the parasitoid was observed. The patches consisted of different numbers of hosts in a viscous

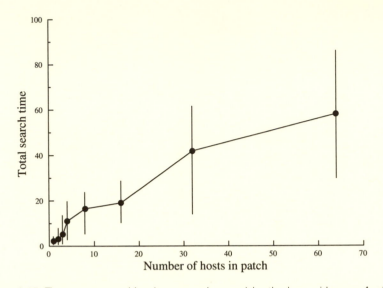

Figure 2.15 Time spent searching (means and ranges) by the braconid wasp *Asobara tabida* on laboratory patches containing different numbers of larval *Drosophila*. (From van Alphen & Galis 1983.)

suspension of yeast on a bed of agar. The patches were carefully constructed so that the numbers of hosts, and the concentration of kairomones, varied while the density of yeast was kept constant. As well as experiments with unparasitized hosts, some wasps were presented with patches containing parasitized hosts and other wasps with patches containing kairomones but with the hosts themselves removed.

The total time spent on a patch increased with host density. Excluding time spent handling hosts, search time also increased with host density (fig. 2.15). Wasps spent less time on patches contaminated with kairomones but containing no hosts. Residence time initially increased with kairomone concentration but soon leveled off (fig. 2.16). Van Alphen and Galis (1983) concluded that successful parasitism increased the likelihood of remaining on the patch. Residence times on patches containing only parasitized hosts were similar to the time spent on patches from which all hosts had been removed. There was thus no evidence that encounters with parasitized hosts act as a strong stimulus to leave the patch (as had been found by van Lenteren 1981, 1991, in another drosophilid parasitoid, the eucoilid *Leptopilina heterotoma*). Giving-up-times were highly variable, increased with host density, and were frequently shorter than intervals between ovipositions. These observations suggest that no simple rule (such as constant time, constant number of ovipositions, or constant giving-up-time) describes *A. tabida* behavior. A further complicating factor is that wasps appear to recognize marks deposited by conspecifics on patches that have just been searched.

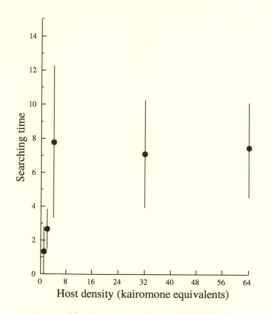

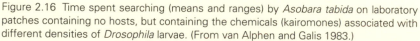

Figure 2.16 Time spent searching (means and ranges) by *Asobara tabida* on laboratory patches containing no hosts, but containing the chemicals (kairomones) associated with different densities of *Drosophila* larvae. (From van Alphen and Galis 1983.)

Foraging in a very different type of patch was studied by Weis (1983). The torymid *Torymus capite* attacks the larvae of the gall midge *Asteromyia carbonifera* which makes blisterlike galls in the leaves of *Solidago canadensis*. Each gall contains between one and four larvae though the large majority of galls (80%) contain just one or two. Weis calculated the number of hosts attacked in each gall discovered by the wasp (his method of calculation is explained in fig. 2.17). The success of the parasitoid increases with the number of hosts in the gall, partly because in large galls not all fly larvae can feed in the center of the gall where they obtain some protection from parasitism. Despite this, parasitoids still seem to leave galls when they contain many unparasitized hosts. The parasitoid cannot assess the number of hosts, and searching a gall is a time-consuming process. The parasitoid moves around the edge of the gall, inserting its ovipositor, and feeling for hosts; each insertion takes about ten minutes, and on average each gall is probed five times. Iwasa et al. (1981) showed that where the variance in patch quality is small, a fixed time-leaving rule may lead to optimal patch exploitation. Noting the low variance in patch quality, Weis (1983) suggested that such a leaving rule might explain patch use in *T. capite*. In laboratory experiments, the time spent on a patch was indeed independent of the number of hosts. Romstöck-Völkl (1990) has also studied patch use by parasitoids attacking concealed hosts. The eurytomid *Eurytoma* (sp. nr. *tibialis*) and the pteromalid *Pteromalus caudiger* attack tephritid flies (*Tephritis conura*) that feed communally in thistle heads (*Cirsium*

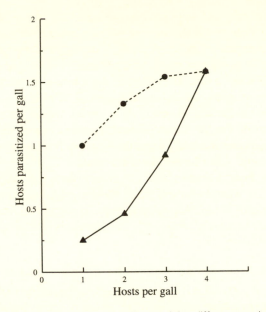

Figure 2.17 Attack by *Torymus capite* on galls containing different numbers of hosts. The dashed line is the number of hosts parasitized per gall, given that at least one host is attacked. Assessing the true risk of parasitism as a function of host density is complicated because the wasp frequently discovers galls but fails to parasitize any of its inhabitants; this is especially true of galls containing a single host, which can hide in the center of the gall. Weis (1983) estimated the numbers of hosts parasitized *per discovered* gall (solid line) as follows. He argued that if a wasp discovered a gall containing four hosts it would always be able to attack at least one host. This enabled him to calculate the fraction of four-host galls that were discovered by wasps. However, laboratory experiments had shown that the probability of gall discovery was independent of the number of hosts it contained. Weis thus assumed that the probability of discovery of any gall was the same as four-host galls, which allowed him to calculate the risk of attack after discovery as a function of host density.

heterophyllum). Many hosts are too deeply buried in the thistle head to be parasitized and it seems unlikely that the parasitoids can assess patch quality. Again, in laboratory experiments, the time spent on the patch was independent of the number of hosts it contained.

Waage's (1979) patch-leaving model envisaged a motivational tendency to remain on a patch that declined in the absence of any reinforcement by oviposition. Sugimoto and his colleagues (Sugimoto et al. 1987, 1990; Sugimoto and Tsujimoto 1988) have investigated a system where Waage's abstract motivational tendency may have a concrete representation. They studied three species of parasitoid (the braconids *Dapsilarthra rufiventris* and *Dacnusa sibirica*, and the eulophid *Chrysocharis pentheus*) which attack a variety of species of leaf-mining agromyzid flies. The wasps search for host larvae in the leaf and, as they search, deposit a water-soluble marker. Wasps tend to spend longer on high-density patches, and experiments excluded the trio of simple leaving rules (constant time in patch, number of ovipositions, or giving-up-time). However, the presence of the marker has a strong effect on patch residence time. Sugimoto et al. suggest that the wasp leaves a patch when the amount of patch marker reaches a critical threshold. As the rate of marker deposition depends on search speed, the cumulative amount of marker is a measure of searching effort on the leaf. A quantitative model fits the data if the further assumption is made that the threshold marker concentration below which dispersal occurs increases markedly when the first host is discovered, and by less dramatic increments when further hosts are found.

To conclude, there is little evidence that parasitoids use simple behavioral rules to determine when to leave a patch. More complicated behavioral rules, based on variants of Waage's hypothesis, provide a better qualitative match to observed behavior. In the next section I discuss an alternative approach to the study of patch-leaving rules.

STUDIES OF BEHAVIORAL MECHANISMS: STATISTICAL MODELING

The behavioral rules discussed so far are a priori hypotheses that are investigated by laboratory experiments. However, there are major problems in statistically assessing the validity of different rules; for example, how good a fit to data constitutes evidence for a particular behavioral rule? An alternative approach is to try and deduce the behavioral rule, with minimal prior assumptions, from the data. The statistical description of behavioral sequences has a long history in classical ethology. Recent advances in statistical theory and computation have provided new techniques that are well suited to the derivation of patch-leaving rules whose performance can be compared with the predictions of classical optimization models. This approach, pioneered by Haccou and colleagues (Haccou and Hemerik 1985; Cuthill et al. 1990; Haccou et al. 1991; Hemerik et al. 1993), is best explained by a specific example.

Haccou et al. (1991) studied patch use in the eucoilid *Leptopilina heterotoma*, the well-known *Drosophila* parasitoid. Wasps were observed parasitizing hosts in the same type of artificial patches used in the experiments of van Alphen and Galis (1983) described above. Parasitoids were presented with patches containing either parasitized or unparasitized hosts, or a mixture of the two. The number and timing of each host attack and rejection were recorded, as well as when the parasitoid left and reentered the patch. If the wasp left the patch for over one minute, the experiment was stopped.

The parasitoid's leaving rule is estimated using a flexible statistical model. It is assumed that each time a wasp oviposits or reenters a patch, a basic leaving tendency is reset; the basic leaving tendency is an unknown increasing function of time. The actual leaving tendency is a product of the basic leaving tendency and a second term which incorporates the effects of past experience on the patch. The aspects of past experience that Haccou et al. reasoned might influence patch leaving were (1) the number of previous ovipositions; (2) the time between each oviposition; (3) the number of rejections between each successful oviposition; and (4) the number of times the parasitoid left the patch.[1] As the overall form of the statistical model must be specified by the investigator, this approach is not wholly without a priori assumptions; however, there are fewer assumptions involved than with previous approaches.

The analysis showed that the total number of host attacks, the time between host attacks, and the number of times the parasitoid had previously left the patch all influenced the tendency to leave; there was no effect of encounters with parasitized hosts. Successful oviposition increases the probability of remaining on the patch. If the intervals between recent ovipositions are small, there is a strong additional tendency to remain in the patch; interoviposition times more than two steps into the past had no discernible influence. The effect of the number of times off the patch is rather complex. The tendency to leave first increases with successive revisits and then decreases again, possibly because the wasp learns there are no nearby alternative patches. Van Lenteren (1991) has also studied patch leaving behavior in *L. heterotoma*, using more traditional statistical methods. His analysis is largely in accordance with Haccou et al. (1991), except that encounters with parasitized hosts appear to increase the wasp's tendency to leave the patch. The explanation for this discrepancy may lie in the experimental design: van Lenteren used much higher frequencies of parasitized hosts than Haccou et al.

Hemerik et al. (1993) have used similar techniques to study the foraging behavior of a related eucoilid, *Leptopilina clavipes*. This species is a mushroom specialist (attacking *Drosophila phalerata*). The artificial patches used in the experiments were 2 cm^2 disks of commercial mushrooms to which were added host larvae and a filtrate containing the chemicals (kairomones) secreted by the appropriate number of hosts. Hemerik et al. observed single parasitoids searching (1) in patches with neither hosts nor kairomone; (2) in patches containing kairomones from four hosts but containing no actual larvae; and (3) in

---

[1] If $\lambda_0(t)$ is the basic leaving tendency, then the actual leaving tendency is

$$\lambda_0 \exp\left[\sum_{i=1}^{n} \beta_i z_i\right],$$

where z_i are the different explanatory variables and β_i are the estimated regression coefficients. The exponential function is used to ensure that the basic leaving tendency is multiplied by a positive number. Interoviposition times are entered as their reciprocals because the parasitoid is likely to be most affected by the rate of discovery of hosts. It was found that the effect of the number of times the parasitoid left the patch was not monotonic, and thus the leaving tendency was made a nonlinear function of this explanatory variable. Note that not all explanatory variables are included in the model at any one time; the time between the last but one and the last but two ovipositions enters only after the parasitoid has attacked three hosts. This type of model is a proportional hazards model, widely used in the analysis of failure times (Cox and Oakes 1984). The unknown regression coefficients and basic leaving tendency are estimated by maximizing the partial likelihood. Test statistics can be derived to allow model criticism.

patches with four hosts plus associated kairomones. They also observed wasps in a series of experiments with four hosts in which the wasp was allowed to oviposit on either one host, two hosts, or to reject a previously parasitized host, before all four hosts were removed—without disturbing the parasitoid. Wasps could leave the mushroom patch, wander around the experimental chamber, and then return to the patch. The effects of a variety of explanatory variables on the tendency (1) to leave, and (2) to return to the patch were studied. While in the patch, the wasp either searched or rested; the tendency to (3) stop and (4) start searching was investigated. Finally, the factors determining the (5) encounter rate during a bout of searching were studied. Although differing in minor details, the methods of analysis were substantially the same as those used by Haccou et al. (1991).

These are obviously very complicated experiments and I shall give only a brief summary of the significant results. The presence of kairomones decreased the tendency to leave a patch but had no effect on the tendency to return. Encounters with hosts reduced the tendency to leave the patch. However, if the first host to be encountered was already parasitized, the tendency to leave sharply increases. Encounters with parasitized hosts after the wasp has begun ovipositing did not affect the tendency to leave. There is an obvious adaptive explanation for this result. Before a wasp begins to oviposit, an encounter with a previously parasitized host is unambiguous proof that the patch has been previously found by another female. After the wasp has begun to lay eggs in the patch, a parasitized host may either have been previously attacked by itself or a conspecific. Encounters with any host, parasitized or unparasitized increased the tendency of the parasitoid to return after leaving a patch. If the last interoviposition time was short, the tendency to return was further increased. The tendency to stop bouts of searching, and the tendency not to resume searching, increased with time during the experiment; this effect was made stronger by encounters with parasitized hosts, and weakened by encounters with unparasitized hosts. The encounter rate with unparasitized hosts during a bout of searching increased after successful oviposition, but declined as the interval between ovipositions increased.

Statistical modeling well illustrates the complexity of the behavioral rules used by parasitoids and explains why early studies on simple rules, such as constant giving-up-times, tended to give ambiguous results. At a qualitative level, most of the patterns revealed in these two studies are in agreement with foraging theory—wasps tend to leave exploited patches and to remain longer on good patches. However, the exciting challenge of a quantitative match between optimality and statistical models has yet to be attempted. It is encouraging that the statistical models suggest new hypotheses for exploration. Hemerik et al. (1993) speculate about why encounter rates and the length of search bouts are initially high but decay during patch exploitation. Previous theory had implicitly assumed that wasps spent all the time on a patch, searching. Possibly there is a trade-off between searching and other basic activities such as cleaning (like most parasitoids, *Leptopilina* spp. keep themselves scrupulously clean). A parasitoid that finds itself on an unexploited patch may be selected to search for hosts as quickly as possible to preempt other females that might arrive subsequently.

PATCHES WITH LINEAR CUMULATIVE FITNESS CURVES

The experiments described so far in this section have involved patches where hosts are hidden and thus not immediately apparent to the parasitoid. In consequence, the curve describing the cumulative gain in fitness rises steeply at first but with decelerating slope (fig. 2.6); in other words the rate of host discovery declines the longer the parasitoid remains in the patch. However, some parasitoids forage in host patches where all hosts are easy to discover and the parasitoid can move from host to host until none remain. If the cumulative fitness gain is plotted against the time in the patch, the result is a linear rise until the last host is discovered (fig. 2.18). The optimal strategy for the parasitoid is both clear and trivial; the insect should parasitize all the hosts in a patch, and then search for a new patch.

A good example of this type of patch is provided by leafhopper egg masses. Many species lay their eggs in neat rows in leaf slits. The eggs are attacked by minute mymarid wasps (often, and rather confusingly, called fairyflies). For example, *Anagrus optabilis* attacks egg masses of the sugarcane hopper (*Perkinsiella saccharicida*). Sahad (1984) describes the oviposition behavior of female wasps which do "not leave the egg mass until all the eggs are parasitised. Even after completion of parasitization of the entire egg mass the parasitoid does not leave the place and repeatedly checks the egg mass to find healthy eggs." A second species of mymarid (*Gonatocerus* sp.) also "never leaves [the egg mass] until all the eggs are parasitised" (Sahad 1982).

However, some mymarids do leave the egg mass before all hosts are attacked (Waloff and Jervis 1987). In particular, another *Anagrus* species, *A. delicatus*, a parasitoid of the eggs of leafhoppers (*Prokelisia*) in North American saltmarshes, regular abandons egg masses when hosts remain to be parasitized (Stiling and Strong 1982; Antolin and Strong 1987; Strong 1989; Cronin and Strong 1990a, 1990b). The wasp leaves the host patch despite carrying apparently mature eggs and may even walk over unparasitized hosts before dispersing. This behavior appears to be maladaptive as the probability of mortality before the discovery of the next host must be quite high.

Strong and colleagues have suggested that premature dispersal may be adaptive because it results in "spreading of risk" or "bet-hedging." The idea is encapsulated in the old adage: don't put all your eggs in one basket. Salt marshes are risky places for insects; very frequently host patches are completely destroyed, for example by high tides. By leaving one host clutch to attack another, the parasitoid increases the probability of at least some of her progeny surviving. In effect, the parasitoid is accepting a lower mean fecundity in order to attain a lower variance in fecundity. The most straightforward

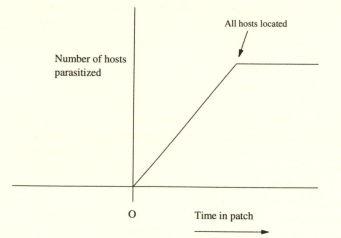

Figure 2.18 The cumulative gain of fitness for a parasitoid searching in a patch where all hosts can be immediately discovered.

circumstance in which bet-hedging strategies can evolve is when there is temporal variation in the fitness of different strategies. The correct measure of fitness is the geometric mean fitness across generations. High variances tend to depress the geometric mean fitness and as a result low variance strategies may be successful, even if their arithmetic mean fitness is lower than some alternatives. However, in the case of *A. delicatus*, spatial rather than temporal heterogeneity appears to be involved. Spatial heterogeneity per se does not lead to selection for bet-hedging; such strategies only evolve if spatial heterogeneity causes temporal variation in fitness. A theoretical demonstration that bet-hedging can occur in this system is lacking. In addition, the costs of dispersal would appear to be very high and it is difficult to believe they could be offset by the advantages of reduced variance.

The evolution of bet-hedging in *A. delicatus* needs to be explored by modeling the geometric mean fitness of different strategies in plausible environments. A different and unusual approach to this problem has been taken by Mesterton-Gibbons (1988). He began with the premise that parasitoid behavior should be an optimal compromise between the total number of hosts attacked, and the number of host clutches in which some eggs are laid. Vector optimization techniques were used to calculate the optimal time to leave a patch as a function of age. The results were consistent with field observations. One problem with this approach is that an arbitrary decision has to be made on how to assess the worth of different compromises. A much more important problem is that the model begins with the assumption that natural selection will favor a compromise. Given this assumption all else follows, but the validity of the assumption is the main question.

At the moment there is no clear alternative hypothesis to risk spreading.

Although Mymaridae are normally considered to be pro-ovigenic, experiments are needed to check that dispersing parasitoids really do have mature eggs because it is sometimes very difficult to distinguish nearly mature from mature eggs (but see Cronin and Strong 1990b). It would also be useful to know the survival of host clutches as a function of the number parasitized; conceivably heavy parasitism might attract predators. But in summary, the behavior of *Anagrus delicatus* is a fascinating unsolved puzzle.

PATCH USE AND SEASONALITY

Parasitoids searching at some times of the year may be at greater risk of death or running out of hosts than those searching at other times. If the parasitoid is able to perceive the time of year, it may be selected to adjust its searching behavior. In particular, if the parasitoid anticipates approaching death or host exhaustion, it may both accept poorer quality patches and poorer quality hosts. These ideas have been modeled by Roitberg et al. (1992) using stochastic dynamic programming. They assumed that parasitoids were able to detect the time of year, perhaps using day length, and examined the optimum host location and oviposition strategies of insects encountering environments composed of patches containing unparasitized hosts (good worlds) and environments where most hosts have already been attacked (bad worlds).

The model predicts that toward the end of the season, all parasitoids, whether they have experienced good or bad conditions, should remain for a relatively long period of time in each patch and should be relatively willing to superparasitize. Near the beginning of the season, parasitoids in the good world should avoid superparasitism and should not linger in the patch. Early in the year, parasitoids in the bad world should remain longer in the patch and should superparasitize, although in both cases less than late-season parasitoids. The results can be understood by considering the costs of spending further time in partially exploited patches and using up eggs for limited fitness returns on previously parasitized hosts: the costs will be small when there is little chance of finding many more patches before death, or when the likely quality of any patches that are discovered is poor.

To test their ideas Roitberg et al. (1992) carried out an experiment with *Leptopilina heterotoma*, the eucoilid parasitoid of *Drosophila*. Wasps were kept under either short or long photoperiods to simulate summer and autumn conditions. The summer wasps were allowed for two days to attack patches of unparasitized (good world) or parasitized (bad world) hosts, although on the day of the experiment all wasps were tested on patches of parasitized larvae. For logistic reasons only one autumn treatment was conducted (good world). The results supported the theory: wasps in the summer good-world treatment spent longer in the patch and were less willing to superparasitize than wasps in the autumn good-world treatment. Summer bad-world wasps spent longer in the patch than summer good-world species, although in this comparison there was no difference in the rates of superparasitism.

These interesting experiments are unusual in linking searching behavior to broader questions of life history and age-specific strategies. It will be interesting to see whether similar behavior is found in other species, especially parasitoids with two or more discrete generations.

COEVOLUTION OF HOST AND PARASITOID DISTRIBUTIONS

It is clear that parasitoid attack leads to many counteradaptations by the host (e.g., sec. 7.4). Is it possible that the distribution of hosts across patches may have evolved as a response to parasitoid attack?

Many herbivorous insects lay their eggs in large discrete clutches which constitute a patch for searching egg parasitoids. In many cases, the number of host eggs present are far in excess of the number that can be attacked by a single parasitoid. Parasitoid attack can lead to selection in favor of the production of large clutches of eggs (Godfray 1986a, 1987a), though the advantage of large clutches depends quite critically on the number of hosts a parasitoid can attack and the relative rate of discovery of different-sized clutches. Parasitoid attack may also influence gregariousness among actively moving hosts. In addition to the same advantages that accrue to large clutches, individual larvae may be selected to associate with conspecifics to reduce their personal risk of parasitism—i.e., to form a selfish herd (Hamilton 1961). There may also be physical advantages of large clutches independent of the searching behavior of the parasitoid. Some moths lay their eggs in irregular mounds, and those eggs at the center are physically protected from parasitoids (Dowden 1961). It is also possible that larval parasitism contributed to the evolution of gregariousness in some butterflies and moths. The larvae of a number of species live together in silken webs which give some protection from parasitoids (Stamp 1981a).

A much more subtle interaction between host distribution and parasitoid attack has been suggested by Thompson (1986), prompted by field observations of a searching parasitoid. *Lomatium dissectum* is a large umbelliferous plant whose flower head (umbel) is divided into discrete clusters of flowers called umbellets. Each umbellet is made up of approximately fifteen flowers, each of which produces two seeds that together form a schizocarp. A small incurvariid moth called *Greya subalba* lays eggs on the schizocarp in which its larvae feed. An undescribed braconid wasp in the genus *Agathis* parasitizes the larvae. The wasp flies between umbellets and then walks among the schizocarps, probing some with her ovipositor to detect larvae. Thompson observed the behavior of searching wasps, recording which schizocarps they probed; afterwards he collected the umbellets visited by wasps to determine which schizocarps contained hosts.

Uninfested and infested schizocarps were equally likely to be inspected, and it appears that the wasps are unable to detect a larva prior to probing with their ovipositor. The time spent by a wasp on an umbellet was independent of host density; there was no evidence that a wasp remained longer on a patch (umbellet) after discovering a host. Thompson found that the distribution of infested schizocarps across umbellets could be described by a truncated geometric distribution. The distribution of hosts across patches strongly affects what a parasitoid can deduce about the number of hosts remaining in a patch after the discovery of the first host. Suppose that hosts have a truncated regular distribution: some patches have no hosts but all the rest have exactly the same number, say for illustration five. If a parasitoid finds one host it can assume four remain; when it finds the fifth host it can assume none remain. Contrast this with a highly aggregated distribution of hosts; the more hosts a parasitoid finds, the greater its estimation of the numbers undiscovered. The truncated geometric distribution is intermediate, the discovery of one host imparts no information whatsoever; the parasitoid cannot assume that the rest of the patch is either better or worse than the rest of the environment. Its optimal strategy is to distribute its search effort across patches using the same truncated geometric distribution as that of its host. Thompson found that the truncated geometric distribution provided a good fit to the number of schizocarps per umbellet probed by the *Agathis* sp. He concluded that the distribution of eggs by the host had evolved as a response to parasitoid attack, and that the searching behavior of the parasitoid was adapted to the distribution of hosts.

This is a fascinating idea and deserves more attention. A number of issues need to be resolved. Although the truncated geometric provides a good fit to the data, other distributions also perform well, for example the truncated poisson. Thus, whether the discovery of a host really provides no information for a searching parasitoid needs to be further investigated. A second problem is theoretical rather than empirical. The distribution of hosts across patches is a property of the population, rather than of individuals. It needs to be demonstrated that the optimal oviposition behavior of individuals results in a populationwide truncated geometric distribution of eggs. In addition to the risk of parasitism, the egg-laying behavior of an individual will be influenced by the availability of oviposition sites, the risks of moving between sites, and the behavior of other individuals in the population.

The tephritid fly *Eurosta solidaginis* makes galls on *Solidago altissima*, where it is attacked by a predacious stem-boring beetle *Mordellistena unicolor* (Mordellidae). The beetle is host to an unnamed braconid in the genus *Schizoprymnus*. In both observational and experimental studies, Cappuccino (1992) found no relationship between either parasitism and beetle density across patches, or between the densities of beetle and tephritid. However, greater levels of parasitism were found in patches of high gall density. Cappuccino speculates that the parasitic wasp finds it difficult to locate the beetle directly and instead uses its prey, the tephritid, in host location. She further suggests that this may result in selection on the beetle to avoid patches of high prey density. Here is another fascinating experimental system that deserves more detailed study.

POPULATION PATTERNS

In addition to the rather few studies that have observed adult parasitoids foraging in the field, a much larger number of studies have recorded the resulting spatial pattern of parasitism (reviews in Lessells 1985; Stiling 1987; Walde and Murdoch 1988; Hassell and Pacala 1990). The motivation for most of these studies was to understand the role of spatial heterogeneity in population dynamics. A number of workers have asked whether optimal parasitoid searching behavior promotes population dynamic stability. This is a difficult question and still not fully resolved. The question can be separated into parts: first, the problem of whether selection on parasitoid searching behavior leads to direct, inverse, or no density dependence in parasitism across patches; and second, the problem of how the pattern of density dependence affects dynamic stability.

The most careful analysis of the first part of this problem is due to Lessells (1985). She showed that a population of parasitoids consisting of individuals each selected to maximize their rate of gain of fitness would, when attacking patchily distributed hosts, cause density-dependent parasitism. The model she considered was equivalent to those of Cook and Hubbard (1977) and Comins and Hassell (1979). This conclusion is independent of handling time and mutual interference. However, domed and inverse density-dependent patterns can occur if the parasitoid is time or egg limited, and if it enters patches at random. In the simplest case, a parasitoid that searches patches which contain many more hosts than it has eggs will attack a constant number of hosts per patch and thus generate inverse density dependence. Patterns other than density dependence can also occur if parasitoids use patch exploitation rules that involve assessment of host density while foraging in the patch (Iwasa et al. 1981; Lessells 1985). There is thus a straightforward answer in the simplest case, but no general result for more complicated situations.

Care must be taken in interpreting the results of surveys of patterns of density dependence in the light of parasitoid searching theory. This theory predicts patterns of patch use over small spatial scales and normally assumes that all patches are equally accessible to all searching females. Many published accounts of parasitism in patchy habitats concern data collected over much larger spatial scales. In these cases, the type of density dependence will be influenced not only by foraging strategy, but by parasitoid dispersal, population dynamics, and by large-scale environmental heterogeneity.

The second part of the problem is to understand how the patterns of parasitoid aggregation generated by optimally searching parasitoids affect dynamic stability. Consider first the effects of arbitrarily chosen patterns of aggregation.

Aggregation has been closely studied in discrete-generation models where parasitoids are assumed to distribute themselves among patches at the beginning of the season where they remain until death (Hassell and May 1973, 1974,

1988; Hassell 1978, 1982, 1984; Hassell et al. 1991; Pacala and Hassell 1990; Pacala et al. 1990; May 1978; Chesson and Murdoch 1986). All forms of aggregation—density dependent, inverse density dependent, and density independent—promote stability. The reason for this is that aggregation causes a reduction in the per capita efficiency of parasitoids as density increases: the reduction happens because parasitoids tend to occur in the same patches and compete for the same hosts. The consequent temporal density dependence in parasitoid efficiency acts to stabilize the interaction. Another way of looking at the increased stability is to note that parasitoid aggregation in certain patches results in a refuge for hosts in those patches where parasitoid density is low. These simple conclusions do depend on limited parasitoid mobility between patches.

A different approach is to assume that parasitoids constantly redistribute themselves so that there is a fixed relationship between the number of searching parasitoids in each patch and the density of unparasitized hosts. Contrast this assumption with the single episode of redistribution in the classical discrete generation models. Murdoch and Stewart-Oaten (1989) analyzed such a model in continuous time and concluded that aggregation normally had little effect on dynamic stability or tended to destabilize the interaction. The nature of the biological assumptions required to justify Murdoch and Stewart-Oaten's model have been the subject of some debate (Godfray and Pacala 1992; Ives 1992; Murdoch et al. 1992). However, what seems certain is that the stabilizing effect of aggregation is reduced by high rates of parasitoid movement among patches. Movement tends to reduce the temporal density dependence in parasitoid efficiency because it causes parasitoids to travel to patches where the hosts have been less exploited. The effect of redistribution will be least marked where aggregation occurs in certain patches irrespective of host density and most marked in cases where parasitoids preferentially move to patches with high densities of unparasitized hosts. At the time of writing, the amount of redistribution required to undermine the stabilizing influence of aggregation is unclear but under investigation.

Studying the population dynamic consequences of optimal patterns of parasitoid distribution across patches is much more difficult. Comins and Hassell (1979) embedded their searching model within a discrete generation host-parasitoid population model and concluded that at least in some circumstances optimal searching behavior may promote population persistence. The explanation for this result is that as parasitoid numbers increase, insects in the optimal set of patches compete more strongly with each other so that parasitoid efficiency is density dependent. However, the optimal redistribution of parasitoids acts strongly to reduce competition, and the stabilizing effect of this form of aggregation is considerably less than when parasitoids are distributed in a clumped manner across patches where they remain throughout their adult lives. To simplify the mathematics, Comins and Hassell analyzed a model in which hosts occur in a single high-density patch and in many low-density patches, a distribution which they point out maximizes the stabilizing effect of the optimal searching behavior. I suspect that with more realistic host distributions, optimal searching behavior will contribute little to the stability of host-parasitoid interactions.

2.6 Conclusions

We now know an impressive amount about the diverse cues that parasitoids employ in locating their hosts. The vast majority of the research in this field is funded by applied agencies, but the results are of immense importance in obtaining a fundamental understanding of parasitoid biology. Paralleling the behavioral advances discussed in this chapter are equally impressive advances in the understanding of the biochemical and physiological bases of host location. Two disadvantages of funding from applied sources is that most attention is naturally concentrated on parasitoids of economically important pests, and most species are studied in an artificial environment. This may lead to a false impression of the types of behavior most important in parasitoids and also leads to few comparative studies. Some of the earliest work on learning in insects was performed on parasitoids, and while parasitoid biologists have never ignored learning, it is probably fair to say that the field is currently enjoying a renaissance. Quite astonishing levels of behavioral opportunism are being demonstrated in many species.

The second major research thrust in this field has concerned parasitoid behavior in patches. The initial euphoria about the application of simple models from mainstream behavioral ecology to parasitoid searching has largely evaporated. The great variance in patch quality normally experienced by searching parasitoids as well as the variation in the numbers of parasitoids searching together put a premium on resource assessment. Parasitoids can also use the presence and frequency of previously parasitized hosts as an additional source of information, a strategy unavailable to most foraging animals. The importance of resource assessment, plus the fact that many parasitoids encounter relative few hosts in a lifetime, conspire to frustrate the employment of simple models. Predicting patch-use behavior in parasitoids using evolutionary models remains an important goal. I suspect the best approach is by using fairly specific models geared to particular host-parasitoid interactions.

An important issue in behavioral ecology is understanding how animals implement adaptive behaviors. I believe that the matching of evolutionary and mechanistic behavioral models is the area of foraging/searching theory in which studies with parasitoids are likely to have the greatest impact on behavioral ecology. Although parasitoid behavior is complicated, it is undoubtedly simpler than that of most vertebrates. This, together with the convenience of working with parasitoids in the laboratory, facilitates the dissection of behavioral mechanisms. Though a priori models of patch-leaving rules have proved useful hypotheses for experimentation in the past, I also believe that the way ahead is through the statistical analysis of behavioral rules. Comparing the predictions of evolutionary searching models with the performance of statistically derived behavioral rules is an exciting prospect in parasitoid biology.

Much of the research on parasitoid behavior in patchy environments has been prompted by population dynamic questions. Theoretical studies have shown that aggregation can be an important factor stabilizing host-parasitoid interactions. However, the exact role of aggregation is unclear, and at the moment subject to considerable debate. In addition, the type of density dependence generated by adaptive patterns of parasitoid foraging is not always certain. Both theoretical and experimental studies are required to resolve these problems; my hunch is that between-patch heterogeneity in parasitoid attack is important for population stability, but that this variability is not generated by parasitoid searching behavior but instead reflect larger-scale environmental heterogeneity that influences the distribution of parasitoids.

3   Oviposition Behavior

After a parasitoid locates a host, it is faced with a sequence of decisions concerned with oviposition. Should the host be used for oviposition, or should it be ignored, or used for food? If the host is accepted, gregarious parasitoids must decide how many eggs to lay. Very frequently, parasitoids encounter hosts that already contain eggs or larvae from previous attacks. Should the parasitoid ignore such hosts or attack them again?

In this chapter I will examine how natural selection has molded the various decisions made by the ovipositing parasitoid. One further question, what sex progeny the parasitoid should produce, is discussed in the next chapter. The first section considers when a parasitoid should accept a host for oviposition and the choice between parasitism and host feeding. The second section discusses the evolution of clutch size, and the final section examines the problem of superparasitism.

3.1 Host Acceptance

The hosts encountered by a parasitoid will often vary in their quality as food for the developing young, in the time needed for their attack and parasitism, and possibly in the risks of mortality to the female during oviposition. Variation in host quality often depends on the age of the host. Most hosts are immature insects and grow in size as they age; larger hosts are often better food sources for parasitoid larvae, though they may be better defended, both physically and physiologically. Many parasitoids attack a number of different host species which vary in their suitability as oviposition sites.

Parasitoids should obviously attack the best hosts and ignore any hosts in which the probability of successful larval development is zero. But how should a parasitoid treat hosts of intermediate quality? I first examine some models of host acceptance in parasitoids and then discuss experimental work designed to test their predictions. In the third section I describe some of the behavioral mechanisms involved in the assessment of host quality. In the final section I consider host feeding, the use of the host as a source of food rather than as an oviposition site.

3.1.1 THEORY

There are two main types of optimal host acceptance models. The most straightforward type of model assumes that the parasitoid's decision is unaffected by its internal physiological state—for example, by the number of eggs in its oviduct. These models are termed "static optimization models" in contrast to "dynamic models" which allow the parasitoid's decision to vary with internal state. Static models are solved using graphs and calculus, while dynamic optimization models require more advanced techniques such as stochastic dynamic programming. Although models of the second kind are obviously more realistic, they are considerably more complicated and it is more difficult to extract generalizations.

STATIC OPTIMIZATION MODELS

The simplest model of host acceptance in parasitoids is formally identical to the optimal diet model of classical foraging theory (MacArthur and Pianka 1966; Stephens and Krebs 1986). In this model, natural selection is assumed to maximize the rate of energy intake. Prey items vary in energy content, in handling time (the time needed to subdue and consume the prey item), and finally in the rate at which they are encountered by searching hosts. If prey are ranked by their profitability (energy content divided by handling time), it is easy to show (e.g., Krebs and Davies 1987) that energy intake is maximized by eating all prey with profitability above a certain threshold value and ignoring all prey beneath this cutoff. Prey should either be eaten or ignored, partial acceptance is never predicted (the zero-one rule). The model also makes the strong prediction that the exclusion of low-ranking prey does not depend on the rate at which they are encountered by the predator, but only on the encounter rate with high-ranking prey. As Krebs and McCleery (1984) put it, when good-quality prey are sufficiently abundant, the predator should never take "time out" from searching for these prey to consume less profitable items.

The optimal diet model is directly applicable to parasitoids whose reproductive success is limited by the time available for searching (Iwasa et al. 1984). Instead of maximizing the rate of energy gain, the parasitoid is assumed to maximize the rate of gain of a quantity such as the number of surviving offspring or the number of grandchildren (host quality frequently affects parasitoid size and hence fecundity). Of course, the underlying assumption is that the parasitoid maximizes fitness, but surviving offspring or grandchildren are used as surrogate currencies. If the profitability of a host is defined as the fitness gain by the parasitoid from oviposition divided by handling time, the model predicts a threshold profitability below which hosts are ignored. As in the diet model, hosts should either be accepted or ignored, and the position of the threshold is unaffected by changes in the abundance of hosts of lower profitability. In general, the range of hosts attacked by a parasitoid should be narrow in rich environments with many good-quality hosts and broad in poorer environments with fewer good-quality hosts.

Like the optimal patch-use model discussed in the previous chapter, the optimal diet model assumes that the animal encounters a relatively large number of prey items (a requirement of renewal theory, the mathematical basis of the model). While this is a fair assumption for most foraging animals, some parasitoids attack only a relatively small number of hosts in a lifetime. A limited life expectancy will tend to increase the range of hosts acceptable to a parasitoid. In the limit, if a parasitoid encounters only a single host, it should be attacked, however poor in quality. Of course, if parasitoids encounter many hosts over a long period of time, a second assumption of static optimization models can be undermined: it may be invalid to assume a single, time-invariant optimum strategy (Mangel 1989a).

Oviposition may be a risky undertaking for a parasitoid if it exposes the insect to increased mortality. If the risks of mortality while ovipositing differ among host types, the optimum choice of hosts to attack may be affected. Iwasa et al. (1984) incorporated differential mortality of ovipositing females into a host acceptance model. They predicted that parasitoids should rank the quality of hosts not by profitability alone, but by a criteria that takes into account the risks of mortality. Not surprisingly, hosts that pose a high risk of mortality to the female are relatively downgraded. Hosts with a long handling time may also be downgraded, as a long handling time increases the period during which the parasitoid is subject to an elevated risk of mortality.

Many parasitoids mature eggs throughout their life (synovigeny). If a parasitoid chooses to attack a wide range of hosts, it must increase its rate of egg production. There are likely to be costs to increasing egg production and these costs may influence the optimal set of hosts acceptable for oviposition. Charnov and Stephens (1988) explored this trade-off using a model in which they assumed that the risks of mortality increased linearly with the rate of egg production. The predictions of their model are identical to the ordinary host acceptance model if the profitability of a host type is calculated in a different way. Instead of dividing the fitness gain from oviposition in a host by handling time, fitness is divided by handling time plus a constant. The constant is a measure of how strongly the rate of egg production affects the risk of mortality (in fact, the slope of the linear relationship). The effect of the redefinition of profitability is to increase the ranking of host types in which oviposition leads to a large gain in fitness. As oviposition reduces longevity, the parasitoid is selected to oviposit preferentially on those hosts where oviposition leads to the highest rewards. Charnov and Stephens also point out that if a parasitoid has a restricted supply of eggs, and no possibility to mature new eggs, it should reduce the set of acceptable hosts. In the limit, a parasitoid with one egg should place it on the best host, even if this host has a long handling time.

The optimal diet model has been a major plank of foraging theory and many more realistic variants have been proposed and analyzed (review in Stephens and Krebs 1986). Some of these variants may be applicable to host acceptance in parasitoids. Van Alphen and van Harsel (1982), Charnov and Stephens (1988), and Janssen (1989), for example, have argued that it is important to include the time taken to recognize different host types in optimal host acceptance models. The major effect of including recognition time is that the position of the threshold quality for oviposition is influenced by the abundance of low-ranking hosts (Hughes 1979; Houston et al. 1980). When poor hosts are common, the time required for their identification and rejection may reduce the maximum achievable average rate of fitness gain in that environment to a point where their inclusion in the set of hosts attacked is no longer disadvantageous. Another factor that may need to be taken into account when studying parasitoid host acceptance is nonrandom encounter rates. Classical foraging theory assumes prey types are encountered randomly, at rates that do not vary over time. However, as was discussed in the last chapter, searching parasitoids frequently locate hosts using attractant chemicals and may learn to associate new stimuli with the presence of hosts as they search. The encounter rate with different host types may thus change over time and complicate the calculation of the optimal set of acceptable hosts.

DYNAMIC OPTIMIZATION MODELS

I now turn to models that explicitly take into account the internal state of the parasitoid. McNamara and Houston (1986), Mangel and Clark (1986, 1988), and Houston et al. (1988) have championed the use of stochastic dynamic programming in the analysis of optimal behavioral sequences. Mangel (1987a, 1987b, 1989a, 1989b) in particular has examined insect oviposition using these techniques, building on the pioneering work of Iwasa et al. (1984). In a dynamic program, the animal is described by one or a number of variables that constitute its "state": for example, the state variables might be the insect's egg complement and/or its nutritional reserves. It is then necessary to specify how different oviposition decisions affect the animal's cumulative gain in fitness, its instantaneous risk of mortality, and the values of each of the state variables. The optimization procedure itself rests on the assumption that given an animal is in a particular state, its optimal policy depends only on future decisions, and not on the past history of its behavior. The optimal decision sequence can then be calculated by working backwards after specifying the final state of the animal.

Mangel (1987a) demonstrated the versatility of this approach by constructing a suite of host-acceptance models embodying different assumptions about insect oviposition. The first model described a pro-ovigenic parasitoid (a species born with a fixed complement of eggs) and contained a single state variable, the number of remaining eggs. The model predicts that the parasitoid should attack a more restricted set of hosts when it begins to run short of eggs and when hosts are particularly common. In addition, the parasitoid should attack a wider set of hosts when the risk of mortality is high, for example toward the end of the season. These predictions can be understood in terms of a trade-off between present and future reproductive success: the insect should be more selective (i.e., should lay eggs only on good hosts) if there is a high probability that opportunities will arise to use the eggs it saves before death. Mangel went on to analyze further models in which the parasitoid was described by two-state variables, egg load and energy reserves. These models are appropriate for synovigenic parasitoids which feed as adults and mature eggs throughout their adult life. Mangel supposed that the parasitoid could either search for hosts or forage for food, the latter used both for maintenance and for maturing more eggs. His models predict that a greater proportion of time will be spent searching for hosts as the parasitoid ages or as the season draws to a close. Before Mangel, Iwasa et al. (1984) had studied a number of rather complicated dynamic programming models of parasitoid host selection. They assumed that the risks of parasitoid mortality were particularly high during the act of oviposition. Parasitoids with a large egg load should restrict oviposition to high-quality, safe hosts; as eggs are used up, the parasitoids should take greater risks because the costs of mortality in terms of lost future ovipositions are now smaller. However, when the parasitoids have very few eggs they should again be more selective in order to avoid running out of eggs when there is still the possibility of finding the best-quality hosts. Parasitoids are thus predicted to attack the widest range of hosts when their egg reserves are at intermediate levels.

The strength and weaknesses of dynamic programming are illustrated by a comparison of the static model used by Charnov and Stephens (1988) to investigate the effect of costly egg production (see above), and a dynamic model written by Mangel (1989a) to investigate the same question. For many parameter values, the predictions of the two models are almost identical, and in these circumstances the relative simplicity and transparency of the static model give it a decided advantage. However, Mangel argues that for other, biologically relevant parameter values, the two approaches differ and the dynamic programming model provides the more realistic predictions. Unlike the static model, the dynamic model allows the parasitoid's oviposition strategy to depend on current egg reserves, and the predictions of the two approaches differ markedly when the size of the egg load is highly variable. An ability to quickly replenish egg reserves, and a large maximum carrying capacity, both lead to marked variation in egg loads. Mangel's model predicts greater selectivity when current egg reserves are low. The two approaches also differ in the case of short-lived parasitoids, which is not surprising as the static model assumes that the insect locates a large number of hosts in its lifetime. Finally, the dynamic model is almost certainly more realistic when there are a number of hosts of nearly similar quality. In these circumstances the position of the threshold in host quality predicted by the static model is particularly sensitive to changes in parameter values.

I think it is fair to say that many of the qualitative predictions from dynamic models are fairly obvious and could be derived by extending static models by verbal reasoning or through simpler approaches that approximate true dynamic optimization (Jaenike 1978; Parker and Courtney 1984). The major advantage of stochastic dynamic programming is that provided it is possible to measure the relevant parameters, strictly quantitative predictions can be made. A second advantage is that the stochastic element allows predictions to be made about behavioral variability. Not only can the optimum strategy of a parasitoid be predicted, given the value of its state variables, but also the expected distribution of the state variables. It is thus possible to suggest, for example, the proportion of parasitoids in the population that will reject a host of a certain quality. Static host acceptance models seldom predict partial preferences. The chief disadvantage of dynamic programming is that although logically straightforward, the mechanics of finding the optimal solution are frequently tedious and cumbersome and it is normally impossible to obtain analytical solutions. Mangel and Clark (1988) provide an excellent introduction to building dynamic models in behavioral ecology.

The two questions of host acceptance and patch use can be studied together within the same modeling framework. Mitchell (1990) and Visser (1991; Visser and Sjerps 1991) have studied the optimal searching strategy for time-limited predators in patches containing good and bad prey (using control theory and simulation studies, respectively). In high-quality environments where good prey are abundant and travel time between patches is short, predators should accept only good prey, irrespective of the number of competitors in the patch. As the quality of the environment falls, a threshold is reached when all predators foraging alone in a patch should switch to accepting both good and bad prey. However, if two or more predators are exploiting the same patch, they should initially concentrate on the best prey and only later switch to accepting both prey types. The reason for this is that competing predators in the same patch are racing against each other to locate the best prey and should not take time out from this competition to handle poor prey until most of the high-quality prey types have been consumed. This model applies to time-limited parasitoids in circumstances when previously parasitized hosts are either not encountered or completely (and instantaneously) ignored. The extension of this model to situations where superparasitism cannot be ignored is discussed later in the chapter (sec. 3.3.1).

3.1.2 EXPERIMENTAL STUDIES

There is abundant evidence that offspring survival and fecundity depend crucially on the size and species of host (see secs. 6.1 and 7.1), and many studies have recorded host preferences in parasitoids. However, relatively few studies have compared quantitatively host preference and host suitability, and even fewer studies have specifically tested host acceptance models.

HOST QUALITY AND ACCEPTABILITY FOR OVIPOSITION

The host selection strategy of the braconid *Asobara tabida*, a parasitoid of drosophilids, has been dissected by van Alphen and colleagues. A searching parasitoid encounters hosts of different age (and hence size), and also of different species. Van Alphen and Janssen (1982) and van Alphen and Vet (1986) compared the behavior of the wasp when offered *Drosophila melanogaster* and one of four other *Drosophila* species. The probability of host rejection was correlated with larval survival and could be broken down into two components: the chance that a parasitoid abandons the host after detection but before the insertion of the ovipositor, and the probability that a pierced host is rejected prior to oviposition. Some unsuitable hosts, particularly small species and those with a thick cuticle, were rejected prior to piercing whereas others were rejected after attack. Further experiments on just two host species have been conducted by van Strien-van Liempt and Hofker (1985) using both *A. tabida* and the eucoilid parasitoid *Leptopilina heterotoma*. Both species preferred the high-quality host, but oviposited more often than expected in the poor host. Van Alphen and Drijver (1982) also examined the behavior of *A. tabida* attacking different-aged *D. melanogaster* larvae. The wasp larvae can only survive when young hosts are attacked. The parasitoid appeared to be more successful at attacking young hosts (larger hosts often escaped) but, after the insertion of the ovipositor, it was as likely to reject old poor-quality hosts as young high-quality hosts. One possible explanation for this apparent failure to discriminate against poor hosts is that the immigrant *D. melanogaster* is a relatively novel host for northern populations of *A. tabida*. Finally, Mollema (1988, 1991) has compared the host acceptance behavior of different strains of *A. tabida*. Populations from Holland and the south of France both successfully parasitized *D. subobscura* but only French strains had a high probability of success on *D. melanogaster* (this species is common in southern Europe but rare in Holland). French wasps accepted *D. melanogaster* as a host much more readily than Dutch wasps. Mollema (1991) obtained evidence for heritability of host acceptance behavior (using mother-daughter regression) in a laboratory strain of *A. tabida* originating from insects collected in Holland.

The survival of another drosophilid parasitoid, the eucoilid *Leptopilina clavipes*, is also strongly affected by host species. In laboratory experiments, Driessen et al. (1991) found a broad association between the acceptability of a *Drosophila* species and its suitability as a host. There were, however, exceptions. One poor-quality species, a close relative of a high-quality host, was accepted much more frequently than expected. Possibly the wasp was unable to distinguish between the two species, or possibly host encounter rates are so low in the field that acceptance of even the poor quality host is always optimal under natural conditions. Curiously, a species in which larval survival was zero, and which was always avoided in the laboratory, was quite frequently attacked in the field.

The pteromalid *Pachycrepoides vindemiae* is also a parasitoid of *Drosophila* sp. but only attacks the fly after it has pupated. Hosts attacked by this species may already contain a larval-pupal parasitoid such as *Asobara tabida* or *Leptopilina heterotoma*. Such hosts can support the development of *P. vindemiae*, which develops as a facultative hyperparasitoid on the younger parasitoid. Previously parasitized hosts tend to be of poorer quality than unparasitized hosts, especially when the larval-pupal parasitoid has consumed the original host and has itself pupated. Van Alphen and Thunnissen (1983) presented individual *P. vindemiae* females with a choice of parasitized and unparasitized hosts of different ages. The pupal parasitoid could potentially develop on all hosts, with or without parasitoids, as long as they were sufficiently advanced that the host had pupated (*A. tabida* and *L. heterotoma* kill the host after pupation). However, *P. vindemiae* was less willing to parasitize hosts containing mature larval-pupal parasitoids and, in the case of *L. heterotoma*, avoided hosts containing mature parasitoid larvae that had not yet entered the prepupal stage. The host stages most often avoided by the wasp were those where offspring survival was low and which produced small adult parasitoids. This study provides good evidence of a correlation between host acceptance and host quality.

The relationship between host acceptance and host quality has also been investigated in aphid parasitoids (especially aphidiine braconids). Stary (1970) argues that in general there is a good agreement between the aphid stage attacked and its suitability as a host. Liu et al. (1984) and Liu (1985) found that *Aphidius sonchi* would attack all stages of the aphid *Hyperomyzus lactucae* (=*sonchi*) but preferred second and third instars where survival was highest. Mackauer (1973) found that *Aphidius smithi* did not distinguish between second, third, and fourth instar aphids despite higher survival on second instars; it did however discriminate against first instars or adults. Care must be taken in interpreting such experiments, as aphid instars differ in their ability to resist parasitoid attack, leading to large differences in handling time.

The braconid *Ephedrus californicus* attacks the pea aphid (*Acyrthosiphon pisum*). Kouamé and Mackauer (1991) offered parasitoids a choice of fed aphids and aphids that had been starved for four or six hours. In one set of experiments all aphids were of the same age but varied in size, while in a second set of experiments aphids were all of the same size but varied in age. In both cases, parasitoids preferred to attack starved hosts, even though smaller adults emerged from these hosts. Two factors in addition to the quality of the host for larval development may have influenced host choice. First, larger and well-nourished hosts are better able to defend themselves against parasitoids;

larger hosts may take more time to subdue, or may get away completely. The preference for starved hosts disappears if all hosts are anesthetized. Second, development time is shorter on starved aphids. In growing populations there is premium on fast development and the wasp might sacrifice higher fecundity for speed of development.

There are a few studies of the relationship between host quality and oviposition preference in other parasitoid groups. For example, trichogrammatid egg parasitoids normally prefer to attack young eggs where survival and reproductive success is highest (Lewis and Redlinger 1969; Marston and Ertle 1969; Hiehata et al. 1976; Juliano 1982; Strand 1986). Nechols and Kikuchi (1985) investigated the encyrtid *Anagyrus indicus*, a parasitoid of the Spherical Mealybug, *Nipaecoccus viridis (=vastator)*. The wasp attacks all three instars and the adult of its host. In no-choice experiments it oviposits most readily on third instar and adult hosts, while in choice experiments it prefers the adults. Part of the explanation for these results is that the wasp locates large hosts more readily, but positive rejection of small hosts is also involved. Host suitability mirrors host acceptance: mortality is lowest on large hosts and development time is shortest on the adult host. In addition, this gregarious species can lay more eggs, with a more female-biased sex ratio on large hosts. In a series of careful experiments, Hopper and King (1984) found a correlation between the preference of the braconid *Microplitis croceipes* for different instars of its hosts (*Helicoverpa zea* and *Heliothis virescens*), and the development time of its larvae. Development time was influenced by host instar but not by host species, and the wasp showed no preference between similar-aged caterpillars of the two species of moth.

RELATIVE VERSUS ABSOLUTE HOST QUALITY

Models predict that the probability of host acceptance is influenced by the quality of other hosts in the environment. Weak support for this prediction is provided by the observation that parasitoids frequently attack unsuitable hosts or host stages in the laboratory when their favored host is absent (e.g., Salt 1935; Schuster 1965; Nechols and Kikuchi 1985). For example, the aphelinid *Aphytis mytilaspidis* will attempt to parasitize quite unsuitable scale insects if given no choice of host; however, the same species are almost completely ignored if presented in conjunction with the normal host (Baker 1976).

Van Alphen and Janssen (1982) noted, in the experiments described above, that *Asobara tabida* was more likely to reject a *D. melanogaster* larva when this species was offered in combination with a higher rather than a lower quality species. Moreover, when *Asobara tabida* encountered mixtures of the good quality *D. subobscura* and the poor quality *D. melanogaster*, it initially accepted a high proportion of the poor-quality host but with time accepted fewer and fewer. In addition, if it had previously experienced mixtures of the two species, it was more likely to reject the poor-quality *D. melanogaster* than if its previous oviposition experience was restricted to this species. Van Alphen and Janssen (1982) and van Alphen and van Harsel (1982) interpret these results to mean that the wasp modifies its host acceptance strategy as it gains information about the overall quality of the environment.

FIELD STUDIES

One of the problems in testing host acceptance models in the laboratory is the difficulty of obtaining realistic estimates of encounter rates with different host types. Janssen (1989) courageously followed individual parasitoids in the field and directly observed the frequency of encounters with hosts. Using a portable binocular microscope attached to a tripod, he watched the two tiny parasitoids *Asobara tabida* and *Leptopilina heterotoma* as they searched for *Drosophila* larvae on fermenting fruit and on the sap fluxes from wounded trees. Nine different species of *Drosophila* larvae fed in these environments, and the handling time and larval survival on each of these hosts were estimated in the laboratory. For both parasitoids, handling time was very short (about thirty seconds) and differed little among host species. However, there was considerable variation in the probability of parasitoid survival in different hosts, and frequently parasitism caused the death of the host (table 3.1). Encounter rates were very low: wasps discovered 0–5 hosts per hour. Substituting these parameter estimates into an optimal host acceptance model (and making corrections for recognition time and encounters with previously parasitized hosts), Janssen was able to predict that a host should be attacked whenever the probability of larval survival was greater than a threshold that varied between 0.002 and 0.03, depending on species and microhabitat. Thus even very poor hosts should be attacked. Janssen observed thirty-three host encounters of which thirty resulted in parasitism. Although he was unable to identify a host species at the time of oviposition, Janssen knew the relative abundance of host species in the different environments (table 3.1), and the very high rate of host acceptance implies that many eggs are laid on species where their probability of survival is small.

Janssen also used his data to investigate the sensitivity of parasitoid fitness to changes in the set of acceptable hosts. The fitness of *Asobara tabida* changed only very slightly as lower ranking hosts were included in the acceptable set. *A. tabida* should always attack *D. subobscura*, a common host on which it does well, but other hosts are either rare or of low suitability and their inclusion has only a marginal effect on fitness. Similarly, the fitness of *Leptopilina heterotoma* changed only slightly with the addition of the third and lower-ranking host types.

In order to predict the set of acceptable hosts, Janssen assumed that the reproductive success of the parasitoids was limited by time. Greater selectivity might be favored if the wasps have a limited supply of eggs and risk running out of eggs before dying. In another courageous field experiment, Driessen and

Table 3.1

The nine species of Drosophila encountered by the parasites *Leptopilina heterotoma* and *Asobara tabida* in field sites studied by Janssen (1989) in the Netherlands.

Host Species	L. heterotoma: Survival after Parasitism of		A. tabida: Survival after Parasitism of		Relative Abundances in the Field (1984)	
	Parasitoids	Hosts	Parasitoids	Hosts	Fruit	Sap-fluxes
Drosophila subobscura	87.3	0.0	87.9	0.0	31.7	72.1
Drosophila immigrans	2.0	22.9	1.4	74.6	34.8	20.5
Drosophila simulans	65.3	10.0	0.0	68.7	31.1	0.0
Drosophila kuntzei	89.5	7.9	5.3	44.4	0.0	2.2
Drosophila phalerata	49.5	2.8	0.0	49.8	0.1	0.9
Drosophila littoralis	10.8	1.1	0.8	74.4	0.1	0.9
Drosophila melanogaster	26.1	3.7	69.4	13.2	0.7	0.0
Drosophila obscura	17.5	0.0	28.5	45.0	1.3	3.5
Drosophila tristis	5.6	0.0	51.9	4.6	0.2	0.0

Notes: For each parasitoid species the table shows the percentage survival of hosts and parasitoids after parasitism (for example, after parasitism of *D. subobscura* by *L. heterotoma*, 87.3% of hosts produced parasitoids, 0% of the hosts survived, and in the remaining 12.7% [100% − 87.3%] of cases both host and parasitoid died). This data was collected in the laboratory. The last two columns of the table show the relative abundance of the host species in two microhabitats in 1984.

Hemerik (1991) attempted to assess whether reproduction by the eucoilid parasitoid *Leptopilina clavipes* was limited by egg supply or by time. *L. clavipes* also attacks *Drosophila* larvae but is more specialized than either *A. tabida* or *L. heterotoma*, chiefly attacking species feeding on stinkhorn fungi, *Phallus impudicus*.

Driessen and Hemerik assumed that wasps hatch with a fixed number of eggs and spend their life flying between patches and foraging for hosts within a patch. Initial egg loads were easily determined by dissecting newly emerged females and varied between about fifty and two hundred. Studying migration between patches and behavior on the patch was considerably more difficult, and a technique for marking individuals wasps with acrylic paint had to be developed. Marked wasps were released at the center of an array of petri dishes baited with fungi, and the time of arrival of wasps at the recapture sites was noted. The behavior of marked wasps on artificial patches of different host density was also studied in the field, and patch residence times were analyzed using the type of statistical analysis described in section 2.5.3. It was found that the time spent on patches was exponentially distributed and did not depend on host density (except that wasps left empty patches more readily than occupied patches). The rate of oviposition while on a patch was estimated from field observations, while life expectancy was measured in large field cages.

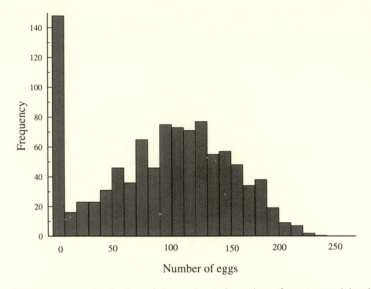

Figure 3.1 Frequency distribution of the estimated number of eggs remaining in the oviducts of dead *Leptopilina clavipes*. Results of Monte Carlo simulations using data collected in the field and laboratory. (From Driessen and Hemerik 1991.)

The results of all these observations and experiments were incorporated into a Monte Carlo simulation model in which wasps were allowed to search in an environment where fungi and hosts were distributed as in the wild. The aim of the model was to find out how many wasps exhausted their egg supply before death. The results obtained using the parameters estimated in the field are shown in figure 3.1. Approximately 13% of wasps were predicted to die with no eggs left. Driessen and Hemerik performed a sensitivity analysis by halving and doubling those parameters, likely to vary in the field. In different simulations, between 3% and 32% of wasps were predicted to die with no eggs left, and the most influential variables were life expectancy and oviposition rate. In conclusion, this study demonstrates that egg limitation should be taken into account when trying to predict the oviposition strategy of *L. clavipes* and similar parasitoids.

3.1.3 BEHAVIORAL MECHANISMS

Parasitoids use a variety of cues to assess the quality of their hosts. Many species, especially those attacking immobile hosts, spend a great deal of time externally examining the host, often by stroking or drumming with their antennae. A number of workers have used model hosts, or have experimentally modified features of the host, to discover the most important stimuli involved in host acceptance. After external examination, many parasitoids insert their ovipositor into the host to obtain additional information about its suitability. In this section I briefly review the behavioral mechanisms involved in host recognition and acceptance.

EXTERNAL EXAMINATION OF THE HOST

One of the first attempts to understand the proximate mechanisms involved in host acceptance was Salt's (1935) study of the polyphagous trichogrammatid egg parasitoid *Trichogramma evanescens*. Salt presented female wasps with a wide variety of egglike objects, some of which only very distantly resembled moth eggs, their normal host. *Trichogramma* initially examine a potential host by "drumming" their antennae over its surface. If accepted at this stage, they attempt to probe the host with their ovipositor. Salt found that size was the most important attribute leading to acceptance. The wasp attempted to probe any vaguely globular object (including seeds, glass beads, and rhombic crystals) between about 0.25 and 4.5 mm in diameter. In another set of experiments, Salt (1958) created artificial eggs with small droplets of mercury. Very small droplets were unacceptable to the wasps and elicited no probing. By adding minute quantities of mercury to the droplet, it was possible to determine the exact range of sizes acceptable to the wasp. Salt (1940) also found that the assessment of size was relative: larger *T. evanescens* females tended to accept larger eggs.

Arthur (1981) and Vinson (1985) review a number of experiments, similar to those of Salt, where the effect of the experimental alteration of different host characters has been studied. Size, shape, and surface texture have all been shown to be important in host acceptance. Occasionally, dramatic mistakes occur in the field. The pipunculid fly *Verralia setosa* has been observed trying to attack the buds of birch trees, apparently mistaking them for their hosts, nymphs of the planthopper *Oncopsis flavicollis* (F. P. Benton, quoted by Waloff and Jervis 1987). Host movement is also frequently a prerequisite for host acceptance (Arthur 1981). The tachinid *Drino bohemica* can be induced to oviposit on a fluttering feather in the presence of sawfly odor, and Monteith (1956) argues that movement is the final stimulus eliciting oviposition. In other cases movement may inhibit oviposition: egg parasitoids are frequently unable to develop on mature eggs and the movement of the developing embryo is used as a signal that the eggs are too old to support a larva (Salt 1938; Jackson 1968). Many parasitoid wasps, especially ichneumonids, have very noticeable white tips or bands on their antenna. It is possible that these markings assist the wasp in assessing the size of the host.

Chemical cues perceived through receptors in the antennae and tarsi are undoubtedly of great importance in host acceptance. Many of the chemicals used in host location (see previous chapter) are also likely to be involved in host acceptance. In addition, nonvolatile chemicals present on the surface of the host may be the final stimuli for oviposition. Arthur (1981) reviews a number of studies which have found that solvent extracts of host cuticle are recognized by parasitoids and produce oviposition behaviour. After the solvent treatment, the actual host may be markedly less attractive to the parasitoid. Chemical cues may act in conjunction with shape, size, and texture. Strand and Vinson (1982b, 1983a, 1983b) found that the scelionid egg parasitoid *Telenomus heliothidis* responded to a protein produced in the accessory gland of its host, the moth *Heliothis virescens*. The wasp would attempt to oviposit on glass beads coated with the chemical, if they were about the same size as a host egg, but they would refuse to oviposit either on uncoated glass beads or on flat surfaces smeared with accessory gland extract. In comparison with the *Trichogramma* studied by Salt, the range of egg-shaped objects that elicit probing by *T. heliothidis* is quite small. Objects must be approximately spherical and between 0.5 and 0.6 mm in diameter. *H. virescens* eggs are spherical when young and become more conical with age. The greater selectivity of *T. heliothidis* is probably explained by its much smaller host range and its preference for young eggs.

INTERNAL EXAMINATION OF THE HOST

Parasitoids frequently insert their ovipositor into a host but do not go on to lay an egg. The ovipositor is normally covered in sensillae and it seems likely that the insect is rejecting the host after perceiving that it is unsuitable for oviposition. The parasitoid may assess the suitability of the host using chemical cues, or possibly by detecting the heartbeat of a healthy host (Fisher 1971). Arthur et al. (1969, 1972; Hegdekar and Arthur 1973) found that the ichneumonid *Itoplectis conquisitor* would oviposit into false hosts made of plastic film containing host hemolymph; the false hosts were hidden in leaf rolls similar to those made by the real host. Wasps would not oviposit into false hosts containing distilled water or saline, but would oviposit into solutions of certain combinations of amino acids and inorganic ions. Some chemical combinations proved more attractive than hemolymph. *Trichogramma* can be persuaded to oviposit in artificial eggs with a paraffin film or wax shell (Rajendram and Hagen 1974; reviewed by S. N. Thompson 1986b). Again, successful oviposition depends quite critically on the chemical makeup of the artificial eggs.

3.1.4 HOST FEEDING

Hosts may be used both for oviposition and as a source of food. Host feeding has been recorded from a large number species of hymenopteran parasitoids, though not from parasitoids in other orders. The taxonomic distribution and physiology of host feeding has been extensively reviewed by Jervis and Kidd (1986), who provide a full bibliography for the subject.

Some parasitoids both feed from and oviposit in the same host (concurrent host feeding). Typically, the wasp may consume hemolymph exuding from the ovipositor wound, or make a small extra incision, but without removing large quantities of fluid from the host. I am unaware of any studies that have compared offspring survival in hosts that have or have not been used for feeding. Other parasitoids use some hosts for food and others for oviposition (nonconcurrent host feeding). In these cases, host feeding typically leads to the death of the host. Again, the parasitoid normally consumes only host fluids, though tissue is also ingested in some groups. Some parasitoids of concealed hosts (chalcidoids in the families Pteromalidae, Eurytomidae, Torymidae, Eupelmidae, and Eulophidae) feed from their host by constructing a feeding tube (Fulton 1933; Askew 1971). The wasp inserts its ovipositor into the cell or cavity containing the host and a viscous fluid is exuded from the tip of the ovipositor. The fluid forms a thin layer around the ovipositor, which dries to form a tube. The wasp extends the tube until it touches the host, which it wounds. The ovipositor is then withdrawn and the hemolymph, which rises through the tube by capillary action, is drunk by the wasp.

Food obtained from the host is used for many purposes, but a simple distinction can be made between using food as a source of energy and using food as a source of nutrients to mature eggs. Host feeding appears to be most important in synovigenic as opposed to pro-ovigenic parasitoids, that is, in parasitoids that mature eggs throughout their life rather than those that emerge with a fixed complement of eggs. Some species are unable to lay eggs until they have host-fed at least once, suggesting that they obtain essential nutrients for egg maturation from the host which they cannot obtain from flowers or honeydew.

Host feeding is one of a number of processes that influence the dynamic balance of energy and nutrient reserves within the parasitoid. The insect is born with certain reserves of energy (e.g., fat bodies) and nutrients that are used up during host location. Once a host is found, it can be used for oviposition which depletes eggs, or host feeding which provides nutrients and energy. Feeding on flowers and honeydew leads to energy intake and probably provides some nutrients that can be used for egg maturation. Finally, in many species eggs in the ovarioles can, if necessary, be resorbed and recycled to provide energy and nutrients. Resorption is a costly strategy because when a food source is finally discovered, the production of fertile eggs can take several days. A few species, when in extremis, resorb wing muscles (Sandlan 1979a). It is against this background of fluxes in physiological variables that the decision whether to oviposit or feed on a host is made.

MODELS OF HOST FEEDING

Jervis and Kidd (1986) constructed a simple model of host feeding based on the assumption that the parasitoid feeds to obtain energy which is used up during host searching and in the production of eggs. Their model assumed that the parasitoid lived for a fixed period of time. When hosts are rare, a large percentage have to be used for food in order that the parasitoid meets its general maintenance requirements. As hosts become more common, the fraction used for host feeding declines. If hosts are very rare, a parasitoid that does encounter a host may be selected to oviposit because even if it used that host for food, the chances of discovering further hosts are minimal. Jervis and Kidd also consider the consequence of various host feeding and resorbtion strategies in a more complex simulation model.

The prediction that the percentage of hosts used for oviposition grows as encounter rate increases probably arises from most energy-based host-feeding models. However, if the assumption is made that the function of host feeding is to provide protein or other nutrients for egg maturation, the fraction of hosts used for oviposition may remain constant. In the simplest case, where egg maturation occurs instantaneously, the parasitoid should lay all the eggs it carries and then host-feed to mature new supplies. Thus if host feeding provides sufficient nutrients to mature, say nine eggs, every tenth host will be used for host feeding irrespective of encounter rate. If egg maturation is not instantaneous, host feeding should occur in anticipation of egg depletion when supplies run low.

The available evidence suggests that the fraction of hosts used for oviposition does increase as encounter rate rises. This has been observed in the encyrtid *Metaphycus helvolus*, a parasitoid of black scale (DeBach 1943), the ichneumonid *Pimpla (=Coccygomimus) turionellae*, a parasitoid of the wax moth *Galleria mellonella* (Sandlan 1979a), and the aphelinid *Aphelinus thomsoni* which attacks aphids (Collins et al. 1981). These results suggest that at least part of the function of host feeding is to provide the resources used by the parasitoid in maintenance and searching.

If parasitoids are given a choice of different-sized hosts, they often choose to feed from relatively small hosts and lay eggs on relatively large hosts (e.g., Bartlett 1964; Abdelrahman 1974; Nell et al. 1976; van Alphen 1980). There are at least two reasons why parasitoids might preferentially feed on small hosts. First, small hosts may be unsuitable for oviposition, so the parasitoid loses nothing by using them for food. Second, even if oviposition is possible on small hosts, the resulting progeny may be small and less fit than those developing on large hosts. Of course, large hosts provide larger meals and so host feeding may also be more profitable on bigger hosts. However, it seems likely that large hosts are relatively more valuable as oviposition sites than feeding sites and so selection will favor concentrating host feeding on smaller hosts. The problem of determining the optimum host-feeding strategy on different-sized hosts is very similar to the problem of finding the optimum sex ratio on hosts of different size. As is discussed more fully in section 4.4, female eggs will be laid on relatively large hosts if daughters gain more than sons from developing on bigger hosts. Models of sex allocation predict an abrupt threshold in host size where the female switches from laying males to females. Although the process has not been specifically modeled, if large hosts are relatively more valuable as oviposition than feeding sites there is likely to be threshold host size below which hosts are used as food. The position of the threshold will be relative, and in an environment with many large hosts the dividing line between feeding and oviposition will be shifted toward bigger hosts.

Static optimization models are likely to be of limited use in studying host feeding because decisions on whether to feed or lay an egg on a host are strongly influenced by physiological state. Dynamic programming has been used by Houston et al. (1991) to study the effect of variability in encounter rates on host-feeding strategy.[1] They assumed that the parasitoid feeds from hosts purely to obtain energy to stay alive and continue searching. If energy reserves fall below a critical threshold, the animal dies of starvation. The optimum strategy for the wasp is to oviposit if energy reserves are above a critical level, but to host-feed when reserves fall below this level. The effect of increased variance in encounter rates is to raise this critical level of energy reserves; the wasp has to guard against the risk of starvation when there is the possibility of long gaps between hosts. Similar predictions are made by models of the fat reserves carried by overwintering birds: larger reserves are predicted when the climate is less predictable (McNamara and Houston 1990).

3.2 Clutch Size

The study of clutch size in birds is one of the oldest areas of evolutionary ecology. In recent years there has been an enormous growth of interest in the study of invertebrate clutch size in general (reviews in Godfray 1987a; Godfray et al. 1991) and parasitoid clutch size in particular (Waage 1986). Clutch size in parasitoids can be defined as the number of eggs deposited on a host in a single oviposition bout. The first part of this section describes the application of clutch size theory to parasitoids, and the second section covers experimental tests of theory. The third part discusses the behavioral mechanisms used by parasitoids to assess how many eggs to lay on a host. The final part reviews the argument that the solitary and gregarious habits in parasitoids are not simply two ends of a continuum, but that the solitary habit is an absorbing state from which it is difficult to evolve to become gregarious.

3.2.1 THEORY

THE LACK CLUTCH SIZE

Lack (1947) suggested that birds are selected to lay the size of clutch which maximizes the number of young fledging from the nest. A modern restatement of Lack's hypothesis, applicable to all animals, is that the mother should lay the number of eggs that maximizes her gain in fitness from the whole clutch. In the simplest case, when all members of the clutch have equal fitness, this

[1] Note added in press: Chan and Godfray (1993) have explored a series of dynamic programming models of host feeding with different assumptions about whether the function of host feeding is to provide energy or nutrients for the parasitoid.

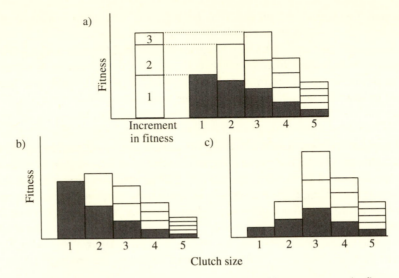

Figure 3.2 Calculation of the Lack clutch size. The shaded bars represent the fitness of individual larvae developing in clutches of different size. The fitness of the parent is obtained by multiplying progeny fitness by clutch size (total height of bar). The highest composite bar defines the Lack clutch size (the parental optimum), while the highest shaded bar is the offspring optimum. (a) and (b): Two examples where offspring fitness declines monotonically with clutch size. (c) An example where the offspring and parental optima coincide. Figure (a) also shows the increments in parental fitness as each egg is added to create the Lack clutch size.

implies that the mother should maximize the product of clutch size and the fitness of individual offspring (fig. 3.2). The clutch size that results in the maximum fitness returns is often called the Lack clutch size.

It is straightforward to apply Lack's hypothesis to parasitoids if the fitness of larvae developing in clutches of different size is known. Figure 3.2 illustrates the calculation of the Lack clutch size when larval fitness declines monotonically with increasing clutch size (fig. 3.2a, b), or when the relationship between larval fitness and clutch size is domed (fig. 3.2c). The Lack clutch size is strongly influenced by host quality, and smaller clutches are predicted on hosts with fewer resources to support parasitoid larvae (compare fig. 3.2a and b).

Animals are selected to maximize their lifetime fitness and thus Lack's hypothesis is only correct if maximizing lifetime fitness is equivalent to maximizing the fitness gains from each clutch. This is true if the animal only ever produces a single clutch, or if the size of the current clutch has no influence on future reproductive success. However, there are several reasons why producing a large clutch of eggs may reduce future reproductive success—in other words, why there may be a cost of reproduction. One reason has been widely discussed by ornithologists. If large broods of offspring require more effort on the part of the parents to feed and protect, then the probability of the parent surviving to reproduce again may be related to clutch size (Williams 1966;

Charnov and Krebs 1974). However, two different costs of reproduction are probably more relevant to parasitoids (Charnov and Skinner 1984, 1985; Parker and Courtney 1984; Waage and Godfray 1985; Wilson and Lessells 1993): the production of large clutches may waste time that could be used more profitably searching for new hosts, or waste eggs that could be placed more profitably in smaller clutches on other hosts.

THE COST OF REPRODUCTION: LARGE CLUTCHES WASTE TIME

Consider first a parasitoid whose reproductive success is limited by the time available for host location and oviposition. Such a parasitoid should maximize its rate of gain of fitness. If oviposition is time consuming, production of the complete Lack clutch size may not be in the parasitoid's best interest. To see why, consider the fitness gain for an ovipositing parasitoid as each egg is added to the clutch (fig 3.2a). In this figure, I assume that the fitness of an individual offspring declines monotonically with increasing clutch size. The parent gains a large increment in fitness when she lays the first egg but then successively smaller increments as she approaches the Lack clutch size. It may not be worthwhile for the parasitoid to add the last egg as it gains only a very small increment of fitness and wastes time that could be used searching for a new host.

There are great similarities in the calculations of optimal clutch size and optimal patch-residence time (sec. 2.5.2) for time-limited parasitoids. In both cases, the solution can be found using the marginal value theorem (fig. 3.3) (Charnov and Skinner 1984, 1985; Iwasa et al. 1984; Parker and Courtney 1984; Skinner 1985). The theory predicts that the optimal clutch size is strongly influenced by the time taken to locate a host and to prepare it for oviposition. If hosts are scarce and rarely encountered, larger clutch sizes should be found. Similarly, if much time has to be spent preparing a host for oviposition, larger clutches are predicted. Such preparation time should only include activities whose duration is independent of clutch size. For example, preparation time includes time spent drilling through wood to reach a concealed host, or time spent drilling through a tough cuticle, as long as a single hole is sufficient for the oviposition of the whole clutch. In the same way that the marginal value theorem can be used to predict residence time on patches that vary in quality, it can be used to predict clutch size on hosts that vary in quality. The parasitoid should remain ovipositing on a host until its rate of gain of fitness drops to the maximum average rate achievable in that environment. This leads to the prediction that at any one time and place, clutch size should be correlated with host quality and that no eggs should be placed on very poor hosts (Charnov and Skinner 1984, 1985; Iwasa et al. 1984; Skinner 1985).

THE COST OF REPRODUCTION: LARGE CLUTCHES WASTE EGGS

Suppose now that the reproductive success of the parasitoid is not limited by time but by its egg supply. More specifically, suppose that there is a finite

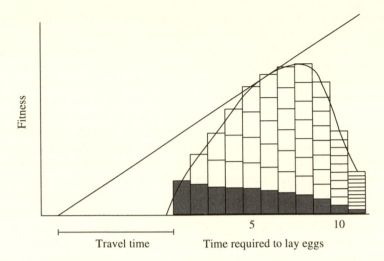

Figure 3.3 Optimal clutch size for time-limited parasitoids. The *x* axis is divided into two parts: the time taken to locate a host (travel time), and the time taken to lay different clutches. The shaded histograms represent offspring fitness, and the composite bars parental fitness, as in figure 3.2. A curve has been fitted by eye to the parental-fitness histograms. The optimal clutch size is found in exactly the same way as the optimal patch residence time in figure 2.6.

chance that the wasp runs out of eggs before it dies. Again, it may not be in the parasitoid's interest to complete the Lack clutch size. Recall that the addition of the last few eggs to make up the Lack clutch size may result in only a relatively small increase in fitness (fig. 3.2a); if eggs are in short supply the parasitoid will be selected instead to use these eggs to initiate new clutches where they will achieve much greater fitness returns. In the limit, when egg exhaustion is a certainty, the parasitoid should produce the clutch size that maximizes its fitness gain per egg. If offspring fitness declines monotonically with increasing clutch size (fig. 3.2a,b), severely egg-limited parasitoids should lay a single egg on each host. Larger clutches may be produced by egg-limited parasitoids if the relationship between larval fitness and clutch size is domed (fig. 3.2c).

The simplest models of clutch size in egg-limited parasitoids assume that the insect has a fixed complement of eggs and lays a constant clutch size throughout her life (Parker and Courtney 1984, Waage and Godfray 1985; Weis et al. 1983 developed a similar model for gall flies). These models predict smaller clutch sizes when there is a risk that a parasitoid will run out of eggs, for example when the risks of mortality are low, or in pro-ovigenic species born with a small complement of eggs.

The assumption that a parasitoid lays the same-sized clutch throughout its life can be relaxed with more sophisticated models using constrained optimization (Parker and Courtney 1984) or dynamic programming (Iwasa et al. 1984; Mangel 1987a, 1987b; Mangel and Clark 1988). These models predict that parasitoids should tailor their clutch size to current egg reserves, reducing the number of eggs laid per host when supplies are low. This suggests that clutch size in pro-ovigenic parasitoids should decline as the wasp ages, as long as the parasitoid does not suffer age-specific mortality. If the risk of mortality (that component unaffected by reproductive effort) increases with time, older parasitoids may produce larger clutches: future egg limitation is unimportant when you have no future. Predictions for synovigenic parasitoids are more complex because the best clutch size depends on the current state of egg reserves and thus on the individual's recent history of oviposition and feeding. Some of the difficulties of this type of model are illustrated by the work of Iwasa et al. (1984). For certain parameter combinations the solutions appear poorly behaved, the optimal clutch size fluctuating wildly as egg reserves vary. This indicates that several oviposition strategies have near equal fitness and the specific predictions of the model are not to be taken too seriously.

Charnov and Skinner (1988) have taken a simpler approach to the study of clutch size in synovigenic parasitoids. They assumed that the parasitoid matures eggs at a constant rate and that it finds hosts sufficiently frequently that on discovering a host it is selected to deposit all its mature eggs. The question they ask is whether the parasitoid, after laying a clutch of eggs, should remain on the host laying further eggs as they mature. The answer depends on whether hosts are encountered at fixed or variable intervals. In the case of fixed intervals, the parasitoid should always deposit its eggs and search for new hosts: a stay on the host always causes a reduction in the rate of gain of fitness below the maximum achievable in that environment. However, if travel times between hosts vary, they showed there was a threshold egg load below which the parasitoid is selected to remain on the host. If hosts are normally encountered relatively infrequently but a parasitoid happens to find a second host very soon after oviposition, its time is most profitably spent maturing eggs on the host rather than attempting to find a fresh host.

IMPRECISION AND UNCERTAINTY

So far, the theoretical models discussed in this section have assumed that parasitoids are able to assess precisely the worth of a host, and that they can always lay exactly the optimum clutch size. In practice, there may be uncertainty about the quality of a host, either because the insect is unable accurately to measure quality, or because the final quality of the host is not yet fixed at the time of oviposition. In addition, the parasitoid may not always be able to lay a precise number of eggs. This is particularly likely to occur in parasitoids that lay a large number of eggs very quickly. The braconid wasp, *Cotesia (=Apanteles) congregatus*, pumps between 30 and 300 eggs into the body of the tobacco hornworm (*Manduca sexta*) in 2–3 seconds (Fulton 1940; Beckage and

Riddiford 1978). If the optimum clutch is, say, 234, it is unlikely that the wasp can achieve this exact figure every time.

The effect of uncertainty and imprecision on insect clutch size has been studied by Godfray and Ives (1988; see also Mountford 1968 and Boyce and Perrins 1988 for similar models applied to vertebrates). Uncertainty or imprecision can affect the optimum clutch size if the penalties of overestimation and underestimation are not the same. Consider an extreme case: suppose the optimum clutch size is x but the parasitoid cannot precisely gauge the number of eggs it lays and is often out by one. If the fitness penalties in being one over, and one under, are the same, the optimum clutch size is unaffected by imprecision. But suppose all larvae die in a clutch of $x + 1$; there would be strong selection pressure on the female to reduce her clutch size to $x - 1$ to avoid the risk of catastrophic mortality. In general, if the fitness penalties for overshooting the optimum clutch size, or overestimating the quality of the host, are greater than undershooting or underestimation, smaller clutch sizes are selected. If the fitness penalties are reversed, larger clutch sizes are favored.

In the absence of any counterbalancing forces, maximum accuracy, both in host assessment and in egg production, will be favored by natural selection. However, there may be physiological and sensory constraints on the maximum achievable accuracy. Alternatively, high precision may be possible, but at a cost, for example in time or increased risk of mortality. The parasitoid's overall fitness might then be maximized by accepting reduced accuracy.

SEXUAL ASYMMETRIES

The fitness of parasitoid larvae normally declines as clutch size increases. However, it is quite likely that the decline in fitnesses of male and female larvae are not identical. For example, a common consequence of increased competition among larvae is a reduction in the size of the resulting adults, and females may suffer more than males from being small (see sec. 7.1.3). What is more, larvae of one sex may be more successful in competition than larvae of the other sex. Thus a particular larva (of either sex) may suffer to a greater extent by sharing a brood with many brothers than with many sisters, or vice versa. These two types of asymmetry were inelegantly dubbed asymmetric density responses and asymmetric composition responses by Godfray (1986b).

Sexual asymmetries in competition can affect both clutch size and sex ratio, the latter discussed in the next chapter. As far as clutch size is concerned, there are two possible outcomes: the parent may lay broods containing individuals of only one sex, or the parent may lay mixed broods. Single-sex broods are most likely to be produced when competition between individuals of different sexes is more intense than competition between individuals of the same sex. The optimum clutch size for all male and all female broods need not be identical; separate Lack clutch sizes can be calculated for the two types of broods.

If mixed broods are produced, the optimum clutch size is a function of a fairly complicated average of the different responses by the two sexes (Godfray 1986b).

SIBLING CONFLICT

Most theoretical studies of clutch size assume a fixed relationship between the number of larvae developing in a host and their individual fitnesses. However, the interactions between larvae will themselves be subject to selection. Later in this chapter, violent conflict among the larvae of solitary parasitoids, often leading to death, will be described. However, more subtle interactions among the larvae of gregarious parasitoids may also influence observed clutch sizes.

Consider the following hypothetical trade-off. Larvae can either develop slowly and utilize host resources with maximum efficiency, or they can develop more quickly by using resources with reduced efficiency. The cost of faster development by the whole brood is that the parasitoids do not fully assimilate all the host's resources and emerge as relatively small, low-fitness adults. How fast will the larvae be selected to feed? If the parent controls the behavior of the offspring, or if the offspring are genetically identical (as in polyembryonic species), then slow feeding will be selected: slow feeding leads to the maximum fitness of each member of the brood (complications arise when the rate of development is also affected by other selection pressures, for example in growing populations and where immature stages are particularly susceptible to predation). However, in a mixed brood of genetically heterogeneous siblings, a rare mutant that causes its bearer to feed faster may be at an advantage. Although its feeding efficiency is reduced, it develops faster and is able to preempt resources in competition with its slow-feeding broodmates. This advantage is lost as the rare mutant spreads to fixation, but nevertheless the advantages when rare can be sufficient for the mutant to displace the resident, slow-feeding genotype.

Competition within broods of siblings has been modeled by Godfray and Parker (1991, 1992; see also Parker and Macnair 1979). In these models, the food obtained by an individual depends on how strongly it competes relative to the other members of the brood; it is also assumed that there are costs to competition. In the present context, the intensity of competition is the speed of development, and the cost of competition is the reduced efficiency of assimilation. The models predict that the evolutionary stable strategy (ESS) is to feed faster than the rate that maximizes brood fitness. The exact rate depends on the costs of increased competition, and on the average relatedness among brood members.

The reduction in brood fitness through larval competition is one aspect of parent-offspring conflict (Trivers 1974). In birds and mammals with parental care, the parent can to varying extents regulate competition among offspring because it is the parent who provides resources for the growing young (Parker and Macnair 1979; Godfray 1991). In species without parental care, it is much harder for the parent to regulate competition. One strategy open to the parent is to manipulate clutch size to allow for the effects of sibling rivalry (Godfray and Parker 1991, 1992). If a parent reduces clutch size, it obviously suffers a reduction in the number of offspring; however, the models suggest that a modest reduction in clutch size can be favored because sibling competition also decreases with clutch size. The reduction in sibling competition occurs because in smaller broods there are more resources to go around and thus fewer advantages to greater competitiveness.

I have discussed the theory of sibling rivalry in some detail because parasitoid wasps may provide some of the best systems in which to test these ideas. In general, tests are difficult because there are major problems in manipulating the competitive ability of offspring. However, the expected intensity of competition between siblings, and hence the predicted clutch size, is strongly influenced by the relatedness of siblings. A prediction can be made that, other things being equal, the least competition and the greatest clutch sizes will be observed in broods of genetically identical individuals (relatedness 1), followed by broods of haplodiploid sisters (relatedness 0.75), and lastly broods of haplodiploid brothers (relatedness 0.5). Mixed broods are intermediate between all-male and all-female broods. The existence of polyembryonic parasitoids, and of parasitoids that produce all male and all-female broods, offer an opportunity to test some of these ideas.

3.2.2 EXPERIMENTAL STUDIES

The fundamental assumption of parasitoid clutch size theory is that offspring fitness is a function of clutch size. As reviewed in section 6.1 (see also below), the chief effect of clutch size is to decrease the size of the emerging adult parasitoid, although sometimes survival is also influenced by clutch size. The relationship between adult size and fitness is well established, at least in the laboratory, although quantitative field data are hard to obtain (sec. 7.1.2). The fundamental assumption is thus well supported by data. In most cases, offspring fitness declines with increasing clutch size, though in a few instances a domed relationship is found because of the poor performance of larvae in smaller clutches.

In this section I review experimental tests of clutch size theory. First, I discuss the evidence that parasitoids adjust clutch size to the quality of their host. Second, I describe attempts to calculate the Lack clutch size and to compare the prediction with observed clutch size. Finally, I discuss the clutch size of polyembryonic wasps, and of wasps that produce all male and all female broods.

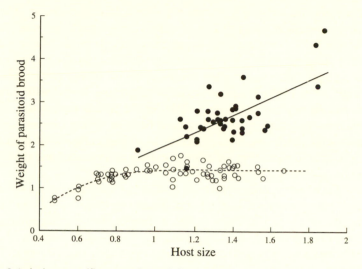

Figure 3.4 An interspecific comparison of clutch size in the genus *Apanteles (sensu lato)*. The logarithm of the volume of wasps emerging from a host (defined as the cube of the length of the adult wasp multiplied by brood size) is plotted against the logarithm of host volume. Solid points represent gregarious species, and hollow points solitary species. The line is a fitted log-log regression. (From le Masurier 1987.)

CLUTCH SIZE AND HOST QUALITY

The prediction that clutch size and host quality should be positively correlated can be tested using both interspecific and intraspecific comparisons. The most common component of host quality that has been studied is host size. Le Masurier (1987) collected data on the host size and brood size (the number of wasps emerging from a single host) of fifty-two gregarious species of the braconid genus *Apanteles sensu lato*. A regression of brood size on host size was highly significant and accounted for 31% of the variation in the data. However, wasps emerging from larger hosts tended themselves to be larger and thus presumably required more resources for their development. To control for this, Le Masurier examined the relationship between the total volume of wasps emerging from a host, and host size (data from forty species). This improved the fit of the regression, which now accounted for 45% of the variation in the data (fig 3.4).

Writing in 1961, Salt listed ten studies of different parasitoid species which had shown that female wasps laid larger clutches on bigger hosts, and today a similar list would probably contain over one hundred species. There is abundant evidence that parasitoids with wide host ranges lay more eggs on larger host species. One of the earliest examples is provided by da Costa Lima (1928), who found that *Telenomus fariai*, a scelionid egg parasitoid of triatomid bugs (the Chagas disease vector), laid 4 eggs in *Triatoma sordida* but 6–8

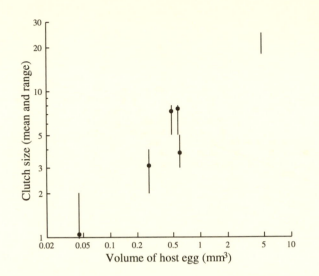

Figure 3.5 The clutch size laid by *Trichogramma embryophagum* on hosts of different size: ranges and means (where available). (From Klomp and Teerink 1967.)

eggs in the larger *Panstrongylus (=Triatoma) megista*. Some of the most detailed work on parasitoids that attack several host species has been on egg parasitoids in the genus *Trichogramma* (Trichogrammatidae). Salt himself (1934, 1936) showed that *T. evanescens* laid more eggs on larger hosts, and this work was extended by Klomp and Teerink (1962, 1967), working with a related species, *T. embryophagum* (fig 3.5).

Many parasitoids that attack the growing stages of a single species lay more eggs on older hosts. In a very careful study of the parasitoids attacking the coconut leaf miner (*Promecotheca reichei*), Taylor (1937) noted that all the gregarious chalcidoid parasitoids laid fewest eggs on first instar hosts and most eggs on the largest third instars. Two more recent examples are shown in figure 3.6. Host age can be important even when host size does not change with age. Marston and Ertle (1969) found that the egg parasitoid *Trichogramma minutum* placed an average of 6.3 eggs in one-hour-old host eggs (the cabbage looper, *Trichoplusia ni*) but just 1.5 eggs in one-day-old eggs; similar behavior has been reported in a number of other trichogrammatids (Pak 1986). Old eggs provide fewer resources for the developing parasitoid larvae than young eggs.

Host species may influence clutch size independently of host size. Figure 3.7 shows the number of progeny emerging from three host species of the pteromalid wasp *Pteromalus puparum* (Takagi 1986). The most variable host species is *Papilio xuthus*, for which the relationship between host size and clutch size is well described by a linear regression. However, the numbers of eggs laid on the two other host species does not fall on the same regression line, both species receiving fewer eggs than expected.

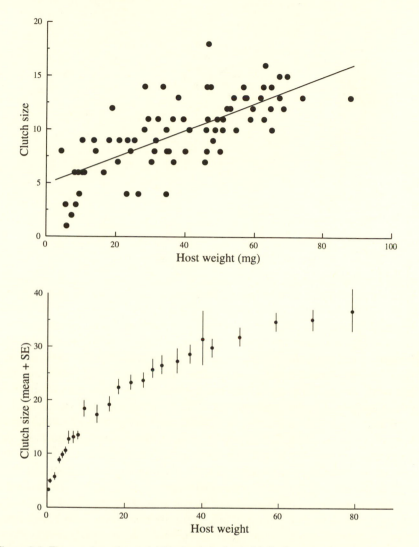

Figure 3.6 The number of eggs laid by (*top*) the bethylid *Goniozus nephantidis*, attacking the pyralid moth *Corcyra cephalonica*, and (*bottom*) the eulophid *Colpoclypeus florus*, attacking the tortricid moth *Adoxophyes orana*, on different-sized hosts. (From Hardy et al. 1992 and Dijkstra 1986, respectively).

ATTEMPTS TO CALCULATE THE LACK CLUTCH SIZE

The first attempt to calculate the Lack clutch size was made by Charnov and Skinner (1984, 1985) using data collected by Klomp and Teerink (1967) on the egg parasitoid *Trichogramma embryophagum*. On different hosts, Klomp and Teerink had measured the relationship between clutch size and adult size, and

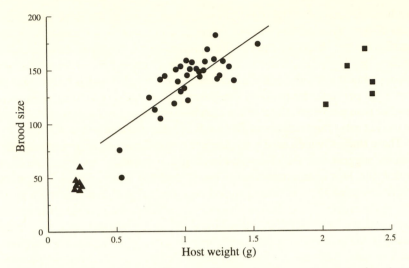

Figure 3.7 The number of pteromalid wasps (*Pteromalus puparum*) emerging from hosts of different size. The hosts are the pupae of three species of butterfly: triangles, the white butterfly *Pieris rapae*; circles, the swallowtail *Papilio xuthus*; squares, the swallowtail *Papilio memnon*. (From Takagi 1986.)

between adult size and lifetime reproductive success. As *T. embryophagum* has a very female-biased sex ratio, Charnov and Skinner concentrated on this sex and calculated the clutch size that resulted in the maximum number of granddaughters (via daughters) for the ovipositing female. On three host types the Lack clutch size was 4, 7, and 9, which compared with observed clutch sizes of 1–2, 5–8, and 5–8, respectively. They also used similar data, collected themselves, to predict the Lack clutch size of the pteromalid *Nasonia vitripennis* on four sizes of host (dipteran pupae). Their estimated Lack clutch sizes were 9, 14, 17, and 21, and the observed values 5, 7, 11, and 11, respectively. Waage and Godfray (1985) attempted to calculate the Lack clutch size in six species of parasitoid wasps using data from the literature on the relationship between clutch size and offspring survival. Observed clutch sizes were in all cases considerably smaller than the calculated Lack clutch size, though these results are not particularly reliable as it was not possible to include the effects of clutch size on the size, and hence fitness, of the adult wasp.

Another trichogrammatid egg parasitoid, *Trichogramma evanescens*, was studied by Waage and Ng (1984). This wasp injects up to six eggs into the eggs of the cabbage moth, *Mamestra brassicae*. Waage and Ng found that the size of the emerging wasps depended strongly on the number of other individuals developing in the host. They calculated the optimal combination of sons and daughters that maximized the parents' reproductive success, making use of two assumptions: (1) the fitness of female wasps of different sizes could be approximated by lifetime fecundity as measured in the laboratory, and (2) all matings are between siblings so only enough males are produced to inseminate their sisters. They concluded that a female wasp attacking a single host egg maximizes her fitness gain per host by producing one male and three females. In fact, they tended to produce a single male and two females. If females attack groups of host eggs, the optimum policy for the female wasp is to lay three to four eggs per host, the majority daughters. Again, observed clutch sizes were smaller than predicted: the average clutch size was approximately two. (Some of the sex ratio implications of this study are discussed in sec. 4.3.3).

These studies, which used a mixture of observational and experimental techniques, suggest that the observed clutch size is normally less than the Lack clutch size. However, ornithologists have found that observational data can be extremely misleading in testing hypotheses about clutch size. The main problem is the frequent correlation between the number of eggs laid by a female and unknown third variables, for example, the ability of the female to rear the young or the quality of the male's territory. In parasitoids, which normally lack parental care, different confounding variables are important. For instance, an experimenter who ignored variation in the size of the host might observe a positive relationship between clutch size and the survival and adult size of the offspring. Parasitoids lay larger clutches on big hosts, and quite often these survive with higher probability to grow into larger hosts (e.g., Hardy et al. 1992). Host size is relatively easy to standardize, but the same problems may arise with other host properties that are less easy for the experimenter to observe and control. Clutch-size manipulation experiments are thus highly desirable to eliminate these problems.

Endoparasitoids and ectoparasitoids pose different problems for experimental manipulation. As yet, no one has directly manipulated clutch size in an endoparasitoid, though indirect manipulation is possible by making use of the behavior of the adult parasitoid. For example, if two parasitoids are placed simultaneously, or in quick succession, on a host, often both lay normal-sized clutches, allowing the experimenter to produce particularly large clutches (Klomp and Teerink 1967). Small clutches can be produced by interrupting a parasitoid before the completion of oviposition. A particularly elegant technique for manipulating clutch size in *Trichogramma* has been developed by Klomp and Teerink (1967). These wasps assess the size of the host by antennal examination prior to oviposition (see below); if a wasp is carefully transferred to a new host, after examination, but before oviposition, the clutch it lays will be appropriate to the size of the first host. It is thus possible to persuade a wasp to lay clutches that are either larger or smaller than those normally laid in a host of a particular size. Manipulation of clutch size in ectoparasitoids is straightforward if the eggs are laid loosely on the surface of the host and can be transferred between hosts without damage. If the eggs are firmly attached, manipulation may require relying on indirect techniques. I now discuss recent attempts to calculate the Lack clutch size using manipulation techniques.

The pteromalid wasp *Pteromalus puparum* is a gregarious endoparasitoid attacking the pupae of many species of Lepidoptera. Takagi (1985, 1986, 1987) studied the wasp and one of its hosts, the swallowtail butterfly, *Papilio xuthus*. Pupae of this host weigh between 0.63 and 1.32 g and the wasp lays on average 150 eggs per gram weight of its host. Takagi artificially produced a range of clutch sizes by allowing between one and seven wasps simultaneously to oviposit into a single host. He found very high immature survival, but that there was a strong relationship between eggs per gram host weight and the size of the emerging wasp. The lifetime fecundity of a female was strongly size dependent, though the precise form of this relationship depended markedly on whether the female was allowed access to honey for feeding. Female longevity was also size dependent, though the relationship appeared to be domed. Substituting the different relationships between female size and fitness into the formula for the Lack clutch size produced a wide range of predictions. Using the fitness measure, lifetime fecundity of honey-fed females, a prediction of 300 eggs g^{-1} was obtained—much higher than observed. Using the measure, lifetime fecundity of non-honey-fed females, predictions in the range 60–70 eggs g^{-1} were found—lower than observed. Finally, using female longevity as a measure of fitness led to a prediction of about 150 eggs g^{-1}.

It is difficult to interpret this study because the three fitness measures give very different predictions; the problem is the lack of field data on the effect of female size on lifetime reproductive success. It is also unclear whether *Pteromalus puparum* is selected to produce the Lack clutch size, so assumptions that lead to the prediction of larger than observed clutch sizes are not necessarily incorrect. Finally, care has to be taken when manipulating clutch size by causing several females to oviposit into the same host. As will be described in the next chapter, many wasps (including *P. puparum*, Takagi 1987), lay more males when superparasitizing and this may influence the optimum clutch size.

Colpoclypeus florus is a eulophid parasitoid of the apple leaf roller *Adoxophyes orana*. The wasp lays its eggs externally on the host and, as shown in figure 3.6b, lays more eggs on larger hosts. Dijkstra (1986) directly manipulated clutch size and showed that adult size was strongly influenced by the number of larvae developing on a host. He was also able to relate female size to fecundity and longevity (in the absence of hosts), although only in the laboratory. Dijkstra reasoned that if the wasp were strongly host limited, the correct measure of a mother's fitness would be the total potential longevity of all her daughters. However, if the wasp were not limited by hosts, but by egg supply, the mother should lay a clutch size that resulted in each of her daughters achieving maximum fecundity. Dijkstra calculated the optimal clutch size based on each of these assumptions. He found that if the mother were truly host limited, clutch sizes greater than those observed should be found, whereas if she were egg limited, clutch sizes less than those observed should occur.

Hardy et al. (1992) studied clutch size in the bethylid wasp *Goniozus nephantidis*, which parasitizes the larvae of microlepidoptera. Unusual for a parasitoid wasp, the female remains with her young until they pupate, protecting them from superparasitism (and possibly hyperparasitism and multiparasitism, Hardy and Blackburn 1991). Hardy et al. argue that because of the long time spent with each brood, the wasp will be selected to maximize her gain in fitness from that brood and produce a Lack clutch size. The argument that large clutches waste time or eggs does not seem to apply in this case, as the time spent in oviposition is negligible compared with the time required to guard the brood, and the wasp attacks relative few hosts in its lifetime and is unlikely to run out of eggs. There is no evidence that the costs of brood guarding are affected by clutch size.

Hardy et al. (1992) created artificial clutches of different sizes on hosts of equal weight. Clutch size had no effect on larval survival but it had a marked influence on the size of the wasps that emerged from the manipulated broods. *G. nephantidis* lays a clutch that normally includes just a single male (see also sec. 4.3.3). The male mates with his sisters soon after eclosion and it seems likely that his fitness is only slightly affected by size. It thus appears that the chief consequence of large clutch size is a reduction in the size, and hence fitness, of female progeny. Hardy et al. (1992) attempted to relate female size to fitness using three different measures of fitness: (1) female fecundity when given unlimited access to hosts; (2) female longevity in the presence of hosts; and (3) female longevity in the absence of hosts. Larger females were at an advantage when fitness was calculated using the first two measures, but not when the third measure was used. The predicted Lack clutch size was about eighteen, using either of the first two measures, but could not be calculated using the third measure (if an optimal clutch size exists it is above the range of manipulated clutch sizes). The wasp laid nine or ten eggs on hosts of the size used in the experiment, considerably below the predicted clutch size using any of the fitness measures.

There are at least three explanations for these results. First, the wasp may be egg limited and able to lay only nine or ten eggs during one oviposition bout. This is unlikely because if the wasp is removed from the host immediately after laying a clutch of eggs and presented with a fresh host, it immediately begins to oviposit. Second, the argument that the natural history of *G. nephantidis* leads to selection for maximizing the fitness gain per clutch may be erroneous. Perhaps there are hidden costs of large clutch size that were not identified. However, Hardy et al. argued that the more likely explanation was a gross underestimation of the fitness penalties of small size. All the experiments were confined to the laboratory, a more benign environment than the field. They concluded that field estimates of the effect of size on reproductive success are essential before quantitative tests of clutch size theory can hope to be successful.

Finally, Le Masurier (1991) attempted to calculate the Lack clutch size of the gregarious braconid *Cotesia* (=*Apanteles*) *glomeratus*, a common parasitoid of white butterflies (*Pieris* sp.). He did not attempt to manipulate clutch size and estimated the number of eggs laid as the sum of wasps that success-fully emerged plus dead pupae and larvae. He was thus unable to control for between-host variation in quality, or to assess egg or young larval mortality. The relationship between adult size and fecundity was estimated by dissection. Working with a British strain of *C. glomeratus* that attacks the Large White butterfly (*Pieris brassicae*), he was unable to calculate a Lack clutch size be-cause the natural variation in egg number per host was insufficient to reveal the density-dependent reduction in fitness in large clutches. However, he also ex-perimented with an American strain that attacks the Small White butterfly (*Pieris rapae*) which as its name suggests is much smaller than *P. brassicae*. Surprisingly, the wasp lays a larger clutch (28) in *P. rapae* than the British race lays in *P. brassicae* (20). If the British race is presented with a *P. rapae* larva in the laboratory, a larger clutch (27) is also laid. Le Masurier estimated the Lack clutch size on *P. rapae* as 23.

Although these results suggest the wasp lays more eggs than the Lack clutch size, calculations from observational studies must be treated with caution, and the most that can be said is that the observed and Lack clutch sizes on *P. rapae* are approximately the same. Le Masurier suggests several explanations for these results. First, the wasp may be adapted primarily to *P. brassicae* and not recognize *P. rapae* as a smaller host. This fails to explain the significantly higher brood sizes on the second host, or why the American race, which al-ways attacks *P. rapae*, has not adapted to its new host in the three hundred generations since its introduction. Second, *P. brassicae* is gregarious and *P. rapae* solitary. As is discussed in the next section, parasitoids that fre-quently encounter hosts will often reduce the number of eggs they lay per host. Finally, Le Masurier was not able to rule out superparasitism as an explanation for the different brood sizes in the two host species. Manipulative experiments with *A. glomeratus* would be valuable in resolving these questions.

THE COST OF REPRODUCTION

Few studies have yet tried to disentangle the different reasons why parasit-oids may be selected to produce smaller clutches than the Lack clutch size. However, there is some evidence for a weak prediction of theory—that clutch size should drop as the frequency of host encounters increases. This relation-ship is predicted both by models that assume reproductive success is limited by time, and by models that assume egg limitation.

Kearns (1934a) remarked of the bethylid *Cephalonomia gallicola*, a parasit-oid of the cigarette beetle (*Lasioderma serricorne*), that "a scarcity of the host seems to encourage the female to lay a greater number of eggs per individual host and often to return to the host upon which she has previously oviposited and lay additional eggs." Jackson (1966) also reported a link between clutch size and encounter rate. She studied an aquatic mymarid wasp (*Caraphractus cinctus*) which oviposits into the eggs of water beetles. On small hosts pre-sented individually, the wasp lays 2–3 eggs, but smaller clutch sizes are found

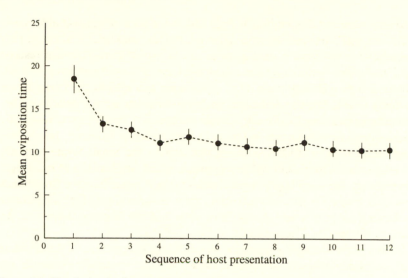

Figure 3.8 Clutch size in *Cotesia (=Apanteles) glomeratus* attacking the white butterfly *Pieris rapae*. The wasps were presented with twelve hosts consecutively and the time taken to oviposit in each host was recorded. Oviposition time is strongly correlated with clutch size. Clutch size is highest on the first host but drops quickly with successive hosts. (From Ikawa and Suzuki 1982.)

when the wasp is presented with a series of hosts. Similar results have been reported in trichogrammatid egg parasites (Glas et al. 1981; Waage and Ng 1984; Hirose et al. 1976; Schmidt and Smith 1985b). Ikawa and Suzuki (1982) found that in the braconid *Cotesia (=Apanteles) glomeratus*, a parasitoid in Japan of the cabbage butterfly (*Pieris rapae*), there is a close correlation between the time taken to oviposit and the number of eggs injected into the host. Making use of this observation, they observed that if the wasp was presented with a series of hosts, it laid a large clutch on the first caterpillar and then smaller clutches on subsequent hosts (fig. 3.8). The authors interpreted this behavior as an adaptive response: the wasp learns that hosts are locally abundant and reduces clutch size to lessen the risks of egg exhaustion.

Theory predicts that clutch size should drop as egg reserves decrease. Unfortunately, manipulating the number of eggs carried by a wasp is difficult and can only be achieved indirectly. In some ingenious experiments, Rosenheim and Rosen (1991) attempted to observe the relationship between clutch size and egg reserves in the aphelinid wasp *Aphytis lingnanensis*, a parasitoid of armored scale insects (*Aonidiella* spp.). This wasp matures eggs throughout its life, but at a rate dependent on ambient temperatures. Egg reserves of individual wasps can be reduced by keeping them at low temperatures. In addition, there is predictable natural variation in reserves; large wasps tend to have more eggs than small wasps. Rosenheim and Rosen found that clutch size was inversely related to egg reserves, which, in a multiple-regression analysis, was the most important factor influencing clutch size. Experience was also important: wasps that had recently encountered hosts laid smaller clutches than those with similar egg reserves that had been deprived of hosts. Low egg reserves and abundant hosts both increase the probability of egg exhaustion and hence favor smaller clutches. Rosenheim and Rosen also found that wasps with larger egg reserves spent less time preparing for oviposition, as well as grooming and resting after the completion of oviposition.

SINGLE-SEX CLUTCHES

Single-sex clutches in hymenopteran parasitoids provide an opportunity to test the prediction that all female clutches should be larger than all male clutches because of the reduced larval competitiveness associated with greater relatedness among sisters. Only a few cases of single-sex clutches are known among parasitoids. The best-studied examples occur among leaf-miner parasitoids in the eulophid genus *Achrysocharoides (=Enaysma)*. There are about fifteen species in the United Kingdom, where they have been studied by Askew and Ruse (1974) and Bryan (1983), including one species in which males are unknown. Of the other species, two or three have mixed broods, about seven produce single-sex broods, and the natural history of the rest is poorly known. In the species with single-sex broods, male broods are always smaller than female broods. Typically, male broods contain one or occasionally two wasps, while female broods contain two or three (table 3.2; see also sec. 4.3.5 for a discussion of sex ratio in this genus). The size of single-sex broods in *Achrysocharoides* is consistent with the clutch size theory outlined above (Godfray and Parker 1991) but obviously requires further investigation. Single-sex broods have also been recorded in the summer (but not the spring) generation of another eulophid *Eulophus ramicornis (=larvarum)* though in this case the sizes of male and female broods are highly variable and not statistically different (Godfray and Shaw 1987).

The pleasing concordance between theory and data provided by *Achrysocharoides* is rather spoiled by another case of single-sex broods. Salt (1931) provides a table (reproduced here as table 3.3) of the sexual composition of 281 broods of the braconid wasp *Bracon (=Microbracon) terebella*, a parasitoid of the wheat-stem sawfly *Cephus pygmeus*. Approximately 90% of the broods are single sex, but in this case the average size of all male broods (3.9) exceeds that of all female broods (2.9).

POLYEMBRYONIC CLUTCHES

In polyembryonic species the female wasp places one or a few eggs in a host; they divide asexually to produce a few to a very large number of embryos (see sec. 1.3.1 for an introduction to polyembryony). The size of polyembryonic broods is thus affected by the number of eggs laid by the parent (the parental clutch size) and the number of asexual divisions by the eggs (which

Table 3.2
Progeny allocation in the eulophid wasp genus *Achrysocharoides*.

Species	Male Broods		Female Broods		Mixed Broods		Overall Sex Ratio
	Sample Size	Average Brood Size	Sample Size	Average Brood Size	Sample Size	Average Brood Size	
(1) *carpini*	2	1.0	545	1.6	4	3.0	0.01
(2) *butus*	4	1.0	5	1.4	0	—	0.31
latreillei	112	1.0	111	2.0	0	—	0.29
zwoelferi	103	1.0	152	2.0	5	3.0	0.28
niveipes	432	1.0	344	2.2	1	2.0	0.40
cilla	406	1.2	445	1.9	2	3.5	0.35
splendens	19	1.5	29	1.9	4	3.0	0.35
splendens 'B'	13	2.4	14	2.6	2	3.5	0.45
(3) *acerianus*	2	2.0	13	2.2	8	2.9	0.24
atys	58	1.2	201	1.5	287	2.6	0.33

Source: Data from Bryan (1983).
Notes: The number of all-male, all-female, and mixed broods are shown with their average brood size as well as the population sex ratio. The species are divided into three groups: (1) an apparently thelytokous species (*carpini*), (2) species in which mixed broods are rare or absent, and (3) species where mixed broods are common.

Table 3.3
The sexual composition of 281 broods of the braconid *Bracon terebella* recorded by Salt (1931).

Number of Individuals in Brood	1	2	3	4	5	6	7
Frequency of male broods (26%)	3	5	25	16	15	6	3
Frequency of female broods (63%)	13	49	61	44	8	1	0
Frequency of mixed broods (11%)	0	2	10	12	6	2	0

Note: A total of 924 wasps were reared, 36.5% of which were males.

determine the offspring clutch size). Is it a coincidence that the largest brood sizes of any parasitoids are found among polyembryonic wasps? Encyrtids in the genus *Copidosoma* (which now includes *Litomastix*) sometimes produce broods of 2000–4000, exceeding, almost by an order of magnitude, their nearest rivals. There are at least three reasons why large brood sizes may be associated with polyembryony:

1. The major clutch size decision in polyembryonic species is made not by the parent, but by the young. The young will not be subject to the type of trade-off that leads the parent to lay clutches smaller than the Lack clutch size. For example, there is no question of wasting time or eggs

while ovipositing. Polyembryonic parasitoids are probably one of the few groups that should invariably produce the Lack clutch size (Godfray 1987a, 1987b).

2. Polyembryonic encyrtids are particularly small wasps, and this in part explains the large brood size. Because polyembryonic broods are genetically identical and thus avoid sibling rivalry, adaptations to improve the efficiency of host utilization will occur far more readily than in sexual broods where each individual has its own genetic opinion. It is possible that small size, and hence large broods, is an adaptation to increase the efficiency of resource use (Godfray and Parker 1991).

3. The previous two explanations have assumed that polyembryony preceded large brood size. However, the opposite argument can be made. Perhaps polyembryony is more likely to evolve in systems where a small parasitoid attacks a very large host. Division of a single egg is an alternative strategy to the oviposition of a large number of eggs.

Without a good phylogeny of different polyembryonic groups, and without more details of their biology and the biology of their nearest relatives, it is difficult to move beyond speculation. One must also remember that polyembryony is not invariably associated with large clutch sizes and cooperation. As discussed in section 1.3.1 there are a number of polyembryonic species with small clutches, and there is also evidence of fighting and competition among siblings.

The size of all-male and all-female offspring clutches should be the sex-specific Lack clutch sizes. If one sex suffers more from small size as an adult, the optimum clutch size for that sex will be smaller. In general, small females are believed to be heavily penalized, so larger male broods would be predicted. In a series of detailed studies, Strand (1989b, 1989c) initially found no difference in the size of male and female broods of *Copidosoma floridanum*, a finding typical of other encyrtids (reviewed by Strand 1989c). However, in his most recent work Strand (pers. comm.) has detected a small but consistent tendency for female broods to be larger. Worse for theory, Strand (1989c) also notes that there is "limited evidence" for smaller male broods in polyembryonic braconids and platygasterids.

Mixed broods are quite common in a number of polyembryonic species, and this has led to some controversy over whether they are derived from one or more than one egg (Clausen 1940a). Whenever a species has been studied in detail, mixed broods have been shown to arise from multiple eggs. The best-studied system is again *Copidosoma floridanum*, where mixed broods constitute 57% of field collections (Strand 1993). Observations in the laboratory have shown that all-male and all-female broods always arise from a single egg, while mixed broods always arise from two eggs (Strand 1989b, 1989c). To create a mixed brood, the female wasp invariably lays a female egg followed by a male egg; mixed broods are of the same size as single-sex broods but are predominantly composed of females. It appears that division by the male egg is affected by the presence of a female egg. Strand (1989b) was able to manipulate the number and sex of eggs laid in a host by using virgin females and by allowing superparasitism. He found that the final number of offspring and the final sex ratio were independent of the number and sex of the eggs laid by the parent (providing, of course, that at least one egg of each sex was present). The reason male eggs produce smaller broods than female eggs is probably due to selection acting on the sex ratio, and I shall return to this in the next chapter (sec. 4.3.3). A final problem is to explain why the female wasp sometimes lays a single egg but more frequently two eggs. Strand (1989b, 1989c) found that two eggs were laid near the beginning of an oviposition bout and single eggs toward the end. It is thus possible that the wasp lays single eggs when its egg reserves are low. The relative advantages of single and double oviposition are likely to depend in a complicated manner on host abundance and the ease with which male and female parasitoids locate mates.

3.2.3 BEHAVIORAL MECHANISMS

What are the proximate cues used by parasitoids to assess the size and quality of their hosts prior to laying a clutch of eggs? Several workers have investigated this question by manipulating different features of the host and observing changes in the number of eggs laid by the parasitoid.

In principle, parasitoids could use either internal or external cues to assess host size. Wylie (1967) partially buried fly pupae in plasticine and exposed them to parasitism by the pteromalid wasp *Nasonia vitripennis*. The clutch size produced by the wasp was unaffected by how much of the host protruded above the surface of the substrate. Wylie suggested that his observations supported the idea that the wasp used an internal cue. For example, the wasp might inject a soluble chemical and then assess host size by sensing the dilution of the marker. However, the experiments did not exclude all external cues; the wasp might still be able to assess host size using the curvature of the fly puparium. Takagi (1986) carried out similar experiments with another pteromalid pupal parasitoid, *Pteromalus puparum*. In this case, clutch size was reduced when the host was partially buried, although the reduction in clutch size was not as great as would be expected if the wasp were purely assessing the volume of the exposed part of the host. The best evidence for the importance of internal cues is Purrington and Uleman's (1972) observation that the eulophid *Hyssopus thymus* adjusts its clutch size to the size of its cryptic host (the gelechiid moth *Metzneria lappella*), which is hidden in plant tissue and cannot be externally examined by the wasp.

Dijkstra (1986) manipulated the length of caterpillars of the tortricid moth

Adoxophyes orana, a host of the eulophid *Colpoclypeus florus*. He produced artificially short larvae by tying ligatures at various distances from the caterpillar's head and cutting off the caudal end. The shortened caterpillars behaved in a reasonably normal manner and were attacked by the wasp which laid clutch sizes appropriate to whole hosts. It appears that the diameter and width of the host are more important than the length in influencing the number of eggs laid by the wasp.

By far the most detailed studies of the behaviors involved in host assessment for clutch size have been carried out on egg parasitoids in the genus *Trichogramma*. On encountering a host egg, *Trichogramma* walk over the surface of the host, drumming continually with their antennae (Salt 1935, 1937a; Klomp and Teerink 1962, 1967). If a wasp is allowed to examine a host of one size, and then quickly transferred to a second host of a different size, she lays a clutch size appropriate to the first host (Klomp and Teerink 1962, 1967). This observation suggests that the clutch size decision is made during the initial external examination, before the insertion of the ovipositor. Visual cues are not important because oviposition proceeds normally in the dark. The precise behaviors involved have been dissected by Schmidt and Smith (1985a, 1985b, 1986, 1987a, 1987b, 1987c, 1987d, 1987e, 1989) using *T. minutum*, a polyphagous species that attacks a variety of hosts that vary both in shape and size. Experiments with real and artificial hosts of different sizes indicated that the curvature of the host egg was an important determinant of clutch size. The wasp keeps its body orientated at a constant angle to the host and measures surface curvature as the angle subtended between the head and the basal segment of the antennae when the tips of the antennae touch the host. However, the surface area of the host exposed to the parasitoid is also very important. If hosts of the same curvature are embedded in the substrate, hosts that have less area protruding receive smaller clutch sizes. Statistical analysis of the movement of the wasp over the host suggested that the specific behavior that most influenced clutch size was the wasp's first walk over the host—the initial transit. On finding a host, the wasp takes a great circle route over the host, including its highest point. If the initial transit is interrupted by placing a barrier in front of the wasp, but the wasp is otherwise undisturbed, smaller clutches are produced. Early observations (Flanders 1935; Salt 1940) had suggested that the assessment of host size was made relative to the size of the wasp: large parasitoids were more likely than small parasitoids to reject small hosts. This seems not to be the case in *T. minutum*, where clutch size is independent of wasp size. If the wasp directly measured the length of the initial transit, then one would expect host assessment to be relative—the only ruler the wasp has is itself. Indeed, there is evidence that the wasp's estimate of host curvature is influenced by its size. However, the time taken to conduct the initial transit is independent of wasp size, small wasps have smaller strides but walk faster, and Schmidt and Smith argue that wasps assess host size by the time needed for the transit. Although external cues seem to be the most important determinant of clutch size, internal cues may also exert some influence. In experiments with artificial eggs, the ionic composition of the egg has been shown to influence clutch size (Nettles et al. 1982).

The insect eggs attacked by *Trichogramma* spp. may be either solitary or laid in loose or compact clumps. Wasps reduce the number of eggs they lay per host when attacking clumped eggs, as predicted by many clutch size models (see above). Schmidt and Smith (1985b) have also explored the mechanistic basis of this observation by examining the behavior of wasps attacking single host eggs, eggs in compact or loose clumps, and eggs arranged in lines. When attacking solitary host eggs, the wasp uses the surface area of the egg to decide how many eggs to lay; Schmidt and Smith argue that the same behavioral mechanism leads to smaller clutches on clumped host eggs as these eggs have less exposed surface area. In support of this idea, host eggs at the edge of clumps, with a greater exposed surface area, receive a larger clutch than those in the interior. Similarly, if host eggs are arranged in lines, eggs at the end receive more parasitoid eggs than those in the middle. However, surface area alone is not the only factor contributing to smaller clutches on clumped host eggs because eggs in loose clumps also receive smaller clutches. It appears that the rate of host encounter is also important (Schmidt and Smith 1987b). In particular, there is a large drop in clutch size if the wasp encounters a second host very soon after attacking a first. This simple rule—reduce clutch size on encountering further hosts—may be sufficient to explain the average reduction in clutch size on loose clumps. There was no evidence that the wasps assessed the size of the egg mass prior to beginning oviposition, or that they returned to "top up" hosts. This last finding contrasts with observations by van Dijken and Waage (1987) on another *Trichogramma* species, *T. evanescens*. At least in the laboratory, when this species attacks a clump of host eggs it frequently returns to hosts it has previously attacked and adds further eggs.

3.2.4 SOLITARY VERSUS GREGARIOUS PARASITOIDS

In many parasitoid species, only one larva is able successfully to complete development on a host. If more than one egg is laid in a host, the two larvae compete and only one survives. Competition may take a variety of forms (see sec. 6.5) and frequently involves physical attack. The larvae of many solitary species have large mandibles which they use to attack other individuals in the host. In the first part of this section I discuss how gregarious parasitoids may evolve from solitary species and vice versa. In the second part I discuss whether solitary parasitoids are ever selected to lay more than one egg in a host.

THE TRANSITIONS BETWEEN SOLITARY AND GREGARIOUS LIFE HISTORIES

Reduction in the brood size of solitary parasitoids often occurs when the immature parasitoids are in their first instar, and the mechanism of brood reduction often involves larval fighting. Early brood reduction is of obvious advantage to the victorious larva because its competitors are destroyed before they have consumed an appreciable amount of host resources. However, the winning larva will suffer a reduction in inclusive fitness if the individuals that it killed were genetically related. When will natural selection favor brood reduction, and will the interests of the parent and the young coincide?

Consider a solitary parasitoid attacking a host large enough to support only a single larva. Supernumerary larvae are destroyed by fighting among first instar larvae. The optimum clutch size for both parent and offspring is clearly one, but let us assume that were a parent to lay two eggs in a host, either individual has an equal probability of killing the other and that the risk of double elimination is negligible. Now suppose that the quality of the host improves so that it can support the development of more than one parasitoid; the optimum clutch size for the mother is now greater than one, providing her offspring do not eliminate one another. Such an improvement in host quality might occur through a host shift, or through a gradual change in the size of either the host or parasitoid over evolutionary time. Selection on the parent to produce a clutch size greater than one can be realized only if selection on the immatures simultaneously favors an abandonment of larval fighting in the first instar.

Using an explicitly genetic model, Godfray (1987b) suggested that it would be difficult for a "nonfighting" gene to spread through a population of "fighters" because a rare gene for "nonfighting" would most often find itself in mixed sibships with fighting individuals and so be eliminated. To offset this high risk of elimination, the minority of nonfighters that shares hosts with other nonfighters must have a marked fitness advantage. Indeed, in the simplest models, the fitness of a nonfighter developing as one of a pair must be *higher* than that of a fighter developing alone. This might occur in cases where successful pupation requires that the host is completely consumed, and where this task is most easily performed by several larvae.

The exact conditions required for the spread of a nonfighting allele depend on the details of gene dominance and penetrance, the costs of fighting, the probability of double elimination, and the sex ratio (Godfray 1987b). However, a robust conclusion is that frequently the parent will be prevented from producing its optimum clutch size by the behavior of her larvae. Figure 3.9 illustrates a possible outcome of this process where the quality of the host has

Solitary   Gregarious

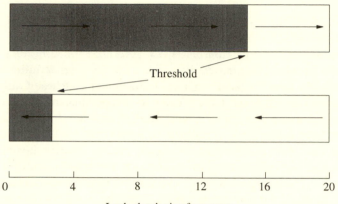

0   4   8   12   16   20

Lack clutch size for parent

Figure 3.9 The evolutionary transitions between solitary and gregarious life histories. Suppose there is a change in host size and hence in quality over evolutionary time: the quality of the host can be measured by the Lack clutch size that it supports (horizontal scale). This figure illustrates an example of the position of the transitions between the gregarious and solitary life histories as the quality of the host increases (top bar) or decreases (bottom bar). Note that there is a hysteresis: the position of the threshold depends on the direction of change. To obtain this figure, it was assumed that the fitness of a larval parasitoid declines exponentially with increasing clutch size, and that if two larvae attack each other there is a probability of 0.1 that both are destroyed.

to increase until the Lack clutch size is fifteen, before selection favors the abandonment of larval fighting.

When will a fighting allele invade a population of nonfighters? Again, the exact answer depends on the details of the model assumptions, but the general result is that a fighting allele can invade when the host is of a size to support a Lack clutch size of about two or three. The fighting allele does not spread in larger clutches because it causes the destruction of too many copies of itself. In smaller clutches, the increased fitness of the survivor outweighs the risk of killing other individuals carrying the same gene.

While gregariousness is favored in large hosts, and single larvae in small hosts, there is a hysteresis in the threshold host size dividing the two types of life history. This is illustrated in figure 3.9, where host quality is measured by the Lack clutch size of the parent. If the parasitoid is initially solitary, host quality must increase dramatically before gregariousness can evolve. However, once the parasitoid is gregarious, host quality must radically deteriorate before the solitary life history reinvades. There is a large intermediate range of host qualities where the life history displayed by the parasitoid will depend on the evolutionary history of the species. The gregarious and solitary life histories are thus not part of a simple continuum, but are better thought of as a dichotomy. The solitary life history also approaches an absorbing state (or as Harvey and Partridge 1987 described it, an evolutionary black hole!); once evolved, it is difficult to lose.

Two predictions arising from this work have been tested by Le Masurier (1987): (1) solitary parasitoids will often be found on hosts of sufficient quality to support more than one larva, and (2) gregarious clutches of the order of two or three should be rare. I have already mentioned that Le Masurier found a correlation between parasitoid clutch size and host volume in a cross-species comparison of *Apanteles, sensu lato* (fig 3.4). If solitary species are included on the same figure, it is clear that they frequently attack hosts large enough to support gregarious clutches of similar-sized congeners. This observation is consistent with the prediction that both life histories should be found on hosts of the same size. Clutches of two, three, or four wasps also seem to be disproportionately rare, again as predicted by theory.

A curious observation from the pesticide literature may also be relevant here. Diflubenzuron is an insecticide that acts as a chitin inhibitor. Khoo et al. (1985) found that if the solitary chalcidid wasps, *Brachymeria intermedia* and *B. lasus*, were treated topically with diflubenzuron and then allowed to attack Gypsy Moth pupae, gregarious broods often emerged. In the laboratory, superparasitism was common, and up to seven parasitoids emerged from the host. Despite the greatly elevated brood size, the adult wasps were able to reproduce although they were quite variable in size. It seems that the chemical allows gregarious clutches to develop by weakening the mandibles of the first instar larvae so that they are ineffective in brood reduction. The fact that large clutches of reasonably normal wasps can develop in the absence of brood reduction suggests that selection on the mother to increase her clutch size may be frustrated by selection on the offspring to retain fighting mandibles.

I should add some qualifying comments to the arguments developed here. The conclusions are based on a simple two-allele model of fighting and tolerance in parasitoids; greater genetic realism would be desirable. The model also ignores the advantages of fighting and killing unrelated larvae, either conspecifics (superparasitism) or larvae of other species (multiparasitism). The argument that the gregarious and solitary life histories are a dichotomy applies only to parasitoids where brood reduction is brought about by larval combat. Larval combat chiefly occurs in endoparasitoids with delayed development (koinobionts), though it is also found in some ectoparasitoids such as pteromalids in the genus *Pteromalus (=Habrocytus)* (Clausen 1940a; Salt 1961). The natural history of some species may preclude the evolution of efficient larval combat. In particular, many ectoparasitoids are relatively immobile and will not encounter other parasitoids on the host. For example, the larvae of

Aphytis species (Aphelinidae) attacking scale insects seem not to interact with each other. Several species have clutch sizes of one, two, or three, and are sometimes referred to as semigregarious. Alternatively, selection may favor very fast growth to the exclusion of all other activities. Klomp and Teerink (1978a) found that larvae of the trichogrammatid egg parasitoid *Trichogramma embryophagum* grew very quickly until the host was completely consumed. Parasitoid growth then stopped and one or a number of larvae died, releasing their gut contents, which were consumed by the surviving parasitoids. A difference of ten minutes in the time between ovipositions affected the probability of larval survival. In such a competitive situation, any mutation that reduced the rate of feeding is unlikely to be favored. Finally, gregarious and solitary strains are known in one species of pteromalid, *Muscidifurax raptorellus*. Gregariousness appears to be a polygenic trait, but the clutch size behavior of female wasps is influenced by the genotype of their mate. This curious finding is discussed more fully in section 5.4.

CLUTCH SIZE IN SOLITARY PARASITOIDS

There are several reasons why a female solitary parasitoid may deposit more than one egg on a host. If there is a finite probability that her first egg is inviable, the deposition of two eggs increases the probability that at least one offspring will survive. Alternatively, the presence of more than one egg in a host may help overcome the host's immune system (see sec. 6.3.1). Against these possible benefits must be set at least two costs. First, there may be a risk that both larvae are eliminated when brood size is reduced to one. Second, laying an extra egg in the host may be a poor use of an egg that can more profitably be placed in another host. Multiple oviposition in solitary parasitoids will be found more often in time-limited than egg-limited species.

In the next section, the behavioral ecology of superparasitism is discussed. If a solitary wasp lays an egg on a previously parasitized host, the survival of the larva will depend on its relative competitive ability and also on the number of eggs laid by the first female. The process of larval elimination may be analogous to a lottery, probably weighted in favor of the offspring of the first female. In such circumstances, one or both females may be selected to lay two or more eggs to increase the probability of one of their progeny surviving (van Alphen and Visser 1990).

There are a few cases where solitary parasitoids have been found to lay more than one egg per oviposition (Beling 1932; van Alphen and Nell 1982; Cloutier 1984; Mollema 1988), while Mackauer (1990) and Chow and Mackauer (1991) state that this behavior occurs most frequently among aphidiine braconids when their hosts (aphids) are rare. However, double oviposition is seldom specifically looked for, and it may be more common than the few reports suggest.

3.3 Superparasitism

There is some disagreement in the literature about the exact definition of superparasitism. Here, I follow van Dijken and Waage (1987) and define superparasitism as the deposition of a clutch of eggs (the clutch may of course consist of a single egg) on a host that has already been parasitized by a member of the same species. Further, I define a clutch of eggs to be the eggs laid during a single oviposition bout. Where the second clutch is laid by a different female, the definition is clear. In self-superparasitism the clutch is laid by the same female and it is possible that there might be some difficulty in defining precisely what is an oviposition bout, though in practice it will normally be unambiguous. A female parasitoid can only avoid superparasitism if it can recognize that a host has been previously attacked, an ability called "host discrimination."

Traditionally, the first observations of superparasitism and multiparasitism were made by Howard (1897) in a study of the parasitoids of the white-marked tussock moth *Orgyia leucostigma*. Writing of an ichneumonid wasp *Itoplectis (=Pimpla) inquisitor*, he stated that "all the apparently anxious soundings and tappings with her antennae, while appearing to satisfy her that everything is all right, do not always result in the depositing of the eggs under just the proper conditions." Superparasitism was an accident and led to "rivalry based upon erroneous instinct." The term "superparasitism" was coined by Fiske (1910), who used it to refer to both superparasitism as understood today (cannibal superparasitism) and to multiparasitism (mixed superparasitism, i.e., the second oviposition is by a different species of parasitoid). H. S. Smith (1916) narrowed the definition of superparasitism and introduced multiple parasitism or multiparasitism for cases where two species were involved. Another important early paper was Pierce's (1910) survey of known cases of multiparasitism and superparasitism which demonstrated the widespread nature of the two phenomena. A spectacular example of superparasitism was recorded by Tothill et al. (1930) who found seventy-two eggs of the solitary tachinid *Ptychomyia remota* on a larva of the zygaenid moth *Levuana iridescens*.

Thus from its first discovery, superparasitism was treated as an error on the part of the parasitoid. From the point of view of the species, there are no benefits to placing an egg in a previously parasitized host. However, as parasitoid biologists have moved to interpreting behavior in terms of selection acting on the individual, there has been a growing realization that superparasitism will often be favored by natural selection (van Alphen and Nell 1982; Iwasa et al. 1984; Parker and Courtney 1984; Charnov and Skinner 1984, 1985; Waage and Godfray 1985; Bakker et al. 1985). The first part of this section describes the theory that has been developed to explain superparasitism, and the second part describes experimental tests of theory. There has been a large amount of work on the behavioral and physiological mechanisms involved in host discrimination which is briefly described in the third part of the section. The final section is a discussion of multiparasitism—oviposition on a host that has previously been attacked by a different species of parasitoid. Superparasitism has been reviewed most recently by van Alphen and Visser (1990).

3.3.1 THEORY

The progeny of a superparasitizing female are normally at a competitive disadvantage in comparison with the progeny of the first parasitoid. In solitary species, the secondary larvae are often more likely to be eliminated, while in gregarious parasitoids they are outcompeted by the older larvae of the first clutch which are able to preempt much of the host resources (sec. 6.5). In the face of such disadvantages, it makes intuitive sense for the parasitoid to be able to distinguish parasitized from unparasitized hosts. This allows the parasitoid to avoid wasting time and eggs on hosts where its progeny have a small probability of survival; in addition, where the host has previously been parasitized by the same female, the ability to recognize parasitism avoids the risk of damaging the host by subjecting it to further attack (van Lenteren 1981). Van Lenteren has stressed that the observation of superparasitism does not necessarily mean that a parasitoid is unable to discriminate between parasitized and unparasitized hosts, a point originally made by Salt (1934). A variety of models have explored the circumstances under which superparasitism will be favored by parasitoids capable of host discrimination (van Alphen and Visser 1990; Spiers et al. 1991). It is easiest to describe separately models for solitary and gregarious parasitoids.

SOLITARY PARASITOIDS

The simplest models of superparasitism in solitary parasitoids are based on the type of host acceptance model described at the beginning of this chapter. Parasitoids are assumed to maximize their rate of fitness gain, and previously parasitized hosts are treated simply as hosts of poor quality. Either the classical optimal diet model is used, which assumes instant host discrimination, or account is taken of recognition time (I. Harvey et al. 1987; Janssen 1989; van Alphen and Visser 1990). These models predict that superparasitism should occur under the same conditions that favor oviposition into poor-quality hosts: (1) superparasitism should not occur when unparasitized hosts are common; (2) parasitized hosts should either be always accepted or always ignored (the zero-one rule); and (3) when recognition is instantaneous, the decision to superparasitize should depend only on the abundance of unparasitized hosts and not on the frequency of encounters with parasitized hosts. When the identification of parasitized hosts takes time, the decision is influenced by the abundance of parasitized hosts. If parasitoids are able to assess the number of eggs in a host, they should be less willing to superparasitize hosts containing two or three eggs than those containing just a single egg.

These simple models of superparasitism suffer from the same problems as simple host acceptance models. Many parasitoids will be egg limited, rather than time limited, and the optimal decision to superparasitize may depend on the internal state of the animal, for example on the number of eggs it carries. The models also assume that the encounter rate with hosts is constant, which will not be the case if hosts are heterogeneously distributed in the environment. There has been some work on superparasitism strategies in egg-limited parasitoids. Price's (1973a) verbal argument that superparasitism will be found most frequently when parasitoids have large supplies of unlaid eggs is confirmed by the dynamic programming models of Iwasa et al. (1984). In the limit, if a parasitoid is certain to die with unlaid eggs, it will have nothing to lose through superparasitism, even if its progeny have only a small probability of survival. Experimentalists who have failed to find host discrimination in parasitoids have also argued that superparasitism avoidance is unlikely to be favored in species with large egg reserves (e.g., Liu and Morton 1986).

Superparasitism poses further problems for these simple models. The strategy adopted by the parasitoid can influence the relative abundance of host types; in attacking a fresh host, the parasitoid not only removes a member of the high-quality class, but adds to the numbers in the low-quality class. This problem is compounded when a number of individuals search together so that the optimum superparasitism strategy depends on what other parasitoids are doing. Instead of a simple static model, these problems suggest that it may be necessary to use dynamic models that allow for variation in physiological state, and game theory models that take into account the behavior of other parasitoids.

Dynamic game theory models are notoriously difficult to analyze. A start toward making superparasitism models more realistic has been made by Visser et al. (1992a) who consider one or more parasitoids searching and parasitizing hosts on a patch. They assume that the parasitoids are time limited, that the patch is depleted by parasitoid searching, and that the parasitoids are selected to quit the patch when the rate at which they discover hosts (weighted by their value as oviposition sites) drops to a fixed value determined by the quality of the environment. When on the patch, the parasitoid may adopt different superparasitism strategies, the benefits of which are influenced by the strategies of other parasitoids on the patch. Parasitoids are assumed to be able to assess the number of times the host has been parasitized; the fitness of an egg declines with the increase in the number of eggs already present in the host. The evolutionarily stable superparasitism strategies on hosts containing different numbers of eggs are calculated by computer simulation. This approach is an advance on static models since it includes interactions between insects on the same patch. However, the quality of the environment is assumed constant and not affected by the superparasitism strategies adopted by the wasp (although this simplification is unlikely to affect the qualitative conclusions). The model also assumes that only one egg is laid during each attack, an assumption that may increase the advantages of self-superparasitism.

A parasitoid searching alone is predicted never to engage in superparasitism. However, if two or more parasitoids search together, superparasitism can be favored. In the latter case, parasitoids should initially accept only unparasitized hosts but then switch to accepting hosts containing one egg, and then later to hosts containing successively more eggs, until they leave the patch. The female is selected to concentrate initially on unparasitized hosts so that it attacks as many as possible before they are found by the other parasitoids in the patch (see also the similar predator model discussed on in sec. 3.1.1). The length of time required to handle the host and the probability that an egg laid in a parasitized host survives are both crucial parameters affecting when the switch to superparasitism occurs. Where one parasitoid searches alone in a patch, superparasitism does not occur because the insect obtains no increment of fitness through the addition of another egg to a host that it has already attacked. However, when other females are present, it may be worth laying an egg in a previously parasitized host because that host may have been attacked by another individual. This of course assumes that the second egg has some chance of survival. When other females are present, a parasitoid may be selected to add eggs to a host already containing its own egg or eggs if this increases the probability that its own offspring eventually triumphs. The increased value of previously parasitized hosts accounts for why females remain longer in the patch when conspecifics are present.

A related stochastic model has been analyzed by van der Hoeven and Hemerik (1990). They assume that a parasitoid can detect the number of eggs already present in a host, and also the number of females searching in a patch. They also assume that the more parasitoid eggs a host harbors, the greater the probability that all parasitoids perish. Parasitoids foraging alone should never superparasitize, but the amount of superparasitism and the number of eggs per host should increase with the number of searching females. The limit to the number of times a host should be attacked is set by two factors: (1) the risk of killing the host by adding more eggs, and (2) the costs of wasting eggs on parasitized hosts that might be placed on fresh hosts, either in the current or a future patch. Whereas Visser et al. (1992a) stress time limitation, this model is more appropriate to egg-limited parasitoids. The quantitative predictions of both models are rather dependent on their detailed assumptions, but the prediction that superparasitism increases when parasitoids search in the presence of conspecifics is both clear and robust. Both models also predict that competition between conspecifics leads to a decrease in the efficiency with which the population of parasitoids exploits the host population, a type of interference that may help stabilize the population dynamics of the interaction (Visser and Driessen 1991).

Self-superparasitism occurs if a parasitoid attacks a host that she herself has previously attacked (Waage 1986). While it is easy to argue that conspecific superparasitism can, under certain circumstances, increase a parasitoid's fitness, self-superparasitism will nearly always be a waste of either time or eggs (Waage 1986; Hubbard et al. 1987). Self-superparasitism may have other costs, for example there may be a risk that two larvae in a host destroy each other. As discussed in section 3.2.4, there are some circumstances when a solitary parasitoid can benefit from placing two eggs in a host—two eggs may saturate the host's defense systems, or increase the likelihood that one of its own offspring defeats a superparasitoid. However, these factors are unlikely to select for self-superparasitism: if it is a good idea to have two eggs in a host, the parasitoid should lay two eggs during the first host encounter. To argue that self-superparasitism is adaptive one has to show that the parasitoid's assessment of the environment changed so that at the first encounter single oviposition was the best policy, but that by the second encounter double oviposition is favored (Waage 1986). A model by Hubbard et al. (1987) does suggest that self-superparasitism can on occasion be selected. However, their model did not allow the parasitoid to lay two eggs on the first encounter. In addition, they assumed parasitoids ignored hosts with two or more eggs, which gave an advantage to self-superparasitism because it "filled up" the host and rendered it immune to further attack. It is doubtful whether such a mechanism would operate in nature. Visser (1993) has used the same modeling framework as Visser et al. (1992a, described above) to investigate the adaptive value of self-superparasitism. He compared the fitness of parasitoids that were able or unable to discriminate between hosts parasitized by itself or by conspecifics. Discrimination was never disadvantageous but in some cases, especially in poor-quality environments, all hosts were accepted for parasitism and discrimination was unnecessary. The advantages of self-superparasitism in Visser's model arise from two sources. First, a parasitoid searching a patch with conspecifics may initially ignore previously parasitized hosts in order to concentrate on locating unparasitized hosts before its competitors. Later, when its assessment of local patch quality (i.e., number of unparasitized hosts) is lower, spending time adding eggs to a host may be favored. The second advantage to self-superparasitism is that it "fills up" a host, increasing the chance of a particular parent's offspring surviving, and reducing the risk of further superparasitism. However, like Hubbard et al. (1987), Visser's model does not allow the parasitoid to lay more than one egg at the first encounter, which generates an inbuilt bias in favor of self-superparasitism. In conclusion, deliberate self-superparasitism is probably rare and most cases are likely to result from the inability to distinguish hosts attacked by self and conspecifics. The ability to distinguish the two types of host will normally be advantageous, especially when the risk of reencountering hosts is high.

The behavior of newly emerged parasitoids on encountering parasitized hosts may differ from experienced insects for two reasons (van Alphen and Nell 1982; van Alphen and Visser 1990; Visser et al. 1992b). First, they are naive and have not yet learned the current absolute abundance of hosts, nor what proportion are already parasitized. Natural selection is likely to equip naive parasitoids with a set of behaviors appropriate for the circumstances they are most likely to encounter (Roitberg 1990). This prior set will be modified by experience. Second, newly emerged parasitoids do not risk self-superparasitism. In species without the ability to identify hosts previously parasitized by themselves, the possibility of attacking the same host again may act as a deterrent to superparasitism. However, a newly emerged wasp can be sure that if the first host it encounters is parasitized, the culprit is another individual, and this may make superparasitism more likely.

In addition to behavioral ecological models of superparasitism, a variety of statistical models describing the distribution of eggs across hosts have been developed (Bakker et al. 1967, 1972; Rogers 1975; D. Griffiths 1977a, 1977b; Meelis 1982; Maindonald and Markwick 1986; Daley and Maindonald 1989). The models differ in the assumptions they make about parasitoid behavior: the simplest models assume an egg is always laid on an unparasitized host and that parasitoids encountering a previously parasitized host lay an egg with a fixed probability (Bakker et al. 1967, 1972; Rogers 1975). In other models, the probability of oviposition is influenced by the number of times the host has previously been discovered by a parasitoid (Meelis 1982), or the number of eggs it contains (Bakker et al. 1972). These models were developed in the hope that a comparison of real egg distributions with model predictions might shed some light on superparasitism behavior. Unfortunately, this hope has not been realized. Parasitoid behavior is generally too complicated to be described by such simple rules; in particular, the probability of superparasitism frequently is influenced by parasitoid experience. Another problem is local variation in parasitoid behavior which masks the true pattern of egg distribution when samples taken at different times or localities are combined for analysis (van Lenteren 1981).

GREGARIOUS PARASITOIDS

The simplest models of clutch size in gregarious parasitoids are based on the classic patch-use models of foraging theory. The parasitoid is assumed to maximize its rate of fitness gain and to add eggs to a host until its marginal gain of fitness drops below the maximum achievable average rate in that environment (sec. 3.2.1). This idea can be extended to predict the clutch size laid by a superparasitizing female (Charnov and Skinner 1984, 1985; Skinner 1985). The second female should also add eggs to the host until her marginal gain in fitness equals the best average achievable in that environment. For most realistic cases, the clutch size laid by the second female is smaller than that laid by the first female. However, it is just possible for the second female to be selected to lay more eggs if her perception of the quality of the environment is very much worse than that of the first female (Skinner 1985). Superparasitism will be most frequent, and second clutches largest, when fresh hosts are rare and the absolute density of hosts is low.

Simple patch-use models of superparasitism ignore the effect of the risk of superparasitism on the optimal clutch size of females attacking unparasitized hosts. If many hosts are attacked more than once, a female ovipositing on a fresh host may be selected to lower her clutch size in order to reduce the amount of competition suffered by her offspring if superparasitism occurs. This decision will, in turn, influence the clutch size produced by the second female. The optimal clutch sizes for the first and second wasps are now functions of each other, and an explicitly game theoretic approach is required. Suzuki and Iwasa (1980) first studied this problem as part of an analysis of sex ratio decisions during superparasitism (see sec. 4.3.3). They assumed no trade-offs between clutch size and future reproductive success and calculated the Lack clutch size for primary and secondary females. Their model predicted that secondary females should lay smaller clutches than primary females, and that primary females should reduce their clutch size in situations where the risk of subsequent superparasitism was high. Parker and Courtney (1984) studied the same question in time-limited parasitoids. They used game theory techniques to calculate the optimal size of primary and secondary clutches and found secondary clutches to be consistently smaller. A problem with their model is that it assumed that there were two distinct classes of parasitoids: primary parasitoids and superparasitoids. The model always predicts superparasitism as this is the only reproductive outlet for the second class of parasitoids. Strand and Godfray (1989) also considered superparasitism in time-limited parasitoids. However, they assumed that all wasps encountered both fresh and parasitized hosts. If the larvae resulting from the first and second ovipositions have identical fitnesses, affected only by the total number of individuals in the host, this model makes the surprising prediction that superparasitism is always favored. In reality, the second clutch is normally at a competitive disadvantage in comparison with the first clutch, and when this is included in the model, superparasitism is only predicted when unparasitized hosts are relatively uncommon and when the members of the second clutch are not too heavily penalized in competition for host resources with the first clutch. In studies of gregarious parasitoids, it is seldom necessary to predict the optimal clutch size of more than two females attacking the same host. There are, however, some groups where many females oviposit together, for example carrion and dung flies, and their reproductive strategies have been modeled by Ives (1989).

Strand and Godfray (1989) also modeled the evolutionary advantages of ovicide, the destruction of the eggs already present on the host by a superparasitizing female. They assumed that the parasitoid was selected to maximize its rate of fitness gain and that the act of ovicide was time consuming. Ovicide was most advantageous when it took little time, when unparasitized hosts were scarce, and when the survival of a second clutch of eggs was very low unless the first clutch was destroyed. Smith and Lessells (1985) have discussed the evolution of ovicide in grain weevils.

Time-limited parasitoids have received most attention by theorists because of the technical simplicity of models based on this assumption. No game theory models of superparasitism by egg-limited parasitoids have been developed, though the dynamic programming models of Iwasa et al. (1984) (which treat previously parasitized hosts simply as hosts of poor quality) predict more frequent rejection of parasitized hosts when egg supplies are low. This makes intuitive sense because a parasitoid with a small egg load will be selected to attack hosts and produce clutch sizes that lead to a high fitness return per egg.

3.3.2 EXPERIMENTAL STUDIES

HOST DISCRIMINATION

The question of whether parasitoids could actively discriminate against previously parasitized hosts was the subject of great controversy in the first thirty years of this century (Salt 1961; van Lenteren 1981; van Alphen and Visser 1990). In a long series of papers, Salt (e.g., 1932, 1934, 1937a, 1961) provided a wealth of evidence that host discrimination was common among parasitoids. Reviewing the field in 1961, Salt listed 25 species from five families in which the avoidance of superparasitism had been demonstrated. Twenty years later, van Lenteren (1981) estimated that host discrimination had been noted in somewhere between 150 and 200 species of parasitoids. Most of these studies demonstrated host discrimination using behavioral observations and experiments. An alternative approach is to deduce the presence of superparasitism avoidance and host discrimination from the statistical distribution of eggs across hosts.

The statistical analysis of egg distribution frequencies presents a number of problems of interpretation. The presence of more than one egg in the host of a solitary parasitoid is strong evidence that superparasitism has occurred, i.e., that parasitized hosts are not always rejected. The only alternative explanation is that a single female laid two eggs during a single oviposition bout, a behavior that appears to be rare (sec. 3.2.4). Equally, a very regular distribution of eggs, with many hosts receiving one parasitoid egg but very few receiving two or more, is strong evidence that host discrimination is common. Problems arise when the distribution of eggs is random among hosts. At first sight this would appear to suggest no discrimination. However, if there is variation in the avoidance of superparasitism, a random distribution of eggs may still be found if the field sample consists of a collection of subsamples which vary in the mean number of eggs per host (van Lenteren et al. 1978; van Lenteren 1981). Clumped egg distributions are also common and present similar problems of interpretation. The absence of at least partial avoidance of superparasitism is very difficult to demonstrate statistically.

In gregarious species, the occurrence of superparasitism can sometimes be demonstrated by showing that the distribution of clutch sizes is bimodal. For example Richards (1940) argued that the occasional very large clutches of the braconid *Cotesia (=Apanteles) glomeratus* on cabbage butterflies (*Pieris rapae*) were caused by superparasitism. Van Alphen (1980) found that the eulophid *Tetrastichus* sp., a parasitoid of the asparagus beetles *Crioceris duodecimpunctata*, could lay only about six eggs in a single oviposition bout. Observed clutches of 12–18 in the field are almost certainly due to superparasitism. Similarly, Werren (1983) was able to recognize superparasitism in the pteromalid *Nasonia vitripennis* because of the resultant unusually large clutches.

DETECTION OF SELF-SUPERPARASITISM

A recent review lists nine studies that have explored whether a parasitoid can distinguish between hosts attacked by itself or by others (van Dijken et al. 1992). Three of these studies failed to demonstrate the detection of self-superparasitism but the other six were positive (table 3.4). The avoidance of self-superparasitism could work in two ways: the wasp might recognize the host itself, or might recognize the patch. Not all the studies in table 3.4 distinguish among these two possibilities, though *E. lopezi*, *V. canescens*, and *L. heterotoma* appear to recognize the host.

The study of *V. canescens* by Hubbard, Marris, and colleagues is particularly interesting because this species is parthenogenetic and individuals from the same strain are either genetically identical or extremely similar. Marris et al. (1992) found that individuals were more likely to superparasitize hosts attacked by wasps from different strains than those attacked by the same strain. Surprisingly, the wasp was also more likely to parasitize hosts attacked by the same strain than those attacked by itself. These results could have been explained by the wasp remembering characteristics of the oviposition site, but the wasp still discriminated among hosts after their position had been changed. It thus appears that the parasitoid marks the host and is able to recognize its own mark. Variation in marks between individuals must have both genetic and environmental components. Because individuals from the same strain are genetically identical or extremely similar, an individual should make no distinction among hosts attacked by itself and by other members of the strain. The observed discrimination may be a constraint due to the wasps inability to disentangle the genetic and environmental components of the mark, or may reflect a behavior that evolved before the species abandoned sex.

Table 3.4
Studies that have explored whether parasitoids can distinguish hosts parasitized by themselves and by conspecifics.

Parasitoid	Reference	Result
Asobara tabida (Braconidae)	van Alphen & Nell 1982	No
Trichogramma evanescens (Trichogrammatidae)	van Dijken & Waage 1987	No
Venturia canescens (Ichneumonidae)	Hubbard et al. 1987; Marris et al. 1992	Yes
Aphelinus asychis (Aphelinidae)	Bai & Mackauer 1990	No
Ephedrus californicus (Braconidae)	Völkl & Mackauer 1990	Yes
Aphidius smithi (Braconidae)	McBrien & Mackauer 1991	Yes
Epidinocarsis lopezi (Encyrtidae)	van Dijken et al. 1991a	Yes
Microplitis croceipes (Braconidae)	F. Wäckers, pers. comm. in van Dijken et al., 1992	Yes
Leptopilina heterotoma (Eucoilidae)	Visser 1993	Yes

Source: Adapted from van Dijken et al. 1992.
Note: There is a tantalizing reference to another possible example in Hill (1926). Hill describes the searching behavior of the platygasterid *Platygaster hiemalis* searching for the eggs of the Hessian fly. The wasps he observed displayed an almost perfect ability to avoid self-superparasitism when searching alone for scattered eggs on a leaf. Hill remarks that a colleague had observed similar behavior and that "he also mentioned the fact that one female did not seem able to recognize the occurrence of oviposition by another female."

FACTORS AFFECTING THE PROBABILITY OF SUPERPARASITISM

The offspring of superparasitizing females are at a disadvantage because they suffer from competition with the parasitoids already present in the host. The extent of the disadvantage depends critically on the time that elapses between the two ovipositions. Actually measuring the relative fitness of the two competitors is difficult and generally requires genetic markers. Visser (1993) used an eye color mutant to study superparasitism in the eucoilid *Leptopilina heterotoma* and found that within the first three hours after initial parasitism, the probability that a second egg successfully develops averages 0.43, approximately the same success rate as the first egg. Superparasitism appeared not to affect the fitness of the winning larva, at least as measured by its adult size. By twenty-four hours, the probability of successful development declines to zero (see also Bakker et al. 1985). Strand and Godfray (1989), also using eye color mutants, showed that the survival of the second clutch of the gregarious braconid *Bracon hebetor* was influenced both by the size of the first clutch and by the time between the two ovipositions.

Parasitoids should be more willing to attack recently parasitized hosts where their larvae have the greatest probability of survival (van Lenteren 1981; Strand 1986; Mackauer 1990). Thus Strand (1986) found that the scelionid *Telenomus heliothidis* which attacks eggs of the moth *Heliothis virescens* would not superparasitize a host after the egg of the first female had hatched. The newly hatched larva causes host necrosis which appears to deter the second female. A number of gregarious species also discriminate against hosts containing parasitoid larvae (Werren 1984b; Tepedino 1988). However, there is surprisingly little evidence that parasitoids discriminate against hosts containing parasitoid eggs, or very young larvae, of different age (Visser et al. 1992c). Cloutier et al. (1984) and Micha et al. (1992) found that the probability that *Aphidius nigripes* and *A. ervi* superparasitizes a host decreased rapidly four to eight hours after the first oviposition, a long time before the first egg hatches. In other species there is some evidence that hosts with older eggs are attacked more frequently (Chow and Mackauer 1986; Hubbard et al. 1987; Hofsvang 1988; Visser et al. 1992c). These differences are most likely related to whether the female marks the host at the time of oviposition. In genera such as *Aphidius* which probably lack this behavior (Micha et al. 1992; but see McBrien and Mackauer 1991), host discrimination can only occur after the parasitoid has in some way altered the behavior or physiology of the host. Where the female does mark the host, discrimination is possible immediately; may wane as the mark evaporates or wears off; and finally it rises again as the parasitoid induces noticeable changes in the host (Mackauer 1990; Visser et al. 1992c). Micha et al. (1992) discuss statistical techniques for estimating the probability of superparasitism as a function of time.

Parasitoids may in some circumstances preferentially attack previously parasitized hosts. Takasu and Hirose (1991) studied the encyrtid wasp *Ooencyrtus nezarae*, an egg parasitoid of the plataspid bug *Megacopta punctatissimum*. Only naive wasps, or wasps that had not oviposited for some time, were willing to superparasitize. Surprisingly, these wasps preferred recently parasitized to unparasitized hosts, even though in these hosts their larvae experienced reduced survival. However, when attacking parasitized hosts, the wasps make use of the hole drilled in the chorion by the first female. This reduces their handling time from about twenty to nine minutes. Takasu and Hirose argue that the wasp can achieve a greater rate of fitness gain by concentrating on parasitized hosts. This is a rare example where the ranking of hosts by profitability (fitness gain divided by handling time) is the reverse of the ranking by fitness gain alone.

There is some evidence that solitary parasitoids are less willing to superparasitize hosts containing two or more eggs than those containing just one. This was first demonstrated in the *Drosophila* parasitoid *Leptopilina heterotoma* (Eucoilidae) (Bakker et al. 1972; van Lenteren 1976) using a statistical argument though van Alphen and Visser (1990) pointed out that these results might alternatively be explained by the avoidance of self-superparasitism. However, recent behavioral evidence supports the original conclusions (Bakker et al. 1990). Another *Drosophila* parasitoid, *Asobara tabida* (Braconidae) is insensitive to the number of eggs in a host (van Alphen and Nell 1982).

The probability of superparasitism is influenced by the encounter rate with parasitized host. It is often observed in laboratory experiments that parasitoids confined with their hosts will resort to superparasitism after all unparasitized hosts have been attacked; it is also often noted that the switch in behavior appears, at least subjectively, to be taken reluctantly, and is associated with a decrease in searching activity (see van Lenteren 1981 for a review of earlier work; van Alphen and Nell 1982; Cloutier 1984; Cloutier et al. 1984; Chow and Mackauer 1986). The significance of these laboratory observations for parasitoids searching under field conditions which can disperse from an exploited host patch is difficult to assess.

The influence of prior experience on the willingness of a parasitoid to superparasitize has been investigated by Visser et al. (1992b) working with the eucoilid *Leptopilina heterotoma*. Parasitoids were allowed to search for hosts on a patch containing either no hosts, all unparasitized hosts, or a high proportion of parasitized hosts. After a variable amount of time, the wasps were transferred to a second patch containing a mixture of unparasitized and parasitized hosts. Wasps were more willing to superparasitize, and spent longer on a patch, if they had previously experienced a patch containing parasitized hosts. Surprisingly, the age of the wasp and the interval between patch presentation had little effect, unless the interval was as short as one hour when superparasitism occurred less frequently. Parasitoids caught in the field and released on a patch of parasitized hosts spent a relatively short time on the patch with little superparasitism, indicating that they had experienced a relatively high-quality environment in the field. The importance of prior experience is also indicated by other experiments with *L. heterotoma* (Visser et al. 1992c) and with another *Drosophila* parasitoid, the braconid *Asobara tabida* (van Alphen et al. 1992).

Mangel (1989b) provides an interesting analysis of a laboratory experiment that illustrates how dynamic programming can be used to construct testable hypotheses. Collins and Dixon (1986) studied the oviposition behavior of the braconid *Monoctonus pseudoplantani*, a parasitoid of the sycamore aphid, *Drepanosiphon plantanoides*. They placed individual parasitoids into tubes containing twenty aphids, some of which had been previously parasitized. Every hour, the aphids were removed and replaced by another twenty aphids. The pattern of parasitoid oviposition was strongly influenced by the cycle of host renewal, peaking in the fifteen minutes after a new batch of aphids was presented and falling as the hosts were located and attacked (fig. 3.10 shows a typical result). However, successive peaks in oviposition activity declined over time. Parasitized aphids were attacked as well as fresh hosts, though significantly less often. Larger parasitoids (with greater egg reserves) showed a higher rate of oviposition and were more willing to superparasitize. Collins and Dixon interpreted these results using the descriptive term "motivation to oviposit." The aim of Mangel's analysis was to show that the same patterns could be generated by an evolutionary model and to use this model to predict

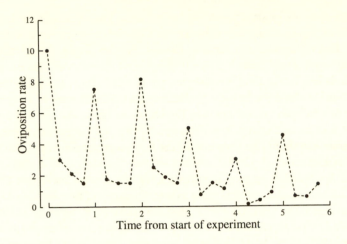

Figure 3.10 The oviposition rate (number of eggs laid every 15 minutes) of the braconid *Monoctonus pseudoplantani* presented with twenty aphids every hour in a laboratory experiment. (From Collins and Dixon 1986.)

the primary determinants of oviposition behavior. To do this, he constructed a suite of three stochastic dynamic programming models designed around Collins and Dixon's experiment. The models assumed that the parasitoid could estimate encounter rate and its own risk of mortality. The first model assumed that the parasitoid had a fixed estimate of the percentage hosts already parasitized and that host discrimination was perfect. The second and third models assumed that the parasitoid started life with an estimate of percentage parasitism which was updated as the animal searched for hosts in the environment; the third model differed from the second in that host discrimination was assumed not to be perfect. The first model could not reproduce the behavior shown in figure 3.10 while the second and third model provided good matches. These models also predicted the observed relationships between parasitoid size (and hence egg capacity), oviposition rate, and the probability of superparasitism.

Mangel concluded that the wasps in Collins and Dixon's experiments learned about the environment as they searched for hosts. The small differences in the predictions of the second and third models suggest that imperfect discrimination is not the key to explaining the pattern of superparasitism. Mangel's models describe the experimental results only if the fitness of an egg laid by superparasitoid is relatively small, and if the wasps live long enough that egg depletion is a real risk. These assumptions can be experimentally verified. Other studies have also noted a correlation between egg load and willingness to superparasitize (Völkl and Mackauer 1990).

There have been few quantitative analyses of superparasitism under field conditions. Janssen's (1989) study of host selection in two *Drosophila* parasitoids, *Leptopilina heterotoma* and *Asobara tabida*, has already been described (sec. 3.1.2). By assuming that the parasitoid is selected to maximize its rate of gain of fitness, Janssen was able to predict that the wasps should attack all hosts in which the probability of offspring survival exceed a threshold that varied between 0.002–0.03 for different parasitoid species and microhabitats. In good-quality host species, the probability that the offspring of a superparasitoid will survive is greater than this threshold, as long as the interval between the first and second attack is not too great. This led Janssen to predict that superparasitism should be common in the field, a prediction supported by his data. Van Dijken et al. (1993) made a similar calculation in a study of the encyrtid *Epidinocarsis lopezi*, a parasitoid of the cassava mealybug (*Phenacoccus manihoti*), and a species that superparasitizes readily in the field. They again assumed time limitation and calculated a threshold for progeny survival of 0.017. The probability that a secondary larva survives frequently exceeds this threshold, and they predicted that superparasitism should be common.

There is clear evidence that solitary parasitoids are more willing to superparasitize in the presence of conspecifics than when they are searching alone (Bakker et al. 1985; van Alphen 1988; Visser et al. 1990, 1992b). The strongest evidence has been obtained by Visser et al. (1990, 1992b) using the *Drosophila* parasitoid *Leptopilina heterotoma* (see also van Alphen 1988). Wasps were allowed to oviposit on artificial patches of hosts. On the first day of the experiment, the parasitoids were placed singly on patches to gain experience of oviposition, while on the second day wasps were placed in groups of one, two, or four. The ratio of hosts to parasitoids was kept constant. The frequency of superparasitism was very low for solitary searching females (although not zero as predicted) but higher in experiments with two or four females (fig. 3.11). Time spent on the patch also increased with the number of searching parasitoids (fig. 3.12). Visser et al. (1990) also found that parasitoids searching alone on a patch were more willing to self-superparasitize if they had been stored the previous day with conspecifics as opposed to alone. It appears that the female uses recent contact with conspecifics as an indicator that there are potential competitors in the environment. The braconid *Asobara tabida* is also more willing to superparasitize when searching for hosts in the presence of conspecifics (van Alphen et al. 1992). These results provide strong evidence that parasitoids alter their searching behavior in the presence of other searching females.

SUPERPARASITISM AND CLUTCH SIZE IN GREGARIOUS PARASITOIDS

Most models predict that gregarious parasitoids should either ignore a parasitized host or lay a reduced clutch size. The first evidence that secondary clutches are smaller than primary clutches was obtained by Wylie (1965) working on the pteromalid *Nasonia vitripennis*, a parasitoid of dipteran pupae. Wylie dissected newly superparasitized hosts and counted the number of eggs

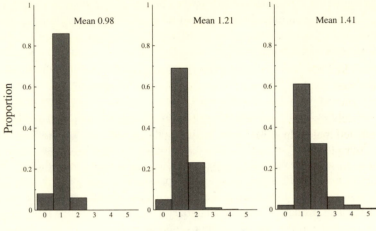

Number of eggs per host

Figure 3.11 The distribution of parasitoid eggs among host larvae when parasitoids search alone (left-hand figure), in pairs (middle figure), or in groups of four (right-hand figure). The ratio of hosts to parasitoids is kept constant. Superparasitism of *Drosophila subobscura* larvae by *Leptopilina heterotoma* increases with the number of searching parasitoids. (From Visser et al. 1990.)

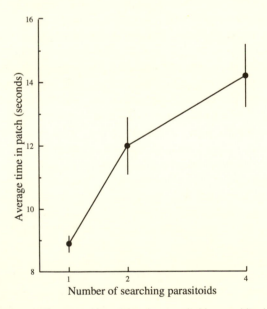

Number of searching parasitoids

Figure 3.12 The average time spent in patches by parasitoids searching in groups of one, two, or four (*Leptopilina heterotoma* attacking *Drosophila subobscura*). The ratio of hosts to parasitoids is kept constant. (From Visser et al. 1990.)

in each clutch. He also showed that smaller clutches were laid if the site of the second oviposition was near that of the first, and that the clutch laid by a third female was smaller still (Wylie 1966, 1970, 1973). Wylie's main results were confirmed by Holmes (1972), Werren (1980, 1984b) (see also fig. 4.9), and King and Skinner (1991b) using genetically marked females, though King and Skinner were unable to repeat the observation that clutch size was affected by oviposition site. Werren (1984b) found that secondary females did not lay larger clutches in hosts previously parasitized by females who had been mated to irradiated males. In these hosts, most female eggs die so that the number of active larvae is much reduced. This result suggests that stimulus causing a reduced clutch size is not associated with the number of feeding larvae.

In some species, there is a strong correlation between the duration of oviposition and the number of eggs injected into a host. Ikawa and Suzuki (1982) used oviposition time as a measure of clutch size in a study of the braconid *Cotesia (=Apanteles) glomeratus*, a parasitoid of cabbage butterflies, *Pieris rapae*. They found that females laid smaller clutches when attacking a parasitized host than when attacking an unparasitized host. Moreover, if a wasp had recently attacked a series of unparasitized hosts, it was more likely to reject a parasitized host and, if it was attacked, to lay a smaller clutch. Smaller clutches were also observed when the wasp was presented with a series of parasitized hosts. Both observations are in qualitative agreement with theory: if unparasitized hosts are common, the wasp should be less willing to waste time and eggs on a parasitized host; if parasitized hosts are abundant, smaller clutches should be laid in each.

The number of eggs laid by *Trichogramma* females can be counted by carefully watching the female oviposit. Van Dijken and Waage (1987) found that the primary clutch size of *Trichogramma evanescens* ovipositing into the eggs of the cabbage moth *Mamestra brassicae* was 2.7 (S.E. = 0.5), but that the clutch laid by a superparasitizing female was smaller (1.3, S.E. = 0.5). They also found evidence of a negative relationship between the size of the first and second clutches. *Trichogramma chilonis* attacks the eggs of swallowtail butterflies (*Papilio xuthus*) injecting about fifteen of its own eggs into each host egg. In contrast to *T. evanescens*, Suzuki et al. (1984) found no difference in the number of eggs laid on parasitized and unparasitized hosts by this species.

There has been some work on the genetics of superparasitism in another species of *Trichogramma*. Wajnberg et al. (1989) created ten isofemale lines of *T. maidis* and counted the number of eggs laid per host by groups of females in experiments where they varied the ratio of hosts to parasitoids. The number of eggs per host varied from one to ten and, as expected, more eggs were laid when parasitoids were common relative to hosts. More interestingly, there was significant variation among lines, both in the average number of eggs per host, and also in the evenness with which eggs were distributed among hosts.

Wajnberg et al. were unable to distinguish primary clutches from subsequent eggs laid during superparasitism, and thus the observed differences among the genetic lines could have arisen from a variety of causes. However, Wajnberg et al. argued that the differences probably reflected genetic variation in super-parasitism behavior.

OVICIDE

Both solitary and gregarious parasitoids have been observed to destroy eggs on previously parasitized hosts. The solitary pteromalid parasitoid *Pachycre-poides vindemiae* attacks *Drosophila* pupae, laying its egg in the space be-tween the pupa and the puparium (the hardened coat of the last larval instar that forms an outer protective barrier). In over half the cases of attempted superparasitism, the second female managed to destroy the egg of the first female (Nell and van Lenteren 1982). Arakawa (1987) also observed that the solitary whitefly parasitoid, *Encarsia formosa* (Aphelinidae) used its oviposi-tor to destroy the eggs of conspecifics inside its host.

Some gregarious parasitoids also practice ovicide. It appears to be most common in the family Bethylidae (Venkatraman and Chacko 1961; Malyshev 1968; Goertzen and Doutt 1975; Griffiths and Godfray 1988; Hardy and Blackburn 1991) but has also been recorded in the ichneumonid *Pleolophus indistinctus* (Price 1970b). One species of bethylid, *Goniozus marasami*, will also destroy larvae as well as eggs. Some, but not all strains of the braconid *Bracon hebetor* are ovicidal, using their ovipositor to destroy eggs already present on the host. Strand and Godfray (1989) found that *B. hebetor* was more likely to destroy eggs at low host density and when unparasitized hosts were relatively rare, observations that support two qualitative predictions of theory.

SUPERPARASITISM BY NAIVE PARASITOIDS

Van Lenteren and Bakker (1975) and van Lenteren (1976) found that naive eucoilid wasps, *Leptopilina heterotoma (=Pseudeucoila bochei)*, were more likely to superparasitize than experienced wasps. The same behavior was also found in another drosophilid parasitoid, the braconid *Asobara tabida* (van Al-phen and Nell 1982). Similar behavior has been reported in trichogrammatid, mymarid, scelionid, and encyrtid egg parasitoids by Jackson (1966, 1969), Rabb and Bradley (1970), Klomp et al. (1980), Strand (1986), and Takasu and Hirose (1991); and in a braconid parasitoid of aphids (Chow and Mackauer 1986). However, at least some species are born with an innate ability to distin-guish parasitized from unparasitized hosts (Nell and van Lenteren 1982; Völkl and Mackauer 1990; Bai and Mackauer 1990; Mackauer 1990). Typically, the parasitoids do not begin to avoid parasitized hosts until they have attacked at least one unparasitized host, though Jackson (1966) found that the mymarid *Caraphractus cinctus* began to show oviposition restraint after attacking a number of parasitized hosts and before encountering an unparasitized one.

The first explanation for this behavior was that naive wasps were unable to tell the difference between parasitized and unparasitized hosts until they had experienced both types of host (van Lenteren 1981). In this view, superparasitism by naive wasps is maladaptive and a constraint imposed by the need for the wasp to learn to discriminate. However, as was discussed above, there are also adaptive explanations for the behavior of naive females: they have yet to learn the current distribution of hosts, and they superparasitize in the certainty that they are not reencountering hosts (van Alphen and Nell 1982). In a series of experiments with *Leptopilina heterotoma* and *Trichogramma evanescens*, van Alphen et al. (1987) showed that although naive females of both species were willing to superparasitize, their behavior on encountering parasitized and unparasitized hosts was not identical: for example, they dispersed more readily after encountering parasitized hosts. This suggests that naive wasps can at least potentially discriminate and makes an adaptive explanation more likely.

If parasitoids are deprived of hosts, they are more willing to superparasitize. Klomp et al. (1980) suggested that parasitoids may lose the ability to discriminate and have to relearn it in the same way as naive individuals. One adaptive explanation for this behavior has already been discussed: host deprivation may be used by the parasitoid as an indication that hosts are rare and thus superparasitism is a more favorable strategy. However, a second explanation is that a parasitoid which has not attacked a host for a long period can be reasonably sure that if it encounters a parasitized host, another female was responsible, and that it can lay an egg without risking self-superparasitism.

3.3.3 BEHAVIOURAL MECHANISMS

THE DETECTION OF PARASITISED HOSTS

The first detailed study of the mechanics of host discrimination was Salt's (1935) work on the lepidopteran egg parasitoid *Trichogramma evanescens*. Salt discovered that after oviposition female wasps deposit an external, chemical mark that deters subsequent females from attacking the host. The chemical is water soluble and if host eggs are washed, the second female will probe the egg with her ovipositor. However, there is a second, internal mark, and probing is seldom followed by oviposition. In addition to deliberate marks left by the ovipositing female, the developing parasitoid causes physiological changes to the host which are also used by subsequent parasitoids in host discrimination (Strand 1986). Some parasitoids mark the surrounding substrate rather than the host itself, and there is a close connection between host marking and the patch-marking behavior discussed in section 2.5.1.

Many examples of the use of marking chemicals are now known in parasitoid wasps. Some species only use an internal mark, for example the pteromalid parasitoid of fly pupae, *Nasonia vitripennis* (Wylie 1965, 1966, 1970; King and Skinner 1991b) and the eucoilid parasitoid of *Drosophila, Leptopilina heterotoma* (van Lenteren 1972). In contrast, other species use only an external mark, for example the scelionid *Trissolcus (=Asolcus basalis)*, which attacks the eggs of the pentatomid bug *Nezara viridula* (F. Wilson 1961), and the aquatic mymarid *Caraphractus cinctus*, which attacks submerged water beetle eggs (Jackson 1966). The use of both internal and external markers, as in *Trichogramma*, is also common (Guillot and Vinson 1972). In egg parasitoids, the host tissue is frequently lysed by chemicals secreted by the parent, the larva, or by teratocytes (see sec. 6.3.2). It appears that parasitoids are able to detect necrotic host tissue and hence identify previously parasitized hosts (Strand 1986).

The source of the internal and external mark in a number of ichneumonoids has been identified as the Dufour's gland (Guillot and Vinson 1972; Harrison et al. 1985). Hubbard et al. (1987) and Marris et al. (1992) suggest that the chemical profile of the Dufour's gland secretion allows wasps to discriminate between hosts attacked by themselves and by other individuals. Dufour gland secretions are known to act as individual markers in some aculeate wasps (Hefetz 1987, 1990).

In addition to chemical cues, some parasitoids may use other mechanisms of host discrimination. It seems likely that ectoparasitoids are able directly to see or feel developing larvae living externally and conspicuously on their hosts. Alternatively, changes brought about by parasitism such as the cessation of movement or death may signal to the parasitoid that the host is no longer acceptable (Ullyett 1936; Thorpe and Jones 1937; Wylie 1965). Ullyett suggested that healthy and parasitized hosts may be detected by differences in the sound produced by drumming the antennae on the host cuticle. It has often been suggested that parasitoids detect the external wounds made by the oviposition of the first female although definite evidence is lacking. A number of encyrtids lay stalked eggs, the stalk protruding from the host after oviposition. Takasu and Hirose (1988) have shown that the encyrtid *Ooencyrtus nezarae*, an egg parasitoid of the bean bug *Riptortus clavatus*, uses the presence of an egg stalk to detect whether a host has been parasitized. Finally, aphids often react to the presence of parasitoids by attempting to defend themselves, typically by kicking. Recently parasitized aphids are thoroughly alarmed and difficult to attack, a factor that may reduce the probability of superparasitism (Gardner et al. 1984).

The recognition of previous parasitism often depends on the time elapsed since the first oviposition. At one extreme, superparasitism may occur if two parasitoids locate a host simultaneously. This is most likely to occur when the parasitoid is small relative to the size of the host. At least in the laboratory, it is straightforward to persuade several *Nasonia vitripennis* (Pteromalidae) to attack the same fly pupa (other examples in van Lenteren 1981). Van Lenteren found that the eucoilid *Leptopilina heterotoma* normally avoided superparasitizing a host except when the second encounter occurred less than seventy seconds after the first oviposition. If the second female attempted to oviposit very near the site of the first attack, then the host was rejected much sooner. It appears that a little time is needed for the mark to diffuse over the body of the host. External marks tend to be most effective soon after parasitism but decay in effectiveness as they get washed off or evaporate. On the other hand, internal markers released by the parasitoid egg, or the physiological changes to the host caused by parasitism, become more noticeable with time. Sometimes, the two effects interact so that the avoidance of superparasitism is initially strong, then drops, and finally rises again (Visser et al. 1992c). Chow and Mackauer (1986) found that superparasitism avoidance in the aphid parasitoid *Ephedrus californicus* (Braconidae) was strongest immediately after the first parasitism but declined after about ten hours as an external marker dissipated. However, after a further four hours, the changes wrought by the developing parasitoid become obvious, and wasps again refrain from superparasitism.

THE EVOLUTION OF HOST MARKING

So far I have discussed why a parasitoid should recognize a marked host but not why the female should mark the host in the first place. Marking takes time and there are likely to be metabolic costs to the production of the marking chemical; so what are the advantages to marking? This problem has been reviewed by Roitberg and Mangel (1988), both in the context of parasitoids and also of herbivorous insects such as tephritid fruit flies which mark fruit after oviposition (Prokopy 1972; Roitberg and Prokopy 1987).

Four explanations have been put forward to explain marking. (1) Marking may improve the efficiency of the parasitoid population by increasing the evenness of host attack. This suggestion relies on population-level selection and is thus hard to justify. (2) The function of marking may be to allow the female to avoid a host she has just attacked. Where females mark patches or trails, the mark functions to alert the female that she is entering an area that has already been searched. (3) The mark may be an altruistic act to help related females forage efficiently. This explanation requires limited dispersal by adult parasitoids so that females are likely to encounter hosts attacked by relatives. (4) The function of the mark may be to alert other females that the host has already been attacked. For this explanation to work, both the original female, and any female that subsequently discovers the host, must benefit from host discrimination and superparasitism avoidance. This double benefit can occur if the offspring of the first female normally triumphs in competition with a superparasitoid's offspring (so that there is an advantage to the superparasitoid in paying attention to the mark) but if the second offspring is occasionally successful (so that there is an advantage to the first female to try to discourage subsequent attack).

The four explanations offer reasons for the maintenance of marking behavior, but to explain its origins other considerations must be taken into account. A rare mutant, marking female will prosper only if her marks are detected. In the case of the second hypothesis—avoidance of self-superparasitism—it may be reasonable to postulate a double mutant that both causes a female to mark a host and allows the same individual to subsequently detect the mark. For the fourth hypothesis to explain the origin of marking, it is necessary to argue that normal individuals in the population have a basic ability to identify previously parasitized hosts, for example by detecting oviposition wounds, and that the mark acts to improve the probability of correct host discrimination.

Roitberg and Mangel (1988) use a variety of modeling techniques to explore the evolution of host marking, chiefly concentrating on the second and fourth of the mechanisms listed above. They consider both the origin of marking through a double mutant, and through the exploitation of a basic ability to detect previous parasitism. They show that both mechanisms can contribute toward the maintenance and origin of marking, their relative contributions depending quite critically on the natural history of the interaction. When there is only a small chance that the progeny of superparasitizing females survive, the avoidance of self-superparasitism is the chief mechanism that promotes marking. As one would predict, it is necessary for females to have a relatively high probability of reencountering hosts before they gain any advantage from marking. When the avoidance of self-superparasitism is the chief function of marking, the mark should be most effective immediately after its application when the risk of reencounter is highest. In fact, most marks are water soluble and lost over a few days, consistent with this hypothesis. On the other hand, by the time the mark wears off, the risk of successful competition from a secondary parasitoid will be small, and the internal changes to the host caused by the developing parasitoid are likely to be obvious to potential superparasitoids, so that long-acting oviposition deterrents may be unnecessary.

3.3.4 MULTIPARASITISM

If a parasitoid attacks a host that has previously been parasitized by a different species, then multiparasitism rather than superparasitism is said to occur (H. S. Smith 1916). Much of the theory developed for superparasitism can be applied directly to multiparasitism although the latter has received considerably less attention. The chief difference between the two phenomena is that competition is likely to be more asymmetric in the case of multiparasitism, where the different biologies of the two species may lead one species to have a consistent advantage over another.

A number of parasitoid species are able to detect that a host has been attacked by a different species and to refrain from further parasitism (Mackauer 1990). According to Salt (1961) the first example of interspecific discrimination is due to Lloyd (1940) who showed that the ichneumonid *Diadegma eu-cerophaga (=Angitia cerophaga)* avoided hosts (the larvae of the diamond-back moth, *Plutella xylostella*) previously attacked by the braconid *Cotesia (=Apanteles) plutellae*. Pupal stages containing *D. eucerophaga* were them-selves avoided by another ichneumonid, *Diadromus collaris*. However, inter-specific host discrimination is considerably less common than the ability to recognize hosts parasitized by conspecifics (van Alphen and Visser 1990). For example, the two sympatric *Drosophila* parasitoids, *Asobara tabida* (Braco-nidae) and *Leptopilina heterotoma* (Eucoilidae) often avoid hosts attacked by their own species, but not hosts attacked by the other (van Strien-van Liempt and van Alphen 1981). *A. tabida* is able to recognize the mark left by its sibling species *A. rufescens* which is sympatric but attacks hosts in different mi-crohabitats. It cannot, however, recognize the mark of two allopatric conge-ners, *A. persimilis* and an unnamed *Asobara* sp. (Vet et al. 1984b).

There are at least three possible reasons why interspecific discrimination is uncommon: (1) parasitoids are more likely to encounter hosts parasitized by conspecifics so there may be stronger selection to identify such hosts; (2) some species may always win in competition against other species and so be under no pressure to avoid hosts containing the weaker competitor (although the weaker competitor will obviously experience strong selection); and (3) it may be harder for a parasitoid to evolve the ability to detect chemical marks left by a different species—marking may have evolved to prevent self-superpara-sitism and so the recognition of conspecific marks is a preadaptation. Tur-lings et al. (1985) model the evolution of interspecific host discrimination for sympatric *Drosophila* parasitoids using parameter values estimated in the lab-oratory and conclude that it is unlikely to evolve. Because their model is a behavior-rich simulation, it is quite difficult to understand the biological basis of the prediction. It appears that the avoidance of multiparasitism has little advantage as the insects are not egg limited, oviposition is not time consum-ing, and because there is always a small chance that eggs laid in a parasitized host will survive.

The complicated competitive interactions between the larvae of three braco-nid parasitoids of aphids (*Aphidius smithi, Ephedrus californicus*, and *Praon pequodorum*) have been disentangled by Chow and Mackauer (1984, 1985, 1986; Mackauer 1990). In all three species the oldest parasitoid normally wins in intraspecific competition. The first individual to reach the larval stage is able physically to destroy its competitors and hence age is of paramount impor-tance (though larvae may later become vulnerable when they enter the amandi-bulate second instar). All three species show intraspecific host discrimination but *A. smithi* alone shows (imperfect) interspecific host discrimination. In in-terspecific competition, *A. smithi* consistently loses to the other two species, and this may account for the evolution of interspecific discrimination in this species.

Aphidius smithi is also an inferior competitor to the congeneric *A. ervi*. The latter is able to kill younger *A. smithi* by physiological suppression and older larvae by direct attack (McBrien and Mackauer 1990). The only circumstance under which *A. ervi* does not have an advantage is if the eggs of the two species hatch simultaneously. Both *A. smithi* and *A. ervi* show interspecific host discrimination and prefer unparasitized aphids. The advantages to *A. smithi* are clear but why has *A. ervi* evolved discrimination? McBrien and Mackauer found that *A. ervi* took longer to develop in hosts parasitized by *A. smithi* and suggest that the costs of slower development may favor discrimination when the wasp is able to locate more favorable, unparasitized hosts. When given a choice of hosts previously parasitized by *A. ervi* and *A. smithi*, both species preferred to lay eggs in hosts containing *A. smithi*, where their larvae have the best chance of survival (McBrien and Mackauer 1991). A similar situation is found in the interaction between *Aphidius ervi* (Braconidae) and *Aphelinus asychis* (Aphelinidae), both of which show intraspecific host discrimination using internal cues. When given a choice of unparasitized hosts and hosts attacked by the other wasp, both species prefer unparasitized hosts (Bai and Mackauer 1991). The benefit of interspecific host discrimination to *A. asychis* is straightforward because unless it has a seven day start over *A. ervi*, it always loses in competition. The advantages of discrimination to *A. ervi* are less clear but again are possibly associated with longer development time or reduced size or fecundity as adults.

The interaction between the scelionid *Telenomus heliothidis* and the trichogrammatid *Trichogramma pretiosum*, both egg parasitoids of the moth *Heliothis virescens*, has also been closely studied (Strand and Vinson 1984; Strand 1986). The winner in multiparasitism depends critically on the relative timing of attack (fig. 3.13). *T. pretiosum* triumphs if it has a head start of at least 6 hours on its competitor. This species is gregarious and is able to consume host resources faster than *T. heliothidis*. However, if *T. pretiosum* has less than a 6-hour start, or if the first attack is by *T. heliothidis*, the latter wins. The egg stage of *T. heliothidis* is shorter than *T. pretiosum* and as soon as it hatches, the *T. heliothidis* larva secretes a cytolytic chemical that both destroys the host tissue as well as any *T. pretiosum* eggs (larvae are protected by their cuticle). Curiously, *T. pretiosum* is the successful competitor if it attacks hosts approximately 84–96 hours after an attack by *T. heliothidis*. At this stage the scelionid is a third instar larva, filling the egg, and the trichogrammatid acts as a hyperparasitoid. The relative competitive advantages of the two species are partially mirrored by the oviposition behavior. *T. heliothidis* never parasitizes hosts attacked 12 or more hours previously by *T. pretiosum*. However, *T. pretiosum* will attack eggs parasitized up to 24 hours previously by *T. heliothidis*. The mechanistic reason for these differences is that both wasps appear to avoid hosts where tissue necrosis has occurred (see also 6.3.2). Necrosis is caused by the *T. heliothidis* larvae, but in *T. pretiosum* it is caused by a substance injected

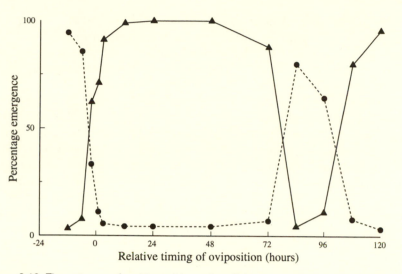

Figure 3.13 The outcome of multiparasitism when *Telenomus heliothidis* (triangles) and *Trichogramma pretiosum* (circles) attack the same host depends on the time between ovipositions. The *x* axis is the number of hours after the *T. pretiosum* attack before parasitism by *T. heliothidis*; negative values mean *T. heliothidis* attacks first. (From Strand 1986.)

by the ovipositing female. Previous parasitism by *T. pretiosum* can thus be detected more quickly than that of *T. heliothidis*.

A number of parasitoids destroy the eggs of other species present on the host. Hardy and Blackburn (1991) obtained indirect evidence that the braconid *Bracon hebetor* destroys the eggs of the bethylid *Goniozus nephantidis*, although ovicide was never actually observed. Some aphid parasitoids in the genus *Ephedrus* (Braconidae) are known or suspected to inject a venom into their host during oviposition which leads to the death of the newly laid eggs of species of *Aphidius*, another braconid genus; the venom has no effect on conspecific eggs (Hågvar 1988; Mackauer 1990).

3.4 Conclusions

Simple models of host acceptance, based on the classical diet models of foraging theory, are probably applicable only occasionally to real parasitoids. The time required to oviposit on a host is normally only a small fraction of the average time required to locate a host. Parasitoids with ample supplies of eggs will thus be selected to lay eggs on nearly all hosts they encounter, including hosts where the probability of the survival of their larvae is small. This conclusion is supported by careful field experiments that have attempted to parameterize simple host acceptance models. More realistic models that incorporate egg limitation are substantially harder to analyze and so far have received little experimental attention. Extending such models to describe the nutrient and energy balance of parasitoids will allow the study of host feeding and of parasitoid time budgets.

There has been great interest in parasitoid clutch size, which in many ways offers an ideal experimental system for manipulative studies. The subject now has a well-developed theoretical foundation, but the major impasse in developing quantitative tests of theory is due to the difficulty in assessing the relationship between adult size and fitness, a difficulty that also frustrates tests of host acceptance models. Laboratory estimates of this relationship almost certainly underestimate the fitness disadvantages of small size, and field studies are essential for the subject to progress. Recent work has started to dissect the behavioral mechanisms involved in clutch size decisions, and this is likely to be an expanding area of research.

The study of superparasitism has occupied a central place in parasitoid biology throughout this century, and we know many details of the physiology of host discrimination. The realization that superparasitism is not always a maladaptive mistake but frequently a strategy that increases individual fitness has led to a reassessment of this phenomenon. Modeling superparasitism presents formidable technical problems, often requiring dynamic game theory, but advances are being made in this field. The willingness of naive females to superparasitize, the influence of other searching females, and the possibility of preferential discrimination against self-superparasitism are all particularly interesting areas of current research.

4 Sex Ratio

In the last chapter, the evolution of host acceptance, clutch size, and superparasitism were discussed. This chapter is about yet another decision faced by a parasitoid when it encounters a host: what sex egg to lay. Sex ratio is a major preoccupation of parasitoid behavioral ecologists because the majority of parasitoids are Hymenoptera and so have a haplodiploid genetic system. Under haplodiploidy, males develop from unfertilized eggs and females from fertilized eggs. The sex of the egg is thus under the direct behavioral control of the mother, and it is relatively easy for natural selection to produce sex ratios adapted to local conditions. Entomologists have known since the last century that parasitoid wasps facultatively alter the sex ratio of their progeny in response to changes in the environment (Kirby and Spence 1816). In other parasitoids which have a diploid (strictly diplodiploid) genetic system, sex is probably determined by the random segregation of sex chromosomes, and it is far less easy for natural selection to change the sex ratio.

The study of sex ratios can claim to be one of the most successful areas of behavioral ecology (Charnov 1982; Leigh et al. 1985; Endler 1986). The subject has a rich and detailed theoretical foundation, and many of the predictions of theory have been confirmed by observation and experiment. Studies with parasitoids have played an important role in testing sex ratio theory, and also in stimulating new theory. Indeed, the only fact many nonentomological behavioral ecologists know about parasitoids is that they have interesting sex ratios. Different aspects of parasitoid sex ratios have been reviewed by Charnov (1982), Frank (1983), Waage and Godfray (1985), Waage (1986), King (1987), and Werren (1987a).

The chapter begins with a discussion of sex determination in haplodiploid parasitoids. The second section reviews Fisher's argument that frequency-dependent selection will often result in equal allocation of resources to sons and daughters, and also discusses cases where sons and daughters require different amounts of parental resources and the curious sex ratio strategies of heteronomous aphelinids. In many parasitoid species, mating often occurs between siblings, and brothers compete with each other for mates. This is normally associated with a female-biased sex ratio. The explanation for this bias, and related phenomena, is the subject of the third section. The fourth section discusses why solitary parasitoids frequently lay male eggs in small hosts, and female eggs in large hosts. The final section reviews a number of other biological and environmental factors that have been suggested to influence parasitoid sex ratios. A variety of elements with non-Mendelian patterns of inheritance are known to affect parasitoid sex ratios, and these are discussed in the next chapter.

Throughout this chapter, the sex ratio is expressed as the proportion of males among the offspring.

4.1 Sex Determination in Parasitoid Wasps

The genetic and physiological mechanisms that cause some individuals to develop as males and others as females can normally be ignored in studies of the evolution of the sex ratio. This is not the case in the haplodiploid Hymenoptera. Although males usually develop from unfertilized haploid eggs and females from fertilized diploid eggs, diploid males are found quite frequently in a variety of different species. Moreover, the occurrence of diploid males is associated with inbreeding, which is also linked to adaptive biases in the sex ratio. Eliminating sex ratio biases caused by the breakdown of the sex determination mechanism is thus important in any study of the evolution of the sex ratio.

Among invertebrates, sex determination has been most closely studied in *Drosophila* and in the nematode *Caenorhabditis elegans*. In both cases, the primary factor determining the sex of an individual is the ratio of X chromosomes to autosomes (a genic balance mechanism). In *Drosophila* the Y chromosome appears not to influence sex determination, unlike in mammals where it has a masculinizing influence; the nematode lacks a Y chromosome. Great strides have been made in recent years in identifying the series of regulatory genes in *Drosophila* and *Caenorhabditis* involved in sex determination (Hodgkin 1990). In both species, the X:autosome ratio influences a master regulatory gene whose activity causes a chain reaction in a cascade of regulatory genes which determines, in different ways, the expression of sex-specific genes in somatic cells and the germ line, and which also controls dosage compensation. Exactly how the X:autosome ratio influences the master regulatory gene is unclear. In *Drosophila* the dosage of a relatively few X chromosome genes in early development, relative to the concentration of general cell machinery products coded for by genes on the autosomes, may be responsible.

In haplodiploid organisms, all chromosomes are present in the same ratio in males and females so it is difficult to see how a genic balance mechanism could work. However, da Cunha and Kerr (1957) (see also Kerr 1974) have suggested a modified genic balance model applicable to haplodiploid animals. In their scheme, sex is determined by a series of alleles with masculinizing and feminizing influences; the effects of the male-determining alleles are non-additive or weakly additive, and thus approximately constant in haploid and diploid individuals, while the effects of the female-determining alleles are additive and thus stronger in diploids. Thus if F and M represent the influences of female-and male-determining alleles, it is possible to have a situation where $M > F$ but $2M < 2F$ and thus haploid individuals become males and diploids females. Crozier (1971) has discussed an alternative genic balance model in which the sex of an individual depends on the relative concentrations of substances produced in the cytoplasm and the nucleus, and where the absolute concentration of the nuclear-derived substance is determined by ploidy. However, Crozier (1971, 1977) has criticized both models on the grounds that they lack a plausible molecular mechanism, and because they have difficulty in explaining the known occurrence of diploid males in a number of hymenopterans. Instead, Crozier favored an allelic diversity model of sex determination, a model originating in the work of P. W. Whiting and his colleagues on the braconid wasp *Bracon hebetor*.[1]

Whiting (1939, 1943; see also A. R. Whiting 1961) showed that sex is determined in *B. hebetor* by the segregation of nine alleles at a single sex-determination locus. If an individual caries only a single allele, either because it is hemizygous (i.e., haploid) or homozygous, it becomes a male, while if it is heterozygous it becomes a female. In *B. hebetor*, homozygous diploid males are infertile. This mechanism is called "complementary sex determination" (CSD) or "allelic-diversity sex determination." CSD has also been demonstrated in another species of *Bracon*, *B. serinopae* (Clark et al. 1963); in honeybees, *Apis mellifera* and *A. cerana* (Mackensen 1951; Woycke 1965, 1979; Hoshiba et al. 1981); in the fire ant *Solenopsis invicta* (Hung et al. 1974; Ross and Fletcher 1985); and in two species of sawfly, *Neodiprion nigroscutum*, (S. G. Smith and Wallace 1971) and *Athalia rosae* (Naito and Suzuki 1991). In all cases, the diploid males either do not develop or are infertile.

Two possible molecular mechanisms have been suggested for CSD. The locus may code for a female-determining molecule consisting of two peptides which is active only as a heteropolymer (Crozier 1971). Alternatively, each allele may code for mRNA containing a different error. The resulting protein is thus invariably nonfunctional in homozygotes, but in heterozygotes a viable gene product can be reconstructed, one allele compensating for the error in the other (Kerr 1974). Haploid mosaics are known in *B. hebetor* and it is found that the borders between the different haploid regions are feminized, indicating that the gene products of different mosaics interact to produce a feminizing substance (Whiting et al. 1934). Supporters of the genic balance model of sex determination suggest that the system found in *B. hebetor* and the other species listed above is atypical and can be explained by a restriction of the additive effects of the female-determining alleles to heterozygous individuals (Kerr 1974).

[1] There is considerable taxonomic confusion surrounding the braconid wasp *Bracon hebetor*. The genus *Bracon* is sometimes divided into smaller units and *hebetor* placed in *Microbracon* or *Habrobracon*. *B. hebetor* itself is highly variable, sometimes called *brevicornis* or *juglandis*, and quite possibly composed of a number of closely related species.

If single-locus CSD is the common method of sex determination in parasit-oid wasps, then diploid males should occasionally be found in the wild and their frequency should markedly increase during systematic inbreeding. Dip-loid males have been detected in electrophoretic studies of wild-caught wasps, for example in the ichneumonids *Diadromus pulchellus* (Hederwick et al. 1985) and *Bathyplectes curculionis* (Unruh et al. 1984), as well as in a variety of nonparasitic bees and wasps. However, systematic inbreedings of a sceli-onid, *Telenomus fariai* (Dreyfus and Breuer 1944); eulophids, *Melittobia* spp. (Schmeider and Whiting 1947); pteromalids, *Nasonia vitripennis* (Whiting 1967; Skinner and Werren 1980); *Muscidifurax* spp. (Legner 1979); *Anisop-termalus calandrae* (Cook 1991); a bethylid, *Gonoizus nephantidis* (Cook 1991); and a braconid, *Heterospilus prosopidis* (Cook 1991) have all failed to detect the expected increase in diploid males. In addition, many species of parasitoids habitually inbreed in the wild and seem to produce consistently female-biased sex ratios. Against this evidence is the frequent observation of increased male biases during mass rearing as part of biological control pro-grams: "It is a well known fact, noticed by many workers engaged in breeding hymenopterous parasites, that there is a tendency in many bisexual species for the sex-ratio to decrease as the breeding work proceeds over a series of gener-ations. In a number of cases the sex ratio decreases to an extent where only males are produced, and the stock dies out" (Simmonds 1947).

Some of these observations can be explained by Crozier's (1971) suggestion that sex determination might involve multiple alleles at two or more loci (Snell 1935 had previously suggested a multilocus model with just two alleles per locus). Haploids, and diploids that are homozygous at all the sex determination loci, develop as males, while individuals heterozygous at one or more loci become females. Single-locus CSD, as found in *Bracon hebetor* and honey-bees, could arise from a multilocus system if all loci except one become homo-zygous. Even with many sex-determination loci, habitual inbreeding will still lead to diploid male offspring. Crozier (1971) suggested that most laboratory strains of wasps considered to be habitually inbred might still contain suffi-cient rare outbreeding to maintain the heterozygosity of at least some sex-determination loci. He also argued that continual inbreeding in nature is very rare and that intermittent episodes of outbreeding are adequate to prevent the breakdown of the sex determination mechanism. However, Cook (1991) sys-tematically sibmated three species of parasitoid wasps for a sufficient number of generations that diploid males would have appeared if fewer than ten sepa-rate loci were involved in sex determination. Similarly, Skinner and Werren (1980) have excluded all forms of CSD in *Nasonia vitripennis* except those with very many loci. It is unlikely that a large number of loci are involved in CSD because diploid males would then be produced so rarely that natural selection on the sex-determining system would be very weak and some loci would become homozygous by genetic drift. Only when a few loci are involved, and some diploid males occur, is natural selection strong enough to counter the effects of drift.

Single-locus CSD results in the continual production of infertile diploid males. The proportion of diploid progeny that are male depends both on the number of alleles at the sex-determination locus and their relative frequencies. If there are x alleles, all with equal frequencies, a fraction $1/x^2$ of diploids will be homozygous for a particular allele and thus, as there are x alleles, overall a fraction of $1/x$ diploids will be male (Laidlaw et al. 1956). This fraction increases as the allele frequencies become less equitable. Whiting found nine alleles segregating at the CSD locus in *B. hebetor*, suggesting that at least 11% of all diploids are infertile males. The production of diploid males does not lead to selection on the sex ratio as it acts as a differential mortality after the production of the primary sex ratio (see next section). With multiple loci, the proportion of diploid males is much reduced. If the number of alleles at k loci are $x_1, x_2, x_3 \ldots x_k$, and all alleles at one locus have the same frequency, the fraction of diploid males is $1/(\Pi_{i=1}^{k} x_i)$ (Crozier 1971).

The few studies of the population genetics of single-locus CSD have been prompted by honeybee genetics (Laidlaw et al. 1956; Woycke 1976; Yokoyama and Nei 1979). In an infinite-sized population, theory predicts that any number of alleles can be maintained and that, at equilibrium, all will have equal frequencies. In finite populations, new alleles will appear by mutation while others will be lost through drift: Yokoyama and Nei (1979) derive approximate formulas for the equilibrium number of alleles and the distribution of allele frequencies in populations of different sizes and with different mutation rates. Fewer alleles can be maintained in small populations. They also make the interesting point that the genetics of CSD are mathematically identical to that of self-incompatibility loci in plants.

To conclude, complementary sex determination is found in a variety of Hymenoptera including sawflies, parasitoids, and aculeates. Among parasitoids, the only proven case is in *Bracon*, but the presence of diploid males and the occurrence of sex ratio biases during mass rearing, suggest its presence in other species, especially in the Ichneumonoidae. Single-locus CSD has been definitely ruled out in many species, and while multilocus CSD may be more widespread, it has yet to be experimentally demonstrated, and has been ruled out in some species. At the present time, there are no alternatives to CSD with experimental support, though several genic-balance mechanisms have been proposed. Where CSD occurs, the production of diploid males acts as a sex-determination load reducing population fitness. The load is reduced the more alleles segregate per locus, and the more loci that are involved in sex determination. The production of diploid males does not directly affect selection on the primary sex ratio, but may influence the observed secondary sex ratio. Diploid male production may, however, act as a selection pressure influencing the evolution of the sex-determination mechanism itself (Bull 1983).

4.2 Fisher's Principle

The modern study of the sex ratio began with Fisher's (1930) explanation of why equal investment in male and female progeny is so often observed in populations of plants and animals. Where male and female progeny require the same amounts of parental resources, equal investment implies an equal sex ratio. In this section, I first describe Fisher's principle and the assumptions on which it is based. I then discuss possible cases of differential progeny costs in parasitoids, and the curious sex ratios of heteronomous aphelinids. Finally, I briefly describe genetic evidence for heritable variation in the sex ratio.

4.2.1 The Theory of Equal Investment in Sons and Daughters

Begin by supposing that sons and daughters are equally costly to produce. The fitness of a mother can be measured by the number of daughters she produces plus the number of females inseminated by her sons (the latter figure may be a fraction if several males mate with a single female). When the population sex ratio is 50:50, a son will on average inseminate a single female and so the mother will gain identical increments in fitness from sons and daughters. Now suppose the population sex ratio is male biased: a son will on average inseminate less than one female and so will be relatively less valuable to a mother than a daughter. Any mutant gene that causes its bearer to produce more daughters than sons would thus be favored by natural selection and, as the mutant increases in frequency, the population sex ratio will move toward equality. Now suppose the population sex ratio is female biased: sons will on average mate with more than one female and so will be more valuable to the mother than daughters. A mutant gene leading to the production of more sons will spread through the population and this again will tend to return the population sex ratio toward equality. The 50:50 sex ratio is the only sex ratio that cannot be invaded by alternative mutants, and is thus the evolutionarily stable strategy (the ESS sex ratio). The mechanism underlying Fisher's argument is a form of frequency-dependent selection: the fitness of a sex ratio strategy depends critically on its frequency in the population.

The argument is only a little more complicated when sons and daughters require different amounts of parental resources. Instead of measuring the fitness increment to the parent per offspring, as was done above, the fitness increment should be measured per unit of investment in offspring. A gene that increases the production of the sex that gives the greater relative fitness return per unit of investment will be favored. Equality of investment in sons and daughters is the only investment ratio that cannot be invaded by alternative mutants, and is thus the ESS. Differential mortality after the young have been abandoned has no effect on the mother's fitness gain per unit of investment and thus has no effect on the optimum sex ratio. For example, suppose 50% of sons die before reproducing but all daughters survive: under these circumstances each son will on average mate with two females. While the differential mortality halves the value of investment in sons, the value is exactly restored by the doubled mating success of the survivors. The sex ratios before and after differential mortality occurs are called the primary and secondary sex ratios, respectively.

Fisher's principle predicts the primary population sex ratio but does not specify how it is achieved. For example, a 50:50 sex ratio occurs if every female in the population produces equal numbers of sons and daughters, or if half the females produce only sons and the other half daughters. In finite populations there is weak selection in favor of all individuals producing sons and daughters with equal probability (Verner 1965; Taylor and Sauer 1980), and this case is by far the most frequently observed.

To obtain Fisher's result, many assumptions have to be made. For parasitoids, the most critical assumptions are that mating is panmictic and that both sexes benefit to the same degree from increased investment. Violations of these two assumptions will occupy most of this chapter. The general assumption required for Fisher's principle to apply is that the fitness returns from increased investment in either sex must be linear, though not necessarily equal (see Frank 1990 for a recent and particularly lucid discussion). Linearity means that if a mother increases her investment in one sex by two units, she gains exactly twice as large an increment in fitness as if she had invested one extra unit.

For Fisher's principle to apply, some genetic assumptions must be met. The mother (assumed to control the sex ratio) must be equally related to both sons and daughters. Relatedness is defined as the probability that a gene present in one individual is also present, identically by descent, in another individual. Equal relatedness occurs in diploid animals and in outbred haplodiploid insects, but in haplodiploid populations with inbreeding, mothers are more closely related to daughters than sons. Greater relatedness implies that daughters are a more efficient means of transmitting genes to the next generation, leading to selection for a female-biased sex ratio. As will be discussed in section 4.3, inbreeding leads to other violations of Fisher's principle in addition to distorting relatedness.

Fisher's principle and other sex ratio models typically make the assumption that there is sufficient heritable variation in the sex ratio that the population can evolve to the optimum strategy. As discussed below, there is some evidence of additive genetic variance in the sex ratio of parasitoid wasps. In the case of other parasitoids, sex is determined by the segregation of sex chromosomes and there may be no heritable variation in sex ratio. Here, the sex-determination system guarantees the production of equal numbers of sons and daughters and acts as a constraint preventing the animals from adjusting their sex ratio, even in environments where biased sex ratios are favored. The sex ratio of parasitoid wasps is so fascinating precisely because they lack these genetic constraints.

4.2.2 THE COSTS OF SONS AND DAUGHTERS AND THE SEX RATIO OF HETERONOMOUS HYPERPARASITOIDS

Even if the assumptions underlying Fisher's principle are met, biased sex ratios will be found if the costs of producing sons and daughters are not identical. Costs are measured in terms of the resource that limits the reproductive output of the wasp. In parasitoids, the most important limiting factors are likely to be the number of eggs and the time available for reproduction.

Consider first a parasitoid whose reproductive output is limited by the number of eggs it carries or can produce. Male and female eggs are unlikely to vary in costs: sex is decided at the last moment before oviposition when the mother chooses whether or not to fertilize the egg, and before that the eggs are indistinguishable. A few parasitoids show a limited amount of parental care (sec. 7.3) but there is no reason to suppose male and female offspring differ in their care requirements. Thus egg-limited parasitoids are likely to invest equally in male and female offspring.

Now consider time-limited parasitoids. Some parasitoids may encounter situations where they are confronted with so many hosts that their reproductive success is limited by handling time. If the time required to oviposit male and female eggs differs, the sex ratio will be biased in favor of the sex with the shorter handling time. The best examples of unequal handling time are found among heteronomous hyperparasitoids, a type of heteronomous aphelinid (see sec. 1.3.2 for a description of the biology of these wasps). Female eggs develop as primary parasitoids and male eggs as hyperparasitoids. A number of workers have observed that it takes longer to lay a male egg than a female egg (J. R. Williams 1972; Buijs et al. 1981; Walter 1986; Donaldson et al. 1986; T. Williams 1991; M. S. Hunter, pers. comm.). Male oviposition may take longer because the wasp has to drill through two layers of cuticle. Though differential handling time can theoretically lead to biased sex ratios, it is probably quite uncommon for a parasitoid to both have sufficient hosts and sufficient eggs for reproductive success to be limited by oviposition time.

The most time-consuming part of reproduction for parasitoids is nearly always host location. Some heteronomous aphelinids lay male and female eggs on different types of host and it may take longer to locate one type of host than the other. Consider heterotrophic aphelinids whose males develops in lepidopteran eggs and females as parasitoids of scale insects (I ignore here the possibility that the site of male development may not be obligatory—see sec. 1.3.2). Suppose the two host types are found in different habitats or microhabitats; how should the female divide her time between the two habitat types? If we assume that the female achieves linear fitness returns from allocating search time to either site, Fisher's principle applies and the wasp is predicted to spend equal time searching for male hosts and female hosts (Godfray and Waage 1990; Godfray and Hunter 1991). If one type of host is more easily found than the other, the observed population sex ratio will be biased in favor of the sex it supports. Note that this argument applies because males and females obligatorily develop in different classes of host; it would not apply to cases where males tend to develop in small hosts and females in large hosts (see sec. 4.4).

Although it is possible that the two host types of heteronomous aphilinids occur in different microhabitats, the more likely situation is that they occur together in the same place. This is especially likely in the case of heteronomous hyperparasitoids (sec. 1.3.2) where males develop as hyperparasitoids of female aphelinids. The sex ratio strategy of time-limited heteronomous hyperparasitoids searching in an environment containing both host types has few parallels in the animal kingdom. As far as I am aware, in all other cases of sex allocation, the production of sons and daughters is mutually exclusive; resources invested in sons are resources lost to daughters and vice versa. This is not the case here. If a searching wasp encounters a host suitable for male progeny, it should lay a male egg, an action that has no effect on the number of daughters she produces. The total reproductive output of the wasp consists of two independent components, sons and daughters. The fitness of the searching wasp is maximized by laying an egg of the appropriate sex whenever a host is encountered. The observed sex ratio will be wholly determined by the relative rates at which the two host types are located by searching wasps (Godfray and Waage 1990; Godfray and Hunter 1991).

These ideas lead to several predictions. First, sex ratios in heteronomous aphelinids should be very variable and correlated with the frequency of the two host types. A number of workers have noted particularly variable sex ratios among heteronomous aphelinids (Chumakova and Goryunova 1963; Kennett et al. 1966; Kuenzel 1975; Viggiani and Mazzone 1978; Donaldson and Walter 1991b) or have suggested that the sex ratio is correlated with the relative abundance of the two host types (Flanders 1967; J. R. Williams 1977; Hunter 1993). The variability of aphelinid sex ratios has also been used to argue that sex ratios are nonadaptive (Donaldson and Walter 1991a, 1991b), a point challenged by Godfray and Hunter (1991). A second prediction is that egg-limited and time-limited parasitoids should adopt different sex ratio strategies. An egg-limited wasp, confronted with many hosts, should produce equal numbers of the two sexes, while a strictly time-limited wasp should lay an egg of the appropriate sex in whatever host it encounters.

Several studies have examined the sex ratio of heteronomous hyperparasitoids presented with mixtures of hosts suitable for male and female development. Hunter, (1989) working with *Encarsia pergandiella*, and Donaldson and Walter (1991a), working with *Coccophagus atratus*, found a correlation between the sex ratio and the ratio of host types. However, the sex ratio tended to be relatively more male biased than the host ratio. Avilla et al. (1991) found that the sex ratio of *Encarsia tricolor* was correlated with the host ratio, but that a relative excess of females was produced. In contrast, T. Williams (1991), also working with *Encarsia tricolor*, found no correlation between sex ratio and host ratio. He considered that the wasps in his experiment had an excess of hosts, and that their reproductive success was limited by egg supply.

It is difficult to interpret these experiments without knowing the factors limiting reproductive output. A critical test of sex allocation theory in heteronomous hyperparasitoids would be to vary both the absolute and relative densities of the two host types so that the response of the wasp can be observed under conditions where it is either host limited or egg limited. It is also important to measure the encounter rate with the two types of hosts (which will be influenced by the density of the host and how easy they are to locate and detect) because the predicted sex ratio of host-limited parasitoids is determined by the relative numbers of the two host types found by the searching wasp.

The sex ratio of heteronomous hyperparasitoids may also be influenced by the possibility that a female is hyperparasitizing one of her own offspring. Colgan and Taylor (1981) showed that where this risk occurs, a more female-biased sex ratio will be favored by natural selection. There is some indirect evidence that this may be a significant factor biasing the sex ratio. *Encarsia tricolor* not only hyperparasitizes females of its own species, but also those of other aphelinids. T. Williams (1991) and Avilla et al. (1991) compared the sex ratio of *E. tricolor* in situations where hosts suitable for male development either contained females of the same species, or females of another species (*E. inaron* or *E. formosa*, respectively). Both studies found a considerably more male-biased sex ratio if the alternative host was of a different species. In choice experiments, Williams found that *E. tricolor* preferred to hyperparasitize females of a different species. There is no evidence that the survival of male larvae is higher when parasitizing females of other species. It is possible that the avoidance of conspecifics may have evolved as a mechanism to prevent parasitism of relatives. However, Buijs et al. (1981) failed to find a preference for females of a different species when they presented *Encarsia pergandiella* with a choice of hosts parasitized by conspecifics and by *E. formosa*.

4.2.3 GENETIC VARIATION IN THE SEX RATIO

Natural selection can lead to changes in the sex ratio only if there is additive genetic variance in the sex ratio. The absence of heritable variation does not imply that the population sex ratio is not well adapted—every individual in the population may have the optimum genotype—but it does imply that the population is unable to respond in the short term to new selection pressures. Sex ratio is an atypical trait because when Fisher's principle applies, and the population is at sex ratio equilibrium, there is no selection against rare sex ratio mutants.

One way to demonstrate additive genetic variance in the sex ratio is through artificial selection. Biological control workers have discussed artificial selection for female-biased sex ratios as a means of increasing natural enemy efficiency. However, where more female-biased sex ratios have been obtained through laboratory breeding (Wilkes 1947; Simmonds 1947), the improvement has probably been due to a reduction in the proportion of diploid males (sec. 4.1).

Parker and Orzack (1985) used the methodology of quantitative genetics to study the sex ratio of five inbred lines of the pteromalid wasp *Nasonia vitripennis* which were initiated from single females collected in the wild or from laboratory stocks. Initial experiments showed that under identical laboratory conditions, the five lines produced different sex ratios suggesting a genetic component to sex ratio variation. This was confirmed by selecting for a male-biased sex ratio in two replicate new lines of wasps (a third replicate formed a control) created by interbreeding individuals from the five original inbred populations. Over twelve to fifteen generations, the mean sex ratio dropped from approximately 80%–90% to 50%–55% females and then remained constant, presumably through exhaustion of additive genetic variance. Other experiments excluded selection for male sterility and brood size as explanations for these results. This experiment reveals genetic variation in the sex ratio, though not within a single natural population. In subsequent experiments, Orzack and Parker (1990; Orzack 1990; see also sec. 4.3.3) created isofemale lines from wasps collected in the field. They found significant between-strain heterogeneity in sex ratio, strong evidence for within-population genetic variance in the sex ratio. A previous experiment designed to study the genetics of sex ratio in *N. vitripennis* failed, though in a very interesting way. Werren et al. (1981) tried to select for a male-biased sex ratio but found that some females produced all male broods caused by a factor subsequently shown to be a supernumerary chromosome inherited only through sons (sec. 5.1). Parker and Orzack (1985) excluded non-Mendelian factors as an explanation for their results.

4.3 Local Mate Competition and Sex Ratio in Structured Populations

Parasitoid wasps frequently have extremely female-biased sex ratios. Such sex ratios are observed most frequently in gregarious parasitoids where there is a high probability that mating occurs between siblings and that brothers compete together for mates. An explanation for these observations was provided by Hamilton (1967), who described the mechanism responsible as local mate competition (LMC).

LMC is now known to be a specific example of a wider class of phenomenon, all involving interactions between relatives in spatially structured populations (Charnov 1982), and capable of generating both male-biased and female-biased sex ratios. Throughout the 1980s, a vigorous debate took place on how best to describe the mechanisms responsible for biased sex ratios in spatially structured populations, a debate that to a lesser extent still rumbles on today. The first part of this section describes Hamilton's original model and the controversy surrounding its interpretation. The second part discusses some of the numerous elaborations of the basic model, many of which have been prompted by work on parasitoid wasps. The third part reviews experimental work on LMC, while the fourth examines the behavioral mechanisms used by wasps to achieve a desired sex ratio. The final part considers other factors in addition to LMC that may bias sex ratios in spatially structured populations.

4.3.1 Local Mate Competition: Hamilton's Model

Hamilton (1967) imagined a species living in a patchy environment where in each generation groups of n mated females (foundresses) colonize patches for reproduction. Each female lays the same number of eggs in the patch, and their progeny develop together and eventually mate among themselves. The females of the new generation then disperse and new groups of n foundresses are chosen at random. Under these assumptions, the ESS sex ratio for each of the n foundresses is $(n - 1)2n$ (fig. 4.1). If only a single foundress colonizes a patch, a sex ratio of zero is predicted. Hamilton interpreted this to mean that a female should lay only enough males to fertilize all her daughters. As the number of foundresses increases, the sex ratio asymptotes at 0.5, the predicted sex ratio when mating occurs at random.

The prediction of a sex ratio of $(n - 1)2n$ is exactly correct only for diploid organisms. In his original paper, Hamilton (1967) conducted computer simulation studies that explicitly incorporated haplodiploid genetics. His numerical studies showed that the ESS sex ratio was slightly more female biased than predicted: when $n = 2$, the ESS was approximately 0.217 instead of the expected 0.25. Later Hamilton (1979), Taylor and Bulmer (1980), and Suzuki and Iwasa (1980) showed that when n is constant throughout the population the true ESS sex ratio for haplodiploids with LMC was $[(n - 1)2n] \cdot [(4n - 2)/(4n - 1)]$ (fig. 4.1). The reason for the discrepancy between the diploid and haplodiploid predictions is that the population structure proposed by Hamilton results in an increase in the equilibrium level of inbreeding which in haplodiploids increases the relatedness of a mother to her daughters in comparison with that of a mother to her sons. Increased relatedness to daughters implies that this sex is a better means of transmitting genes to future generation and leads to a sex ratio bias in favor of females. In diploid populations, the changes in relatedness caused by inbreeding are sexually symmetric and do not affect the optimum sex ratio. A useful way of incorporating the effects of inbreeding in an expression for the sex ratio is to define g as the ratio of the mother's relatedness to her daughters and her sons. The ESS sex ratio is then given by $(n - 1)/(1 + g)n$ (Werren 1987a).

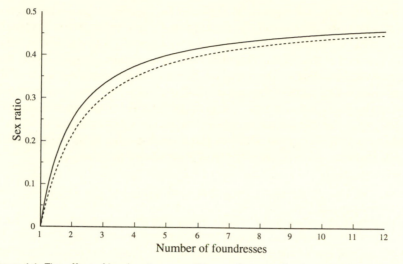

Figure 4.1 The effect of local mate competition on the sex ratio: the prediction of Hamilton's model which assumes *n* foundresses reproduce together on a patch and their offspring mate among themselves. The solid line is the prediction for diploid species, the dashed line for haplodiploids.

There are two main ways to interpret the female-biased sex ratios found in Hamilton's model. The first uses the methodology of individual and kin selection and arises naturally from Hamilton's original treatment. The second approach is to analyze the sex ratio using hierarchical selection theory, an approach that also has its roots in papers by Hamilton (1975, 1979) and has been particularly developed (though in different ways) by Colwell (1981) and Frank (1986). Both methods, when applied correctly, predict exactly the same sex ratio and are thus equally "right." The choice of method should thus be determined by the convenience of analysis. The next two subsections briefly describe the two approaches and can be skipped by readers uninterested in sex ratio modeling.

THE INDIVIDUAL AND KIN SELECTION APPROACH

Consider first why Fisher's principle does not apply when sons compete for mates (including their sisters). Increased investment in sons will not bring linear fitness returns for the mother, thus violating the major assumption of Fisher's argument. Sons compete among themselves for mates, and there will be diminishing returns from the production of sons at the expense of daughters. In the most extreme case, when sons compete only with their brothers, the mother will obtain no extra increment in fitness by producing more sons than are necessary to fertilize her daughters.

There are actually two reasons why a female-biased sex ratio is advantageous when local mate competition occurs: (1) fewer sons mean less wasteful competition for mates among brothers, and (2) more daughters mean more potential mates for sons (Taylor 1981, 1988; Grafen 1984; Bulmer 1986). In haplodiploid species there is a further advantage: (3) when inbreeding occurs mothers are more closely related to daughters than sons. Sometimes the first process alone is referred to as "local mate competition" while the second process is called "sibmating" or (confusingly) an "effect of inbreeding" (as the extra potential mates are sisters). More normally, and in my view preferably, local mate competition refers to the joint action of all processes. The ESS sex ratio is defined as the sex ratio at which the marginal gain in fitness from a further reduction in sons is exactly balanced by the normal frequency-dependent disadvantage of producing more of the commoner sex. Note that as n gets large, changes in the mother's sex ratio have no effect on either the competition experienced by her sons, or on the number of their potential mates, and the sex ratio returns to equality.

Nunney (1985; also Nunney and Luck 1988) has pointed out that in Hamilton's original model, a decrease in competition among sons is always accompanied by a greater number of potential mates. The two advantages of a female-biased sex ratio act in concert and can be treated together as a single process. Nunney uses the term "local parental control" to describe the joint process because the sex ratio bias is found only when the action of the parent is able to alter the sex ratio conditions experienced by her progeny. However, there does seem to be some advantage in retaining the dual categorization of the cause of biased sex ratios with LMC, as circumstances can be envisaged where the two processes are disassociated (see below). On a more semantic note, adaptations to LMC can still occur when the sex ratio is not under parental control, for example in polyembryonic species.[2]

THE HIERARCHICAL SELECTION THEORY APPROACH

Consider now the interpretation of Hamilton's model based on hierarchical selection theory (Hamilton 1975, 1979; Colwell 1981; D. S. Wilson and Colwell 1981; Frank 1983, 1985, 1986). This method has its roots in population genetics and in particular a technique devised by G. R. Price (1970, 1972) to study changes in gene frequency in hierarchically structured populations where selection can act at a number of different levels. The population structure described in Hamilton's model is hierarchically structured, and the essence of this method is to examine changes in gene frequency at two levels, within the group of n foundresses (*intra*demic selection where in this case the

---

[2] Nunney (1985) also argues that the female-biased sex ratio found in Hamilton's model occurs in an artificial sex ratio model in which competition among brothers for mates (which he equates with local mate competition) is explicitly excluded. However, this artificial model includes a new factor that biases the sex ratio, absent in other models: a dependence of total number of clutches on sex ratio (Godfray 1986b). It is this new factor that substitutes for competition between brothers and results in the same female-biased sex ratio.

foundresses form the deme) and within the "population" of groups (*inter*demic selection).

Following Colwell (1981), suppose the population is made up of two alleles: "Fisher," producing a sex ratio of 0.5, and "Hamilton," producing a sex ratio of $(n - 1)2n$. The numbers of each allele in a group of n foundresses will depend on the overall frequencies of the alleles in the population and on the statistical process of random sampling. Given the composition of a particular group, the frequency of each allele in the progeny of the n foundresses can be calculated. The average change in allele frequency across all groups is a measure of intrademic selection. It turns out that in groups containing both alleles, the Fisher allele always increases in frequency at the expense of the Hamilton allele: thus intrademic selection favors a sex ratio of equality.

Now consider the contribution of all groups to the change in gene frequency in the whole population. It is important to consider not only the gene frequencies in particular groups, but also the relative numbers of inseminated females produced by different groups. A group that produced only enough males to fertilize all the females in the group would maximize its production of inseminated females. Groups that contain a large proportion of individuals expressing the Hamilton allele produce many more inseminated females than groups that contain a large proportion of individuals expressing the Fisher allele. Thus interdemic selection favors the Hamilton allele.

Whether the Hamilton or Fisher allele increases in frequency in the whole population depends on the relative strengths of the interdemic and intrademic selection pressures. In this case the Hamilton allele is favored as the interdemic selection pressure is strongest. Even though the frequency of the Hamilton allele is always less in the progeny of a mixed group of foundresses than in their parents, this is more than compensated for by the greater productivity of groups containing a high proportion of Hamilton alleles. In a sense, the Hamilton allele loses all the battles, but still wins the war.

A more complete analysis of Hamilton's model using this methodology would consider all candidate sex ratios and calculate which was the ESS. Hamilton (1979) and Frank (1986) found that the ESS sex ratio for diploids was $0.5P$, where P is the genetic variance within a group (V_g) divided by the genetic variance within the population (V_p). Now the group is composed of n individuals chosen at random from the population and the variance of a sample of size n is simply $(n - 1)/n$ times the variance of the sampled population. Thus $V_g = [(n - 1)/n]V_p$, and so $P = (n - 1)/n$ and the ESS sex ratio is $(n - 1)/2n$—exactly that calculated for diploids by the alternative method.

Hierarchical selection theory allows the importance of selection at different levels of the population to be identified and combined to predict the overall selection acting on an allele. In the analysis of Hamilton's model, two levels were involved: individuals among groups, and groups within the population. The method can be extended upwards to examine "supergroups," aggregations of groups within the population, or downwards to examine the effect of biased segregation of chromosomes within the individual.

4.3.2 MODIFICATIONS TO HAMILTON'S MODEL

Hamilton's (1967) model has been modified and adapted in many different ways. Some of the modifications to the basic model have been deliberately unrealistic, designed to reveal the underlying mechanisms responsible for sex ratio biases (Maynard Smith 1984; Werren 1983; Charnov 1982; Nunney 1985). However, other modifications have been prompted by the natural history of real systems. A summary of the most important modifications of Hamilton's model, and the predictions they make, is given in table 4.1.

4.3.3 EMPIRICAL STUDIES OF LOCAL MATE COMPETITION

Hamilton (1967) supported his original theory of LMC by listing a number of examples of animals with strong local mate competition which also showed a female-biased sex ratio—his "biofacies of extreme inbreeding." Of the twenty-six examples he quoted, not fewer than eighteen were parasitoid wasps (a nineteenth was a fig wasp). Today, many more examples could be added to Hamilton's list and it does appear that female-biased sex ratios are particularly common in parasitoid wasps.

Parasitoid wasps are both genetically and ecologically predisposed to show female-biased sex ratios caused by local mate competition. Their haplodiploid genetic system allows the flexibility in sex ratio necessary to respond to LMC, while their ecology frequently results in competition for mates among brothers. LMC is found most often in two categories of parasitoids: gregarious parasitoids, and parasitoids attacking gregarious hosts. Families of gregarious parasitoids frequently pupate together in the vicinity of their exhausted hosts. Adult males emerge and begin looking for mates at around the same time as their sisters emerge, and so brother-sister mating and competition between brothers are both highly likely. The same thing happens if a female of a solitary species attacks a gregarious host; for example, solitary egg parasitoids attacking clumps of host eggs, or solitary larval parasitoids attacking hosts in discrete resource units such as fungi, dung, or rotten fruit. Although LMC is common in parasitoid wasps, markedly female-biased sex ratios are still the exception rather than the rule. It is quite common to read in nonspecialist books statements that imply that all parasitoid wasps have female-biased sex ratios, a distortion that reflects the influence of Hamilton's pioneering work.

Local mate competition provided an evolutionary explanation for a variety of earlier observations on the influence of parasitoid density and superparasitism on sex ratios in parasitoid wasps. Salt (1936) had observed more male-

Table 4.1
Modifications of Hamilton's Local Mate Competition model.

Modification	Effect	Reference
Haplodiploid genetics	In inbred populations, mothers are more closely related to daughters than sons and the sex ratio is thus female biased in comparison with diploids.	Taylor & Bulmer 1980; Hamilton 1979
Sex ratio control exerted by male in haplodiploid species	Sex ratio twice as male biased $(=(n-1)/n)$. If the sex ratio is controlled by other nonautosomal genes, biased sex ratios may occur.	Hamilton 1967
Variable foundress number, females able to assess the number of other foundresses in the patch	The prediction of sex ratio as a function of n applies as a conditional strategy for females experiencing variable foundress numbers. In haplodiploids, variable foundress number increases inbreeding and the conditional strategy is slightly more female biased.	Herre 1985; Frank 1985
Variable foundress number, females unable to assess the number of other foundresses in the patch	A good approximation of the predicted sex ratio is obtained by substituting the mean value of n in the expression for sex ratio.	Nunney & Luck 1988
Individuals differ in brood size and adjust their sex ratio accordingly	Small clutches will be relatively male biased and large clutches relatively female biased with the overall sex ratio approximately unchanged.	Werren 1980; Herre 1985; Frank 1985
Individuals differ in brood size: individual sex ratio adjustment does not occur, though clutch size variance may influence sex ratio strategy adopted by all females	Leads to an increase in sibling interaction and (in haplodiploids) to closer mother daughter relatedness: sex ratio becomes more female biased.	Hamilton 1979; Charnov 1982; Frank 1985
Patch composed of hosts which vary in size; solitary parasitoids	Male eggs tend to be laid on small hosts, female eggs on large hosts (see also sec. 4.4); overall sex ratio unchanged or less female biased.	Werren 1984a; Werren & Simbolotti 1989
Sequential discovery of patch by foundresses (females unaware of their position in sequence)	Leads to an increase in the average level of sibling interaction and thus to a more female-biased sex ratio.	Nunney & Luck 1988

(Table continues on following page)

Table 4.1 (*continued*)

Modification	Effect	Reference
Male dispersal: some males disperse from natal patch and obtain matings on other patches	Leads to a decrease in the average level of sibling interaction and thus to a more male-biased sex ratio. Even moderate levels of dispersal can strongly affect the sex ratio. If hosts vary in size, extra males are placed on smaller hosts.	Nunney & Luck 1988; Werren & Simbolotti 1989
Subdivided population: females do not disperse randomly throughout the population but are more likely to settle in patches with other females from the same subpopulation	Increases the likelihood of interactions between relatives and thus leads to a greater female bias in the sex ratio.	Frank 1985
In gregarious species, individuals suffer greater competition from larvae of one sex than the other	If male competition is stronger, sex ratio becomes more female biased while the reverse occurs if female competition is stronger.	Godfray 1986b
Erratic selection pressures leading to an advantage for recombination	Females become more valuable because only in this sex is recombination possible: female-biased sex ratio selected.	Hamilton 1979
Unequal investment in the sexes	As before, except ratio of sons to daughters replaced by ratio of investment in sons and daughters.	Uyenoyama & Bengtsson 1981, 1982

Note: The brief summaries presented here can only caricature the conclusions of sometimes extensive modeling.

biased sex ratios in *Trichogramma evanescens* when a number of female wasps oviposited together, and similar observations had been made by Kanungo (1955), Wiackowski (1962), Jackson (1966), Velthuis et al. (1965), Wylie (1965, 1966), Walker (1967), Viktorov (1968), Viktorov and Kochetova (1971, 1973b), Kochetova (1972), Shiga and Nakanishi (1968), and Holmes (1972). Previous interpretations of these observations had invoked differential mortality (Salt 1936; Kanungo 1955) or mechanistic explanations such as the effects of contacts with, or disturbance by, other females during oviposition (Jackson 1966; Wylie 1966; Kochetova 1978).

More recently, there have been a variety of quantitative studies of the predictions of LMC theory. In the following subsections I discuss comparative cross-species studies, experiments and observations relating sex ratio and foundress number, studies of sex ratio and superparasitism, and finally work on polyembryonic species.

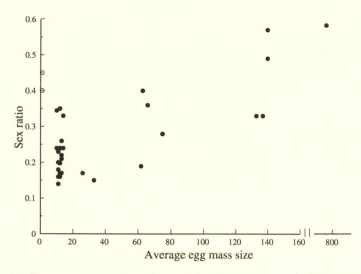

Figure 4.2 The average sex ratio, and average host egg mass size, for thirty-one species of Scelionidae. The unfilled circles represent the sex ratio of two species attacking hosts which lay solitary eggs. (From Waage 1982a.)

COMPARATIVE STUDIES

In cross-species comparisons, the average sex ratio should be related to the average level of local mate competition. Waage (1982a) tested this suggestion in a comparative study of the Scelionidae, a family largely composed of solitary egg parasitoids. Different scelionid species attack hosts that lay their eggs in large clumps, in small clumps, and hosts with solitary eggs. Waage argued that LMC was likely to be greatest in wasps that attacked hosts that laid their eggs in small clumps. A small clump would normally be found by only a single or a few wasps whose progeny would develop on the egg mass and mate among themselves after emergence. Wasps attacking larger clumps would experience less LMC as these clumps would be discovered by many foundresses; wasps attacking solitary eggs would have to disperse prior to mating and thus would be relatively unlikely to mate with siblings. Female scelionids often fight to defend clumps of eggs from other females and so reduce foundress number, though this behavior appears to occur most frequently in species attacking small egg batches (sec. 7.3). Waage collected data from the literature on sex ratio and host egg distribution in thirty-one species of scelionid (fig. 4.2). There is a significant relationship between sex ratio and egg mass size for species attacking clumps of more than one egg. The two species that attack solitary eggs have sex ratios close to 0.5. The data thus broadly confirm Waage's hypothesis.

A different prediction of LMC theory was tested by Griffiths and Godfray (1988) in a comparative study of sex ratio and clutch size in the family Bethylidae, gregarious parasitoids of moth and beetle larvae. The clutch size of different bethylids varies from one (found in a single species) to over a hundred. A number of arguments suggest that the majority of matings occur between brothers and sisters from the same clutch. Bethylids tend not to disperse away from the pupation site prior to mating, and superparasitism leading to mixed parentage is rare because females normally stay with the host to guard their young (see sec. 7.3). If brother-sister mating is the rule, the female wasp is predicted to produce just enough sons to mate all her daughters, leading to a correlation between sex ratio and clutch size. Griffiths and Godfray obtained data from the literature on clutch size and sex ratio in twenty-six species of bethylids. As predicted, sex ratio declined with increasing clutch size, a regression explaining 45% of the variation in the data (fig. 4.3). They also compared their data against the specific hypothesis that a female lays one son which then inseminates all the females in the clutch (i.e., sex ratio is the reciprocal of clutch size). This provided a rough fit to the data though more males tended to be produced than expected in larger clutches. Possibly the assumption that one son can inseminate the rest of the clutch is untrue in these species. An alternative explanation is that in large clutches the mother is selected to produce an extra son as insurance against the death of the first son before mating (Hartl 1971).

The results of both these studies must be treated with some caution as the authors assumed that data from different species could be considered as independent in statistical analysis. Nonindependence arises because species are phylogenetically related; species characters may be evolutionarily conserved and inherited from a common ancestor, or may be highly correlated with an unknown, phylogenetically inert third variable (Harvey and Pagel 1991). The argument that sex ratio and clutch size are likely to be evolutionarily labile (Griffiths and Godfray 1988) is only a defense against the first of these problems (Ridley 1988a).

SEX RATIO AND VARIABLE FOUNDRESS NUMBERS

Hamilton's model predicts that the sex ratio of an individual wasp is influenced by the number of other females ovipositing in a patch. This prediction has been tested in two ways: by using natural variation in foundress number in the field, and by experimental manipulation in the laboratory. The tests are based on the assumption that the wasp can both assess the number of other females in the patch, and can adjust her sex ratio accordingly. A problem occurs in testing LMC models using wasps from populations with variable foundress numbers. The standard expression for sex ratio in haplodiploid populations with LMC is $[(n - 1)/2n$ y $[(4n - 2)/(4n - 1)]$, the second term in square brackets representing the effects of asymmetric relatedness caused by

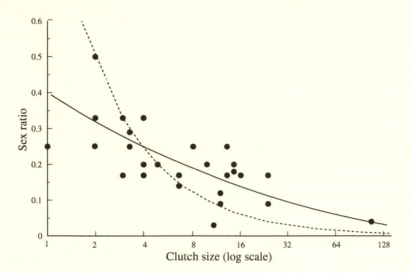

Figure 4.3 Sex ratio and clutch size from twenty-seven studies of twenty-three species of bethylid wasp. The solid line is a fitted weighted regression through the data (in some of the original papers, clutch size is given only as a range and not a mean, and these points are given less weight in the regression). The broken line is the reciprocal of clutch size, the predicted sex ratio if every clutch contains a single male. (From N. Griffiths & Godfray 1988.)

inbreeding. As Herre (1985) and Frank (1985) point out, this expression is exactly correct only if n is constant throughout all patches. To obtain an accurate prediction for the sex ratio, it is necessary to measure the distribution of foundress numbers in the wild.[3]

The first experiment designed to investigate this hypothesis was performed by Werren (1983) working with the gregarious pteromalid *Nasonia vitripennis*, a parasitoid of the pupae of blowfly and related Diptera. Werren placed different numbers of female wasps (foundresses) in containers with host pupae; wasps were allowed to leave the experimental arena at will because unnatural confinement tends to lead to excess superparasitism and high mortality. In the field, wasps frequently attack patchily distributed hosts and have a female-biased sex ratio. Male *N. vitripennis* have reduced wings and do not fly, increasing the likelihood that mating takes place between related insects. Werren showed that wasps ovipositing alone laid a very female-biased sex ratio, and that the sex ratio became more male biased as foundress number increased (fig. 4.4). Field observations on *N. vitripennis* reinforce these findings. The wasp attacks the puparia of a number of species of cyclorraphous

[3] The equilibrium sex ratio in a patch with n foundresses is $[(n - 1)/2n].[(4\bar{n} - 2)/(4\bar{n} - 1)]$ where $\bar{n}$ is the harmonic mean of the number of foundresses per patch (Herre 1985), or equivalently $(n - 1)/(1 + g)n$, where g is the expected ratio of a mother's relatedness to her daughters and sons (Werren 1987a).

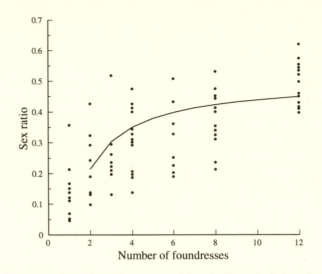

Figure 4.4 The sex ratio produced by groups of female *Nasonia vitripennis* ovipositing together in the laboratory (from Werren 1983), with the predicted sex ratio from LMC theory (Hamilton 1979).

Diptera, in a wide variety of habitats. In birds' nests, both hosts and parasitoids are rare and LMC is likely to be common. On animal carcasses, the number of foundresses varies but is likely to be roughly proportional to the size of the carcass. Finally, in rubbish dumps, both hosts and parasitoids are abundant and brother-sister mating is likely to be rare. Werren made collections of parasitized pupae from a variety of natural sites and argued that foundress number would be roughly proportional to the number of parasitized pupae in each collection. Excluding some sites where most hosts were superparasitised (recognizable by their large brood size), Werren found the most female-biased sex ratios occurred in sites with few foundresses, while in rubbish dumps the sex ratio approached equality.

Laboratory experiments manipulating foundress number in *Nasonia vitripennis* have also been conducted by King and Skinner (1991a). Their results are qualitatively similar to Werren's (1983) except that they found that the sex ratio became very male biased (>0.5) as foundress number increased. The reason for this difference probably lies in the details of the experimental method. As mentioned above, Werren allowed his wasps to leave the experimental arena at will; in contrast, King and Skinner deliberately confined wasps with their hosts. The advantage of Werren's method is that it prevents superparasitism (which causes male-biased sex ratios) although, as King and Skinner argue, it is possible that wasps leave the experiment without oviposition so that real foundress numbers are smaller than expected. In the same study, King and Skinner manipulated foundress number in a closely related species, *Nasonia giraulti*, which has a similar natural history. They hypothesized that sex ratios in this species should be less female biased because, unlike *N. vitripennis*, male *N. giraulti* are fully winged and thus likely to disperse farther and mate with unrelated females. In fact, they found the reverse. *N. giraulti* produces a more female-biased sex ratio than *N. vitripennis*, although it too lays relatively more males as foundress number increases. Some possible explanations for this surprising result are listed by King and Skinner (1991a). Another species with a similar natural history is *Pteromalus puparum*, a pupal parasitoid of a variety of butterflies. Males are fully winged, but most mating appears to occur on the host. Takagi (1986) found that sex ratio increased with foundress number in a laboratory experiment where one host pupa was enclosed with one to ten parasitoids. One problem in interpreting this experiment is to disentangle the influence on sex ratio of the increased superparasitism and possible differential mortality that will inevitably accompany high foundress number in this experimental design.

Waage and Ng (1984) and Waage and Lane (1984) studied LMC in the trichogrammatid egg parasitoid, *Trichogramma evanescens*. This wasp attacks the eggs of many Lepidoptera, including species that lay solitary eggs and species that lay their eggs in clumps. Typically, the wasp places a clutch of one to four eggs in each host egg. Waage and Ng compared the sex ratio produced by indvidual wasps ovipositing on single eggs and on small clumps of eggs (the host was the cabbage moth, *Mamestra brassicae*). Adult wasps mate on the host eggs immediately after emergence, suggesting that a female ovipositing by herself should lay just enough male eggs to ensure all her daughters are inseminated. Solitary females attacking a single-host egg normally laid one male egg and two or three female. If the female was presented with small clumps of eggs, she produced a highly female-biased sex ratio with just one or a few males. In a second experiment, Waage and Lane placed between one and twelve wasps on a patch of twelve host eggs and recorded the sex ratio of the progeny emerging from the patch. The sex ratio became progressively more male biased as foundress number increased (fig. 4.5), in qualitative agreement with LMC theory, although at all wasp densities the sex ratio was significantly more male biased than predicted by Hamilton's model. Waage and Lane suggested that part of the reason for this discrepancy might be superparasitism that leads to differential mortality of females. Similar results have been obtained in experiments with the scelionid egg parasitoids *Telenomus remus* (van Welzen and Waage 1987) and *T. heliothidis* (Strand 1988). The ideas of Waage, Strand, and colleagues about the behavioral mechanisms that generate the response to foundress number are discussed in section 4.3.4.

Some of the best examples of experimental tests of LMC come from pollinating fig wasps (see sec. 1.3.3 for a description of fig wasp biology). Hamilton (1967) included one species of fig wasp in his list of insects displaying sibmating, haplodiploidy, and female-biased sex ratios. In 1975 he studied the

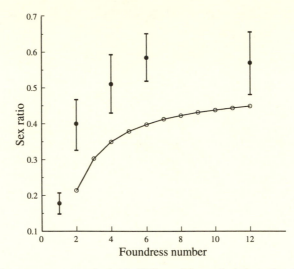

Figure 4.5 The average sex ratio (with 95% confidence limits) produced by different numbers of *Trichogramma evanescens* (foundresses) ovipositing into host egg masses containing twelve eggs in the laboratory (from Waage and Lane 1984). The predicted sex ratio from LMC theory (Hamilton 1979) is also shown.

biology of fig wasps inhabiting two species of fig in Brazil (Hamilton 1979). After a female wasp enters and oviposits in a fig, it dies but does not completely decompose. By dissecting figs at the right time, it is possible to count the number of foundresses and to measure the sex ratio of their progeny. Hamilton compared foundress number and sex ratio in one of the pollinating fig wasp species he studied. While sex ratio increased with foundress number, it remained substantially more female biased than predicted. Hamilton suggested unequal brood size and fluctuating selection pressures (see table 4.1) as possible explanations for the poor fit.

Herre (1985) used Hamilton's technique to assess foundress number and sex ratio in three species of fig wasp in Panama. He also surveyed the natural distribution of foundress numbers and calculated the average level of inbreeding for each of the three species of wasp. In all three species, sex ratio increased with foundress number, as predicted by Hamilton's theory. Herre also found that in figs colonized by the same number of foundresses, more inbred species tended to have more female-biased sex ratios (fig. 4.6). Thus local mate competition and inbreeding both influence observed sex ratios. In a later paper, Herre (1987) extended his study to thirteen species of fig wasp from Panama. Again, sex ratio increased with foundress number in all species and, in figs colonized by the same number of foundresses, more inbred species produced more female-biased sex ratios. However, deviations from the predicted optimum sex ratio were not random. A species was less likely to pro-

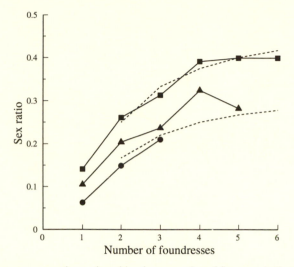

Figure 4.6 Average sex ratio produced by three species of fig wasp in figs colonized by different numbers of foundresses. The three species differ in their average level of population inbreeding. The proportion of sibmating in *Tetrapus costaricensis* (■) is 0.50, in *Blastophaga* sp. 1 (▲) is 0.71, and in *Blastophaga* sp. 2 (●) is 0.88. Higher average levels of inbreeding are associated with more female-biased sex ratios for the same foundress numbers. The upper dotted line is the theoretically predicted sex ratio when the average level of sibmating in the population approaches 0%, and the lower line is the sex ratio when sibmating approaches 100%. (From Herre 1985.)

duce the "correct" sex ratio in a situation it encountered rarely in the field (fig. 4.7). This makes sense as natural selection will be weaker and so adaptation poorer in circumstances that occur only occasionally. This is a very important, and almost unique, quantitative demonstration of the limitations of adaptation. Herre's two papers on fig wasps rank as some of the most elegant studies in the whole field of behavioral ecology.

Finally, Frank (1985) also studied fig wasps, this time in Florida. Unlike Hamilton and Herre, he experimentally manipulated foundress number by placing wasps on young figs and watching them enter. He too found that sex ratio increased with foundress number, but that it was considerably more female biased than predicted by theory, even after allowing for variable foundress numbers (fig. 4.8). Frank suggested a number of factors that might contribute to the extra female bias (see table 4.1). One possible explanation is that foundresses are not picked at random from the population but tend to be genetically similar, perhaps because of temporal or spatial genetic subdivisions within the population. If genetic similarity among foundresses is important, then it is possible that fig wasps might produce a more female-biased sex ratio if they found themselves in a fig with closely related foundresses. Frank tested

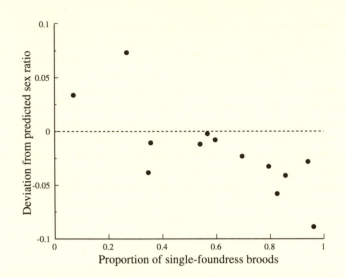

Figure 4.7 The difference between the observed sex ratio of thirteen species of fig wasp in figs with two foundresses and the prediction from theory. The differences are plotted against the proportion of single-foundress broods found in the field. Sex ratios are consistently more female biased then expected in species where the majority of figs are colonized by single foundresses, and thus two-foundress figs are rare. Species that rarely experience single-foundress conditions tend to deviate in the direction of more males. (From Herre 1987.)

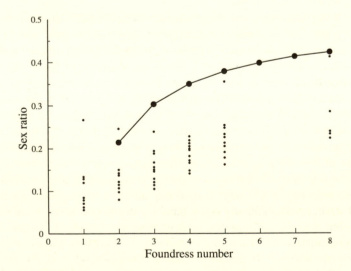

Figure 4.8 The sex ratios produced by the fig wasp *Pegoscapus assuetus* in figs containing different numbers of foundress (from Frank 1985). The theoretical prediction from Hamilton (1979) is shown as a solid line.

this hypothesis by comparing the sex ratio produced in figs colonized by related and unrelated foundresses but found no significant differences. Another possibly important explanation is variation in brood size by different foundresses. If certain foundresses contribute a disproportionate share of offspring to the mating pool, the effective number of foundresses is reduced and a more female-biased sex ratio is predicted (Werren 1980; Frank 1985; Herre 1985).

The most important factor that complicates experimental studies of LMC is the possibility that differential mortality affecting females also increases with foundress number. As more females oviposit together on a patch, clutch size or superparasitism may increase, leading to greater competition among larvae for host resources. This is particularly likely to occur if foundress number is manipulated while the number or size of hosts in the patch remains constant. There is evidence that increased larval competition typically leads to an increase in female deaths (Wilkes 1963; Wylie 1966; Benson 1973; Suzuki et al. 1984; see also sec. 4.5.1). Of course, the absence of a response to foundress number in a species with a female-biased sex ratio does not necessarily mean that LMC is not responsible (Werren 1987a). The wasp may be unable to assess foundress number, or foundress number may be sufficiently constant in nature that selection for a facultative response has never existed. Nevertheless, mating between siblings is probably too rare in the majority of parasitoids for LMC to have a detectable influence on the sex ratio. Van Dijken et al. (1989) provide an example of an outbred species, the encyrtid parasitoid *Epidinocarsis lopezi*, which shows no response to foundress number in laboratory experiments (see also King 1989b).

I finish this section by describing an interesting but puzzling experiment by Galloway and Grant (1989) on the gregarious braconid *Bracon hebetor*. This experiment provides a counter example to the rule that more male-biased sex ratios are associated with increased foundress number. Galloway and Grant argued that *B. hebetor* is normally an outbred species, an argument supported by Antolin and Strand's (1992) recent observations on mating in this wasp (see sec. 7.2.1). Even though outbred, the wasp usually produces a sex ratio of about 0.4, a female bias that lacks an explanation. Galloway and Grant placed either one or two *B. hebetor* females with a host caterpillar and recorded the sex ratio of the emerging wasps. Two different eye-color strains were used to enable the sex ratio of individual females to be measured. One strain produced the same sex ratio (c.0.4) when ovipositing alone or when one of a pair. The second strain produced nearly equal numbers of the two sexes when isolated (0.485) but a significantly more female-biased sex ratio (0.442) when paired. The approximately 50:50 sex ratio of this strain is probably a result of the loss of all but two alleles at the sex determination locus (see sec. 4.1). An initial 2:1 ratio in favor of diploid offspring would be reduced to a 1:1 sex ratio by the death of all diploid males.

One consequence of two females attacking the same host is increased larval competition. In this species, intense larval competition often leads to very small or dwarf individuals. Grosch (1948) studied dwarfism in *B. hebetor* and found that while dwarf males survived more often than dwarf females, they were unable to mate with normal-sized females and thus had zero fitness. Galloway and Grant (1989) suggested that the avoidance of dwarf males may provide an explanation for the increased female-biased sex ratio when two wasps attack the same host.

These results are interesting but difficult to interpret. The statistics used (a *G* test on the total eggs laid by all solitary and paired females of one strain) may overestimate the significance of the sex ratio bias because the sex ratio produced by a parent, rather than the sex of each egg, is the basic statistically independent unit. The high rate of diploid male mortality may also obscure other sex ratio patterns. Galloway and Grant used a very large and unusual host species for *B. hebetor* in their experiments in order to reduce competition and minimize differential mortality. Similar experiments with other strains of *B. hebetor* on hosts of normal size failed to show a sex ratio shift with foundress numbers, or to produce reproductively impaired dwarf males (Cook et al. 1993).

SEX RATIO AND SUPERPARASITISM

Superparasitism by gregarious parasitoids has proved to be a valuable tool in investigating local mate competition. The progeny of a gregarious brood of parasitoids frequently mate among themselves at the pupation site. If the gregarious brood has a single mother, then Hamilton's model with a single foundress applies. If superparasitism has occurred, this situation is similar to Hamilton's model with two foundresses, though with the difference that the two females oviposit sequentially rather than simultaneously. Hamilton (1967) considered some of the sex ratio complications that arise with superparasitism in what was the first explicit application of the theory of games to an evolutionary problem. However, he assumed that the clutch sizes produced by the two wasps were the same, whereas normally (see sec. 3.3.2), the superparasitizing wasp lays fewer eggs.

The optimal sex ratio for a superparasitizing wasp, with variable clutch sizes, was found independently by Werren (1980) and Suzuki and Iwasa (1980). Werren derived a formula that predicted the optimal sex ratio given the sizes of clutch produced by the two wasps, while Suzuki and Iwasa simultaneously solved for optimal clutch size and sex ratio. In most cases, the second wasp lays a smaller clutch of eggs and produces a more male-biased sex ratio. The optimal sex ratio of the second wasp is quite sensitive to the ratio of the two clutch sizes, and is also influenced, though less strongly, by the sex ratio of the first clutch. Early observations on the sex ratio of superparasitizing wasps supported this hypothesis. Wylie (1965, 1966), Walker (1967), and Holmes

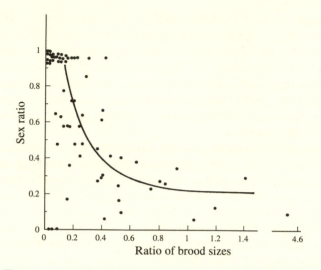

Figure 4.9 The sex ratio produced by superparasitizing *Nasonia vitripennis* as a function of the relative sizes of the first and second clutches (for example, a value of 0.2 indicates that the second clutch is one-fifth the size of the first). The line is the theoretical prediction assuming that the sex ratio of the first clutch is consistently 0.87, the overall average. (From Werren 1980.)

(1972), working with the pteromalid *Nasonia vitripennis*, Shiga and Nakanishi (1968), working with the ichneumonid *Gregopimpla himalayensis*, and Jackson (1966) in a study of the mymarid *Caraphractus cinctus* all observed the production of relatively male-biased clutches during superparasitism.

The first quantitative test of theory was by Werren (1980) using *Nasonia vitripennis*. Genetic eye-color mutants were used to identify the progeny of primary and secondary wasps, and clutch size and sex ratio were recorded. When the second wasp laid relatively few eggs, it produced a very male-biased sex ratio. As the relative size of the second clutch increased, the sex ratio became more female-biased (fig. 4.9; see also Werren 1984b). Although sex ratios were very variable, there was a strong trend in the data in agreement with theoretical predictions. In a separate set of experiments, Werren (1984b) found that secondary wasps produced the same sex ratio in hosts previously attacked by virgin females. There thus seems to be no effect of the sex ratio of the first clutch on the sex ratio laid by the second female. Suzuki et al. (1984) also compared the sex ratio of primary and secondary parasitoids using the trichogrammatid wasp *Trichogramma chilonis*. In contrast to Werren's results they found no significant difference between the clutch size or sex ratio in primary or secondary clutches.

The optimal sex ratio of an egg parasitoid attacking a clump of host eggs, some of which have been previously parasitized, is formally identical to that of LMC with superparasitism if all matings take place around the exhausted hosts. This phenomenon has been investigated in two egg parasitoids: the trichogrammatid *Trichogramma evanescens* (Waage and Lane 1984), and the scelionid *Telenomus remus* (van Welzen and Waage 1987). Wasps were placed on a group of host eggs, some of which had been previously parasitized by another individual. Theory predicts that a more male-biased sex ratio should be produced as the number of eggs laid by the second wasp declines relative to the numbers laid by the first. In *T. evanescens*, the sex ratio produced by the second wasp was variable but it was more male-biased, with relatively small second clutches (see analysis in Orzack 1990). With *T. remus* the sex ratio of primary and secondary females was indistinguishable, except that when only a small number of hosts were left unparasitized the second wasp produced proportionately more males. Experiments with egg parasitoids thus provide some support for the predictions of theory.

Orzack and Parker (1986, 1990; Orzack 1986, 1990) have also studied the sex ratio of *Nasonia vitripennis* in parasitized and unparasitized hosts. The aim of their work was threefold: to test the predictions of local mate competition theory, to assess genetic variation in the sex ratio, and to use the system as a model with which to evaluate the biological and philosophical foundation of the behavioral ecological research program. Their first set of experiments, reported in the 1986 papers, used laboratory strains of *N. vitripennis*, some maintained in culture for very many generations. Their second set of experiments, reported in the 1990 papers, used isofemale lines derived from females recently collected in birds' nests in Sweden.

The results of most of these experiments agree well with the previous work of Werren (1980). The sex ratio produced by most *N. vitripennis* strains is more male biased on previously parasitized hosts than on fresh hosts, and there is normally a correlation between sex ratio and the relative size of the second brood. An example of one of Orzack's (1986) results is shown in fig. 4.10. Note that compared with Werren (fig. 4.9), the relative sizes of the second brood are larger and often exceed that of the first brood. The reason for this difference is probably experimental technique: Werren allowed his wasps to remain with their hosts until they left of their own accord while Orzack removed the wasps after a fixed time. As a result, the first wasps in Orzack's experiments sometimes laid an unnaturally small clutch. Both Werren (1984b) and Orzack (1986) have suggested that wasps that accept a host containing very few eggs for oviposition may treat it as if unparasitized. In the first set of Orzack and Parker's experiments, one strain (MI) produced similar, very female-biased sex ratios on parasitized and unparasitized hosts. However, this appears to be the same strain that other workers have shown to possess a maternally inherited factor that causes a female-biased sex ratio (*msr*, see sec. 5.1), and this observation must be interpreted with caution.

Orzack (1990) suggests a number of interesting ways to compare the predictions of LMC theory with the observed data. In one set of analyses, he fits a

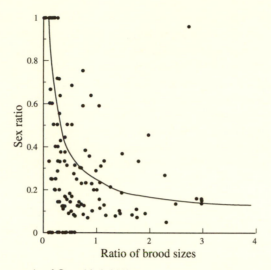

Figure 4.10 An example of Orzack's (1986) experiments with *Nasonia vitripennis* (his fig. 1b). The results are plotted in the same way as for Werren's experiment (see fig. 4.9); the theoretical prediction assumes a consistent sex ratio in the first clutch of 0.82.

generalized linear model (McCullogh and Nelder 1983) with logistic regression to the data (using the GLIM statistical package). The advantage of this procedure is that the predictions of LMC theory can be included as an explanatory variable in the analysis, and each observation can be accorded a different weight. In nearly all analyses, incorporation of the LMC predictions markedly improved the explanatory power of the model. However, as Orzack repeatedly stresses, in no case does the model precisely predict the exact sex ratio of individual wasps. Orzack explores the effects of a number of different weighting procedures which allow points that have the largest effect on fitness to have greatest influence in determining the parameters of the statistical model. In this case, however, the weights make little difference to the explanatory power of the model. If data were available on the distribution of primary clutch sizes in the field, it would be possible to give added weight to data points representing situations that the wasp is likely to encounter frequently. As Herre (1987) has argued, the predictions of optimality theory are more likely to be met in circumstances that are often experienced by the animal.

In nearly all comparisons of different isofemale strains, Orzack and Parker found significant between-strain variation in both the first and second sex ratios. In many cases there was also significant within-strain variation in sex ratio. Much of this between-strain variation is likely to be genetic, and if the laboratory results reflect what is happening in nature, and if differences in sex ratio strategy are not correlated with differences in other traits, these observations imply heterogeneity in fitness within the population.

Orzack (1986, 1990) uses his results to mount an angry attack on modern behavioral ecology. Phenotypic modeling of problems such as this, based on optimality theory, predicts that the population is monomorphic with each individual possessing the optimal phenotype—yet this assumption is contradicted by the observed genetic heterogeneity in sex ratio. The theory purports to predict the behavior of individuals, yet can do so only approximately, and tells us nothing about behavioral variability. However, few behavioral ecologists would argue that the precise assumptions underlying optimality arguments are ever met or need be met, or that the *precise* predictions of theory will ever be observed in nature. Indeed, given the inherent variability in most biological systems, and the number of intervening variables affecting character evolution, it is rather unrealistic to expect precise quantitative fits to evolutionary models. The role of optimality theory is to generate hypotheses about how natural selection molds phenotypes. It has been particularly successful in the case of LMC and superparasitism: we have an explanation for the relatively more male-biased sex ratios of superparasitizing wasps, and the correlation between sex ratio and brood size, where before we had none. Orzack (1990) accepts that the "theory is a success in a more general sense" but because its fundamental assumptions are only approximately correct, "the role [of] theory is problematic." I believe Orzack and Parker's work is important, not because it challenges the fundamental tenets of behavioral ecology, which I think they misunderstand, but because it provides one of the few experimental investigations into genetic variation in the sex ratio. To explain the origin and maintenance of this variation we shall require further experiments and further theoretical analysis (building on work such as Uyenoyama and Bengtsson 1981, 1982, and Karlin and Lessard 1986)—but for an understanding of phenotypic patterns in parasitoid sex ratios the behavioral ecological approach has yet to be bettered.

SEX RATIO IN POLYEMBRYONIC WASPS

Polyembryonic wasps (see sec. 1.3.1 for an introduction to their biology) are able to produce large broods of one sex by the asexual division of a single egg. However, mixed broods, usually with a female-biased sex ratio, are common in many species. Despite early controversy, there is now a strong consensus that mixed broods occur only if both a male and a female egg are laid in the same host. Why are mixed broods female biased, and is the sex ratio controlled by the parent or the offspring?

Only one species has been sufficiently well studied that answers to these questions can be attempted. In a series of papers, Strand (1989a, 1989b, 1989c, 1993) investigated the reproductive strategy of the encyrtid wasp *Copidosoma floridanum*, an egg-larval parasitoid of plusiine moths (Noctuidae). In the field, about 57% of broods are mixed, but with a strongly female-biased sex ratio. In the laboratory, Strand showed that female wasps either laid one or two eggs into each host egg. Single eggs might be of either sex, but when two eggs were placed in a host, a female egg was always followed by a male egg. Single-sex and mixed broods of *C. floridanum* are of the same size, approximately 1200 wasps, but in mixed broods only about fifty males develop. There are at least two possible explanations for the reduced number of males in mixed broods: males may suffer higher mortality in mixed broods than females, or the asexual division of male eggs may be reduced in the presence of females. Detailed serial dissections of single-sex and mixed broods showed there was insufficient immature mortality for male deaths to account for the observed sex ratio biases. By carefully manipulating the behavior of individual wasps, Strand (1989b) increased the number of eggs laid into a host from one or two to a maximum of eight. However, the number of wasps emerging from a host was not influenced by the initial clutch size, though the sex ratio was a little more male biased in broods initiated by four or more eggs.

Wasps emerging from mixed broods frequently mate among themselves around the remains of their host (Strand 1989c). Local mate competition is thus a likely explanation for the female-biased sex ratio in mixed broods. Although it is possible that the ovipositing wasp controls the rate at which her male and female eggs divide, it is far more likely that the extent of division is controlled by the eggs themselves. The male eggs constitute a single genetic unit and will be selected to maximise the total number of females inseminated by male brood members. If most matings take place among siblings in mixed broods, the male egg should divide until enough males are produced to fertilize all the females in the brood. Because one male will certainly be able to fertilize many females, a very female-biased sex ratio results . However, many single-sex broods are produced in the field, and there may be selection on the male egg to divide further to produce individuals that can seek out broods of females. A further complication is that males and females share the same host, which is essentially a fixed-value resource; more males will mean more competition among the brood and this will affect the number or fitness of the female brood members which are both sisters and potential mates. Selection will also influence the number of individuals produced by the female egg, and, a final twist, the greater relatedness of brothers to sisters compared with sisters to brothers introduces an asymmetry into the evolution of sibling competition.[4]

---

[4] Note added in press: Grbic et al. (1992) have recently shown that the largely female precocious larvae found in the encyrtid *Copidosoma floridanum* attack and kill male eggs in hosts containing mixed broods. If males can obtain at least some matings with nonsibling females developing in other hosts, there will be sibling conflict over the sizes of male and female clutches in the host. They speculate that because hymenopteran sisters are less closely related to their brothers than males are to their sisters (i.e., the coefficient of relatedness between different-sexed siblings is asymmetric in haplodiploids), selection may favor siblicide by females but not by brothers. They also suggest that the reason male eggs develop in the abdominal fat body is to obtain some protection from female precocious larvae.

4.3.4 ACHIEVING THE ESS SEX RATIO

P. E. King (1961, 1963) suggested that the sex ratio of *Nasonia vitripennis* (approximately 33% males) could be explained simply in terms of the morphology of the reproductive tract. If the egg traveled down the oviduct with a random orientation, one-third of the time the micropyle through which the sperm enters the egg will be placed away from the mouth of the spermathecal duct, and so the egg will fail to get fertilized. Recent work, described in the last section, suggests that the sex ratio strategy of this species is more complicated, but King's suggestion is an example of a proximate mechanism that allows a wasp to produce a particular sex ratio. In this section I first discuss the degree of sex ratio precision achieved by parasitoids and then the behavioral mechanisms involved in producing precise sex ratios and in assessing foundress number.

PRECISE SEX RATIOS

Consider a wasp subject to LMC and selected to produce a sex ratio of 0.25. If it lays exactly one male for every three females, it is said to produce a precise sex ratio. Alternatively, it could average a sex ratio of 0.25 by laying a series of eggs, each of which is male with probability one in four. In the second case, some clutches of eggs will have more males than expected and some less, and the chances of different outcomes will be determined by the binomial probability distribution.

There may be costs to producing sex ratios with binomial variance, and these costs may influence the optimum sex ratio. The first explicit treatment of this problem was by Hartl (1971). Consider the case where just a single foundress colonizes a patch. Theory predicts that the wasp should produce just enough sons to inseminate her daughters. In a clutch of ten eggs, a wasp capable of producing a precise sex ratio might lay just a single male and nine females. But what about a wasp constrained to produce a sex ratio with binomial variance? Should she aim for an average sex ratio of 0.1? If she does, a fraction 0.35 (i.e., 0.9 to the tenth power) of her broods will contain no sons and her daughters will remain unmated. Hartl argued that such a wasp should lay more males than required to inseminate her daughters in order to guard against the possibility of female-only clutches: it is much more disastrous to produce an all female brood than one with slightly too many males. A similar argument applies, although less strongly, if foundress number is larger than one: sex ratios above Hamilton's prediction will be favored if the female is unable to produce a precise sex ratio.

Where LMC occurs, natural selection will favor the production of sex ratios with less than binomial variance. However, even if a wasp has precise control over the primary sex ratio, mortality among her offspring may affect the secondary sex ratio. Thus continuing the example of the last paragraph, if the wasp were always able to ensure she laid one son, her daughters would still be unmated if that son happended to die. Green (1982) have modeled the optimum sex ratio in species with LMC where there is a risk of offspring mortality. If whole broods tend to die together, the optimum sex ratio is unchanged, but if mortality affects individual brood members, selection favors the production of more sons.

Although many earlier workers had implicitly assumed that wasps were able to produce sex ratios with less than binomial variance (e.g., Clausen 1940a; Hamilton 1967), the first explicit demonstration of precise sex ratios was by Green et al. (1983) in two species of *Goniozus* (Bethylidae). Sex ratios with less than binomial variance have now been shown to occur in many wasp species and seem to be the rule in groups such as scelionids, trichogrammatids, and bethylids (reviewed by Hardy 1992). One exception is the braconid wasp *Asobara persimilis*, studied by Owen (1983). This *Drosophila* parasitoid seems to experience little LMC in the field and produces equal numbers of sons and daughters. However, if brought into the laboratory and individual females are caged with hosts, it switches to a female-biased sex ratio of about 0.25. Of relevance to the present argument, the variance in the brood sex ratios was significantly higher than binomial. Owen speculates that precise sex ratios might not evolve in a species that only infrequently encounters LMC. The significance, if any, of the suprabinomial variation in the sex ratio is unclear.

SEX RATIO AND EGG SEQUENCE

Many parasitoids achieve precise sex ratios by laying male and female eggs in a fixed sequence (reviewed by Hardy 1992). For example, the scelionid egg parasitoid *Gryon pennsylvanicum (=atriscapus)* typically lays a male egg first in the ovipositional bout, followed by a long series of females; another male may be laid later in the sequence on large clumps of host eggs (fig. 4.11; Waage 1982a). Males are also laid early in the ovipositional bout by a number of other scelionid and trichogrammatid egg parasitoids (Hokyo et al. 1966; Waage and Ng 1984; Waage and Lane 1984; Suzuki et al. 1984; van Dijken and Waage 1987; van Welzen and Waage 1987; Strand 1988; Noda and Hirose 1989; Braman and Yeargan 1989). The early production of males may be a good strategy for small egg parasitoids if they are unable to assess the size of the egg mass prior to oviposition because it ensures that they never run out of host eggs to parasitize before they have laid at least one son to inseminate their daughters. This strategy may also be advantageous for gregarious parasitoids because if the wasp is disturbed before oviposition is completed, the host will already contain males that can inseminate the females in the clutch.

While Suzuki et al. (1984) also found that *Trichogramma chilonis* laid males early in the oviposition sequence, the first egg was nearly always female and the second male (this is consistent with earlier observations on *Trichogramma* spp. by Flanders 1935 and E. P. Jones 1937, and also by Jackson 1966

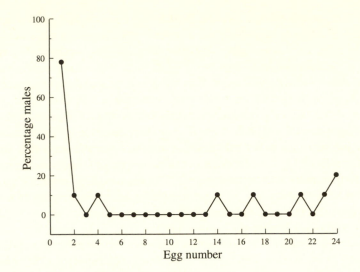

Figure 4.11 The sequence in which the scelionid *Gryon pennsylvanicum* lays male and female eggs when attacking host egg masses of twenty-four eggs. (From Waage 1982a.)

on the mymarid *Caraphractus cinctus*). After producing a son, the wasp typically laid a long string of females, often with more sons every eight or so eggs. Suzuki et al. suggested that the strategy of producing a female followed by a male was adaptive for *T. chilonis* because of this species' large host range, which includes both solitary host eggs that can just support a single wasp and host eggs laid in large masses. Mating almost invariably occurs around the host egg mass so that individuals developing by themselves in solitary host eggs are destined never to mate. This may be more serious for males than females: an unmated female can still produce sons, and if she lays eggs on a host egg mass colonized by other females, her sons will be able to compete for mates. Laying a female first ensures that any progeny developing alone are daughters. The scelionid *Gryon japonicum* also has a strong tendency to lay a female egg first followed by a male egg (Noda and Hirose 1989). This is surprising, since the host of *G. japonicum* (the bean bug *Riptortus clavatus*) typically lays solitary eggs and this is not a species in which one would expect LMC to be important (Waage 1982a). Noda and Hirose (1989) found that a delay of three hours between host attacks was sufficient to reset the sequence of son and daughter production.

The opposite strategy, the production of males late in the oviposition sequence, has been recorded from a number of gregarious bethylids (Powell 1938; Finlayson 1951; Mertins 1980, 1985), from the gregarious eulophid *Colpoclypeus florus* (Dijkstra 1986) and from the gregarious pteromalid *Dinarmus vagabundus* (Rojas-Rousse et al. 1988). For example, Mertins (1985) noted in a study of the bethylid *Laelius utilis*, a parasitoid of the dermestid beetle *Anthre-*

nus fuscus, that "almost invariably, only the last egg is male." In these species, all eggs are laid in a single host, and the female wasp can estimate the resources available for her offspring before beginning to lay eggs. Yet another pattern has been discovered in the eulophid *Nesolynx albiclavus*, a parasitoid of the tsetse fly (*Glossina* spp.). This wasp lays gregarious clutches of up to sixty eggs in the puparium of its host. Putters and van den Assem (1985) observed that the wasp produced a highly female-biased sex ratio with much less than binomial variance. Though they were not able to characterize precisely the sequence of sex allocation, it appeared that the wasp first laid five females followed by a male. After the first male more females were produced with further males later in the sequence. The sex allocation sequence was reset by a break of twelve hours in oviposition. Whether these nonrandom oviposition sequences in gregarious parasitoids have any adaptive value beyond their ability to provide precise sex ratios is not yet clear.

Finally, the sequence of son and daughter production may vary in different circumstances. Werren (1984b) has preliminary data suggesting that when the pteromalid *Nasonia vitripennis* attacks an unparasitized host it tends to lay male eggs late in the oviposition sequence, but that when it attacks parasitized hosts the males are produced at a constant rate or near the beginning of the sequence.

ASSESSMENT OF FOUNDRESS NUMBER

Females may assess the number of foundresses present in a patch by the frequency of contacts with other searching parasitoids, or by the frequency of contacts with parasitized hosts or marks left by other females. Foundress number may also influence the sex ratio produced by an individual female by a more indirect root. When many foundresses are present, there will be greater competition for hosts and each female is likely to lay fewer eggs. If females tend to produce male eggs early in an oviposition sequence, a reduction in the number of eggs laid will result in a more male-biased sex ratio.

Viktorov and Kochetova (1973a) found that female *Trissolcus grandis* (Scelionidae) produced a more male-biased sex ratio after being stored in vials that had previously contained other wasps. Chemical traces of the previous occupants were presumably responsible for the change in sex ratio. Strand (1988) observed a related phenomenon in the scelionid egg parasitoid *Telenomus heliothidis*. Newly emerged wasps produced a sex ratio of about 0.23 when presented with ten consecutive hosts. Wasps kept in large containers with many females and then presented with hosts produced a more male-biased sex ratio. The bias increased with the length of time that the wasps were stored together, until after fifteen days the sex ratio remained constant at about 0.5. However, if a wasp was removed from the large container and placed in isolation for a number of days before being given hosts, the sex ratio again was about 0.23. A likely explanation for these results is that the wasps are using contacts with other females as an index of the number of foundresses likely to colonize a patch. Contacts with other females have no effect on the sex ratio in the pteromalid *Spalangia cameroni* (King 1989a).

There is little evidence that females use the number of encounters with parasitized hosts to assess foundress number. In his experiments with *T. heliothidis*, Strand (1988) found that alternating parasitized with unparasitized hosts had no effect on the sex ratio produced in the latter. Similarly, Waage and Lane (1984) found that contacts with parasitized hosts appeared not to influence the sex ratio of *Trichogramma evanescens*. In this species as well, wasps did not to alter their sex ratio after contact with conspecifics.

One important consequence of the presence of multiple foundresses is that the number of hosts parasitized by each wasp will normally drop. If the wasps always produce males early in their oviposition sequence, a reduction in the number of eggs laid will automatically increase the sex ratio. Waage and Lane (1984) and Strand (1988) illustrated the workings of this mechanism in *Trichogramma evanescens* and *Telenomus heliothidis*. In both wasps, sex ratio increased with foundress number, in broad agreement with Hamilton's model. To assess the effect of the reduced number of hosts attacked, they then confined individual wasps with ½, ⅓, ¼, etc., the number of hosts used per patch in the main experiment (fig. 4.12). They discovered that the reduction in the number of hosts attacked could largely explain the observed changes in sex ratio with foundress number. It should be stressed that these observations do not provide an alternative explanation to Hamilton's LMC theory, but suggest a proximate mechanism that allows the wasp to approximate the optimal sex ratio. Indeed, the strategy of placing males early in the oviposition sequence may partly have arisen because of its adaptive value in the presence of other foundresses. How well a fixed sequence of male and female egg production approximates the LMC sex ratio will depend quite critically on the number of hosts available for parasitism and how competition affects the number of eggs laid. In another species of scelionid, *Telenomus remus*, van Welzen and Waage (1987) discovered that although part of the response to increased foundress number could be explained purely by sex allocation sequence, there was in addition a switch to the production of more males independent of oviposition sequence. The proximate cue responsible for the shift toward males was not definitely identified, but was probably a response to chemical traces left by other wasps.

Gregarious parasitoids that lay smaller, more male-biased clutches when superparasitizing must of course respond to signs of previous parasitism. It would seem likely that the same type of cues used in the detection of superparasitism (see sec. 3.3.3) are used as a signal to produce a more male-biased sex ratio. However, King and Skinner (1991b) found evidence that different cues may be employed in clutch size and sex ratio decisions (see also Werren 1984b). Female *Nasonia vitripennis* always reduced their clutch size on previ-

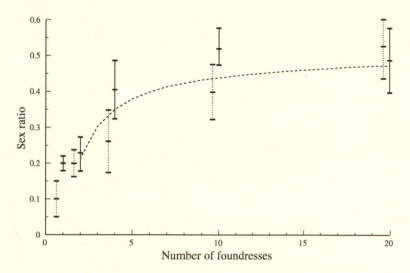

Figure 4.12 Sex ratio and foundress number in the scelionid wasp *Telenomus heliothidis* (from Strand 1988). The solid bars show the average sex ratios (with 95% confidence limits) produced by different numbers of foundresses confined on patches of twenty host eggs. The dashed bars show the sex ratios produced by individual females confined with the average number of hosts they might expect to parasitize when searching with different numbers of foundresses. Thus when two foundresses attack twenty hosts, each is expected to parasitize ten, and the dotted bar shows the sex ratio an individual wasp produces when attacking ten hosts. Note that the dotted bars are all displaced slightly to the left. The dotted line is the prediction from LMC theory (Hamilton 1979).

ously parasitized hosts, but a male-biased sex ratio was found only if the site of the second oviposition was near the first oviposition. The independence of the two reproductive decisions suggests that the sex ratio response is not merely due to the production of male eggs at the beginning of the oviposition sequence. Working with the same parasitoid, Wylie (1966, 1976) found evidence that contact between females may be a proximate cue leading to a more male-biased sex ratio during superparasitism, though this effect alone was insufficient to explain the observed sex ratio shifts.

4.3.5 OTHER FORMS OF SIBLING INTERACTION IN STRUCTURED POPULATIONS

The sex ratio biases caused by LMC can be interpreted as due to stronger competition among siblings of one sex—specifically competition among brothers for mates. Sex ratio biases may also occur if there is differential competition for other resources. Clark (1978) first suggested that resource competition influenced the sex ratio in a study of primates. In some species daughters remain near their natal site and compete to inherit their mothers' territories.

Sons, on the other hand, disperse widely and are likely to compete with unrelated males. As daughters compete among themselves to a greater extent than sons, the optimal sex ratio will be male biased. This phenomenon is known as "local resource competition" (Charnov 1982).

In parasitoids, it is possible that adult sisters compete with one another for access to hosts, a process that would bias the ESS sex ratio toward more males. However, a more likely source of sex ratio biases through local resource competition is sexual asymmetries among larval, gregarious parasitoids.

SEXUAL ASYMMETRIES IN LARVAL COMPETITION

Sexual asymmetries in larval competition have already been discussed in the context of selection on clutch size (sec. 3.2.1). Consider a gregarious parasitoid where the members of a clutch compete together for host resources and where this competition determines either the probability of survival until the adult stage or fitness as an adult. There are two types of sexual asymmetry in competition that may influence the sex ratio: (1) the fitness of males and females may be affected to different degrees by the strength of competition; (2) males and females may differentially affect the level of competition experienced by other members of the brood. It is the second of these asymmetries that can cause a sex ratio bias. If, for example, an increase in the number of males has a greater negative effect on the fitness of brood members in comparison with an increase in the number of females, there will be selection for a female-biased sex ratio (Suzuki and Iwasa 1980; Godfray 1986b). In calculating the optimum sex ratio, both types of asymmetry need to be taken into account. However, biased sex ratios will be found only if the second asymmetry is present; differential responses to competition, like differential mortality, cannot alone lead to a bias in the primary sex ratio.

There is some evidence in parasitoid wasps for the type of sexual asymmetries in competition that lead to sex ratio biases. Chacko (1969) compared the fitness of female *Trichogramma evanescens* developing in host eggs containing one other parasitoid, either a male or a female. Females reared with brothers were smaller, less fecund, and lived for a shorter period of time in comparison with females reared with sisters. This suggests that males cause a greater reduction in the fitness of clutch mates than females, a factor that would contribute toward a female-biased sex ratio. There is also evidence that males exert stronger competitive effects in the braconid wasp *Bracon hebetor*. Benson (1973) found that as clutch size increased, the brood sex ratio becomes relatively more male biased. It appears that males hatch first and are liable to increase female mortality by either damaging unhatched eggs or by preempting host resources. Rotary and Gerling (1973), working with the same species, suggested that male cannibalism of their sisters was responsible for the differential mortality. The faster development of males may explain a general tendency for females to suffer greater competition from brothers rather than from sisters.

SEGREGATED BROODS

As discussed in the last chapter, asymmetric competition between the sexes can result in selection for the segregation of the sexes in separate broods. There may also be differences in the size of male and female broods. If the reproductive success of the parasitoid is limited by the number of hosts it can locate, the optimum reproductive strategy is to produce equal numbers of male and female broods. The overall sex ratio will then be biased in favor of the sex produced in the largest broods. This hypothesis can be tested with data collected by Askew and Ruse (1974) and Bryan (1983) on the eulophid genus *Achrysocharoides* in Britain (table 3.3). A number of species produce single sex broods with female broods larger than male broods and with an overall female-biased sex ratio. Consider the four species that produce almost exclusively single-sex broods and where over one hundred broods have been reared. A χ^2 test of the hypothesis that male and female broods are produced with equal frequency is rejected for two species (*zwoelferi*, $\chi_1^2 = 4.70$; *niveipes*, $\chi_1^2 = 4.98$: 5% critical value 3.84) and accepted for two species (*latreillei*, $\chi_1^2 = 0.006$; *cilla*, $\chi_1^2 = 0.89$). The data are in partial agreement with the hypothesis though they fail to explain the excess number of female broods in *A. zwoelferi* and male broods in *A. niveipes*. Single-sex broods have also been found in the summer generation of another eulophid, *Eulophus larvarum* (Godfray and Shaw 1987). Both male and female brood sizes are very variable (no statistically significant difference in size) and the overall sex ratio is close to 50:50.

If broods tend to contain a predominance of either males or females, the sex ratio is termed "overdispersed." It is possible that the same factors that lead to segregated broods may also cause over-dispersion. Pickering (1980) observed this phenomenon in an ichneumonid wasp (*Pachysomoides stupidus*), which attacks social wasps in the genus *Polistes*. The larvae of this species are gregarious ectoparasitoids on host pupae, commencing feeding on the abdomen and working their way up to the head. As they feed, the group of larvae grows to fill the cross-sectional area of the host's cell, and this often leads to the physical exclusion of one or more larvae from access to the diminishing host resources. The excluded larvae pupate in the rear of the host cell and are more likely to be male. Pickering found that the average sex ratio of fifty broods in which no mortality had occurred was approximate 0.5 but that it was significantly overdispersed. He suggested that this was adaptive for the parent wasp, as an overdispersed sex ratio would increase the genetic relatedness of individuals within a brood and, in particular, prevent daughters from monopolizing resources that the parent would prefer to see used by sons (recall that in Hymenoptera sisters are more closely related to one another than to brothers). While an interesting idea, this suggestion remains far from proved. Further theoretical work is needed to show that such a conflict of interests between parents and daughters can lead to overdispersed sex ratios rather than to the complete segregation of the sexes. In addition, the interpretation of the experimental data is complicated by the fact that twenty nests where differential mortality occurred were not included in the analysis. If mortality occurred more frequently in nests with nearly equal numbers of the two sexes, an overdispersed sex ratio would be found in the remaining broods.

4.4 Sex Ratio and Host Quality

4.4.1 THEORY

Fisher's sex ratio argument implicitly assumes that all females in the population have the same amount of resources to invest in offspring production, and that the relationship between fitness and resource allocation is identical for sons and daughters. These assumptions were first relaxed by Trivers and Willard (1973) who considered the sex ratio strategy of mammals in either poor or good condition in circumstances where one sex gained more from extra investment than the other. They predicted that females in good condition, with ample resources for reproduction, should invest in the sex that benefits most from increased resources, while poor-quality females with limited resources should invest in the other sex. Charnov (1979) and Bull (1981) showed that the Trivers and Willard hypothesis can be applied to a wide range of biological situations where mothers vary in the resources available for reproduction, and one sex benefits more from increased investment.

Charnov et al. (1981) put forward a very convincing argument that this idea explains an old observation that solitary parasitoid wasps tend to lay male eggs in small hosts and female eggs in large hosts (e.g., Chewyreuv 1913; Brunson 1937; Clausen 1939). Here the resource available for reproduction is the host, which may vary in size. Charnov et al. argued that females would benefit more than males from developing in large hosts. In solitary wasps the size of the adult is normally strongly correlated with the size of the host in which it develops. The fecundity and hence fitness of female wasps is usually strongly correlated with adult size (sec. 7.1). Male wasps may also benefit from being large, but small size probably impairs mating less than it impairs oviposition. In addition, there is some evidence that females suffer higher mortality in small hosts than males (Kishi 1970; Sandlan 1979b).

Charnov et al.'s suggestion that females suffer more from being small is eminently reasonable although exceedingly difficult to prove. While there is much evidence that female longevity and fecundity drop with decreasing body size, it is much more difficult to assess male lifetime reproductive success (see sec. 7.13). A second problem is that lifetime reproductive success has normally been measured in the laboratory where conditions are more benign than in the field, almost certainly leading to the underestimation of the true costs of being small.

In addition to the qualitative prediction that males should be placed in small hosts and females in large hosts, models of sex allocation with variable host quality make a number of further predictions (Charnov et al. 1981). First, wasps should lay male eggs in all hosts smaller than a critical threshold and female eggs in all hosts above this size: the relationship between sex ratio and host size should be a step function. Second, relative rather than absolute host size is important. A wasp should lay a male egg in a medium-sized host if most other hosts in the environment are large, while the same-sized host should attract a female egg if the environment is full of small hosts. Finally, the primary sex ratio will be male biased (Frank and Swingland 1988; Bull and Charnov 1988; Charnov 1993), though if mortality in small hosts is greater than that in large hosts, the observed secondary sex ratio of adults may be near equality, or even female biased.

Some hosts are so small that the best policy is to ignore them completely. Green (1982) used a simple foraging model to study sex ratio and host acceptance. He assumed that parasitoids were time-limited and that they would not be selected to waste time laying an egg on very small hosts for low fitness returns. His model predicts two thresholds in host size: hosts smaller than the lower threshold are ignored; hosts of an intermediate size receive male eggs, and hosts larger than the upper threshold receive female eggs. The positions of the two thresholds are fixed relative to the local distribution of host sizes.

4.4.2 Tests of Theory

There is abundant evidence that many parasitoid wasps place male eggs in small hosts and female eggs in large hosts (Charnov 1982; King 1987). However, only in the case of an undescribed ichneumonid wasp in the genus *Dolichomitus*, a parasitoid of weevil pupae (Kishi 1970), is a step function even approached (fig 4.13). There are a number of reasons why we should not expect to find a perfect step function (Charnov et al. 1981). First, the value of the threshold may vary both spatially and temporally, and combining measurements taken from different places or at different times may lead to a blurred transition. Individual wasps may also differ in their estimate of the host size distribution, again leading to a smooth rather than a stepped transition. Finally, it is possible that the wasp and the experimenter may be ranking hosts using subtly different criteria; if the experimenter measures host weight and the wasp host length, a perfect step function when sex ratio is plotted against host length will be obscured when sex ratio is plotted against host weight.

An important prediction of Charnov's hypothesis, that sex allocation strategy should reflect the host size distribution, can be tested in several ways (Charnov 1982). First, populations of the same species attacking hosts with different size distributions can be compared. The two populations are predicted to evolve sex allocation strategies appropriate to their different hosts.

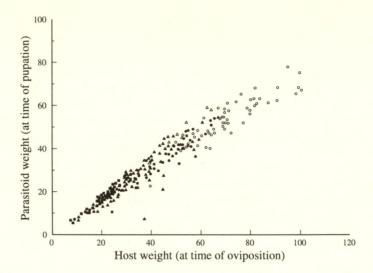

Figure 4.13 The weight of adult male and female *Dolichomitus* sp. (Ichneumonidae) emerging from hosts (pupal pine bark weevils, Curculionidae) of different weight and species. Solid symbols are male wasps, hollow symbols are female wasps; circles, *Niphades variegatus*; triangles, *Shirahoshizo* sp.; squares, *Pissodes nitidus*. From Kishi 1970.)

Second, individual wasps can be presented with different host size distributions. As long as the sex allocation strategy of the species is phenotypically plastic, the insects should respond by adjusting their sex ratio strategy. Examples of both tests are given below. Finally, should a parasitoid possess an inflexible sex allocation strategy, it might be possible to observe evolutionary change in the sex ratio if separate cultures were to be maintained on hosts with different size distributions. Such an experiment has yet to performed. In testing Charnov's hypothesis, it is important to exclude differential mortality of the two sexes on different-sized hosts as an explanation for observed sex ratio biases. Of course, as Charnov points out, such differential mortality would itself contribute to selection for host-size specific sex ratios.

A strong test of the host size hypothesis is to compare sex allocation strategies in populations of a single parasitoid species which experience different host size distributions. Hails (1989) studied three species of pteromalid wasps in the genus *Mesopolobus* which attack a range of cynipid gall wasps, of varying size, in Britain. The density of one of the smallest species, *Andricus quercuscalicis*, varies considerably between different sites. In some localities, this species is a very minor component of the gall fauna, while in other localities it constitutes the majority of available galls. Hails argued that if Charnov's hypothesis applied to this species, mostly males should be laid on *A. quercuscalicis* in sites where it was rare, but a more equal sex ratio in sites where it was very abundant. In fact, she found a constant, very highly male-biased sex ratio in all localities, for all three species of parasitoid. There are several possible explanations for this result. *A. quercuscalicis* has only relatively recently colonized the British Isles and the parasitoids may not have had time to adapt to this host species. Dissections of developing galls showed little immature wasp mortality although the very early differential mortality of female eggs, while unlikely, could not be excluded. The most likely explanation is that the wasps use a fixed rule of thumb in allocating sons and daughters to hosts of different size, and that this rule occasionally leads to very male-biased population sex ratios.

The second way to test Charnov's hypothesis is to manipulate the size distribution of hosts and observe whether the wasp adjusts its sex ratio accordingly. In their original paper, Charnov et al. (1981) reported the results of two experiments of this kind with parasitoids of stored-product pests. In both experiments, differential mortality was ruled out as an important factor biasing the results. The pteromalid wasp *Lariophagus distinguendus* attacks weevils (*Sitophilus granarius*) living inside cereal grains. Hosts measuring 1.4 mm received a sex ratio of 15% males when presented to the wasp alone. However, hosts of the same size produced a sex ratio of 30% males when presented alternately with larger hosts, but 2% males when presented alternately with smaller hosts (see also van den Assem 1971). The same effect was observed for a wide range of host weights, thus providing very strong evidence for the relative host size hypothesis (fig. 4.14). However, the prediction that in the absence of choice, hosts of all sizes should receive the same sex ratio (0.5) was not confirmed. Sex ratios were strongly male biased on small hosts and female biased on large hosts. Werren and Simbolotti (1989) suggested that *L. distinguendus* experiences local mate competition and developed a model which incorporated both LMC and host size effects (see also Werren 1984a). When local mate competition is strong (two foundresses per patch), the model predicts that females are exclusively placed on large hosts unless small hosts are very uncommon. Small hosts, when rare, receive only male eggs but as the percentage of small hosts increases, more and more female eggs are laid on hosts of this size. Werren and Simbolotti compared their predictions to data from experiments in which female *L. distinguendus* had been presented with different ratios of large and small hosts (Simbolotti et al. 1987). As the authors stress, the experiments were designed to investigate other questions, and while their results might be consistent with their model, they could not provide a critical test. The observed sex ratios on large and small hosts provided a good qualitative fit to their predictions if LMC was assumed to be high and if certain other were assumptions made about the relative fitness of sons and daughters and large and small hosts. However, the model incorporating LMC still predicted that the wasp should produce an identical sex ratio on hosts of different size when they are presented without choice. One possible explanation is that while LMC is important, some males disperse from the patch to mate with

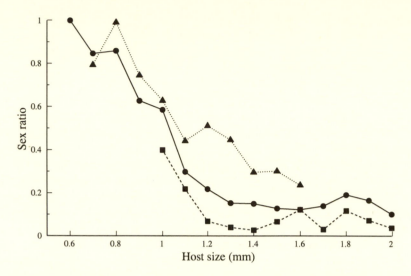

Figure 4.14 Sex ratio and host size in the pteromalid *Lariophagus distinguendus*. Hosts (larvae of bean weevils) of different size were either offered alone (solid lines) or as the larger (dashed lines) or smaller (dotted lines) of mixtures of hosts differing by 0.4 mm in size. More female-biased sex ratios were found when the host was the larger of the two classes offered, while more male-biased sex ratios were found when it was the smaller. The size of the host was estimated by measuring the width of the feeding tunnel inside the bean using X-ray photography. (From Charnov et al. 1981.)

other females in the population (or perhaps to remate previously inseminated females). A female confronted with a patch composed purely of small hosts is then selected to lay male eggs in the hope that they might find mates in other patches, rather than to lay female eggs that would develop into adults with very low fitness (Werren and Simbolotti 1989).

The second parasitoid studied by Charnov et al. (1991), the braconid *Heterospilus prosopoidis*, attacks bruchid beetles (*Callosobruchus* spp.). Preliminary experiments revealed that the wasp laid a strongly male-biased sex ratio in small hosts and a female-biased sex ratio in large hosts. Charnov et al. compared the sex ratio produced by the wasp on two sizes of hosts presented either without choice or in different proportions (fig. 4.15). The sex ratio was largely unaffected by the size of other hosts present. It thus appears that *H. prosopoidis* does not show a facultative response to changing host densities (see also W. T. Jones 1982).

Similar experiments have been conducted on a variety of other species. Van Dijken et al. (1991b) offered different proportions of second and third instar cassava mealybugs (*Phenacoccus manihoti*) to the encyrtid *Epidinocarsis lopezi*. The wasp always laid a highly male-biased sex ratio in small second-instar hosts, irrespective of their abundance. Third-instar hosts received a fe-

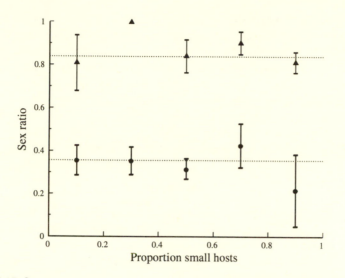

Figure 4.15 Sex ratio and host size in the braconid *Heterospilus prosopoidis*. Wasps were offered mixtures of large and small hosts in varying proportions. The observed sex ratios on large (circles) and small (triangles) hosts remain roughly constant and independent of the distribution of host sizes. The dotted lines are the observed sex ratios when large and small hosts are offered without choice. (From Charnov et al. 1981.)

male-biased sex ratio, except where they constituted the majority of available hosts when the sex ratio was near equality. *E. lopezi* has been introduced to control the cassava mealybug in west and central Africa. In cassava fields, young mealybugs often make up the majority of the population when hosts are abundant, and in these circumstances *E. lopezi* produces a very male-biased sex ratio (van Dijken et al. 1991b). In its natural range, the hosts of *E. lopezi* tend to be at low density and its sex ratio strategy under high-density conditions may not be adaptive. Heinz and Parella (1990) studied a eulophid, *Diglyphus begini*, released in glasshouses to combat the leafminer *Liriomyza trifolii*. Over the course of a season, females were consistently produced on relatively larger hosts than males, despite the fact that the distribution of host sizes showed marked fluctuations. Heinz and Parella excluded differential mortality and any effects of superparasitism, and concluded that the wasps were responding to relative host size. Sex ratios were consistently male biased, varying from about 0.55 to 0.75 over the season.

Curiously, as Charnov et al. (1981) noted, the first sex ratio experiment in which a wasp was presented with the same host as the larger and smaller of a pair, was performed as long ago as 1913 by Chewyreuv. An ichneumonid wasp (*Pimpla* sp.) produced more males on *Pieris brassicae* (Large White butterfly) pupae in the presence of larger *Sphinx ligustri* (Privet Hawk moth) pupae than in the presence of smaller *Cynthia cardui* (Painted Lady butterfly) pupae.

Failure to find a facultative sex allocation response to changes in the host size distribution does not exclude Charnov's hypothesis as an explanation for the oviposition of male eggs in small hosts. Indeed, if a wasp always experiences the same host size distribution, a response to absolute host size will be identical to a response to relative host size. Alternatively, if the distribution of host sizes fluctuates markedly over the lifetime of the searching wasp, it may not be selected to respond to short term changes (Charnov 1982). Finally, the wasp may not be able to assess the host size distribution, or assessment may be expensive in time and energy and so has never have been selected.

4.4.3 BEHAVIORAL RULES

In order for a parasitoid to make adaptive sex ratio decisions on hosts of different size, it must be able to measure the size of the host. In addition, those species that adjust their sex ratio to the current distribution of host sizes must have some means of assessing this distribution. A number of studies have examined the behavioral mechanisms involved in these decisions.

The ichneumonid wasp *Pimpla (=Coccygomimus) turionella* shows a strong response to host size as predicted by the size advantage hypothesis. Aubert (1961) and Sandlan (1979b) investigated the cues used by the wasp to assess host size. The wasp attacks lepidopteran pupae concealed in silken cocoons. Artificial cocoons of different sizes can be constructed from balls of cotton and real pupae hidden within them. Aubert (1961) discovered that the wasp used the size of the cocoon, rather than the pupae, in deciding whether to lay a male or female egg. Sandlan (1979b) concealed parts of pupae and cocoons in wax blocks and found that the sex ratio was most sensitive to the length of the longitudinal axis of the cocoon. He also found that the wasp was unable to assess host size after removal of the antennae, and that the cocoon was only recognized as part of the host if it had absorbed host kairomones.

In the same study, Sandlan found that if a single wasp was presented with a series of hosts, all of the same size, the wasp initially produced a female-biased sex ratio on large hosts and a male-biased sex ratio on small hosts. However, as more hosts were presented, the sex ratio on hosts of all sizes reached an asymptote at the same value (a sex ratio of 0.25, the female bias possibly due to LMC). This suggests that the wasp is using information about the distribution of host sizes gained during foraging to update its sex allocation strategy.

Based on a series of very detailed experiments, van den Assem et al. (1984a) studied the behavioral rules used by the pteromalid *Anisopteromalus calandrae*, a parasitoid of stored product pests, to respond to local variation in the distribution of host sizes. In their first experiment, wasps were presented on successive days with batches of either large or small hosts. In both cases, the wasp produced a sex ratio of about 0.25, though more time was taken for the sex ratio to settle at this average with successive batches of small hosts. When the wasp was given alternating batches of small and large hosts, it tended to produce many more males on the small hosts and more females on the large host, as Charnov et al. (1981) predicted. Interspersing successive batches of small hosts with a few large hosts was sufficient to prevent the wasp switching to the production of mostly females on small hosts, as they had done when presented with this category alone. The wasp showed a strong tendency to reject small hosts, and in mixtures of large and small hosts the latter were largely avoided. Van den Assem et al. suggested a behavioral rule to explain this and related observations. They proposed that the wasp laid one male every x hosts encountered. If the intervening hosts are good quality, females are placed in them, otherwise they are ignored. The estimation of host quality is relative to the local environment and is updated through experience. However, while this model describes, at least qualitatively, the experimental results, it can lead in more natural situations to a poorly adaptive sex ratio strategy.

In a related study, Simbolotti et al. (1987) proposed behavioral rules to describe the sex allocation behavior of the pteromalid *Lariophagus distinguendus* in the face of changes in both the size distribution and density of hosts. They suggested that each wasp ignores hosts below a certain size threshold, lays male eggs on hosts above this threshold but below the second threshold, and lays female eggs on hosts larger than a second threshold value. The actual values of the two thresholds are determined by the number of mature eggs carried by the wasp: when hosts are scarce, and the female thus has a greater egg load, the wasp is predicted to lower the value of both threshold host sizes.

4.4.4 How Widespread Is the Correlation between Host Size and Sex Ratio?

Since Chewyreuv's (1913) first observation that host size influenced parasitoid sex ratios, this pattern has been observed in a very large number of parasitoid wasps (e.g., Clausen 1939; Charnov 1982; King 1987, 1989a). A parasitoid may find that its hosts vary in size for at least two reasons. Some wasps attack the growing stages of hosts, while others attack a range of host species that vary in size. Although rigorous experiments have been conducted only on a minority of species, it does not seem unduly incautious to suggest that Charnov's size advantage hypothesis explains the large majority of these observations.

Can any broad statement be made about the type of wasp in which this pattern of sex allocation will occur? First, it is probably largely confined to solitary species. In gregarious species, the fitness of the adult wasp is determined chiefly by the number (and possibly the sexual composition) of the clutch, rather than directly by host size. In what Werren (1984a) has called "semigregarious species," which produce clutches of the order of one, two, or three eggs, there may be interesting interactions between clutch size, sex ratio, and host size with single males produced in small hosts, females in bigger hosts, and then particular combinations of males and females in still larger hosts. This possibility has yet to be explored either theoretically or experimentally (but see Williams 1979 for a study of the same problem in the context of sex allocation in mammals).

Second, Waage (1982b), Werren (1984a), and Wellings et al. (1986) have argued that size-dependent sex allocation is most likely to be found in species that either attack nongrowing host stages, or prevent their hosts from continuing to grow (idiobionts). Species in which the host continues to develop and feed after parasitism (koinobionts) are less likely to show size-dependent sex allocation because it will be difficult for a female parasitoid to estimate the eventual size of the host at the time of oviposition. King (1989a) tested this hypothesis with data collected from the literature on twenty solitary idiobionts and twelve solitary koinobionts. Seventeen out of the twenty idiobionts (85%) showed size-dependent sex ratios, but only five out of the twelve koinobionts (42%) did so. Although these results must be treated with caution—the literature may be biased, differential mortality is only seldom ruled out, and there is no control for taxonomic artefacts—they do support Waage's hypothesis. King also questions whether the estimation of final host size is impossible for all koinobionts and provides examples where host size at parasitism is correlated with final host size (see also Mackauer 1986). Koinobionts that attack several host species which vary in size may be able to recognize the species and hence the final size of their host at oviposition, and adjust their sex ratio accordingly.

An interesting comparison can be made between two species of *Spalangia* (Pteromalidae) studied by Donaldson and Walter (1984, *S. endius*) and King (1988, *S. cameroni*). Both species are parasitoids of fly pupae and thus attack a host that does not change in size after parasitism. The size of adult *S. cameroni* is influenced by host size, but not the size of adult *S. endius*. Charnov's hypothesis assumes host size affects adult parasitoid size and so should not apply to *S. endius*. Donaldson and Walter found no host-size specific sex allocation in *S. endius* whereas it was observed by King in *S. cameroni*. However, King was unable to relate adult female size to fitness (see sec. 7.1.3). It was suggested in the last chapter (sec. 3.2.4) that many solitary parasitoids develop in hosts large enough to support several parasitoids. If this is correct, the larvae of these species are not resource limited and Charnov's hypothesis should not apply.

4.4.5 OTHER ASPECTS OF HOST QUALITY

Charnov's hypothesis was originally formulated to apply to situations where hosts vary in size and where the fitness of male and female parasitoids was correlated with adult size. However, other aspects of host quality may influence parasitoid fitness; and other components of male and female fitness in addition to adult size may be differentially affected by host quality.

Host species or host age, independent of size, are two possible components of host quality that might influence the sex ratio (King 1987). Kochetova (1978) provides examples of scelionid and trichogrammatid egg parasitoids that place male eggs in unfavorable hosts, both host species and developmental stages not normally attacked by the wasps. King (1990) found that *Spalangia cameroni* females laid slightly more male eggs on old hosts than young hosts when the two were offered in combination. No difference was found when the two host types, which differ in weight but not size, were offered without choice. Differential mortality was not responsible for these results. King suggests that older hosts are of poorer quality, but she was unable to detect any significant difference in the size or development time of wasps developing on the two host types. There are examples of a correlation between the sex of the parasitoid and the sex of its host (Clausen 1939; McGugan 1955; Lyons 1977). However, in these cases female hosts are larger than male hosts and there is no evidence that, other in their size, males are worse hosts than females.

A host that already contains a parasitoid might be thought of as a low-quality host. Waage (1982b) suggested that the assumptions of Charnov's hypothesis might apply here and lead to the oviposition of male eggs by superparasitizing solitary parasitoids. Exactly when supernumerary larvae are eliminated is quite critical. If elimination occurs very early, before substantial host resources are consumed, a successful superparasitoid may inherit a host of normal quality. Van Dijken et al. (1993) found no evidence that the encyrtid *Epidinocaris lopezi* laid more male eggs on parasitized hosts than on unparasitized hosts, perhaps because the elimination of supernumerary larvae occurs early in this species. Similarly, sex ratio and oviposition order appears to be independent in the eucoilid *Leptopilina heterotoma* (Visser et al. 1992c). However, the facultative hyperparasitoid *Pachycrepoides vindemiae* (Pteromalidae) tended to lay male eggs on poor-quality hosts where the primary parasitoid had already consumed much of the host resources (van Alphen and Thunnissen 1983). Similarly, the encyrtid egg parasitoid *Ooencyrtus telenomicida* tends to lay more male eggs in cases where its progeny develop as a secondary parasitoid (Kochetova 1978).

King (1988) has suggested that host quality might influence parasitoid fitness through an effect on development time. The problem here is to explain why development time has different effects on male and female fitness. One possibility is that faster development is more important for males because they gain a mating advantage from emerging early as adults (protandry, see sec. 7.2.2). It is quite likely that there is a trade-off between development time and adult size, which may be different in males and females, leading to a very complicated interaction between the two components of fitness.

4.5 Other Factors

4.5.1 DIFFERENTIAL MORTALITY

Most of the discussion in this chapter has concerned the primary sex ratio, the sex ratio at oviposition. The observed adult sex ratio may differ from the primary sex ratio if one sex suffers greater mortality while immature. It is important to exclude the effects of differential mortality in experimental tests of hypotheses concerning the primary sex ratio. Differential mortality has been recorded in a number of species of parasitoid wasp, in most cases females suffering greater mortality (e.g., Salt 1936; Grosch 1948; Kanungo 1955; Wilkes 1963; Wylie 1966; Kishi 1970; Benson 1973; Rotary and Gerling 1973; Suzuki et al. 1984; Sandlan 1979b; Wellings et al. 1986) although greater male mortality has also been recorded (Jenni 1951).

Despite the empirical observation that differential mortality tends to penalize females, one source of mortality will affect only males. Smith and Shaw (1980) point out that haploid males will die due to the unmasking of lethal recessive mutants. Nearly all lethal alleles in haplodiploid species will be eliminated in the male sex and females will inherit only nonlethal alleles from their father. Suppose the average rate of mutation to a lethal allele per haploid genome per generation is m. Mutations in males will be eliminated immediately, but those in females will be protected by the second allele and will be lost only when transmitted to males. This argument implies that the rate of mortality due to unmasked lethals in haploid males each generation is $3m$ (the three represents the two haploid chromosome sets in the female part of the population plus the single set in the male part). As mutation is likely to be a Poisson process, the survival of haploids, relative to diploids, will be $\exp(-3m)$ (the zero term of a Poisson distribution with mean $3m$). Smith and Shaw estimated m by comparing the survival of immatures in virgin and mixed broods of braconid wasps in the genus *Apanteles* (*sensu lato*). Their estimate of 0.035 (95% confidence limits, 0.015–0.059) is very similar to estimates of m from *Drosophila*. This figure suggests that male hymenopterans suffer a differential mortality of approximately 10% relative to females. Thus the expected secondary sex ratio when Fisher's principle applies is not 1:1 but slightly female biased. Further estimates of m for parasitoid wasps are needed, especially as

Smith and Shaw's experimental technique could not detect lethal recessives expressed very early in development. Werren (1993) has also estimated the relative genetic load of haplodiploid males and females. For outbred species and recessive lethal genes, haplodiploid males are predicted to suffer about 150% greater mortality than diploids while haplodiploid females almost completely escape mortality. He estimates that about 3.5% of males die due to the unmasking of recessive lethals in both outbred and inbred populations. Werren goes on to study the mutational load arising from nonlethal genes, and from dominant or partially dominant genes. This "detrimental load" also affects males more heavily than females and may contribute to increased mortality in this sex.

Detection of differential mortality is difficult because the sexes can normally be distinguished only in the late pupal stage. However, there are at least three approaches that allow the direct measurement of primary sex ratios and hence the estimation of differential mortality.

1. Cole (1981), working with the ichneumonid wasp *Itoplectis maculator*, observed that the act of fertilization caused a noticeable pause in activity during oviposition, thus allowing the sex of the egg to be identified. A similar technique had previously been shown to work for honeybees (Gerber and Klostermeyer 1970) and has now been used on a variety of parasitoid species (Suzuki et al. 1984; van Dijken and Waage 1987; Strand 1989b). The observation of a pause in oviposition does not guarantee that a female egg has been laid, since virgin females will also pause during a proportion of ovipositions, as if they were fertilizing an egg (Pallewatta 1986; Strand 1989b).

2. The most direct techniques for estimating primary sex ratios are cytological or molecular. The use of karyotypes has the advantage that one need only tell if the organism is haploid or diploid to know its sex. Problems may arise as Hymenoptera chromosomes are notoriously small, and nonreproductive tissue is frequently polyploid. Ryan and Saul (1968), Dijkstra (1986), Nur et al. (1988), and van Dijken (1991) describe techniques that have been successfully applied to pteromalids, eulophids, and encyrtids. Molecular techniques have not been used, but DNA densitometry may prove valuable in species where somatic polyploidy is not a problem.

3. In the absence of more direct techniques, it is possible to estimate sex-specific mortality by comparing the survival of all male broods laid by virgin females, with the survival of mixed broods laid by mated females. This technique is based on the assumption that the male progeny of virgin females suffer the same mortality as the male progeny of mated females which, though untested, is probably a reasonable first approximation. Wellings et al. (1986) have developed statistical methods based on maximum likelihood arguments that allow this technique to be used on solitary parasitoids attacking hosts of different size.

4.5.2 Environmental Influences

The sex ratio of a number of parasitoid wasps is known to be influenced by light, temperature, and humidity. Photoperiod influences the sex ratio of the pteromalid *Pteromalus puparum* (Boulétreau 1976) and the ichneumonid *Campoletis perdistinctus* (Hoelscher and Vinson 1971), and relative humidity the sex ratio of a number of braconids and pteromalids (Legner 1977; Odebiyi and Oatman 1977). King (1987) lists fourteen examples of temperature affecting sex ratio although in the majority of cases this results from rather gross disruption of reproduction by extremes of hot or cold.

The evolutionary significance of environmentally induced sex ratio changes has received little attention. In some cases, the environment may be correlated with factors in the wasp's biology that are important in determining optimal sex ratios. Thus if LMC reaches a peak in the summer, the wasp may be selected to respond to higher temperatures by producing a more female-biased sex ratio. However, in other cases, environmental conditions, especially when extreme, will have a direct physical action on the production of sons and daughters and will cause nonadaptive disruption of the sex ratio.

4.5.3 Constrained Sex Allocation

Haplodiploid females, unlike diploids, are still able to reproduce by producing males if they are unmated or have run out of sperm. In fact, as described in section 7.2.2, there is a variety of mechanisms, of varying importance, that can lead to what I call constrained sex allocation: the enforced production of one sex (normally males) by haplodiploids. In this section I discuss some of the sex ratio consequences of constrained sex allocation. In particular, I shall ask (1) what is the ESS sex ratio of unconstrained females in the presence of constrained, ovipositing females; (2) will natural selection favor the investment of time and energy in oviposition or in activities leading to the removal of the constraint (for example finding a mate); and (3) is constrained oviposition important in wild populations of haplodiploid parasitoids?

THE SEX RATIO OF UNCONSTRAINED FEMALES

Consider a population of parasitoids with panmictic mating where the assumptions necessary for the application of Fisher's principle apply. In the absence of constrained oviposition, a 50:50 sex ratio will evolve. Now suppose some constrained females begin to produce all male broods. The population sex ratio will become biased toward males and the relative value of daughters will increase. This will lead to selection on unconstrained mothers to produce a female-biased sex ratio which will continue until the population sex ratio returns to equality. It is straightforward to show that the ESS sex ratio for unconstrained females is $(1 - 2p)/2(1 - p)$, where p is the fraction of con-

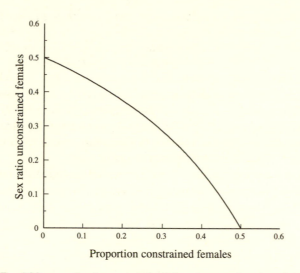

Figure 4.16 The ESS sex ratio of unconstrained females as a function of the fraction of constrained females ovipositing in the population.

strained, ovipositing females in the population (Godfray 1990). As p rises, the sex ratio of unconstrained females becomes progressively more female biased until when 50% of females produce only sons, all unconstrained females produce exclusively daughters (fig. 4.16). It is possible, though difficult, to study the effect of constrained oviposition on the sex ratio of parasitoids subject to LMC. The presence of constrained females among the foundresses colonizing a patch leads to a reduction in the sex ratio of unconstrained females (i.e., to a further female bias), though the effect is small compared with the panmictic case (Godfray 1990).

Ideally, unconstrained females should adjust their sex ratio to the current proportion of constrained females in the population. This will be impossible in most cases and the sex ratio produced by unconstrained females will reflect the average fraction of constrained females over evolutionary time. However, there may be some indirect cues that a female can use to assess the proportion of constrained females, for example the time required to find a mate or the number of courting males encountered. Hoelscher and Vinson (1971), working with the ichneumonid *Campoletis perdistinctus*, and Rotary and Gerling (1973), working with the braconid *Bracon hebetor*, have shown that more female-biased sex ratios are produced if wasps mate late in life. These observations are consistent with theoretical expectations, though other age-related factors that may influence sex ratio are not excluded.

The population sex ratio of parasitoids is often measured in the laboratory using standardized females, each exposed to a male for mating. However, this standardization may bias the estimated sex ratio toward females if unconstrained wasps always produce more daughters than sons in the expectation that there will be some constrained females ovipositing in the population. Another consequence of the presence of constrained, ovipositing females is that the clutches of eggs produced by unconstrained females of gregarious species will be relatively more female biased than the population sex ratio. This increases the likelihood that sisters interact with each other and, as sisters are closely related in haplodiploid genetic systems, may increase the predisposition of haplodiploid insects to evolve eusociality (Godfray and Grafen 1988).

WILL CONSTRAINED FEMALES BE SELECTED TO OVIPOSIT?

If there is no possibility of removing the constraint, so that the only reproductive outlet of constrained females is the production of males, constrained oviposition is always selected. However, the wasp may be faced with a choice of either investing time and energy in the production of sons, or in the removal of the constraint, perhaps by searching for a mate. Now, if the population is at sex ratio equilibrium, the value of a son and a daughter is the same. Thus, when constrained oviposition is rare, there will be no selection pressures preventing its initial spread in the face of the obvious advantage of immediate reproduction. Because constrained oviposition leads to a male-biased population sex ratio, the advantage of being able to produce daughters will increase as the trait spreads, and this advantage may be strong enough to select for the postponement of oviposition until after mating. However, if unconstrained females respond to the increasing male bias of the population sex ratio by producing more females, the relative values of sons and daughters remain the same and there will be no disadvantage to the production of only males. In this case, there would seem to be no bar to the spread of constrained oviposition until its frequency reached 0.5.

Could the processes outlined in the last paragraph lead to runaway selection for the production of all male broods? Probably not; there are always likely to be some benefits in having the capability to produce offspring of both sex. As discussed earlier in this chapter (sec. 4.2.1), in finite populations each individual is selected to produce the population sex ratio. In addition, constrained wasps are unable to increase their fitness by adjusting sex ratios to take account of local mate competition and variation in host size. This may be sufficiently important for some species that they refrain from oviposition and actively search for a mate. Other species may immediately begin to oviposit and only mate when discovered by a male. Here the equilibrium level of constrained oviposition will be determined by the ability of males to locate females and the average number of eggs laid by a female prior to mating.

THE IMPORTANCE OF CONSTRAINED SEX ALLOCATION

There are a number of ways to obtain estimates of constrained oviposition in field populations. (1) *Laboratory oviposition*: females can be collected in the act of oviposition and confined with unlimited hosts in the laboratory to determine the proportion that produces only males. (2) *Dissection*: the spermathecae of females caught in the act of oviposition can be examined to see if they contain sperm: unmatedness and sperm depletion are perhaps the most important causes of constrained sex allocation. The usefulness of this technique depends on the ease with which the spermathecae can be dissected, and the transparency of the spermathecal walls. At least with small chalcids, this technique works very well. (3) *Gregarious broods*: the frequency of all male broods in gregarious species can be measured. Gregarious species often have female-biased broods and in these species all-male broods stand out as unusual and are thus probably the result of constrained sex allocation. However, sex ratio diseases (chapter 5) need to be excluded as an alternative cause. (4) *Non-dispersing males:* in groups where males do not disperse from the site of development, females developing in a patch with no males, or leaving the patch unmated, are unlikely to find mates. For example, the males of many fig wasps never leave the fig. A female developing in a fig with no males, or leaving the fig unmated, is destined never to mate.

The available evidence on constrained oviposition is summarized in table 4.2 (based on Hardy and Godfray 1990; Godfray and Hardy 1993). The percentage of constrained females ranges from 0% through values so low as to have no detectable effects on the observed sex ratio, to values as high as 12% or 23%, large enough to act as a significant selection pressure on individual sex ratios. These data indicate that while constrained oviposition will frequently not be an important factor affecting sex ratios, it should not be dismissed without consideration. However, it should be noted that table 4.2 is a very biased selection of parasitoids, weighted in particular to fig wasps, *Drosophila* parasitoids, and the genus *Apanteles* (*sensu lato*), and much more information is required before the importance of constrained oviposition can be fully assessed.

4.5.4 Sex Ratios with Overlapping Generations

Most sex ratio models assume discrete generations and so ignore the complications of overlapping generations and differences in the longevity of the two sexes. Werren and Charnov (1978) and Charnov (1982) have made a start at exploring these complexities.

Consider first a partially bivoltine species with spring and autumn generations. Females born in the autumn survive until the spring when they reproduce and die, while those born in the spring survive until the following autumn, and then die. Spring-born males also survive until the autumn but autumn-born males survive for a full year and reproduce in both the following spring and autumn. This natural history is found in a number of solitary wasps. Female wasps laying eggs in the spring will be selected to produce a female-biased sex ratio as their sons will have to compete not only with males of their

Table 4.2
Estimates of the prevalence of virginity in field populations of haplodiploid insects (all Hymenopterans).

Species	Method (see text)	Percentage Virgin Males	Total Insects Examined	Reference
Ceratosolen dentifer (Agaonidae) [pollinating fig wasp]	Nondispersing males + dissection	2%	n=100 from 10 figs	Godfray 1988
Philotrypesis sp. (Torymidae) [parasitic fig wasp]	Nondispersing males + dissection	2%	n=100 from 10 figs	Godfray 1988
Apocryptophagus sp. (Torymidae) [parasitic fig wasp]	Nondispersing males + females in groups with no males	4%	n=1212 from 98 figs	Godfray 1988
Apocrypta mega (Torymidae) [parasitic fig wasp]	Nondispersing males + females in groups with no males	23%	n=154 from 46 figs	Godfray 1988
Cotesia (=Apanteles) glomeratus (Braconidae)	Gregarious broods	12%	n=535	Tagawa 1987
Cotesia (=Apanteles) glomeratus (Braconidae)	Gregarious broods	0%	n=18	M. R. Shaw, pers. comm. *
Cotesia (=Apanteles) glomeratus (Braconidae)	Gregarious broods	19%	n=63	le Masurier 1987
Cotesia (=Apanteles) abjectus (Braconidae)	Gregarious broods	18%	n=11	M. R. Shaw, pers. comm. *
Cotesia (=Apanteles) bignelii (Braconidae)	Gregarious broods	6%	n=16	M. R. Shaw, pers. comm. *
Cotesia (=Apanteles) melitaearum (Braconidae)	Gregarious broods	8%	n=12	M. R. Shaw, pers. comm. *
Cotesia (=Apanteles) spurius (Braconidae)	Gregarious broods	0%	n=13	M. R. Shaw, pers. comm. *
Cotesia (=Apanteles) zygaenarum (Braconidae)	Gregarious broods	0%	n=24	M. R. Shaw, pers. comm. *
Cotesia (=Apanteles) flavipes (Braconidae)	Gregarious broods	5%	n=20	A.I. Mohyuddin, pers. comm. *
Glyptapanteles (=Apanteles) fulvipes (Braconidae)	Gregarious broods	0%	n=39	M. R. Shaw, pers. comm. *
Asobara tabida (Braconidae)	Dissection	0%	n=58	Hardy & Godfray 1990

Table 4.2 (continued)

Species	Method (see text)	Percentage Virgin Males	Total Insects Examined	Reference
Asobara tabida (Braconidae)	Laboratory oviposition	0%	n=46	J. M. Cook, pers. comm. *
Tanycarpa punctata (Braconidae)	Dissection	6%	n=97	Hardy & Godfray 1990
Leptopilina heterotoma (Eucoilidae)	Dissection	2%	n=119	Hardy & Godfray 1990
Encarsia pergandiella (Aphelinidae)	Dissection	0%	n=48	M. S. Hunter, pers. comm.*
Habrocytus sp. (Pteromalidae)	Dissection	3%	n=124	J. Wilson & H.C.J. Godfray, unpub.
Trioxys indicus (Braconidae)	Laboratory oviposition	1%	n=199	Singh & Sinha 1980
Trioxys indicus (Braconidae)	Laboratory oviposition	0%	n=47	R. Singh, pers. comm. *
Lysiphlebus mirzai (Braconidae)	Laboratory oviposition	0%	n=78	R. Singh, pers. comm. *
Lysiphlebus delhiensis (Braconidae)	Laboratory oviposition	0%	n=112	R. Singh, pers. comm. *

* I am very grateful for permission to quote unpublished data.

own generation, but also with males from their parents' generation (Werren and Charnov 1978; Seger 1983). The model can be extended to species with different life histories, the mother always producing more offspring of the sex that enjoys a temporary reproductive advantage relative to the other sex (Seger 1983; Grafen 1986).

Werren and Charnov (1978) consider a species with completely overlapping generation that experiences a temporary episode of strong mortality that affects one sex but not the other. If the developmental period of the young is short compared with the adult life span, the current dearth of one sex will also be experienced by the progeny of reproductively active females. In these circumstances, females should respond by producing more of the rare sex.

There is good evidence to suggest that partial bivoltinism influences the sex ratio of solitary wasps (Seger 1983) and it may be important in some species of parasitoid wasps. Werren and Charnov (1978) suggested that the sex ratio response of some parasitoid wasps to day length may be an adaptation to seasonal variation in the survival of males and females, but, as they themselves stress, more work is needed to exclude alternative hypotheses. Mohamed and Coppel (1986) found that chalcidid wasps (Brachymeria intermedia) increased the proportion of daughters among their progeny if ovipositing in the presence of many conspecific males. They suggested that the wasp was responding to a local abundance of males by producing more of the rarer sex.

4.6 Conclusions

Sex ratios in parasitoid wasps show some of the clearest examples of adaptive patterns in animal behavior. Experimental studies with parasitoids have been important both in testing theory and in suggesting new subjects for theoretical investigation.

Hamilton's theory of local mate competition provides a compelling explanation for the common occurrence of highly female-biased sex ratios in species where mating frequently occurs between siblings. A consensus seems to have been reached that no single theoretical approach to LMC enjoys a natural primacy, but that different approaches offer varying advantages in ease of analysis and as useful verbal metaphors. Some experimental work has provided detailed quantitative confirmation of theoretical predictions, though more normally qualitative trends consistent with theory are discovered. Exactly how the latter type of data should be interpreted is a problem common to most areas of behavioral ecology, and has been particularly debated in the context of tests of LMC theory. The behavioral rules that allow parasitoids to produce adaptive sex ratios has been, and I believe will continue to be, a fertile area of research. The production of nonrandom sequences of male and female eggs is an obvious trait whose functions have not yet been fully resolved. It is also not clear whether local resource competition influences the sex ratio of gregarious parasitoids.

The second widespread sex allocation pattern shown by parasitoids is the placement of male eggs on small hosts and female eggs on large hosts. Charnov's size-advantage hypothesis provides a clear explanation for these observations. The greatest problem with this theory is the difficulty of comparing the relationship between size and fitness of the two sexes in the field; laboratory experiments almost certainly underestimate the disadvantages of being small. Tests of the size-advantage hypothesis have almost exclusively involved offering parasitoids different combinations of large and small hosts. The two other techniques suggested by Charnov—artificial selection and the study of different geographical populations—have been used hardly at all by parasitoid biologists.

It is frustrating that the mechanism of sex determination in Hymenoptera is still so poorly understood. Studies of sex determination in *Drosophila*, nematodes, and man have made great strides in the last few years and hopefully some of the techniques developed in these fields will be applied to haplodiploid insects.

There are a number of other factors that may select for biased sex ratios in parasitoids. The presence of some females in the population producing only male offspring will lead to selection on unconstrained females to produce more daughters. It is unclear whether this is of any importance in the field. I believe a major unexplored source of sex ratio biases involves differential mortality in populations with overlapping generations. This is a difficult subject to study theoretically, and the lack of theory has hindered experimental investigation. Work in this area may clarify whether correlations between sex ratio and environmental factors such as temperature and day length have an adaptive basis.

5

Selfish Genetic Elements

The fusion of evolutionary and molecular biology has revealed some remarkable cases of individual genetic elements distorting normal sexual reproduction to their own advantage. For example, a number of chromosomes or chromosomal regions have been identified in *Drosophila* and other animals that bias the random segregation of chromosomes to ensure that they are overrepresented in the gametes (Lyttle et al. 1991). Similarly, many cytoplasmically inherited elements distort the sex ratio in favor of daughters because they are not transmitted through sperm (Cosmides and Tooby 1981). The action of these genetic elements is often described as selfish because by increasing their own fitness, they normally decrease the individual fitness of their host (Dawkins 1976; Doolittle and Sapienza 1980; Orgel and Crick 1980; Werren et al. 1988). One of the most interesting recent discoveries in parasitoid biology is of a series of selfish genetic elements that have a variety of different effects on the biology of their carriers. The nature of many of these elements is unknown, but one has been identified as a supernumerary chromosome, and others are known to be microorganisms with maternal inheritance.

Without doubt, the best-studied selfish elements in parasitoids are found in pteromalid wasps in the genus *Nasonia*, described in section 5.1. Section 5.2 discusses recent work which suggests that many examples of parthenogenetic (thelytokous) reproduction in parasitoids are caused by non-Mendelian microorganisms. Very recently, selfish genetic elements have also been implicated in the biology of heteronomous aphelinids, the subject of section 5.3. In section 5.4 I discuss curious aspects of the evolution of gregariousness in the pteromalid genus *Muscidifurax*.

5.1 Non-Mendelian Genetic Elements in *Nasonia*

The small pteromalid wasp *Nasonia vitripennis* has been intensively studied in the laboratory and has played a major part in the development of sex ratio theory (see the preceding chapter). *N. vitripennis* is a gregarious endoparasitoid of dipteran pupae, typically laying twenty to fifty eggs in a single host, with a highly female-biased sex ratio. In 1979 Werren and his colleagues set up an experiment to investigate heritable variation in the sex ratio. To their surprise, wasps in lines selected for a male-biased sex ratio frequently produced families composed only of males (Werren et al. 1981; Werren 1991). Genetic crosses quickly established that the male-only trait was inherited through males. This was unexpected because in haplodiploids, the "father" of an all-male brood is effectively sterile and does not contribute normal chromosomal genes to future generations. The element responsible for all male broods was named *psr* (paternal sex ratio; in the first papers it was termed *dl*, daughterless). Two early hypotheses about *psr* were that it was (1) a venereally transmitted factor that prevented fertilization, or (2) an element carried by the sperm that converted fertilized eggs to males. The second hypothesis received strong support from a series of experiments by Werren and van den Assem (1986) that revealed a strong correlation between *psr* transmission and the rate of egg fertilization in experiments with different genetic strains of *N. vitripennis*.

Recent work using both cytological and molecular genetic techniques has identified the *psr* element as a small supernumerary or B chromosome (Werren et al. 1987; Nur et al. 1988; Werren 1991). B chromosomes are widespread but uncommon in animals and plants, and often have difficulty surviving meiotic cell divisions (Werren et al. 1988). Spermatogenesis in haplodiploid wasps does not involve a reductive division and it is thus likely that any B chromosomes would be transmitted more efficiently through sperm. Ideally, a B chromosome in a fertilized gamete would wish to convert its carrier from an egg-producing female to a sperm-producing male.

The *psr* chromosome achieves this feat. It is transmitted via sperm with near 100% efficiency and its normal fate would be to end up in fertilized, and thus female, eggs. However, in a manner that is not yet well understood, the *psr* element interferes with the normal functioning of the paternal chromosome set. The chromosomes carried in the sperm, with the exception of *psr*, remain in the condensed state, are unable to participate in normal cell division, and are quickly lost. The maternal chromosome set is unaffected and the fertilized zygote develops as a haploid male. The *psr* chromosome is thus transmitted from generation to generation, temporarily associated with different sets of chromosomes which are destroyed after each new fertilization, without doubt one of the most "selfish" of all known genetic elements.

Molecular investigation of *psr* has shown that it is a short chromosome containing a large fraction of middle-length repetitive DNA with individual repeats probably arranged in single tandem arrays (Nur et al. 1988; Werren 1991; Eickbush et al. 1992). Some repetitive DNA sequences are shared with *N. vitripennis* autosomes (and also with other species of chalcid) while others are unique to *psr*. The unique sequences share a palindromic conserved sequence that may act as a protein binding site or might be involved in sequence amplification. One hypothesis is that these numerous protein binding sites mop up a chemical essential for the proper functioning of the paternal chromosome set, perhaps a protein that allows it to "unpack" from its condensed state. Alternatively, *psr* might contain genes coding for a product that prevents successful processing of the paternal chromosome (Eickbush et al. 1992). These ideas are being investigated using nonfunctional *psr* mutants obtained by exposing males to radiation (Beukeboom and Werren 1993; Beukeboom et al. 1993). Nonfunctional *psr* mutants are able to survive ovigenesis, though as expected the efficiency of transmission is much lower than through spermatogenesis. Crosses can be set up that lead to males with more than one B chromosome, allowing mutant complimentation and dosage experiments to be performed.

The production of all male broods obviously has profound consequences for the population biology of the wasp (Skinner 1986; Werren 1987b, 1991; Werren and Beukeboom 1993). Analysis of this question is complicated by the highly subdivided population structure and female-biased sex ratios found in *N. vitripennis*. Consider first the conditions under which a rare *psr* chromosome will invade a population with panmictic mating. The *psr* chromosome will spread if more *psr* males are produced as a result of a mating between a *psr* male and a female than the number of uninfected males that result from the union of two normal wasps. Now if we assume that the transmission of *psr* in sperm and the efficiency of sex ratio conversion are both 100% (as they very nearly are), then the number of infected males produced by a female carrying *psr* sperm is proportional to the fraction of fertilized eggs, while the number of uninfected males produced by a female carrying uninfected sperm is proportional to the fraction of unfertilized eggs. Thus the *psr* chromosome spreads when females fertilize more than 50% of their eggs, i.e., when they attempt to produce a female-biased sex ratio.

At first glance, it would appear that the female-biased sex ratios found in *Nasonia* would enhance the spread of *psr*. However, the above argument applies only to panmictic populations, and the equivalent argument for subdivided populations is substantially more complicated (Werren and Beukeboom 1993). The *psr* chromosome is penalized in subdivided populations because if most foundresses colonizing a patch happen to carry *psr* sperm, their infected sons will have relatively few opportunities to mate, and will suffer intense competition for whatever females are available. Quantitative analysis of a model based on the population structure assumed by Hamilton (1967; sec. 4.3.1) suggests that the *psr* chromosome can invade a population with a female-biased sex ratio caused by LMC, as long as at least four foundresses colonize each patch (Werren 1991; Werren and Beukeboom 1993). However, when it can invade, the equilibrium frequency of *psr* is very low, and never exceeds 3%. In the wild, the *psr* chromosome has been found only in Utah, but there it may reach frequencies of up to 11%. It thus appears that the theory described so far is insufficient to account for the pattern of *psr* abundance observed in nature.

Werren and his colleagues have suggested that the distribution of *psr* may be influenced by a second non-Mendelian sex ratio factor named *msr* (for maternal sex ratio) discovered by Skinner (1982). As its name suggests, this factor is transmitted maternally, in the cytoplasm. Infected females produce a heavily female-biased sex ratio, something that is obviously advantageous to *msr* which cannot be transmitted via males. Females containing the *msr* factor fertilize all, or nearly all, their eggs, suggesting that *msr* somehow influences the sex allocation behavior of the wasp, though the mechanism is not known. The actual nature of *msr* is also at present unknown. The *msr* factor has been discovered in many North American populations of *N. vitripennis*, and reaches frequencies of up to 17%.

It is unclear how *msr* is maintained at low frequencies in natural populations. A maternally inherited cytoplasmic factor that causes all-female families is predicted to spread through a population to fixation, at the same time driving the population to extinction (Hamilton 1968; Cosmides and Tooby 1981). In populations where both *psr* and *msr* exist, the two factors may interact to facilitate their mutual coexistence. By increasing the fraction of fertilized eggs, *msr* assists the spread of the *psr* chromosome through a population and allows it to equilibrate at relatively high frequencies. Interestingly, the presence of *psr* can prevent *msr* from going to fixation, at least in populations divided into small mating groups. This occurs because when *psr* is common, nearly all males carry the *psr* chromosome and females able to produce uninfected sons that can mate with their own daughters are at a premium. Normal females gain an advantage over *msr* females as they produce more uninfected sons (Werren and Beukeboom 1993).

Some of these ideas have been investigated in cage experiments with *N. vitripennis* (Beukeboom and Werren 1992). Populations were reared in the laboratory under conditions that mimicked strong (three foundresses) and weak (twelve foundresses) local mate competition. The *psr* chromosome could not be maintained under strong LMC, but did persist with weak LMC. However, if *msr* was present in addition to *psr*, then the latter did persist at high frequencies in both populations with three and twelve foundresses. The frequencies of the two factors were sometimes sufficiently high to imperil population persistence, especially with twelve foundresses. When both elements were present and LMC was strong, the presence of *psr* led to a reduction in the frequency of *msr* and in consequence to a reduction in its own fitness. Beukeboom and Werren (1992) found a generally good quantitative match between the observed dynamics in the laboratory and their model predictions. However, the situation in which *psr* is most likely to promote the coexistence of *msr*—very low foundress number—occurs infrequently in the field and thus *psr* is unlikely to be the main factor allowing *msr* to persist. The fact that *msr* persists in populations outside Utah where *psr* is absent also proves other factors are involved.

A third factor that biases the sex ratio, *sk* (son killer, a bacterium), is also known from *N. vitripennis* (Skinner 1985; Huger et al. 1985; Werren et al. 1986). The bacterium, recently described in a new genus (*Arsenophonus nasoniae*: Enterobacteriaceae, Gherna *et al.* 1991), is an infection of female wasps and causes the death of unfertilized eggs. The reproductive tract of females is infected by *sk* which is injected into the host puparium during oviposition and taken up by the wasp larvae as they feed on the host. If the host is superparasitized by an uninfected female, her progeny are also infected and so both horizontal and vertical transmission occur (Skinner 1985). The efficiency of both modes of transmission is about 95%. The bacterium has been reported from a number of *N. vitripennis* populations but at low frequencies (2%–4%).

It is not clear why only males succumb to *sk*, or what is the evolutionary advantage, if any, of male destruction to the bacteria (Skinner 1985). The death of males does not directly increase the transmission of *sk* because death occurs after oviposition and mothers do not compensate for the loss of some of their progeny by producing more eggs. However, there may be an indirect advantage as male larvae compete for resources with female larvae in the host, and their death may lead to larger and fitter infected females. Against this advantage must be set the risk of infected females failing to find mates (Skinner 1985; Werren et al. 1986).[1]

The three elements, *psr, msr*, and *sk*, do not exhaust the menagerie of nonmendelian factors found in *Nasonia*. Saul, Ryan, and colleagues (Ryan and Saul 1968; Ryan et al. 1985; Conner and Saul 1986; Richardson et al. 1987) discovered a cytoplasmic factor responsible for unidirectional incompatibility among laboratory strains of *Nasonia vitripennis*. In forbidden crosses, the paternally derived chromosome set fails to condense properly. The agent responsible could be eliminated by an antibiotic (tetracycline) suggesting a microorganism to be responsible. The incompatibility is termed unidirectional as sperm from a male carrying the microorganism cannot successfully fertilize eggs from strains without the microorganism, although the reverse cross results in healthy female offspring. Unidirectional incompatibility is quite widespread in insects (see Breeuwer and Werren 1990) and the causative symbiont can invade a population and quickly spread to fixation (Caspari and Watson 1959). For this reason, it is unlikely that unidirectional incompatibility is important in speciation. The mechanism of action is unknown, but possibly the

---

[1] There is a curious observation in the literature that may have a similar explanation. Jackson (1958, 1966, 1969) conducted a long series of experiments on the aquatic mymarid *Caraphractus cinctus*, a parasitoid of water beetle eggs. Virgin females oviposit normally, their offspring consisting only of sons. However, virgin females of one strain, derived from a single wild-caught female, oviposited as usual but nearly all their eggs died (Jackson 1958, p. 544). When this strain was mated, it produced healthy offspring, although nearly all were female. One explanation for these results is that the wasp carried a male-lethal factor that caused the death of most of its male offspring.

paternal chromosomes are chemically imprinted by the symbiont and are only able to function properly in females also containing the symbiont.

Breeuwer and Werren (1990) have studied the role of cytoplasmic symbionts in interspecific crosses. Until recently, *Nasonia* was considered a monotypic genus. However, two sibling species of *N. vitripennis* have recently been discovered (Darling and Werren 1990). This was surprising—and slightly alarming: *Nasonia vitripennis* is perhaps the most thoroughly studied of all chalcids. However, the two new species are confined to bird nests, while most laboratory stocks are derived from blowflies breeding in artificial habitats, such as rubbish dumps, where *vitripennis* alone occurs.

Bidirectional incompatibility was found in the case of *N. vitripennis* and *N. giraulti*. If a male mates with a female of the other species, his sperm are unable to successfully fertilize the female's eggs. Individuals without microorganisms can be produced using antibiotics and interspecific crosses are successful (at least partially) as long as the male is uninfected. Bidirectional incompatibility is found in relatively few insects (Breeuwer and Werren 1990) and can be a potent force maintaining reproductive isolation. Other isolating mechanisms also operate in the three species of *Nasonia*; there is a variable amount of premating isolation at the courtship stage, and hybrids have considerably lower fitness (Breeuwer and Werren 1990).

The nature of the incompatibility factors in *Nasonia* has recently been resolved by Breeuwer et al. (1992). Sequencing of a 16S ribosomal gene extracted from wasps using a primer specific to prokaryotes, and amplified using the polymerase chain reaction, identifies the causal factor as an α-Proteobacteria, related to *Rickettsia*. The incompatibility factor in *Nasonia* is very similar to incompatibility factors in other insects which are placed in the bacterial species *Wolbachia pipientis*. O'Neill et al. (1992) have recently conducted a comparative analysis of incompatibility from many different insects (but not including *Nasonia*). They find remarkable sequence homogeneity and no correlation between host and bacterial phylogeny, suggesting that the bacteria is transmitted laterally between different insect lineages. The closest known relatives of *W. pipientis* appear to be intracellular vertebrate parasites (*Ehrlichia* and *Anaplasma*) which are transmitted by arthropods. Other bacteria previously included in *Wolbachia* are more distantly related (O'Neill et al. 1992)

The breakdown of chromosomes through hybrid incompatibility offers a possible explanation for the evolutionary origin of the *psr* chromosome. Ryan et al. (1985) noted that a fragment of the largely destroyed paternal chromosome set could sometimes be inherited. The rate of transmission of the fragment was low through females, but much higher through males. Werren (1991) has suggested that *psr* might have arisen in this way, either as a result of an unsuccessful cross between two *N. vitripennis* strains, or between two *Nasonia* species. Any such fragment would be under strong natural selection to ensure that it found itself in a female carrier as often as possible.

5.2 Microorganisms and Thelytoky

Thelytoky, parthenogenetic reproduction in which females give birth to other females without mating, is rather common among parasitoid wasps and has been reported from at least 20 families, 103 genera, and 205 species of Hymenoptera (R. Stouthamer, quoted in Stouthamer et al. 1990b). In common with the distribution of parthenogenesis in other sexual organisms, thelytoky is normally restricted to individual species within sexual taxa—completely asexual genera are rare and asexual higher taxa do not occur. Before discussing recent work on thelytoky and microorganisms, I will briefly describe other suggestions for why parthenogenesis is common among hymenopteran parasitoids.

Discussion of the evolution of thelytoky is particularly frustrating because one constantly runs up against one of the greatest unsolved problems in evolutionary biology—the evolutionary advantage of sex. Maynard Smith (1978), Bell (1982), and Michod and Levin (1988) review a wide variety of hypotheses that have been proposed to account for the maintenance of sex, yet none is wholly satisfactory. If parthenogenesis is particularly common among parasitoids, then a consideration of their biology may furnish some clue to the selection pressures maintaining sex. A number of factors have been associated with thelytoky in parasitoids. (1) *Haplodiploidy.* Many cytogenetic forms of parthenogenesis lead to either immediate or steadily increasing homozygosity at all loci and the unmasking of recessive lethal or deleterious genes may penalize any parthenogenetic mutant. In species with a haplodiploid genetic system, recessive genes are exposed to selection in the male sex (with the exception of genes whose expression is restricted to females), thus reducing the numbers of harmful recessives. (2) *Inbreeding.* It is often suggested that the major advantage of mating with a male is that it leads to offspring that are genetically diverse and able to face novel environments (both physical and biotic). In highly inbred species, potential mates are so genetically similar that this advantage disappears (Price 1980; Cornell 1988). However, there may be substantial advantages to occasional outbreeding (Crozier 1977). The cost of sex may also be quite low in inbred species which normally have female-biased sex ratios. (3) *Allee effect.* Solitary parasitoids with widely dispersed hosts may have problems finding a mate. One advantage of thelytoky may be the ability to produce females without having to wait until a mate is located (Price 1980; Cornell 1988). However, the ability of haplodiploid organisms to produce sons without mating means that some reproductive activity is possible in the absence of a mate.

No clear answers emerge to explain the frequency of thelytoky among parasitoids. Recent work by Stouthamer et al. (1990a) suggests that the wrong questions are being asked: there may be no advantage of thelytoky to the wasp, but an advantage to non-Mendelian elements that control reproduction.

Stouthamer et al. (1990a; Stouthamer 1990) studied a variety of strains and species of minute egg parasitoids in the genus *Trichogramma* (Trichogrammat-idae). They discovered that apparently asexual strains could be induced to produce males if fed certain antibiotics (Stouthamer 1991), or subjected to high temperatures. Further investigation suggested the presence of a micro-organism, probably a bacteria, in the cytoplasm of asexual but not sexual wasps. The microorganism is vertically transmitted in the cytoplasm via eggs, but not via sperm. Using cytological stains, Stouthamer and Werren (1993) were able to see microorganisms in the eggs of "curable" thelytokous strains of wasps. The micro-organisms were absent in sexual strains of wasps and in laboratory strains of cured wasps. Very recently, Stouthamer et al. (1993) have shown by analysis of the 16S ribosomal DNA gene that the organism respon-sible is very closely related to the bacteria that cause cytoplasmic incompati-bility in *Nasonia*.

Genetic and cytological studies reveal that the microorganism has no effect on fertilized eggs destined to become females, but "converts" unfertilized male eggs to females. It does this by preventing chromosome segregation at the first mitotic division, thus converting a haploid to a diploid organism. If the micro-organism has little effect on fecundity, it should spread through the population to fixation. The condition for spread is that infected individuals produce more females than uninfected insects. As it spreads, the number of males produced by the population dwindles to zero so that all female progeny are the result of conversion by the microorganism. However, Stouthamer and Luck (1993) have shown that when the wasps are presented with unlimited hosts, thelytokous strains produce fewer total offspring then sexual strains, and either the same number or fewer female offspring. Why the presence of the microorganism influences fecundity is not known. However, under natural conditions wasps may not be egg limited and so may not be disadvantaged by the infection. Most thelytokous *Trichogramma* are found with sexual individuals and typi-cally constitute less than 5% of the population (one population of *Tricho-gramma* on the island of Kauai, Hawaii, is 100% thelytokous). The factors that determine the relative frequencies of parthenogenetic and sexual individuals are not yet clear (Stouthamer and Luck 1993).

For a microorganism that induces thelytoky to be successful, its host must be willing to oviposit when unmated. Also, the doubling of the haploid chro-mosome set and the consequent diploid homozygosity must not lead to male development, as it would in species with complementary sex determination (see sec. 4.1). The fact that chromosome doubling leads to female develop-ment suggests that sex determination in these wasps is due to a balance of nuclear and cytoplasmic effects.

Because the microorganism can spread only through female progeny, there is an obvious advantage to the conversion of sons to daughters. Sex ratio bi-ases caused by maternally inherited agents are common in plants and animals, but the complete suppression of male production is very unusual.

Stouthamer et al. (1990b) have used their ability to "cure" asexuality in *Trichogramma* to investigate an outstanding problem in the taxonomy of the genus: how to classify thelytokous strains. Apart from the perennial problem of deciding what constitutes a species in a parthenogenetic taxon, there are particular problems in this genus because the most important characters are only displayed in males. The taxonomy of *Trichogramma* is of more than academic interest as these wasps are widely, and lucratively, used in inundative biological control, and several institutes operate reference libraries of species and strains maintained as live cultures. Stouthamer et al. found they were able to cure forty-four out of forty-seven thelytokous strains available for study. The vast majority of asexual strains were morphologically similar to sympatric sexual species and hybridized freely. There is probably gene flow in the wild from sexual to asexual strains as males are able to mate with and fertilize the eggs of "asexual" females. Stouthamer and Werren (1993) were unable to detect microorganisms in the eggs of the three "uncurable" strains, suggesting that thelytoky may have a different cause in these cases.

Since Stouthamer et al.'s work on *Trichogramma*, antibiotics have also been used to restore male production in the aphelinid *Encarsia formosa*, an important, obligatorily parthenogenetic parasitoid of whiteflies, widely used for biological control in greenhouses (Zchori-Fein et al. 1992). The genus *Encarsia* contains heteronomous hyperparasitoids (see secs. 1.3.2 and 5.3), though male *E. formosa* are produced as normal primary parasitoids of whitefly. Males are nonfunctional, suggesting that this species has been thelytokous for a sufficient period of time for male function to be lost through the accumulation of mutations. Finally, Stouthamer (1990) quotes unpublished work by H. Nadel and T. Unruh, who have used antibiotics to convert to sexuality thelytokous strains of an aphelinid and encyrtid, respectively.

There are good grounds to suspect that microorganisms may be implicated in thelytoky in many other groups. Stouthamer et al. were able to induce male production in *Trichogramma* by subjecting wasps to elevated temperatures. There are many reports in the biological-control literature of males appearing in thelytokous species after exposure to high temperatures, not only in Trichogrammatidae, but also in Encyrtidae, Aphelinidae, Signiphoridae, and Pteromalidae, all families of Chalcidoidea (reviewed in Vinson and Iwantsch 1980a; Stouthamer et al. 1990a, 1990b). A particularly interesting example is the pteromalid *Muscidifurax uniraptor* studied by Legner (1985a, 1985b, 1987a; see also below, sec. 5.4). Material collected in the field is normally thelytokous, but males appear if the wasps are exposed to high temperatures. When different strains of the species are cultured in the laboratory, males appear and increase in frequency over successive generations. What is more, if male *M. uniraptor* are mated to females of certain strains of the congeneric *M. raptor*, this normally sexual (arrhenotokous) species is able to reproduce by thelytoky. The most likely explanation for these results is that an unidentified agent causing asexual reproduction is transferred between the two species. Very recently, Stouthamer et al. (1993) have found a bacterium in *M. uniraptor* that is closely related to the agent causing thelytoky in *Trichogramma*. There is thus strong circumstantial evidence that microorganisms are widely associated with thelytokous reproduction in parasitoids, at least among the Chalcidoidea. However, non-microbe-induced parthenogenesis also occurs. Only further research will tell whether microorganisms are responsible for the majority of instances of parthenogenesis in parasitoids.

5.3 Primary Male Production in Heteronomous Hyperparasitoids

Heteronomous hyperparasitoids in the family Aphelinidae produce females as primary parasitoids of whiteflies and other homopterans, and males as hyperparasitoids (secondary parasitoids) through females of the same or another species of wasp (see sec. 1.3.2 for a wider discussion of the biology of these wasps). *Encarsia pergandiella* is a typical heteronomous hyperparasitoid attacking whitefly (Hunter 1989). However, in one population from Ithaca, New York, some males were observed to develop as primary parasitoids (Hunter et al. 1993). These males were termed primary males to distinguish them from secondary males, which developed in the normal way as hyperparasitoids.

Hunter et al. (1993) found that primary males could be produced only by mated females, while secondary males were produced by both virgin and mated females. Mated females produced primary males only if they had been mated by a primary male. This was true even if the female was derived from a family that had produced primary males. These observations suggested that the ability to produce primary males was a trait inherited only through males. Females mated with primary males produced 0% to 100% primary males among their offspring.

Cytological studies showed that all eggs laid as primary parasitoids are fertilized. In a proportion of eggs from females mated with primary males, one chromosome set condenses and does not take part in cell division. Such eggs are functionally haploid and develop as primary males. If primary males are irradiated and then mated with females, the females produce fewer daughters, but the number of primary males among their progeny is not reduced. The irradiation damages the paternal chromosome set and affects the survival of daughters in which the paternal chromosomes are incorporated and expressed. The fact that primary sons are unaffected suggests that it is the paternal rather than the maternal chromosome set that is condensed and lost.

The observation that primary males are caused by a paternally inherited factor that leads to the condensation and loss of the paternal chromosome set is strongly reminiscent of *psr* in *Nasonia vitripennis* (sec. 5.1). Hunter et al.

(1993) conclude that the most likely explanation for primary males in *E. per-gandiella* is the presence of an analogous non-Mendelian element that converts females to males. The physical basis of the primary male factor is unknown although there is no sign of a supernumerary (B) chromosome as in *N. vitri-pennis*. Another difference between this element and *psr* is the considerably lower transmission efficiency. The primary male factor will spread if an infected female produces more primary males than the number of secondary males produced by a normal female. The low rate of transmission decreases the likelihood of spread but, at least in the Ithaca population, this is countered by a very high rate of egg fertilization. Most eggs are fertilized because of the superabundance of unparasitized whitefly (hosts for females) in comparison to whitefly containing wasp immatures (hosts for males) (M. S. Hunter 1993; see sec. 4.2.2 for a discussion of sex ratios in heteronomous hyperparasitoids). Finally, if the element responsible for primary males has a higher transmission frequency through sperm than through eggs, there is a clear evolutionary advantage to sex conversion.

5.4 Gregarious Oviposition in *Muscidifurax*

Parasitoids in the pteromalid genus *Muscidifurax* attack a variety of muscoid flies inhabiting cow sheds, stables, and other similar habitats. *Muscidifurax raptor* is a widespread species found in both the old and the new worlds while the other four described species are more local and known only from the new world (Legner 1969; Kogan and Legner 1970). It is possible that the four species with restricted geographical range have evolved from *M. raptor* since the European discovery of the New World. Legner (1979, 1985a, 1985b, 1987a, 1987b, 1988a, 1988b, 1988c, 1988d, 1989a, 1989b, 1989c, 1991 and included references) has conducted a long series of very detailed breeding experiments with a number of strains of the five species of wasp. He has revealed different levels of within- and between-species genetic variation in a variety of reproductive traits, and also increased vigor (heterosis) in hybrid lines. Mention has already been made in this chapter (sec. 5.2) of thelytoky in *M. uniraptor* and the possibility that it is caused by a microorganism. Here I briefly describe experiments on the inheritance of gregariousness in another species, *M. raptorellus*; for much greater detail see Legner (1987b, 1988a, 1989a, 1989b, 1989c, 1991). While selfish genetic elements are probably not involved here, I am at a loss where else in this book to describe these results!

Muscidifurax raptorellus is recorded from a number of localities in South America, but most of Legner's experiments have involved two strains, one from Peru and the other from Chile. Females of the Peruvian strain are solitary, nearly always laying one egg per host and with only one larva successfully completing development in a host. On about two-thirds of the hosts they at-

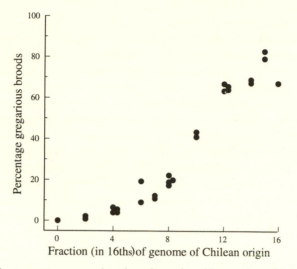

Figure 5.1 The percentage gregarious broods produced by a variety of hybrid strains of the pteromalid *Muscidifurax raptorellus* as a function of the fraction of the genome of Chilean origin. (From Legner 1991.)

tack, virgin females from the Chilean strain lay more than one egg and their larvae develop gregariously. When gregarious, an average of just over three parasitoids develop per host. Virgin females of F_1 hybrids between the two strains show intermediate behavior, they lay more than one egg on about 20% of hosts, and the mean number of progeny developing in a gregarious brood is a little over two. More extensive experiments suggest that the pattern of inheritance of gregariousness is consistent with a polygenic, chromosomal trait. For example, the fraction of hosts receiving more than one egg is strongly correlated with the proportion of the genome derived from Chilean stock in a variety of different hybrid stocks (fig. 5.1). Legner (1991) provides rough estimates for the heritability of gregarious oviposition and of gregarious brood size, and also an estimate of the number of loci involved in each case.

The most curious aspect of the inheritance of gregariousness in *Muscidifurax* is that the oviposition behavior of a female changes after mating. Thus Chilean females significantly reduce the percentage of hosts in which they laid gregarious clutches from about 75% to 55% after mating with a Peruvian male. Peruvian females normally never produce gregarious broods, but do so about 10% of the time if mated with a Chilean female. Virgin F_1 hybrids produce about 20% gregarious broods; those mated with Peruvian males produce 10%, while other females mated with Chilean males produce 45% (Legner 1989a). The size of gregarious broods shows parallel changes. The shift in behavior is permanent and unaffected by additional matings. In further experiments, Legner has shown that the extent to which behavior changes after mating depends critically on the genetic composition of the two partners.

How does the male influence the behavior of the female he mates? Presumably some substance or agent is transmitted in the seminal fluid at the time of mating. Because the response of the female is so predictable, Legner favors the transmission in some form of genetic material that is expressed in the mated female and alters her behavior. As yet there is no evidence for or against this hypothesis. It is also not clear whether this phenomenon has any adaptive significance, or whether it only occurs in laboratory breeding experiments. Legner has considered the possibility that the extranuclear factor that changes the behavior of a mated female is also responsible for the genotype of the progeny. However, the pattern of inheritance is so typical of a polygenic trait that he has rejected this idea (Legner 1991). There is much yet to be discovered about the evolution of gregariousness in *Muscidifurax*.

5.5 Conclusions

Nearly all the work described in this chapter has been published in the last ten years, most in the last five years. This is a rapidly advancing field, and one is left with the strong impression that only the tip of an iceberg has so far been revealed. It is clear that in interpreting observed patterns of sex ratio, the possibility of non-Mendelian elements distorting the sex ratio must be borne in mind. It is also clear that it may be necessary to completely revise ideas about some subjects, for example the evolution of thelytoky. The lack of sex determination constraints that allow parasitoids to produce locally adaptive sex ratios, and that make the behavioral ecology of parasitoid sex ratios so rewarding to study, may also make their reproductive system particularly vulnerable to distortion by rogue genetic elements.

6

The Immature Parasitoid

There is no certainty that a parasitoid egg laid on a host will develop success-fully to the adult stage. Some hosts are of such poor quality that the parasitoid dies during development, or death may be caused by a defense response mounted by the host. Parasitoids are frequently a major source of host mortal-ity and there is intense selection pressure on the host to evolve defenses against parasitoid attack. The parasitoid itself is selected to develop counteradapta-tions against these defenses. The parasitoid may also evolve the ability to manipulate host behavior and physiology to improve its chances of survival. Larvae may have to compete against other parasitoids, both of their own and other species; the larvae of solitary parasitoids often have large mandibles and fight until just a single individual remains, while the larvae of gregarious spe-cies must compete for the resources that constitute the host.

This chapter discusses the behavioral and evolutionary ecology of imma-ture parasitoids. An understanding of the factors that influence the growth and development of the young insect is essential in interpreting the patterns of oviposition behavior discussed in chapters 3 and 4. In addition, the evolution-ary conflict between hosts and parasitoids, and the competitive interactions between larval parasitoids, present a variety of interesting evolutionary prob-lems. There is a very large literature on the physiology of immature parasit-oids, and their molecular and cellular interactions with the host. Here, no attempt is made to review this field except where it is of direct relevance to evolutionary problems under discussion. Two important sets of recent reviews of the physiology of host-parasitoid interactions are the collections of papers in *The Journal of Insect Physiology* (1986, 32, 249–423) and *Archives of In-sect Biochemistry and Physiology* (1990, 13, 1–265).

In the first section of this chapter I will discuss the factors that influence larval fitness in hosts of different quality. The second and third sections de-scribe adaptations by the host to defend itself against parasitism, and counter-adaptations by the parasitoid. The manipulation of host behavior and physiol-ogy by the parasitoid is described in the fourth section, while the fifth section treats competition between parasitoid larvae in the host. The first instar larvae of a few parasitoid groups are highly modified for host location. This aspect of the behavior of immature parasitoids has already been discussed in the chapter on host location (see sec. 2.2.5).

6.1 Host Quality and the Juvenile Parasitoid

When discussing the evolution of oviposition behavior and sex ratio (chapters 3 and 4), I assumed that hosts vary in quality for the developing parasitoid, though the mechanisms responsible were not specified. An understanding of these mechanisms can help disentangle the process of host acceptance, and also suggests new evolutionary problems about the utilization of host resources that have received rather limited attention.

Three main components of parasitoid fitness are influenced by host quality: (1) parasitoid survival until the adult stage; (2) parasitoid size and fecundity as an adult and; (3) parasitoid development time. The first component is obviously important and there is abundant evidence that the second component, parasitoid size, has a strong effect on lifetime reproductive success, although the effect of adult size on male and female fitness is not normally identical. The relationship between adult size and fitness is discussed in section 7.1. The last component is less straightforward. If the population sizes of hosts and parasitoids are constant, a female parasitoid should not care when her offspring emerge (I assume that development time does not influence juvenile mortality). However, suppose the parasitoid population is increasing in size, perhaps at the beginning of the season. Fast development is now favored, as the sooner offspring become sexually mature, the sooner they themselves can reproduce. The advantage of early reproduction in growing populations with overlapping generations (analogous to the advantage of investing money in a bank as soon as possible to maximize interest) is a potent evolutionary force favoring early reproduction (Fisher 1930; Lewontin 1965). There are other consequences of development time: the number of hosts available for parasitism will change over time, perhaps in a predictable manner during the course of a season. Females emerging at different times may thus experience poorer or better conditions for reproduction. Similarly, male parasitoids emerging at different times may find it harder or easier to locate mates. The wealth of subtle interactions that may arise with differential development times and temporal variation in fitness has been very little studied.

6.1.1 HOST SIZE

Host size has a major influence on parasitoid fitness as it determines the maximum amount of food available for the developing parasitoid. Host size is especially critical for idiobiont parasitoids which kill or immobilize the host at parasitism. The hosts of koinobiont parasitoids continue to grow after parasitism although their final size may be related to their size at parasitism. The first major review of the effect of host size on the components of parasitoid fitness was by Salt (1941), although as early as 1844 Ratzeburg had noted that the great variability in the size of the ichneumonid *Pimpla examinator* was due to variation in the size of its host.

The importance of host size in influencing the survival and adult size of idiobiont parasitoids has been demonstrated numerous times (Salt 1941). In gregarious species, fitness is affected not only by host size but also by the number of parasitoids developing in the host (Charnov and Skinner 1984; Waage and Godfray 1985). Hosts may vary in size for many reasons, for example because they belong to different species or are of different instars of the same species. Occasionally, the distribution of host sizes is bimodal, leading to a bimodal size distribution of parasitoids. For example, Mickel (1924) found a strongly bimodal size distribution of both sexes of the mutillid wasp *Dasymutilla bioculata* caused by female oviposition into two species of host (sphecid wasps).

When the host is very small, resources for parasitoid development may be so scarce that the parasitoid fails to mature and dies. If resources are just plentiful enough to prevent death, malformed runts may be produced. Runts are particularly common in the egg parasitoids of the genus *Trichogramma* (e.g., Salt 1941). The smallest individuals tend to be males with misshapen or absent wings. Male *Trichogramma* may be able to mate with females from the same egg batch even if apterous, and it is possible that they have been selected to abandon wings under conditions of extreme food stress (see sec. 7.5.1). Jackson (1958) gives a detailed account of the occurrence of brachypterous runts in another egg parasitoid, the aquatic mymarid *Caraphractus cinctus*.

In koinobiont species, the size of the host at parasitism may be unrelated to its size when killed by the parasitoid. Indeed, as will be discussed below, the parasitoid may manipulate the growth physiology of its host causing premature or delayed maturation at abnormally small or large sizes. However, in some species there is correlation between size at parasitism and size at death (King 1989a; Gunasena et al. 1989). For example, Mackauer (1986) found that when the braconid *Aphidius smithi* attacked aphids weighing between 26 and 146 fg, the size of the adult wasp was linearly related to host size at the time of parasitism. Above 146 fg, wasp size and host size were unrelated.

The capability of the host's cellular immune system to defend itself against parasitoid attack often depends on the size of the host (Salt 1968). For example, several encyrtids in the genus *Metaphycus* are encapsulated by the later instars of soft brown scale (*Coccus hesperidium*) but not by very young scales (Blumberg and DeBach 1981). Similarly, the noctuid moth *Peridroma saucia* cannot mount an effective immune response against the ichneumonid *Hyposoter exiguae* in its first and second instar (Puttler 1961; other examples in van den Bosch 1964, Walker 1959).

Development time is also often influenced by host size, and cases are known in which development is fastest on large, small, or medium-sized hosts (King 1987; Sequeira and Mackauer 1992b). In idiobionts, development often proceeds faster on small hosts, probably for the simple reason that it takes less time to consume a host of this size. In koinobionts, the picture is more complicated. The eggs or larvae of many species remain quiescent or develop only slowly until the host matures and either enters its final instar or pupates. In these species development time is greatest on small hosts (Salt 1941; Vinson and Iwantsch 1980a). However, in some koinobiont braconid parasitoids of aphids, development is fastest on smaller hosts (Kouamé and Mackauer 1991) while in other species the time required for development is relatively constant (Mackauer 1986). In the braconid *Aphidius ervi*, development is slowest on intermediate-sized hosts (Sequeira and Mackauer 1992a, 1992b). The influence of host size on development time may be different for males and females (King 1988).

Larger hosts need not always be associated with increased fitness. Small clutches of gregarious species may have problems developing in large hosts. DeLoach and Rabb (1972) found that immature mortality of a tachinid fly parasitoid (*Winthemia manducae*) attacking the tobacco hornworm (*Manduca sexta*) was very high in small clutches of one to three compared with normal clutches of ten to twenty or higher. Small clutches of the braconid wasp *Apanteles glomeratus* suffer heavy mortality from host encapsulation, which may set a lower limit to viable clutch size (Kitano and Natatsuji 1978; Ikawa and Okabe 1985; Kitano 1986). However, the small clutches in these experiments were produced by disturbing oviposition, and it is possible that this may have interfered with the injection of a venom, known to reduce encapsulation (Kitano 1982; Wago and Kitano 1985). Small clutches of *Trichogramma* tend to drown in large hosts or are forced to overeat and emerge deformed (Flanders 1935; Schieferdecker 1969; Strand and Vinson 1985). Jackson (1958) observed that large clutches of the aquatic mymarid *caraphractus cinctus* could cause the host to split open, leading to the death of the whole brood. Similar observations have been made of *Pediobius (=Pleurotropis) painei*, a eulophid endoparasitoid of a leaf-mining beetle (Taylor 1937), and of *Aprostocetus (=Tetrastichus) hagenowii*, a gregarious eulophid egg predator of cockroach oothecae (Heitmans et al. 1992). Boulétreau (1971) found that small clutches of the gregarious pteromalid wasp *Pteromalus puparum* took longer to develop than larger clutches in the same host (butterfly pupae). Small clutches failed to consume all the host resources, and possibly their longer development time reflected an attempt to eat as much host tissue as possible. Alternatively, faster development may be selected in large clutches through competition among siblings to preempt host resources (sec. 3.2.1).

6.1.2 HOST AGE

The age of nongrowing host stages may influence their suitability for parasitoids. As the development of the egg or pupa proceeds, host tissue is converted into the morphological structures of the following stage. It is likely that the developing egg parasitoid finds it easier to metabolize yolk than the tissue of the embryonic larva, and this may explain why successful parasitism by some egg parasitoids occurs only if newly laid eggs are attacked (Strand 1986). Similarly, pupal parasitoids may find it easier to feed from relatively undifferentiated young pupae and prefer to attack this stage (e.g., Chabora and Pimentel 1977). If parasitoid attack occurs very late in the host egg or pupal stage, the host may hatch or emerge before the parasitoid egg has had time to hatch.

Host age and host size are intimately related for parasitoids that attack growing stages such as the larva. However, host quality may be influenced by aspects of host age independent of size such as the stage of the moulting cycle. Many tachinid flies lay their eggs externally on the surface of insect larvae. If their eggs have not hatched before a molt occurs, they are sloughed off with the discarded skin. Hosts near to moulting are thus not susceptible to parasitism. The endocrine cycles associated with moulting and pupation may also influence host suitability. The ichneumonid *Hyposoter exiguae* suspends development as a first instar larva until a hormonal cue from its host (the cabbage looper, *Trichoplusia ni*) triggers the parasitoid to moult into the second instar and continue development. Parasitism cannot occur successfully after the final ecdysone peak that signals pupation because the parasitoid larvae then fails to receive the correct hormonal stimuli (Smilowitz 1974).

6.1.3 HOST CONDITION AND DIET

Host condition, independent of size and age, can also influence parasitoid fitness. As with host size, host condition may have both positive and negative consequences for the parasitoid. Hosts in poor condition contain reduced nutritional resources for the growing parasitoid, but may also have a weakened cellular defense system (Muldrew 1953; Salt 1956; van den Bosch 1964). For example, the survival of the eucoilid *Leptopilina boulardi* may be twice as high in starving *Drosophila melanogaster* larvae as in well-fed larvae (Wajnberg et al. 1985; Boulétreau 1986). Wasps from starved hosts are smaller and develop faster (Wajnberg et al. 1990). The related eucoilid *L. heterotoma* similarly survives better in poorly nourished hosts (Boulétreau 1986). Kouamé and Mackauer (1991) found that the braconid parasitoid *Ephedrus californicus* developed faster on aphids that had been starved of food. In other koinobionts whose development is synchronized with their host, starvation reduces host growth and hence increases parasitoid development time (Pierce and Holloway 1912). Reduction in host condition may occur at high host densities due to competition for limiting resources (Vinson and Iwantsch 1980a), leading to direct or inverse density-dependent mortality depending on whether parasitoids perform better or worse on stressed hosts.

Parasitoid survival may be influenced by the ecology of the host. *Drosophila* species, for example, live in habitats that vary in ethanol concentration.

Species and strains that live in ethanol-rich habitats show a marked tolerance to this chemical (David and van Herrewege 1983). Parallel adaptations are found in their parasitoids, with strains and species attacking *Drosophila* in alcohol-rich microhabitats showing greater metabolic tolerance of ethanol (Boulétreau and David 1981; Boulétreau 1986).

Secondary chemicals in plants are thought to be used as protection against herbivores (e.g., Feeny 1976; Rhoades and Cates 1976). These compounds may have a variety of effects on larval parasitoids (their use by adult parasitoids in host location has already been discussed, in sec. 2.2.1). Parasitoids may suffer the same detrimental effects as their hosts or they may benefit from developing in a stressed host less capable of mounting an immune defense. Some herbivores sequester plant chemicals and use them in defense against predators and possibly parasitoids (e.g., Roeske et al. 1976); parasitoids themselves may sequester the same compounds for their own defense (Reichstein et al. 1968; Benn et al. 1979; Slansky 1986). Larvae of moths in the genus *Zygaena* (Zygaenidae) sequester cyanogenic compounds from their food plants (*Lotus* and related Fabaceae) but their specialist parasitoids possess enzymes that detoxify these compounds (Jones 1966).

There are a number of examples of the effect of host plant on the performance of juvenile parasitoids (Flanders 1942; Vinson and Iwantsch 1980a). For example, larger adult chalcids (*Brachymeria intermedia*) are reared from gypsy moth (*Lymantra dispar*) that have fed on the leaves of red oak in comparison with alternative foodplants (Greenblatt and Barbosa 1981). Red scale (*Aonideiella aurantii*) is immune to the encyrtid *Habrolepis rouxi* when feeding on sago palm, but not when feeding on citrus (Smith 1941). These results might be due to the action of specific chemicals or to general differences in the size and physiology of hosts fed different foodplants. It is easier to pinpoint the importance of particular chemicals if hosts can be fed on artificial diets with or without secondary plant compounds. When nicotine is added to the artificial diet of the tobacco hornworm (*Manduca sexta*), the proportion of parasitoids failing to pupate successfully increases (Thurston and Fox 1972; Barbosa 1988). However, the effects of nicotine on this specialized herbivore and parasitoid are relatively small compared with the detrimental effects experienced by several parasitoids of *Spodoptera frugiperda* when nicotine, a chemical they would not normally encounter, is added to their diet (Barbosa et al. 1986; Barbosa 1988). The chemical tomatine from tomatoes has been shown to have numerous detrimental effects on survival, speed of development, adult size, and morphology of the polyphagous ichneumonid parasitoid *Hyposoter exiguae* (Campbell and Duffey 1979, 1981). The weight of *Heliothis virescens* and its ichneumonid parasitoid *Campoletis sonorensis* both increase when small quantities of the important cotton secondary compound, gossypol, are added to its artificial diet. The two species occur frequently on cotton and are probably well adapted to this food plant, although high concentrations of gossypol lead to reduced growth and survival (Williams et al. 1988). These data support Barbosa's (1988) hypothesis that specific secondary compounds will have least effect on parasitoids that specialize on hosts feeding on plants containing the chemical and most effect on generalist parasitoids (see also sec. 8.2.5).

6.1.4 HOST SPECIES

Many parasitoids lay eggs on a variety of host species and their larvae suffer different levels of immature mortality. A good example of this is provided by Janssen's (1989) study of the braconid *Asobara tabida* and the eucoilid *Leptopilina heterotoma* attacking nine species of *Drosophila* (see table 3.1). The probability of larval survival varies from 0% to 90%. The major factor influencing parasitoid survival is probably how well the larvae are adapted to cope with the host's cellular encapsulation system (see below), and this is likely to be the main factor influencing host suitability for most endoparasitoid species. The response of *Drosophila* parasitoids to a spectrum of different host species has also been studied by Jenni (1951) and Carton and Kitano (1981). Carton and Kitano found evidence that the ability to encapsulate the eucoilid *Leptopilina boulardi* "appears in some way to parallel the phylogenetic relationships" among the different host species.

Parasitoids probably only rarely meet different strains of the same host while foraging in nature. However, in the laboratory, parasitoid fitness can be shown to vary on host strains from different localities. For example, Chabora (1970a, 1970b) and Chabora and Chabora (1971) found marked differences in the performance of the pteromalid *Nasonia vitripennis* on housefly strains (*Musca domestica*) collected in different places. Understanding the relative performance of parasitoids on geographical strains of one species is of great · importance for planning biological control programs.

6.2 Host Defenses

The first line of host defense is to avoid discovery by the parasitoid. When this fails, it may be possible to fight off the parasitoid before an egg is laid. These host defenses, primarily against the adult parasitoid, are discussed in the next chapter (sec. 7.4). For hosts that are killed or permanently paralyzed at the time of oviposition, the prevention of parasitoid attack is the only defense. Other parasitoid species, however, do not paralyze their host, or the paralysis is only temporary. The host recovers and continues feeding while the immature parasitoid enters a stage of arrested or slowed development, and delays killing the host until it reaches a larger size (koinobionts). Hosts suffering this type of parasitism may be able to destroy the parasitoid before it does any permanent damage. The vast majority of koinobionts are endoparasitoids, and the main mechanism of host defense is through a cellular immune reaction called encapsulation.

6.2.1 ENCAPSULATION

Insect blood contains cells called hemocytes that move around the body in the hemocoel. On the basis of cell morphology, a number of hemocyte types are recognized in the literature (Gupta 1985; Brehélin and Zachery 1986). When the insect is infected by a parasitoid (or a parasite or an inorganic contaminant), a type of cell called a plasmatocyte recognizes the infection as foreign and adheres tightly to the parasitoid egg or larvae. The cells form a multilayered capsule that probably causes the death of the parasitoid through asphyxiation (Salt 1963, 1970). The recognition of the parasitoid as foreign may be mediated by another cell type, the granular hemocyte, which ruptures on contact with the parasitoid, releasing the contents of granular inclusions onto the surface of the egg or larva (Götz and Boman 1985; Götz 1986; Gupta 1986). The capsule is normally melanized and may be visible through the wall of the host larva. In *Drosophila*, the presence of a parasitoid egg causes some hemocytes to change shape and become very flattened (lamellocytes). The lamellocytes form the capsule around the egg or larva which is rapidly melanized, a process that involves the disintegration of another type of hemocyte, the crystal cell or oenocytoid (Nappi 1975; Rizki and Rizki 1984a; Nappi and Carton 1986). After a capsule has been formed and the parasitoid destroyed, some cells may leave the outer layers of the capsule, which appears to decrease in size. The molecular basis of foreign-body recognition and capsule formation is not yet understood.

In comparison with the vertebrate immune system, the immune system of invertebrates is only weakly specific. There is also little counterpart to learned immunity (but see Karp 1990). There are no records of the response to a second parasitoid attack being affected by previous experience. Encapsulation is found in chelicerate and crustacean arthropods as well as insects, and similar defense responses occur in annelids and molluscs (Götz 1986). Encapsulation thus evolved long before the genesis of parasitoids as a generalized defense against pathogens, parasites, and other foreign bodies.

Although encapsulation is the chief physiological defense against parasitism, in some species other mechanisms may be involved. When the ichneumonid *Banchus flavescens* attacks the cabbage looper *Trichoplusia ni*, the larvae are expelled during moulting in a cyst that forms between the cuticle and the epidermis (Arthur and Ewen 1975; Ewen and Arthur 1976). Cell-free (humoral) immunity involving prophenyloxidase cascade may also play a role. Melanin can be deposited directly onto the surface of foreign objects, though the importance of this reaction in parasitoid biology is unclear (Vinson 1990a).

6.2.2 BEHAVIORAL DEFENSES AGAINST JUVENILE PARASITOIDS

In a few cases, the host may be able to defend itself against parasitism by physically removing eggs. A tachinid described as *Winthemia quadripustulata* places its eggs on the last thoracic segments of *Cirphis unipuncter* (Noctuidae) because eggs placed farther back are groomed off by the caterpillar (Allen 1925). Several hosts which are attacked by the first instar larvae of tachinid flies destroy the planidial larvae as they attempt to burrow through the host integument (Muesebeck 1922; Strickland 1923), but "the great majority of hosts, however, show no discomfort during the time the maggots are penetrating the body" (Clausen 1940a). The relatively few ichneumonids with external eggs often place them out of the reach of the host mandibles (Baltensweiler and Moreau 1957). The eggs of tryphonine ichneumonids have elaborate anchors that enable them to survive moulting and possibly also more active attempts by the larvae to rid themselves of their parasitoids (Mason 1967; Kasparayan 1981). Vinson and Iwantsch (1980b) report that the boll weevil (*Anthonomus grandis*) can remove the eggs of the braconid ectoparasitoid *Bracon mellitor.*

6.2.3 SUICIDE

If a host is attacked by a parasitoid and is unable to mount a physiological or behavioral defense, it is doomed and will pass no genes to future generations (the only exceptions are hosts that are attacked as adults or mature nymphs and which may mate or reproduce before death). However, selection may act on the host to commit suicide because by killing the parasitoid it protects conspecifics from attack (Shapiro 1976; Smith Trail 1980). It is unlikely that a host would be selected to commit suicide for the general good of the species or population. Such an explanation requires group selection, which is at best a very weak force in most biological populations (see Grafen 1984 for a discussion). However, suicide might be favored if the act promoted the well-being of relatives. Here kin selection is responsible (although the argument could be rephrased in terms of intrademic selection); a conditional suicide gene perishes when expressed, but its overall frequency increases because of the greater survival of copies of the gene in relatives.

Shapiro (1976) suggested that some nymphalid butterfly caterpillars commit suicide, although his observations can also be explained by the manipulation of host behavior by the parasitoid and are discussed in section 6.4.3. A more recent study of host suicide concerns the pea aphid *Acyrthosiphon pisum* and its braconid parasitoid *Aphidius ervi* in British Columbia (McAllister and Roitberg 1987; McAllister et al. 1990). *A. pisum* is found in both the wet coastal and the hot dry interior regions of British Columbia. When colonies of aphids are disturbed by predators, individuals adopt one of three strategies: they remove their stylets from the plant and remain rigid; they run from the predator; or they drop from the plant. Dropping from the plant is the most effective defense against predation but may lead to desiccation and death; this is especially likely to happen in the dry interior, and Roitberg and Myers (1978) had shown previously that aphids from this region were less likely to attempt to avoid predation by dropping. McAllister and Roitberg (1987) studied the behavior of parasitized aphids and found that in contrast to healthy aphids, parasitized aphids were more likely to drop from the plant if they came from dry areas where the risk of subsequent death was high (fig. 6.1a). They interpreted this behavior as adaptive suicide. McAllister et al. (1990) also compared the behavior of newly parasitized fourth instar aphids and aphids of the same age that had been parasitised four days previously. The latter group die without reproducing, and suicide is thus without personal costs, while the former group normally produce a few offspring before succumbing to the wasp larva, offspring that would be lost if the aphid commits suicide. They found that aphids which still possessed some residual reproductive value were less willing to drop from the plant than aphids with no future reproductive prospects (fig. 6.1b).

McAllister and Roitberg's (1987) first study, published in *Nature*, prompted some lively correspondence in the same journal criticizing the conclusions (Tomlinson 1987; Latta 1987). It was suggested that the observations might be explained by changes in host behavior caused by different strains of parasitoid, rather than by host suicide. In their second paper, McAllister et al. (1990) explored the effect of wet region and dry region parasitoids on hosts from both localities. They found that the increased propensity to drop from the plant was largely a property of the host and not the parasitoid. However, it is possible that in the absence of selection for suicide, the dry region biotype of aphid may react to parasitism in a way that makes falling off the plant a more likely accident of attempted escape from the predator.

Some other criticisms are harder to answer. An increase in the propensity to drop from the plant when a predator approaches seems an odd way of committing suicide: why not just stop feeding, or drop from the plant without waiting for a predator to appear? Better still, why not walk toward the predator? Of course natural selection will modify existing behavior, and there may be counteradaptations by the parasitoid, but it does seem strange that such a roundabout method of suicide should evolve. Other objections might be countered by further work. Tomlinson (1987) and Latta (1987) point out that natural selection does not distinguish between the actual death of the parasitized aphid and the removal of the parasitoid from the vicinity of kin. Might not falling off the plant always be the best strategy, because even if the parasitoid survives it is unlikely to attack the aphids kin? Pea aphids are parthenogenetic during the summer and it would be interesting to know the spatial distribution of different clones. Latta (1987) raises the possibility that it may be in an aphid's interest to carry the

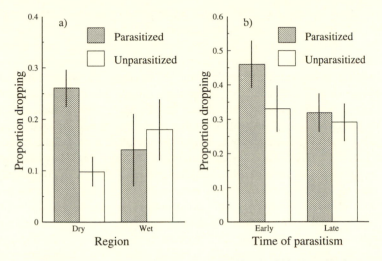

Figure 6.1 The proportion of parasitized and unparasitized aphids dropping from a plant when their colony is approached by a coccinellid predator. (a) A comparison of the response of aphids from the dry interior and wet coastal regions of British Columbia. (b) The effect of time of parasitism on the likelihood of dropping from the plant. Aphids from the dry interior were parasitized during their second instar (early) or during the fourth instar (late). The bars represent one standard error. (From McAllister and Roitberg 1987, and McAllister et al. 1990.)

parasitoid, like a Trojan horse, into the camp of an another clone. Natural selection could favor this behavior because it increases the competitive advantage of one clone over another. Behavior of this type ("spite") has been invoked and studied on other occasions in behavioral ecology, though models suggest that it is unlikely to be a strong force in evolution (Knowlton and Parker 1979).

I am not convinced that the behavior observed by McAllister and Roitberg (1987) is adaptive suicide. Adaptive suicide has also been invoked to explain the change in behavior of bumblebees parasitized by conopid flies (Poulin 1992), a phenomenon I again think has a more plausible alternative explanation (Müller and Schmid-Hempel 1992; see sec. 6.4.3). Nevertheless, these studies raise fascinating questions, and I see no reason why under certain conditions adaptive suicide cannot evolve, especially when hosts live in small isolated patches that survive for several generations.

6.3 Countermeasures

Parasitoids have evolved a variety of stratagems to overcome host defenses (Salt 1968; Vinson 1990a). Here I make a broad division between the avoidance of host defenses by passive means, and mounting an active attack on the host's defenses. I also consider the possibility of successive counteradaptations leading to an arms race between host and parasitoid.

6.3.1 AVOIDANCE

Parasitoids may hide their eggs in tissues where they are protected from encapsulation or coat their eggs or larvae with a protective layer that prevents their recognition as foreign bodies. Some parasitoids tolerate the host's encapsulation response, or even turn it to their own advantage. It has also been suggested that the oviposition of several eggs is a means of passively exhausting the host's immune system.

HIDING

Many parasitoids lay their eggs in specific organs or tissues. By placing their eggs in sites away from the cells circulating in the hemocoel, they are probably protecting their offspring from encapsulation (Salt 1963, 1968, 1970). For example, many species of *Platygaster* (Platygasteridae) which chiefly parasitize gall midges, lay their eggs in nerve ganglia where they are protected by a cyst of host tissue. Any eggs laid in the body cavity are encapsulated (Clausen 1940a, Salt 1970). The ichneumonid *Amblyteles subfuscus* places its eggs in the salivary glands of larvae of the noctuid moth *Euxoa ochrogaster* (Strickland 1923), while several parasitoids of the geometrid *Bupalus piniaria* develop in gut tissue (Klomp and Teerink 1978b).

The larvae of many tachinid parasitoids that emerge from eggs ingested by the host migrate from the gut to take up temporary residence in ganglia, salivary glands, gonads, muscle fibers, or in the fat body, although some species remain in intestinal tissue (Clausen 1940a; Salt 1968). Strickland (1923) records that *Gonia capitata*, a parasitoid of various cutworm larvae (noctuid moths), remains in the intestines for some days and then makes a dash for the brain. On its way, it evokes a vigorous encapsulation reaction which destroys many parasitoid larvae. After some days in the brain, the larva is able to enter the hemolymph, apparently without adverse effects.

It is thought that endoparasitoids normally evolve from ectoparasitoids, and that koinobionts normally evolve from idiobionts (see sec. 8.3.3), although transitions in both directions are possible. Feeding externally or permanently paralysing or killing the host is one strategy for avoiding the host immune response. Host eggs generally lack hemocytes capable of mounting an immune response, and it is possible that egg parasitism may have arisen as a means of avoiding host defenses. A few parasitoids complete development extremely quickly and it has been suggested that this is an adaptation to outpace the host's immune response (Salt 1968). For example, the ichneumonid *Phaeogenes nigridens* takes three days to complete three instars in pupae of its host, the European corn borer (*Ostrinia (=Pyrausta) nubilalis*) (H. D. Smith 1932).

PROTECTIVE COATING

The hemocoel is bounded by a basement membrane which presumably has surface properties that allow the hemocytes to recognize it as self and not attempt to mount an immune response. Parasitoids may protect themselves by mimicking the surface properties of the host. Alternatively, the surface of the parasitoid may resist adhesion by encapsulating cells.

The importance of the surface of the parasitoid in escaping encapsulation is suggested by a number of lines of evidence. The ichneumonid *Venturia (=Nemeritis) canescens* normally avoids encapsulation in the flour moth *Ephestia kuehniella*. However, Salt (1965) found that if the surface of the egg was gently abraded (or chemically treated) and then the egg injected into the host, the parasitoid was encapsulated. *V. canescens* eggs, as they pass through the calyx region of the female's reproductive tract, are coated in viruslike particles (Rotheram 1967). Antibodies raised against the proteins contained in the particles cross-react to proteins from the unparasitized host, suggesting that the viruslike particles may be involved in molecular mimicry (Schmidt and Schuchmann-Feddersen 1989). However, the host proteins are implicated in the immune system and thus the interaction may be more complex than just mimicry (Schmidt et al. 1990). The braconid *Cardiochiles nigriceps* oviposits into the moths *Heliothis virescens* and *Helicoverpa (=Heliothis) zea* although the wasp's egg is encapsulated in *H. zea*. The egg has an outer flocculent layer in *H. virescens* which is lost in *H. zea*, possibly as a result of enzymes circulating in the hemolymph. The layer appears to protect the egg from encapsulation (Lewis and Vinson 1968; Vinson 1990a). If a parasitoid is injured during competition between larvae (see below), hemocytes can be seen aggregating around the site of the wound where its outer protective coating has been compromised (Salt 1970).

The reason why endoparasitoids remain as first instar larvae for long periods of time while their hosts grow and increase in size is normally assumed to be to avoid damaging the host. Salt (1968) has suggested an alternative, or supplementary, advantage: the first instar larva retains a protective covering that would be lost by moulting into the next instar. He also argues that an egg laid in a very young larva (or, in the case of egg-larval parasitoids, the egg) often elicits a very mild immune response and gives the larva time to become acceptable to the host by developing a protective coating. Oviposition in poorly defended stages may also provide time for the more active forms of parasitoid countermeasure discussed below to permanently destroy the immune system.

TOLERANCE

Encapsulated larvae often die through asphyxiation. The larvae of many tachinid flies have posterior spiracles armed with hooks which they use to penetrate the skin of their host, or an air sac or major trachea. The host responds to the injury by laying down wound tissue that forms a funnel around the posterior region of the fly larvae (the respiratory funnel). Once the maggot has tapped into a source of oxygen, it cannot be destroyed by asphyxiation and it is often able to prevent the hemocytes from forming a rigid capsule, or to use the capsule formed by the hemocytes as a respiratory sheath. Salt (1968) contrasts this deflection of the host's immune response with more active means of defense as "ju-jitsu rather than slogging"!

Dryinid wasps in the subfamily Aphelopinae, parasitoids of typhlocybine Cicadellidae (leafhoppers), have a curious life history that also seems to involve the subversion of host defenses. The egg and first instar larva are endoparasitic but the second instar larva pushes its way through an intersegmental membrane so that its head remains in the host but the rest of the body is external. As the larva grows, its cast skins form a sac that protrudes from the side of the leafhopper. Host tissue of unusual histology grows around the head of the dryinid to form a large cyst with a hole in front of the larva's mouthparts. The physiological interactions between host and parasitoid are unclear but it does appear that the wasp is tolerating and manipulating a host defensive response. Dryinids in other subfamilies have similar biology but without cyst formation (Clausen 1940a; Waloff and Jervis 1987; Olmi 1993).

The formation of another type of capsule has recently been described in the aphelinid *Eretmocerus mundus*. Eggs are laid on small whitefly nymphs and the parasitoid larva burrows into the host where it remains in a state of suspended development until the whitefly enters the fourth instar. During this period it is surrounded by a capsule that appears to be derived from host material, although of different structure to that formed by the cellular immune response (Gerling et al. 1990). The formation of the capsule appears to be a normal part of the development of this wasp and may serve to protect it from other host defenses.

OVIPOSITION OF SEVERAL EGGS

A number of authors have noted that the probability of at least one egg surviving encapsulation in solitary parasitoids is greater when the host contains several eggs. (Askow 1968a) For example, Puttler (1974) and Berberet et al. (1987) found that the presence of several parasitoid eggs of the ichneumonid *Bathyplectes curculionis* increased the chance of successful parasitism of the weevil *Hypera postica*. Similarly, while only 8% of solitary eggs of the ichneumonid *Hyposoter exiguae* survive encapsulation in the noctuid moth *Laphygma exigua*, 97% of larvae with two eggs give rise to a (single) parasitoid adult (Puttler and van den Bosch 1959). It is not known in these examples whether a single female lays more than one egg during a single oviposition bout, or whether superparasitism occurs. The increased avoidance of encapsulation may favor multiple oviposition during a single oviposition bout.

Several parasitoid eggs may exhaust the cellular immune response of the host: the multiple target hypothesis. Salt (1968) has gone further and suggested that gregariousness and polyembryony also have evolved to saturate the host immune system. However, some hosts are able to encapsulate many parasitoid eggs, and insect larvae can frequently encapsulate very large numbers of inorganic particles injected into the hemolymph. These observations led Vinson (1990a) to conclude that the multiple target hypothesis is unlikely to be responsible for increased successful parasitism in hosts with several eggs. A possible alternative explanation is that substances that attack the host immune system, either injected by the parasitoid at oviposition or by the immature parasitoid, act in a dosage-dependent manner. In some parasitoids there is evidence of within-population variability in susceptibility to encapsulation (sec. 6.3.3), and Streams (1971) argued that the increased avoidance of encapsulation in superparasitized individuals can be a purely statistical effect: with two eggs there is a greater chance that at least one is able to resist encapsulation.

6.3.2 ATTACK

Parasitoid larvae wriggle vigorously in the host and this probably allows many species to shake off a mild encapsulation attack. However, Salt (1968) doubts whether physical resistance by itself can ever save a parasitoid from a full immune response in a strange host. Chemical warfare appears to be the most important strategy in overcoming host defenses. Here I briefly describe the effects of chemicals injected by the adult wasp at oviposition, the importance of teratocytes and other larval products, and finally the action of viruses associated with ichneumonoid wasps.

CHEMICALS INJECTED BY THE ADULT WASP

At oviposition, many parasitoids sting the host to cause permanent or temporary parasitism. The chemicals used by Hymenoptera to sting parasitoids and prey have been reviewed by Piek and Owen (1982) and Piek (1986). In general, hymenopteran venoms consist of one or more low molecular weight proteins which typically disrupt membrane function or act at nerve synapses. The best-studied parasitoid, the braconid *Bracon hebetor*, injects two proteins into its host which act presynaptically at neuromuscular junctions. The venom paralyzes the host, which may live for several weeks after attack but without moving or moulting. Occasionally a potential host species is hypersensitive to parasitoid venom, causing the death of both the host and the parasitoid (Gerling and Rotary 1973; Vinson and Iwantsch 1980a).

Egg parasitoids are faced with the problem of a rapidly developing host that may hatch before their own eggs finish development. The egg parasitoid *Telenomous heliothidis* (Sceliondae) injects a substance into the host (*Heliothis virescens*) during oviposition that causes the arrestment of embryogenesis (Strand et al. 1985, 1986; Strand 1986). In *T. heliothidis*, a substance produced by the teratocytes (see below) acts in concert with the arrestment chemical and causes host necrosis. The trichogrammatid egg parasitoid *Trichogramma pretiosum* attacks the same host. This species lacks teratocytes, but a substance injected by the adult both arrests development and causes tissues necrosis (Strand 1986). Arrestment of development and tissue necrosis are common features of the development of egg parasitoids.

Unlike egg parasitoids, larval parasitoids must face the full host immune response. Timberlake (1912) first proposed that the adult wasp injects substances that disable the immune system. Rizki and Rizki (1984b) suggested that the eucoilid *Leptopilina heterotoma*, which parasitizes *Drosophila melanogaster*, injects at oviposition a chemical (lamellolysin) that renders ineffective the cells responsible for cellular encapsulation (the lamellocytes). The related parasitoid *L. mellipes* is normally encapsulated by *D. melanogaster* although survives if the same host has been attacked by *L. heterotoma* (Streams and Greenberg 1969). (A similar observation was made by Pemberton and Willard (1918), who found that the eulophid *Tetrastichus giffardianus* could survive only in Mediterranean Fruitflies that had been attacked by the braconid *Psyttalia (=Opius) fletcheri*). Recently it has been demonstrated that lamellolysin is probably a virus (Rizki and Rizki 1990; see below) and it is unclear whether chemicals injected by the parent are involved in these interactions. However, some ichneumonids that lack teratocytes and polydnaviruses are believed to inject venom into their hosts to suppress the immune response (Vinson 1990a). In braconid wasps, venom is also important in preventing encapsulation but appears to act in a complicated manner in concert with teratocytes and polydnaviruses (Kitano 1982, 1986; Wago and Kitano 1985; Tanaka 1987; Kitano et al. 1990; Strand and Dover 1991; Strand and Noda 1991; Strand and Wong 1991; and see below).

Moulting by larval hosts obviously has not evolved as a parasitoid defense mechanism though it does result in the loss of any ectoparasitoid eggs that happen to be cemented to the host integument. Eulophid wasps in the genus *Euplectrus* sting lepidopteran caterpillars at oviposition and induce temporary paralysis. They lay a variable number of eggs externally on the host, which recovers and continues feeding until ready for the next moult. However, the caterpillar fails to moult and the eggs hatch and the parasitoid larvae consume the host (Neser 1973; Gerling and Limon 1976). Coldron and Puttler (1988) and Coldron et al. (1990) studied *E. plathypenae* attacking the cabbage looper *Trichoplusia ni* and found the failure to moult was caused by a substance injected by the female at the time of oviposition that probably acted directly on the epithelial cells that respond to increased ecdysone titers during moulting. The related *Eulophus larvarum*, another gregarious eulophid, stings hosts (a variety of moth larvae) immediately after they enter ecdysis. The timing of attack appears to be crucial, and the parasitoid remains with hosts near to moulting, waiting for the precise moment to sting. Stung hosts continue to feed at a depressed level but fail to moult again (Shaw 1981a).

TERATOCYTES AND THE ACTION OF THE LARVAE

In a number of families of parasitoid wasps (many Braconidae, Scelionidae, and some Trichogrammatidae) the serosal membrane that surrounds the parasitoid embryo within the egg shell breaks up into individual cells that continue to grow, but not to multiply, within the host body. These cells, which may grow to become very large, are called teratocytes (in one species the teratocytes increase in volume 3000-fold; O. J. Smith 1952). Between fifteen and eight hundred teratocytes may be produced by a single parasitoid. The surface of teratocytes is covered in microvilli and bears a strong resemblance to the cells of the mid-gut epithelium (Stoltz 1986).

At least three functions have been proposed for teratocytes (reviewed by Vinson and Iwantsch 1980b; Dahlman 1990). First, they may have a trophic function. Teratocytes are normally ingested by the parasitoid and they may take up substances from the hemolymph of the host. Second, they may indirectly weaken the host's immune system, either by providing a large number of targets for encapsulation, or by the uptake of important nutrients from the host hemolymph, causing stress and a reduction in the efficiency of parasitoid attack (Salt 1968, 1971). However, the injection of a few teratocytes into a host does not elicit an encapsulation response which weakens the multiple target hypothesis (Vinson and Iwantsch 1980b; Vinson 1990a). Finally, the prime function of teratocytes may be secretory, perhaps secreting digestive enzymes, substances that manipulate host physiology to the benefit of the parasitoid, substances that interfere with the immune system, or even substances that reduce fungal infection. The ultrastructure of teratocytes (microvilli, extensive granular endoplasmic reticulum, cellular outpocketings, etc.) strongly suggests a secretory function. Although true teratocytes are restricted to certain groups of parasitoids, it is likely that some of their functions in other groups are carried out by intact serosal membranes, for example those found in alysiine and opiine braconids (Salt 1968; Lawrence 1990).

If eggs of the braconid *Cardiochiles nigriceps* are removed from the female, washed, and injected into a host they are normally encapsulated. However, if teratocytes are also injected, encapsulation is much less effective (Vinson 1972b). Similar results have been obtained from a number of other species, and there is evidence that products secreted by the teratocytes of braconids interfere in some way with the proper functioning of the immune system, probably acting synergistically with the action of venom and polydnaviruses (Tanaka and Wago 1990; Kitano et al. 1990; Strand and Wong 1991; and see next section). It is likely that teratocytes have more than one function, and there is evidence that old and young teratocytes may not behave in the same way

(Tanaka and Wago 1990). Scelionids (and trichogrammatids) are egg parasit-oids and thus do not have to contend with a cellular immune system. As dis-cussed above, teratocytes in these families secrete substances that help digest host tissue. Teratocytes may also be involved in the endocrine manipulation of host physiology (see below).

Parasitoid larvae themselves may secrete substances into the host. For ex-ample, the ichneumonid *Pimpla turionellae* secretes a substance from its anus with antimicrobial activity (Willers et al. 1982; Führer and Willers 1986). In hosts with disabled immune systems, microbial infections may be a major prob-lem for the developing parasitoid.

POLYDNAVIRUSES

It has been known for some time that the reproductive tract of many species of Ichneumonidae and Braconidae, particularly those attacking Lepidoptera, contain particles that are injected into the host at parasitism (Rotheram 1967; Salt 1968, 1970). However, it is only relatively recently that many of these particles have been recognized as viruses (Vinson and Scott 1974; Stoltz and Vinson 1979). One type occurs in many different ichneumonids and braconids, although there are differences in viruses isolated from the two families. Viruses from braconids are rod-shaped and of variable length, are enveloped by a sin-gle unit membrane, and often have a tail. The better-studied viruses of ichneu-monids are spindle-shaped (fusiform) with a double membrane (Blissard et al. 1986). In all cases, the genetic material is DNA, which is organized into a number of double-stranded, circular molecules that are coiled into super-helices. The characteristic and unique arrangement of the genetic material of these viruses has led to their placement in a new family called Polydnaviridae (poly-DNA-viruses) (Stoltz et al. 1984). The molecular biology and physiol-ogy of polydnaviruses has recently been reviewed by Fleming (1992).

Mature eggs drop from the ovarioles into the calyx and from there into the ovipositor. Under the electron microscope, the fluid in the calyx can be seen to be filled with virus particles. At least in the Ichneumonidae, the virus genetic material is present as a provirus incorporated in the wasps' chromosomes and found in all cells, both in the male and female (Fleming and Summers 1986; Stoltz 1990). However, viral replication occurs exclusively in the calyx cells of the ovary. In some species viral particles containing the correct combination of DNA molecules bud from the cell wall while in other species the cell dies releasing viral particles into the calyx. In the best-studied system, the virus of the ichneumonid *Campoletis sonorensis* (CsV), there are at least twenty-eight different DNA types, ranging in size from 6 to 21 kilobase pairs (kbp). The total viral genome of 190 to 240 kbp is made up of a combination of these molecules in nonequimolar ratios (Blissard et al. 1986). After injection into the host, the virus invades many cell types, and transcription products can be found after about two hours (Fleming et al. 1983). If a parasitoid egg is extracted from a wasp, washed clean of calyx fluid, and then injected into the host, the egg fails to develop and is normally encapsulated. However, if calyx fluid or purified virus is also injected, the parasitoid survives (Edson et al. 1981). Injection of the virus without the egg results in many of the symptoms of parasitism (Vinson et al. 1979). Viral products are thus important in compromising the host's immune system and allowing the parasitoid to escape encapsulation (Stoltz and Guzo 1986; Davies et al. 1987; Fleming 1992). Viruses may also influence the host's nutritional and endocrine physiology to the benefit of the parasitoid (see below). In a number of braconids, venom is necessary for polydnaviruses to prevent encapsulation (Kitano 1982; Tanaka 1987; Strand and Noda 1991), and teratocytes may also be required (Kitano et al. 1990; Tanaka and Wago 1990; Strand and Wong 1991). It is possible that the major role of polydnaviruses (which do not replicate in the host and whose effectiveness may decline with time) in braconids is to protect the egg from encapsulation, a role that after hatching is partly or wholly taken over by the larva or its teratocytes.

There is no evidence of viral replication in the host, and transmission of the virus appears to be through the parasitoid genome alone (Fleming and Summers 1986; Stoltz 1990; Fleming 1992). There are three possible evolutionary origins of polydnaviruses (Whitfield 1990): (1) The viruses may have originated as pathogens of the host. There is some similarity between the structure of polydnavirues and some baculoviruses, common pathogens of Lepidoptera. Baculoviruses can be transmitted passively between hosts by parasitoids (Young and Yearian 1990), and this may represent the first step in the evolution of a more obligate relationship. However, the fact that polydnaviruses do not replicate in the host is evidence against this hypothesis. (2) The viruses may originally have been pathogens of the parasitoid. At least one baculovirus is known to attack a parasitoid (Hamm et al. 1988), and it is possible that such a virus might evolve from being a pathogen to a mutualist. One problem with both these theories is to explain how the virus incorporated itself into the host genome. Incorporation is common among retroviruses, but these are RNA rather than DNA viruses. (3) The third explanation avoids this problem. Perhaps the virus originated in the host genome and is nothing more than a means of delivering venom and other parasitoid products more efficiently to the sites in the host where they are required. In support of this idea there is evidence that some of the genes carried by polydnaviruses are similar to genes expressed in parasitoid venom glands (Webb and Summers 1990). It is possible, however, that genes for venom or other parasitoid products could have been transferred to a virus derived from a host or parasitoid pathogen.

Polydnaviruses from different wasp species vary in their profile of different DNA molecules. They also appear to function in a restricted range of hosts. Within ichneumonids and braconids, polydnaviruses appear to have a restricted taxonomic distribution though as yet too few species have been examined for the taxonomic and ecological correlates of the presence of viruses to become clear. Other types of viruses have been observed in parasitoid calyx fluid (Stoltz 1981; Stoltz et al. 1988; Styer et al. 1987; Hamm et al. 1990; Fleming 1992) although their functions are more obscure. The possible functions of virus-like particles which coat the eggs of *Nemeritis canescens* and some other ichneumonids have already been described (sec. 6.3.1). Recently, a viruslike particle has been found in the eucoilid *Leptopilina heterotoma* which appears to cause the destruction of the host encapsulation response previously attributed to the substance lamellolysin (sec. 6.3.1; Rizki and Rizki 1990). The identification of a possible virus in a cynipoid is particularly interesting in view of Cornell's (1983) suggestion that cynipid gall wasps (which are derived from parasitoids) induce the production of gall tissue in plants by in some way injecting DNA that is expressed in the plant. If this explanation is true, it would be interesting to see if gall wasps used a homologous mechanism to eucoilid parasitoids.

6.3.3 THE EVOLUTION OF DEFENSE AND COUNTERMEASURE

It may be possible to observe the evolution of host defenses and parasitoid countermeasures. There is some evidence for within-population genetic variation in the two traits, and the genetic variation may be maintained by frequency- and density-dependent selection resulting from the population dynamic interaction between host and parasitoid. To understand the maintenance of genetic variation, it is important to know the costs of defenses and countermeasures. An important way of studying the reciprocal selection pressures imposed by hosts and parasitoids is by setting up new interactions, either in the laboratory, or in the field making use of biological control introductions.

The ability of *Drosophila melanogaster* to encapsulate the eucoilid *Leptopilina boulardi* has a genetic basis, and there is evidence of both between- and within-population genetic variability. Boulétreau and colleagues (Boulétreau and Fouillet 1982; Carton and Boulétreau 1985; Wajnberg et al. 1985; Boulétreau 1986) compared hosts and parasitoids from France, Tunisia, and the Congo. Strains from the Congo showed a much higher rate of encapsulation than those from France, which in turn were higher than strains from Tunisia. If a host is parasitized by *L. boulardi*, both host and parasitoid may die (this occurs about half the time), the parasitoid may survive, or the host may successfully encapsulate the parasitoid. Using isofemale-line techniques, within-population genetic variation was found in parasitoid survival in all three host populations, and in encapsulation rate for the Congo population (the frequency of encapsulation was too low in the other population for variance to be discernible). Crosses between populations revealed an inverse relationship between the ability of wasp strains to avoid encapsulation, and the efficiency of the hosts' encapsulation response: Tunisian flies were the worst defended and Tunisian parasitoids the best at avoiding encapsulation; the reverse was true for the Congolese strains. This was an unexpected result and may be explained by the presence of alternative hosts in the same locality (see also Carton and Nappi 1990). Recently, Carton and Nappi (1990) have used isofemale-line techniques to show within-population genetic variability in encapsulation ability in a laboratory strain of *Drosophila melanogaster* established from African stock. The hosts were tested with *L. boulardi* collected at the same locality. Only relativly few major genes appear to be responsible. Carton and Nappi (1990) used the same technique to demonstrate within-population genetic variability in the ability of *L. boulardi* to avoid encapsulation (by *Drosophila simulans*). Between-population differences in encapsulation rate have also been found in other *Drosophila* parasitoids (Walker 1959; Hadorn and Walker 1960). There have been few studies of non-Drosophilan hosts apart from Morris (1976), who found that different genetic strains of the fall webworm (an arctiid moth, *Hyphantria cunea*) differed in the number of day degrees required for eclosion and also in their ability to encapsulate a range of parasitoids.

What maintains within-species additive genetic variance in the traits discussed here? One explanation is that there is local adaptation to the host or the parasitoid over the geographical range of the species plus migration between different subpopulations. Alternatively, there may be temporal variation in selection pressures. For example, if successful host defenses evolve, the density of specialist parasitoids will fall and, in consequence, the selection pressures exerted by parasitoid attack will decline. If host defenses have costs, there may be selection on the host population to reduce defenses in the absence of significant parasitism, and this, in its turn, may lead to an increase in parasitoid numbers. The possibility of interesting population dynamic and population genetic interactions has been little explored (Godfray and Hassell 1991).

What are the costs to host defense and parasitoid countermeasures? It seems likely that the diversion of resources into circulating hemocytes and biochemical armory may slow down the development of the host, or lead to fewer resources that can be used for reproduction by the adult. It also seems likely that there are metabolic costs for the parasitoid female in the production of venom, polydnaviruses, and other host defense countermeasures. Though these costs are plausible there is very little evidence to support these conjectures. Carton and David (1983) have shown that *Drosophila melanogaster* which have successfully encapsulated a *Leptopilina boulardi* larva are smaller and produce fewer eggs. However, what is of evolutionary significance is not the costs of successful encapsulation, but the costs of being competent to encapsulate a parasitoid. In his study of the encapsulation ability of different strains of the fall webworm (*Hyphantria cunea*), Morris (1976) concluded that "larvae of the genetic strains that have the highest survival rates under constant rearing conditions without parasites are the ones least capable of encapsulating." Zareh et al. (1980) found that an increased resistance to the pteromalid *Nasonia vitripennis* by the housefly *Musca domestica* (see below) was associated with a reduction in adult female fecundity.

There are a few reports of changes in encapsulation ability and avoidance after new host-parasitoid interactions have been created in biological control programs. However, firm evidence that these changes are the result of selection pressures exerted by the other member of the interaction is lacking.

One of the most often quoted examples of a change in the ability to encapsulate a parasitoid concerns the larch sawfly (*Pristiphora erichsonii*) and its ichneumonid parasitoid *Mesoleius tenthredinis* (Muldrew 1953; Turnbull and Chant 1961; Messenger and van den Bosch 1971). The parasitoid was introduced into Manitoba (Canada) in 1912–1913 and initially appeared to act as an effective biological control agent. Subsequently, the larch sawfly became a problem in British Columbia and the parasitoid was released there in 1934–1936. However, by the 1940s it appeared to have lost its effectiveness in Manitoba and other areas of central Canada; only about 5% of parasitoids developed successfully without encapsulation, compared to 96% in British Columbia. It was tentatively concluded that *P. erichsonii* had evolved to encapsulate *M. tenthredinis*. More recent research suggests a different explanation (Maw 1960; Wong 1974; Ives and Muldrew 1981). *M. tenthredinis* was imported from Europe and released by placing parasitized cocoons in the field. The European strain of *P. erichsonii* is able to encapsulate *M. tenthredinis* and it appears almost certain that some were released at the same time as the parasitoid. Perhaps because of its superior resistance to the ichneumonid, it increased in frequency and range and has displaced the original strain from large areas of Canada.

Bathyplectes curculionis is a parasitoid of the two weevils *Hypera postica* and *H. brunneipennis* (van den Bosch 1964; Puttler 1967; Salt and van den Bosch 1967; Messenger and van den Bosch 1971). The parasitoid was introduced into America from Europe to combat *H. postica* and then subsequently introduced into southern California, where *H. brunneipennis* had become a problem. Initial study showed that *H. brunneipennis* was able to encapsulate a large fraction (c. 40%) of *B. curculionis* eggs, but when the same population was studied again fifteen years later, far fewer parasitoids succumbed to the host defenses (c. 5%). An increase in the ability to survive in *H. brunneipennis* was accompanied by a modest decrease in the ability to avoid encapsulation in *H. postica*, the original host. Van den Bosch (1964) and Salt and van den Bosch (1967) suggested that *B. curculionis* had evolved to overcome the host defenses of *H. brunneipennis* but stressed the speculative nature of this interpretation and the difficulties of comparing encapsulation frequencies over time. Nevertheless, *B. curculionis* provides the best example of the possible evolution of parasitoid countermeasures to host defense.

There is some evidence from laboratory experiments of changes in encapsulation frequency. Different strains of *Drosophila melanogaster* vary in their ability to encapsulate the eucoilid *Leptopilina heterotoma* (=*Pseudeucoila bochei*). Hadorn and Walker (1960) and Walker (1962) reared one strain of the fly with the parasitoid for twelve generations in the laboratory and found a marked increase in the concentration of dispersed melanized hemocytes which they equated with the ability to encapsulate. However, Carton and Kitano (1981) have argued that this is only a poor measure of encapsulation ability and state that selection was responsible for just a 2% increase in the number of parasitoids destroyed. Boulétreau (1986) cultured *D. melanogaster* and *L. boulardi* in the laboratory and selected hosts for high and low parasitoid developmental success. Unsuccessful development normally resulted in the death of both host and parasitoid rather than encapsulation and host survival. There was considerable variation within each line, although a clear response to selection was not found.

In a long series of laboratory experiments with the pteromalid *Nasonia vitripennis*, Pimentel and colleagues (Pimentel 1968; Pimentel and Al-Hafidh 1965; Pimentel and Stone 1968; Pimentel et al. 1963, 1978; Olson and Pimentel 1974; Zareh et al. 1980) demonstrated that the wasp could select for increased resistance to parasitism by houseflies (*Musca domestica*). In most experiments, populations of wasps and flies were cultured together in the laboratory for a number of generations and then their ability to parasitize or resist parasitism was compared with insects from control lines where evolutionary change had been minimized. The most radical changes occurred in the host, which quickly evolved partial resistance to parasitism. *N. vitripennis* is a pupal parasitoid, and experimental hosts appeared to have heavier, stronger pupae, and to spend less time in the vulnerable pupal stage. When selection on the houseflies was relaxed by removing the parasitoid, the resistance of the host tended to diminish. The evidence for evolutionary change in the parasitoid is much more equivocal. Experimental parasitoids tended to have lower fecundity than controls. However, these differences are probably not due to genetic change: parasitoids that develop on partially resistant hosts are likely to be poorer-quality adults with reduced fecundity and longevity. Weis et al. (1989) have recently highlighted the likelihood of reciprocal changes in hosts and parasitoids where genetic change occurs only in the host, the changes in the parasitoid being purely a phenotypic response to modifications of the host.

The main aim of Pimentel's experiments was to study how evolutionary change affected population regulation. In the laboratory cultures, evolution in the host was typically associated with increased host densities and a reduction in population fluctuations. A trend for host densities to increase has also been observed in some other long-term laboratory experiments with hosts and parasitoids (Utida 1957; Takahashi 1963) and may have a similar explanation. In early commentaries, Pimentel (e.g., 1968) interpreted these results as an example of "ecological homeostasis brought about by genetic feedback." He argued that in general host-parasitoid and other victim-exploiter systems would tend to evolve toward ecological stability. As a general principle I feel this is hard to justify, and some of Pimentel's arguments appear to rely on population-level selection. However, in some cases selection acting on one or both partners of the interaction will tend to change demographic parameters in a way that increases population dynamic stability. For example, the evolution of partial resistance by the host will decrease the intrinsic rate of increase of the parasitoid, which may reduce the likelihood of host exploitation and cyclic population dynamics. However, as Pimentel et al. (1978) point out, the precise population dynamic consequences of evolution will depend quite critically on the details of the interaction.

Based on the hypothesis that evolution or coevolution tends to decrease the destructive power of natural enemies, Pimentel (1963) proposed that the release of natural enemies with no evolutionary experience of a pest would be more likely to result in economically significant biological control. In a survey of biological control programs, Hokkanen and Pimentel (1984) concluded that the creation of new associations was linked with a higher probability of successful control. Their statistical analysis attracted widespread criticism (Goeden and Lok 1986; Greathead 1986; Waage and Greathead 1988; Waage 1990; see Hokkanen and Pimentel 1989 for a reply). In my opinion, the statistical and methodological problems in testing this hypothesis with the quality of the available data base probably precludes a firm empirical answer to this question. I also agree with Kareiva (1987a) that the theoretical basis of this prediction is open to question. While evolution acting on the host will often, but not always, lead to an increase in the average density of the host, it is far from clear how evolution acting on the parasitoid will affect host densities, and the outcome of selection acting on both parties of the interaction is even more obscure.

In this section I have only discussed evolutionary interactions concerning defenses against the immature parasitoid. In chapter 8 (sec. 8.2.4) I discuss how parasitoids may lead to selection for changes in the host's ecological niche, and the possible coevolutionary response of the parasitoid.

6.4 Host Manipulation by Endoparasitoids

Endoparasitoids, especially those species that allow their host to feed and grow after parasitism (koinobionts), must live in intimate physical contact with their host for an extended period of time. Over this period, parasitoids frequently avoid damaging important organs that would lead to the death of the host. However, in addition to these passive adaptations, parasitoids may be selected to manipulate the endocrine physiology, nutritional biochemistry, or behavior of the host to their own advantage. Vinson and Iwantsch (1980a) called active manipulation of the host's physiology and behavior "host regulation," a slightly unfortunate term as ecologists use host regulation in a different context. Of course, distinguishing active host manipulation from the passive traumatic effects of parasitism may be very difficult.

6.4.1 HORMONAL INTERACTIONS

The growth and metamorphosis of larval insects is controlled by a complicated and intricate series of hormonal interactions. There is a vast literature on this subject, and the characterization of hormonal influences on development is one of the major achievements of insect physiology (e.g., Kerkut and Gilbert 1985). The two most important hormones are juvenile hormone (JH), secreted by the corpora allata, which maintains juvenile characters, and ecdysone, secreted by the prothoracic gland, which causes moulting. When the JH concentration is low, the presence of ecdysone causes metamorphosis in holometabolous insects. Hormonal interactions between parasitoids and their hosts have been reviewed by Beckage (1985).

Some parasitoids do not manipulate the levels of hormone in the host, but make use of host hormones to regulate their own development and to insure that it is synchronized with that of the host (Lawrence 1986 calls these species "physiological" conformers). A number of species of parasitoids lay their eggs in very young hosts, and their larvae develop only as far as the first instar. There is good evidence that the stimuli for the parasitoid to moult to the second larval instar, and to enter into its rapid feeding stage, are the hormonal changes in the host associated with metamorphosis or the preparation for metamorphosis (lepidopteran larvae, for example, often go through a wandering phase prior to pupation whose onset and length are under hormonal control) (e.g., Baronio and Sehnal 1980; Lawrence 1982). The synchronous switch from proliferation to differentiation in broods of polyembryonic wasps (see sec. 1.3.1) also appears to be influenced by changes in host hormone titers (Strand et al. 1991a, 1991b).

Other parasitoids actively manipulate the host endocrine system (called "physiological regulators" by Lawrence 1986). Some parasitoids induce the host to remain longer in the larval feeding stage. This phenomenon has been studied in the gregarious braconid *Cotesia (=Apanteles) congregatus*, which attacks the tobacco hornworm *Manduca sexta*. The parasitoids emerge from and kill the host at the beginning of the final instar. The parasitoids can cause up to six supernumerary larval instars and prevent metamorphosis by suppressing the normal drop in host JH levels and by blocking the conversion of ecdysone into its active form 20-hydroxyecdysone (Beckage and Riddiford 1982, 1983a, 1983b; Beckage and Templeton 1986). The parasitoid may manipulate host hormone levels indirectly by interfering with their production or metabolism. One possibility is that levels of JH-esterase, the enzyme that degrades JH, are reduced in parasitized *M. sexta*; another that the parasitoid causes increased JH production be interfering with its hormonal or neurohormonal regulation (Beckage 1985). Starved, unparasitized larvae frequently go through supernumerary instars, and thus some of the effects of parasitism may simply be a consequence of the trauma of parasitoid attack. Finally, the parasitoid itself may secrete JH (Beckage and Riddiford 1982). Other parasitoids delay host pupation, possibly by manipulating host hormones. For example, the larch casebearer (a moth, *Coleophora laricella*) normally pupates in the spring but remains as an active larva throughout the summer if parasitized by the ichneumonid *Diadegma (=Angitia) nana* (Thorpe 1933).

The precocious initiation of metamorphosis is found in braconids in the genus *Chelonus* sp. (D. Jones et al. 1981; D. Jones 1987) which are egg-larval parasitoids of moths such as noctuid *Trichoplusia ni*. Parasitized caterpillars initiate wandering and other behaviors associated with the metamorphic moult one instar earlier than unparasitized hosts. However, metamorphism itself is prevented, probably because of the reduced concentration of ecdysone. Venom and polydnaviruses injected by the adult wasp appear to redirect host development. The ichneumonid *Campoletis sonorensis* also prevents pupation of mature *H. virescens* larvae and reduced ecdysone titers are involved. Here the polydnavirus appears to be responsible for the degeneration of the prothoracic gland, the source of the hormone (Dover et al. 1988). There is evidence that the ichneumonid *Homotropus(-Diplazon) fissorius* causes premature pupation of its syrphid host by secreting ecdysone (Schneider 1950, 1951; Beckage 1985), and that the retardation of wing development in aphids parasitized by the braconid *Aphidius platensis* may be caused by elevated JH (Johnson 1959; Beckage 1985). Two dramatic examples of precocious metamorphosis are the braconid *Polemochartus(-Polemon) liparae* that induces its host, the gall-forming chloropid *Lipara lucens*, to pupate six months early, and the eurytomid *Eurytoma curta* that causes its host, the knapweed gall fly *Urophora soltitialis*, to pupate eight months prematurely (Varley et al. 1933). In neither case has the interaction been studied physiologically, but parasitoid manipulation of the host endocrine system is a likely explanation for these observations.

The advantage of prolonging the feeding stage of the host is that it contains more resources when eventually destroyed by the parasitoid. This advantage will be greatest when host resources are relatively scarce, for example when a relatively large solitary parasitoid attacks a small host or when many gregarious larvae feed together on a host. In support of the latter suggestion, Beckage and Riddiford (1983a) noted that large clutches of *Cotesia congregata* induced more supernumerary instars than small clutches. Offsetting these advantages are several potential disadvantages. A longer feeding period may increase the risk of predation or hyperparasitism, a possibility that may be exacerbated if hosts in supernumerary instars are large and obvious, or if their normal behavior is affected by physiological manipulation. A longer larval period also increases generation time, which may be significant when the parasitoid population is increasing rapidly.

I argued in chapter 3 (sec. 3.2.4) that some species of solitary endoparasitoid develop in relatively large hosts that are not completely consumed. Selection acting on the larvae to attack other parasitoid larvae prevents the evolution of gregarious clutches. These species are particularly likely to accelerate host maturation, because a full-sized host contains more resources than required for parasitoid growth. Accelerated maturation has the advantages of decreased generation time and a possible reduction in the risk of predation. Another advantage is the decrease in the amount of unconsumed host tissue that may putrefy, causing a risk of infection for the parasitoid pupa.

Parasitism of adult female insects is often associated with the cessation of oviposition. This occurs even when the ovaries themselves have not been attacked by the developing parasitoid. Ovigenesis is under complicated neural and hormonal control in insects, and it seems likely that parasitoids interfere with this control to prevent egg production. The evolutionary advantage of reduced adult fertility is clear: nutrients that would have been expended on eggs are retained to be available to the parasitoid (S. N. Thompson 1983; Beckage 1985). Parasitism can sometimes affect the reproduction of male insects. Male typhlocybine leafhoppers parasitized by dryinid wasps frequently have reduced external genitalia and may lose their stridulatory organs (Olmi 1993). However, male castration is probably less widespread than its female counterpart. Sperm contain fewer parental resources than eggs and from the parasitoid's point of view are probably a less serious drain on host resources.

An interesting example of host manipulation which, although it has not been studied physiologically, probably has a hormonal explanation, is Kornhauser's (1919) observation that the dryinid *Crovettia (=Aphelopus) theliae* causes male hosts (brightly colored adults of the membracid *Thelia bimaculata*) to assume drab female colors. The advantage of this change to the parasitoid is probably a reduction in the risk of predation.

6.4.2 NUTRITION AND METABOLISM

It is unclear to what extent parasitoids manipulate host metabolism to their own benefit. It has been suggested that parasitoids may disrupt the uptake of nutrients from the hemolymph in order that more are available for themselves (Fisher 1971; Vinson and Iwantsch 1980a, 1980b; Slansky 1986). Some gregarious parasitoids increase not only the duration of the host feeding stage, but also the rate at which food is ingested (Slansky and Scriber 1985). However, in studying parasitoid nutrition, it is particularly difficult to disentangle direct and adaptive changes brought about by the parasitoid, and indirect changes caused by the trauma of parasitism (Vinson and Iwantsch 1980b; S. N. Thompson 1983, 1986a, 1986b; Slansky and Scriber 1985; Slansky 1986). For example, the fatty acid profile of polyphagous parasitoids has been shown to mirror the profile of the host (Thompson and Barlow 1974). Whether "duplication" is adaptive and allows better physiological integration in different hosts, or whether it is purely a neutral consequence of developing in different host species, is unclear (S. N. Thompson 1986a). The braconid *Chelonus* sp. nr. *curvimaculatus* causes a premature rise in the levels of hemolymph storage proteins which normally increase in concentration just prior to pupation. This species emerges from its host (*Trichoplusia ni*) after cocoon formation and, because it only ingests hemolymph, the increase in storage proteins may be beneficial to the parasitoid (Kunzel et al. 1990). Again, it is not clear whether the increase in storage protein is an indirect consequence of the hormonally induced premature maturation caused by the parasitoid, or whether it is an independently evolved adaptation.

6.4.3 BEHAVIORAL MANIPULATION

There are some spectacular examples of the manipulation of host behavior by true parasites (e.g., Holmes and Bethel 1972) but much less information about behavioral manipulation by endoparasitoids. Fritz (1982) collected a number of examples of where parasitized hosts had been found to suffer lower mortality compared to unparasitized hosts, and suggested that some of the differences may be due to behavioral changes induced by the parasitoid. The difficulty with this argument is that the trauma associated with parasitism may cause many effects such as sluggishness which indirectly reduce the risk of predation. Although the end results are the same, the selection pressure (if any) responsible for the change in host behavior is not reduced predation. It must also be asked why, if the parasitoid can reduce the predation suffered by the host, selection has not acted directly on the host. The most likely explanation is that there are trade-offs between the risks of predation and other components of both host and parasitoid fitness; if the trade-offs are not identical, then behavioral manipulation by the parasitoid may be favored. Parasitoids will not always be selected to change host behavior to reduce predation. Some parasitoids may benefit from rapid feeding by the host, even if this results in a greater risk of predation.

Stamp (1981b) studied the braconid wasp *Cotesia* (=*Apanteles*) *euphydryidis* which attacks larvae of the checkerspot butterfly *Euphydryas phaeton*. Parasitized larvae crawl from dense vegetation into more exposed areas before being killed by the parasitoid. Smith Trail (1980) had previously suggested that parasitized larvae of aposematic species such as *E. phaeton* should advertise their presence to predators so that their kin are protected from parasitism. Stamp argued that the more likely explanation for her observations is reduced hyperparasitism in exposed areas. *C. euphydryidis* is attacked by several hyperparasitoids, especially a wingless ichneumonid (*Gelis* sp.) and a pteromalid (*Pteromalus* (=*Habrocytus*) sp.) and mortality is less on cocoons formed high in vegetation. A possible additional advantage of pupating high is that when the adult wasps emerge they are in a good position to attract mates using sex pheromones. Attack by hyperparasitoids is likely to be a more potent selection pressure than predation in leading to behavioral manipulation of the host. Whereas both the host and parasitoid are selected to avoid predation, only the parasitoid is at risk from hyperparasitism (R. E. Jones 1987).

A number of other parasitized lepidopteran larvae die in conspicuous positions. Shapiro (1976) suggested that parasitized caterpillars of the nymphalid butterfly *Chlosyne harrissii* advertise themselves to predators. This species lives in large colonies, and Shapiro (1976) has argued that this behavior may be a result of kin selection. However, the risk of hyperparasitism was not considered. Sato et al. (1983) describe the behavioral manipulation of common armyworms (*Leucania separata*; Noctuidae) by the gregarious braconid *Apanteles kariyai*. The nocturnal caterpillars normally hide in leaf litter during the day, but individuals with mature parasitoids leave their hiding place and climb vegetation where they remain motionless while the parasitoid larvae leave the host and spin silk cocoons. After the parasitoids have quit the host, which takes about ten minutes, the perforated caterpillar suddenly resumes activity and walks away from the pupae, although it quickly dies. A plausible interpretation of these observations is that the wasp causes the host to move to a favorable position for pupation, and prevents the host from dying in the vicinity of the pupae where it might rot or attract predators or hyperparasitoids.

The behavior of aphids (*Macrosiphon euphorbiae*) parasitized by the braconid *Aphidius nigripes* has been studied by Brodeur and McNeil (1989, 1990, 1992). Aphids containing diapausing parasitoids leave the colony and move to concealed microhabitats where the parasitoid consumes the host, pupating under the eaten-out shell of its host (the mummy). This behavior probably acts to reduce hyperparasitism and predation of the parasitoid pupae, and may also protect the pupae from severe weather. Aphids containing nondiapausing parasitoids do not move to concealed sites but tend to leave colonies on the underside of the leaf to pupate on upper surfaces, especially near the top of the plant. Experimental manipulations suggested that this behavior again leads to a reduction in the rate of hyperparasitism. Hyperparasitoids may use cues associated with the aphid colony to locate their hosts. Development may also be faster if the parasitoid pupates in a warm and sunny site near the top of the plant. This work is a particularly clear example of behavioral modification by parasitoids. A slightly different behavior appears to be induced by other parasitoids of homopterans prior to pupation. Leafhoppers (typhlocibine Cicadellidae) parasitized by dryinid wasps cease movement and embed their rostrum in the plant prior to being killed by the parasitoid (Olmi 1993). It is probably in the parasitoid's interest that the host is not dislodged from the plant.

Bumblebees suffer from parasitism by conopid flies and are frequently attacked and parasitized in mid-air. In central Europe, the common bumble bee *Bombus pascuorum* is parasitized by the two conopids *Sicus ferrugineus* and *Physocephala rufipes*. In some Swiss populations, parasitism is often between 30% and 70% (Schmid-Hempel et al. 1990; Schmid-Hempel and Schmid-Hempel 1991). The fly develops within the host which continues foraging until the parasitoid reaches the end of its third instar. A mature parasitoid within the host's body cavity affects the foraging success of the host in several ways. First, it is a metabolic drain on the host, reducing its general condition; second, the weight of the parasitoid may affect the aerodynamics of flight; and third, the bulk of the parasitoid restricts the honey crop in the abdomen and reduces the amount of nectar that can be collected. However, the parasitoid does not affect the bee's ability to collect pollen, which is stored in corbiculae on the hind legs.

Schmid-Hempel and Schmid-Hempel (1990, 1991) found some marked differences in the behavior of parasitized and unparasitized bees. They tended to visit different types of flowers, and bees with nearly mature parasitoids tended to collect nectar rather than pollen. In populations where levels of parasitism are high, these differences may have significant ramifications for the economics of the bee colony, and also for floral pollination. Some of the behavioral changes are probably just due to the metabolic and physiological consequences of parasitism. However, behavioral manipulation by the parasitoid may also be involved. A bee containing a nearly mature fly is doomed to an early death; however, it can still aid the colony by collecting pollen which is used to feed the bee larvae. From the point of view of the parasitoid, foraging for nectar is more worthwhile as this is used as food by the adult bee and so can benefit the parasitoid. Schmid-Hempel and Schmid-Hempel (1991) suggest that the preponderance of heavily parasitized bees collected while foraging for nectar may be a consequence of behavioral manipulation by the parasitoid.

A further difference in the behavior of parasitized and unparasitized bees was described by Schmid-Hempel and Müller (1991). Parasitized insects tended to spend most of their time foraging in meadows and did not return, or returned only rarely, to the nest. This change in behavior may again be a nonadaptive consequence of being parasitized, but Schmid-Hempel and Müller point out that there may be benefits to the fly. When the parasitoid reaches the end of its third instar it consumes vital organs of its host, which literally drops dead, the conopid pupating *in situ* within the husk of its host. If the bee were to die in the nest, its body would be subject to fungal and bacterial infections that occur on the decomposing combs after the nest has been abandoned. By ensuring the host is unlikely to die in the nest, the parasitoid increases its chance of pupating in a safe site. This interpretation was challenged by Poulin (1992), who argued that selection acting on the host may favor the change in behavior. He suggested that parasitized bees might deflect the attention of parasitoids and predators from their nest mates. In addition, in avoiding the nest they would not consume stored food collected by the colony. He also argued that the costs to the host of abandoning the nest would be small if parasitized individuals are inefficient foragers. In replying to Poulin, Müller and Schmid-Hempel (1992) stress that the definite attribution of cause and effect is difficult in a system where experimentation is impossible and without detailed information about costs and benefits. Nevertheless, they show that Poulin's assumptions are unlikely to be correct; bumblebees tend not to forage with their nest mates, and any advantage of the dilution of parasitism and predation would be very weak.

Parasitised bees also seem able to forage for themselves and hence do not require to use stored resources in the colony. Finally, while it is true that bees with mature flies are inefficient foragers, bees with smaller parasitoids can contribute to rearing offspring, and their absence from the nest probably leads to a reduction in their inclusive fitness.

6.5 Interactions between Immature Parasitoids

Parasitoid larvae not only have to contend with host defenses, but also with competition and attack from other parasitoid larvae, both of their own and other species. In some cases, the ovipositing parasitoid physically destroys competitors already present on the host (sec. 3.3.2) or injects a substance that kills larvae of other species (sec. 3.3.4). In most cases, however, the outcome of competition is determined by the behavior of the immature parasitoids. Most parasitoids can be classified as either solitary or gregarious. Only one solitary parasitoid is able to develop on a host, and if more than one egg is laid, all but a single parasitoid perishes. In the first part of this section, the physiological and behavioral mechanisms leading to brood reduction in solitary parasitoids are described. The second section discusses competition between the larvae of gregarious parasitoids.

6.5.1 Solitary Species

Reduction of brood size in solitary parasitoids can occur through physical conflict, chemical attack, or resource competition. Which mechanism occurs, and who wins, frequently depends on the relative ages of the two parasitoids.

The first instar larvae of many solitary parasitoids are equipped with relatively large opposable mandibles and a rigid head capsule that forms a firm base for muscular attachment (fig. 6.2) (Clausen 1940a; Salt 1961). The mandibles are used to attack other larvae in the host, both conspecifics and individuals of other species. When the larvae moult into the second instar and begin their rapid growth phase, the large mandibles are usually lost. Actual conflict has been observed in many parasitoid species while the frequent presence of scars and wounds on moribund larvae is strong inferential evidence for its presence in others (Salt 1961). Fighting mandibles have been recorded from most families of hymenopteran parasitoids, and from a few dipteran species (Salt 1961; Vinson and Iwantsch 1980a). Parasitoid larvae appear to move actively around the host hemocoel, searching for other larvae. Often, the death of a competitor does not result directly from wounds received during conflict, but from attack by host defenses. Wounding appears to compromise the surface integrity of the parasitoid so that it is recognized as foreign by host hemocytes.

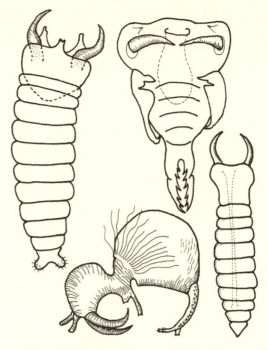

Figure 6.2 The mandibles of first instar solitary parasitoids. (Drawn by I.C.W. Hardy after Salt 1961.)

Physical conflict can result in one parasitoid larva successfully eliminating seven or eight rivals in the same host (Salt 1961). However, occasionally two parasitoids simultaneously wound each other and both die. The fate of the host then depends on whether it can survive the other consequences of parasitism. For example, if two larvae of the ichneumonid *Cardiochiles nigriceps* manage to destroy each other, the host (a larva of the moth *Heliothis virescens*) still perishes because of the action of the polydnavirus injected at oviposition (Vinson 1972b),

Physical conflict can occur between the larvae of ectoparasitoids as well as endoparasitoids. For example, many pteromalid wasps feed externally on their hosts in concealed locations such as inside stems or seeds. The larvae move actively over the surface of the host and will destroy any eggs they find. If more than one larva hatches, the younger usually triumphs because of its greater mobility (Clausen 1940a). Ectoparasitoids are normally at a competitive advantage to endoparasitoids which they destroy as they consume the host. Ectoparasitoids are also frequently facultatively hyperparasitic and feed on the larvae of other ectoparasitoids already present on the host.

Eggs and young larvae often appear to cease development or degenerate in the presence of older larvae (Fisher 1971). Early this century, Timberlake (1910, 1912) suggested that the older larvae secrete a chemical that inhibits or destroys younger competitors, a suggestion that has often been repeated (e.g., Pemberton and Willard 1918; Spencer 1926; Tremblay 1966). The best evidence for the chemical suppression of competitors comes from studies of aphid parasitoids in the braconid subfamily Aphidiinae. When the first egg hatches into a larva, other parasitoid eggs present in the host die and become opaque, as do the aphid's embryos (Mackauer 1986).

Older larvae can also eliminate younger eggs or larvae through competition for limiting resources. Of course, very much older larvae can consume the host before a competitor hatches (Fiske and Thompson 1909), but more subtle forms of resource competition also occur. Fisher (1961, 1963), working with the ichneumonids *Venturia canescens* and *Diadegma (=Horogenes) chrysostictos*, showed that older larvae could reduce the oxygen available in the host's hemolymph to so low a level that eggs or younger larvae perished. In these species, eggs and young larvae are particularly susceptible to lack of oxygen while older larvae are more tolerant. Similar physiological suppression appears to operate in the braconid *Microplitus croceipes* (Edson and Vinson 1976). Fisher (1971) argues that physiological suppression by asphyxiation may be a common mechanism by which established larvae eliminate young competitors.

To conclude, endoparasitoid larvae of approximately the same age tend to eliminate each other by physical conflict. Older larvae are normally able to eliminate younger competitors by physical suppression although the exact mechanism of suppression is seldom understood. In general, there is probably an advantage to being the oldest endoparasitoid in a host, especially in intraspecific competition. There are, however, exceptions; some first-instar larvae of aphidiine braconids are able to use their mandibles to kill older amandibulate second instars (Chow and Mackauer 1984, 1986). In contrast to endoparasitoids, there is often an advantage to being the youngest ectoparasitoid on a host, especially in species that are facultatively hyperparasitic.

6.5.2 GREGARIOUS SPECIES

Gregarious larvae do not possess fighting mandibles, and competition between individuals is normally restricted to exploitation competition for host resources. As was discussed in chapter 3 (sec. 3.2.1), there may be selection on individual larvae to feed swiftly in order to consume resources before their brood mates. In some cases, members of a brood may suffer physical exclusion from the host resources, Pickering (1980) studied an ichneumonid wasp (*Pachysomoides stupidus*) that attacks the larvae of social wasps in the genus *Polistes*. The brood of parasitoid larvae begin feeding on the abdomen of a host larva which occupies one of the cells of the *Polistes'* nest. As the parasitoid larvae grow, they move up the host and come to fill the width of the cell. In medium to large clutches, some larvae are often excluded from the diminishing host resources and die before pupation (see also sec. 4.3.5).

In gregarious parasitoids, defense against super- and multiparasitism is normally through resource pre-emption and possibly asphyxiation or physical suppression of younger larvae. One exception is among polyembryonic wasps in the family Encyrtidae (see sec. 1.3.1 for an introduction to polyembryony). A fraction of eggs develop precociously to mandibulate larvae which defend the main body of eggs against competitors. The precocious larvae do not mature further and thus sacrifice themselves for their (genetically identical) siblings (Cruz 1981; Strand 1989a).

6.6 Conclusions

Although it is possible to study the reproductive behavior of adult parasitoids without worrying about the details of the physiological interactions between immature parasitoids and their host, an understanding of how host quality affects parasitoid fitness greatly helps in interpreting oviposition behavior. It is clear that host quality affects parasitoid fitness by increasing the risk of immature mortality and through its effects on adult size. Host quality also often influences development time, an aspect of parasitoid fitness that has received very little study. Development time is a difficult component of fitness to study because its importance depends on the population dynamics of the parasitoid and possibly the host.

Recent advances in parasitoid physiology have revealed a series of complex adaptations that allow endoparasitoids to overcome the defenses mounted by the host. The next decade is likely to see great progress in our understanding of the molecular basis of these interactions. It is possible that the direct injection of chemicals into the host, the secretion of substances by the larvae or their teratocytes, and the expression of genes on polydnaviruses are each alternative routes for delivering substances to knock out host defenses. Polydnaviruses are a wonderful example of either a highly specialized mutualism, or of an extended host phenotype. The prospect of more information on their evolutionary origins in the next few years is very exciting.

It is clear that some parasitoids can manipulate the physiology and behavior of their hosts to their own advantage. Greater understanding of this manipulation is hampered by the difficulties in distinguishing adaptive changes brought about by active intervention by the parasitoid from nonselected consequences of the trauma of parasitism. These problems are particularly acute in studies of parasitoid physiology. Perhaps the best way to resolve them is by an increased understanding of the molecular basis of changes in host physiology after parasitism so that the role of specific parasitoid gene products can be identified.

Finally, the population biology of host-parasitoid interactions at the physiological level has received disappointingly little attention. There is evidence for within-population variation in encapsulation ability, and also in the ability to avoid encapsulation. However, little is known about the maintenance of variability, or about the costs of the two traits. Few biological control releases have been studied in sufficient detail to detect evolutionary or coevolutionary changes after new interactions have been created. Host-parasitoid interactions are likely to make good laboratory systems to study reciprocal evolution, while biological control releases offer unrivaled possibilities for replicated field experiments.

7

The Adult Parasitoid

In this chapter I examine the behavioral ecology of the adult parasitoid. I begin by considering the relationship between adult size and fitness. Because adult size is often strongly influenced by host quality, a knowledge of this relationship is crucial in interpreting many aspects of parasitoid oviposition behavior. The second section describes parasitoid mating systems and also discusses mating behavior and mating-strategy polymorphisms. A few parasitoid species defend hosts from conspecifics and other parasitoids and are the subject of the third section. The next section discusses host defenses against the adult parasitoid and the countermeasures adopted by parasitoids. The fifth section describes parasitoid dispersal and the evolution of winglessness, while the sixth section discusses adaptations by adult parasitoids to avoid predation. The final section considers the evolution of temporal synchronization between parasitoids and their hosts. Two important activities of the adult female parasitoid, host location and oviposition, have already been discussed in chapters 2 to 4.

7.1 Size and Fitness

The size of the adult parasitoid is largely determined by the amount and quality of food consumed by the larva. Decisions made by the ovipositing parent about the choice of host, clutch size, superparasitism, and sex ratio thus have a major effect on adult size. To understand the evolution of oviposition behavior, it is crucial that the relationship between adult size and fitness is known. In most cases, this relationship will be different for males and females.

How might adult size influence fitness? In a population of constant size, the fitness of a female parasitoid is strongly correlated with the number of hosts attacked and, in the case of gregarious parasitoids, with the number of eggs laid per host. The size of the adult female may influence fitness by affecting searching efficiency, longevity, or egg supplies. It is also possible that parasitoid size may influence the quality of host attacked: a small female may be unable to subdue a large, good-quality host. Male fitness will be largely determined by the number of matings achieved. The size of a male may affect his longevity and also his ability to locate females. More speculatively, small males may suffer in intrasexual competition for mates. In populations that are increasing in density, the timing as well as the number of hosts attacked (or matings achieved) will be significant because early reproduction is favored.

7.1.1 LABORATORY STUDIES

Ideally, the relationship between adult size and fitness should be measured in the field. This is a formidable challenge and, as yet, nearly all studies have been carried out in the laboratory. The following fitness components have received most study.

1. *Potential fecundity*. The female wasp is dissected and the number of eggs or the number of ovarioles counted. In pro-ovigenic species, the number of eggs represents maximum lifetime fecundity while in synovigenic species it represents the current egg load.

2. *Maximum realized fecundity*. Parasitoids are provided in the laboratory with a surfeit of hosts and the numbers attacked are recorded. In the case of gregarious parasitoids, the maximum numbers of progeny emerging from the hosts are counted. Parasitoids may or may not be supplied with additional food sources such as honey.

3. *Longevity*. The lifespan of male or female parasitoids is measured in the laboratory. Insects may be kept with or without hosts, and may or may not be provided with honey or other sources of food.

4. *Mating success of individual parasitoids*. Males are exposed to a large number of females which subsequently are allowed to reproduce. Experiments have been conducted largely with haplodiploid hymenopteran parasitoids, and the total number of daughters produced by all mates is used as a measure of male fitness (sons can be produced without mating).

5. *Mating success in competition with other parasitoids*. Males of different size are allowed to compete for mates. The winner may be identified by direct observation or indirectly using genetic markers.

Most studies have found that female fitness increases with adult size (King 1987). In pro-ovigenic species, dissection of newly emerged females normally shows that larger individuals have more eggs and often more ovarioles (e.g., Iwata 1966; fig. 7.1a). In synovigenic species, maximum egg load is normally correlated with body size (e.g., Rosenheim and Rosen 1991; fig. 7.1b). Greater potential fecundity translates into greater realized fecundity when large insects are supplied with excess hosts. There are some exceptions to this pattern; Rotheray and Barbosa (1984) found no relation between size and fecundity in the chalcidid *Brachymeria intermedia*, while Corrigan and Lashomb (1990) found that while large *Edovum puttleri* (Eulophidae) could store more eggs, there was no relation between size and lifetime fecundity when the wasps were provided with a surfeit of hosts.

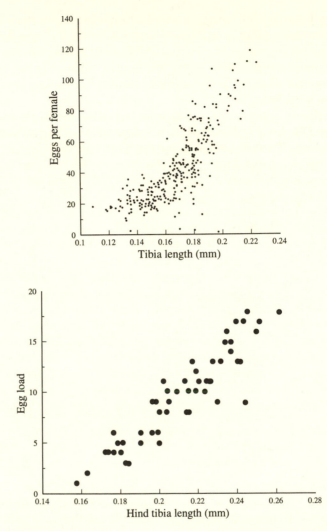

Figure 7.1 The relationship between adult size and fecundity in (*top*) pro-ovigenic and (*bottom*) synovigenic wasps. *Top*: The trichogrammatid egg parasitoid *Trichogramma evanescens* (Waage and Ng 1984). *Bottom*: The aphelinid *Aphytis lignanensis* (Rosenheim and Rosen 1991). In both cases, hind tibia length is used as a measure of adult size.

Larger insects tend to live longer, but both maximum longevity and lifetime reproductive success can be strongly influenced by the amount and type of food provided. For example, larger individuals of the bethylid wasp *Goniozus nephantidis* lived longer than small individuals when supplied with food, but the reverse was true when wasps were starved (Hardy et al. 1992). Some other more specialized advantages of large size in parasitoids have been summarized by Hurlbutt (1987): large female *Trichogramma* spp. travel faster than smaller individuals and may locate more hosts (Boldt 1974); larger females of species that attack protected hosts may find it easier to oviposit through the protective covering (Salt 1940, 1941); and species that have to defend themselves from their hosts, or from conspecific females, may gain a physical advantage by being large.

Male longevity often increases with body size (King 1987), although King (1988) found longevity and male size to be independent for the pteromalid *Spalangia cameroni*. There is some evidence that large males have greater mating success in the pteromalid *Nasonia vitripennis* (Grant et al. 1980) and the braconid *Heterospilus prosopoidis* (Charnov et al. 1981; W. T. Jones 1982), although not in *Spalangia cameroni* (King 1988).

Laboratory studies clearly show a correlation between adult size and several components of fitness. How valuable are laboratory-based experiments in estimating the importance of size in the field? The answer depends on the factors limiting the reproductive success of male and female parasitoids. If the most important limiting factor is egg or sperm supply, then laboratory-based experiments may be quite successful as both these quantities are relatively easy to assess. However, if the parasitoid's reproductive success is limited by the number of hosts or females discovered, laboratory-based experiments are far less likely to be successful. In these circumstances the relationship between size and longevity will be crucial in determining fitness, but estimates obtained in the relatively benign environment of the laboratory are unlikely to match those in the field. How size influences the risk of death from predators and raindrops may be of critical importance in the field, but almost impossible to measure in the laboratory.

7.1.2 FIELD STUDIES

How might field estimates of the relationship between adult size and fitness be obtained? The most direct way is to mark and release parasitoids and to relate the probability of recapture to size. Techniques are available for individually marking all but the smallest parasitoids (e.g., Driessen and Hemerik 1992), but the success of this method is limited by the feasibility of recovering a sufficient number of marked insects. If parasitoids are mass-reared in the laboratory, large numbers of marked individuals can be released into the field, although there may be some danger of atypical behavior because of postemergence conditioning (sec. 2.3.1) prior to release.

A less direct method is to compare the size distribution of parasitoids at emergence with collections of parasitoids caught ovipositing or mating. This procedure might be applied to parasitoids such as those attacking *Drosophila* where large collections can be made at emergence and oviposition sites; it is also possible to collect pairs *in copula* to estimate the reproductive success of males. Problems may occur if the size distribution of parasitoids changes over time, or if the oviposition site attracts parasitoids from unsampled emergence sites.

7.1.3 COMPARATIVE STUDIES OF MALES AND FEMALES

Charnov's size-advantage hypothesis (sec. 4.4) suggests that parasitoids lay male eggs in small hosts and female eggs in large hosts because larger females enjoy a proportionally greater increase in fitness than larger males. The logic behind this assumption is that female size has a major effect on fitness through egg load and longevity, while the ability of a male to inseminate a female is less dependent on size. The argument does not deny that larger males are often fitter than small males, it just states that the advantages of being big are relatively less for males than females.

A number of workers have attempted to compare the relative fitnesses of males and females of the same species of wasp. W. T. Jones (1982) found that the largest females of the braconid *Heterospilis prosopoidis* were able to lay in the laboratory twenty-one times more eggs than the smallest individuals. The difference in male insemination capacity between large and small wasps was only threefold. Van den Assem et al. (1989) carried out a similar study on the pteromalid *Lariophagus distinguendus* which, like *H. prosopidis*, attacks bean weevils (*Callosobruchus* spp.). They found that large females had a higher rate of oviposition and lived longer than small females and that overall they laid approximately three times more eggs. Male reproductive success also increased with body size, but large males enjoyed only a twofold advantage in their capacity to inseminate females. Large males were a little more successful than small males in circumstances when several males competed for the same female, and also in sperm competition. King (1988), in contrast, found that longevity, fecundity and mating success were largely independent of adult size in the pteromalid *Spalangia cameroni*. The only correlation between size and a component of fitness that she observed was faster development by females in larger hosts. Finally, Heinz (1991) compared the relationship between size and fitness in males and females of the eulophid *Diglyphus begini*, a parasitoid of agromyzid leaf miners such as *Liriomyza trifolii*. He used the number of eggs laid by mated females as a measure of male fitness and the total number of eggs laid as a measure of female fitness. Reproductive success was measured under conditions of high and low availability of hosts and mates (in a factorial experimental design). Female reproductive success always increased with body size, whereas male body size was only important when mates were easy to find. It was not clear whether large males were more efficient at locating and courting females, or had greater sperm reserves. Under field conditions, mate accessibility is likely to be low.

Overall, these results suggest that the assumptions of Charnov's size advantage hypothesis are often met. However, as stressed above, it is very important that studies of the relationship between size and fitness should move from the laboratory to the field.

Hurlbutt (1987) has surveyed the sexual dimorphism of 360 species of parasitic wasps and has found that in most cases females are larger than males. The main exception is the family Ichneumonidae, and in particular the subfamily Ichneumoninae, where the trend is reversed. It is not clear why male Ichneumoninae should be relatively larger than females.

7.2 Mating

The spatial location of hosts and other resources in the environment influence the distribution of adult parasitoids and the type of mating systems that evolve. The first part of this section discusses parasitoid mating systems while the second part reviews a variety of different aspects of mating behavior, including courtship, sexual selection, mating frequency, sperm precedence, protandry, and factors leading to constrained sex allocation.

7.2.1 Mating Systems

Parasitoid mating systems are influenced by the spatial and temporal predictability of females in the environment, by the nature of male-male competition, and by female reproductive biology. The most important aspects of female reproductive biology are (1) whether mating is possible immediately after emergence; (2) the frequency of mating; and (3), in cases of multiple mating, whether the first or last male enjoys sperm precedence.

Perhaps the simplest type of mating system occurs when males search randomly in the environment for females. Such a system is unlikely to be common because males will nearly always be able to use some cue or stratagem to increase the probability they encounter a mate. Females will be distributed nonrandomly in the environment and in particular will tend to be found at their emergence sites, and also where they feed and oviposit. Females may also emit sex pheromones which assist mate location by males. Finally, males may be selected to swarm or aggregate at certain sites in the environment to which females are attracted.

Below I discuss some of the main mating systems found in parasitoids, although precise categorization is impossible because the different systems blend into one another. Mating systems have recently been reviewed by N. D. Davies (1991), while Thornhill and Alcock (1983) provide an important account of insect mating systems.

ASSISTED MALE SEARCH

Many female insects release volatile sex pheromones that attract males from long distances. Quite often, relatively large quantities of chemicals are produced by special glands, while specialized organs have evolved to assist in broadcasting the pheromone. The chemistry of many pheromones has been resolved and artificial analogues synthesized. Pheromones are routinely used for monitoring a variety of different pest species. It is clear that chemical cues are very important in mate recognition by hymenopterous parasitoids, and males often show arrestant and short-range attractant responses to chemicals associated with females (Gordh and DeBach 1978; Eller et al. 1984; van den Assem 1986). However, there are rather few examples of long-range sex pheromones among parasitoid wasps.

Female wasps in the tiphiid subfamily Thynninae emerge from underground pupation sites and crawl up vegetation to elevated perches. There they adopt a characteristic calling position and emit a pheromone that attracts males which conduct long patrolling flights over suitable habitat (Ridsdill Smith 1970; Alcock 1981). Lewis et al. (1971b) showed that male *Cardiochiles nigriceps*, a braconid parasitoid of *Heliothis virescens*, were attracted over long distances to female wasps. They used this observation to construct pheromone traps which allowed the density of the wasp to be monitored. Curiously, females were also attracted to females while males showed some attraction to other males. However, females were not attracted to males. The two sexes fly at different heights in the crop, the males near to the ground where the females emerge, and the females at crop level where hosts occur. Robacker et al. (1976) and Robacker and Hendry (1977) studied mate finding in the ichneumonid *Itoplectis conquisitor*. Males orient toward chemicals produced by newly emerged females and, to a lesser extent, males. They were able to show that the substances nerial and geraniol were components of the pheromone. Associative learning has been found in female *I. conquisitor* searching for hosts (sec. 2.3.2), and similar behavior is shown by males looking for mates. Male wasps will inspect a blank "cup dispenser" if they have previously encountered cups containing the pheromone. Eller et al. (1984) have shown that the sex pheromone of the ichneumonid *Syndipnus rubiginosus* (a parasitoid of *Pikonema alaskensis*, the yellow-headed spruce sawfly) is ethyl(z)-9-hexadecenoate. There is no cross reaction between *S. rubiginosus* and the congeneric *S. gaspesianus*, and the attraction appears to be species specific.

Eller et al. (1984) review other evidence for sex pheromones in parasitoid wasps although most of the examples they give refer to short-range attraction over distances of a few centimeters. In some cases the origin of the chemical is known or suspected, for example Dufour's gland (Weseloh 1976c) or a "gland at the base of the second valvifer" (Tagawa 1977). In other cases the attractant chemical appears to be distributed all over the surface of the female.

Very often males are also attracted to insects of their own sex, which suggests that the chemical attractant may not be produced specifically as a sex phero- mone but may be a chemical produced for another reason that happens to be useful in mate location.

It is difficult to say whether the scarcity of long-range sex pheromones in parasitoid wasps is real or just reflects the lack of attention the subject has received from entomologists. However, I would argue that there is a reason why sex pheromones should be less common in haplodiploid species com- pared with other groups. The production of long-range sex pheromones pre- sumably incurs metabolic costs, and thus they are unlikely to evolve in species where females gain little advantage from mating. Unmated females of haplo- diploid species are able to lay male eggs which, if the population is at sex ratio equilibrium, are of the same evolutionary value as female eggs (Godfray 1990; see sec. 4.5.3 for a full discussion). Thus hymenopteran females can begin to reproduce before mating and the advantages of investing in sex pheromones may be relatively small. Of course, if the population is not at sex ratio equilib- rium, or if there are advantages in having the capability of producing both sons and daughters, then natural selection may favor sex pheromone production. However, I would expect sex pheromones to be (1) comparatively less fre- quent among haplodiploid species; (2) when found, to be produced in rela- tively smaller quantities than in diploid species; and finally (3) to be most uncommon in species where investment in the two sexes is approximately equal (i.e., when the assumptions of Fisher's sex ratio argument are met).

A good way of testing these ideas would be to compare the distribution of sex pheromones in parasitoid wasps and flies. The two groups share a similar natural history, but flies are unable to reproduce without mating and should be willing to invest more resources in finding a mate.

MALES REMAIN AT EMERGENCE SITE

Mating at the emergence site is very common among gregarious species and among solitary species that attack gregarious hosts or hosts living in discrete, well-defined patches. Males tend to emerge first and then wait for females to emerge. Male *Trichogramma papilionis* (Trichogrammatidae) help females emerge (from a butterfly egg) by pulling their bodies through a hole chewed in the chorion (Suzuki and Hiehata 1985; Kurosu 1985). It is not uncommon for males to mate with females before they emerge. For example, bethylid wasps typically pupate in a loose aggregation around the remains of their hosts. Newly emerged males may bite into the cocoons and mate with the unemerged females (Bridwell 1920, 1929; Kearns 1934a, 1934b; Gordh 1976; Mertins 1980). The egg parasitoid *Trichogramma dendrolimi* typically develops gre- gariously in clutches of about twenty wasps, of which one or two are males. The wasps pupate inside the eaten-out eggshell, and mating takes place within the host egg before the females see the light of day (Suzuki and Hiehata 1985).

Often, males never leave the host egg, and previous workers, observing successful reproduction without mating, had assumed *T. dendrolimi* to be parthenogenetic. The congeneric *T. papilionis* which has an overlapping host range will also mate within the host egg, but whereas 100% of female *T. dendrolimi* are mated before leaving the host, Suzuki and Hiehata estimated the figure for *T. papilionis* to be 87%. Interestingly, *T. papilionis* males which left the egg remained in the vicinity and courted emerging females, while those few *T. dendrolimi* that did leave the host seemed uninterested in courtship.

Competition among males at the emergence site may be direct or indirect. Direct competition between males ranges from the mild to the lethal. Male *Nasonia vitripennis* (Pteromalidae) emerge from the puparia of cyclorraphous flies and await their sisters to follow. The females emerge from the same holes in the puparium as their brother and males may jostle each other to be nearest to the hole (P. E. King et al. 1969; van den Assem et al. 1980a). When few males are present, a single individual can defend an emergence hole against other males present on the puparia, but when male density is high, territoriality breaks down and the males scramble for emerging females (van den Assem et al. 1980a). Scelionid wasps are egg parasitoids that often attack hosts which lay their eggs in clumps. The first male to emerge frequently defends the egg mass from other males and obtains most of the matings (e.g., F. Wilson 1961). In a survey of descriptions of male behavior in scelionids, Waage (1982a) found that fighting among males was recorded in fourteen out of fifteen species attacking clumps of up to fifty host eggs. There is some evidence that fighting is less common in species attacking larger egg masses—in only one out of three cases was male aggression observed in species attacking host clutches of more than fifty eggs—possibly because the greater number of males present on larger clutches makes territorial defense uneconomical. The egg masses of the pentatomid bug *Antiteuchus tripterusi* are attacked by two species of scelionid wasps (*Trissolcus bodkini* and *Phanuropsis semiflaviventris*). On emergence, males fight not only conspecifics of the same sex, but also males of the other species (Eberhard 1975). This may result in some females leaving the egg mass unfertilized. Grissell and Goodpasture (1981) describe fighting among male *Podagrion mantis* (Torymidae), which attacks mantid ootheca; this species is strictly a predator because each wasp feeds on several mantid eggs.

Fighting takes a more vicious form in the tiny eulophid wasps of the genus *Melittobia*, parasitoids of solitary bees and wasps, and occasionally of their prey as well. Approximately one hundred wasps develop gregariously on a host, with a sex ratio strongly biased toward females. The few males hatching from a host fight among themselves until just a single individual remains (Balfour Browne 1922; Buckell 1928; Schmeider 1933, 1938). Sometimes males are killed in the act of emerging from their pupae (van den Assem 1986). It is worth repeating Buckell's much-quoted description of fighting in *Melittobia chalybii*: "A dead male, or even a small piece of one, will be fiercely pounced upon by another male, and dragged around and thrown about with a great show of anger, like a terrier with a rat." Males of several species of *Melittobia* show dispersal polymorphisms which are discussed below (sec. 7.5.1).

Some of the most extreme forms of male-male competition at emergence sites occur in fig wasps (see sec. 1.3.3 for a description of their natural history). The males of many species never leave the fig and have become morphologically highly adapted for competitive mate-searching in the specialized environment of the interior of the fig. Males of different species have adopted two main strategies: speed and aggression. In the first case, males compete to inseminate as many females as quickly as possible. Normally, males do not wait for the female to emerge but bite through the wall of the female's gall. They then either insert their abdomen into the gall or crawl in themselves. The abdomens of the pollinating fig wasps (agaonids) are highly elongated and reflexed under the thorax to facilitate mating with the unemerged females (see fig. 1.2). Parasitic fig wasps also have elongate and telescopic abdomens, although not normally reflexed under the abdomen. Some species of parasitic fig wasp are highly elongate and snakelike to assist full entry into the female gall. The speed at which males can locate and mate females is very impressive. To give one example, the nonpollinating fig wasp *Apocryptophagus* sp. forms large galls in the figs of *Ficus hispidioides* in New Guinea (Godfray 1988). If a mature fig is broken open, the rush of air causes the males to hatch and immediately begin searching for female galls. The serpentine males are highly active and can bite into, enter, and mate a female in forty-eight seconds. Males never enter galls containing unemerged males, parasitoids, or females mated by themselves or their competitors (see Joseph 1958 and Murray 1987 for related observations on other species). Three or four males can mate fifty females in about ten minutes, although the process is possibly slower in the less oxygen rich atmosphere of a closed fig. Only when all females are mated do males begin to re-enter galls and then only briefly, suggesting mating does not occur. The rush of air also causes males of the specific parasitoid of *Apocryptophagus* to hatch (unfortunately called *Apocrypta mega*). The males of this species are also serpentine and display similar behavior to their host. However, they never enter galls containing hosts, or males of their own species. Murray (1990) has observed similar behavior in the related *Apocrypta bakeri*. However, this species exhibits multiple mating, mate guarding, and mild fighting, behaviors never observed in *A. mega*. Hamilton (1979) describes how two males of an unnamed South American fig wasp became firmly jammed in a single gall as they competed to mate with its occupant. Not all wingless male fig wasps mate with females prior to their emergence; Hamilton mentions one species which mates in the cavity of the fig, the male assisting the female to quit the gall.

A substantial number of fig wasp parasitoids adopt a different strategy. In its most extreme form, males hatch first and fight among themselves, the winner mating all the females. Males of other species only fight if two individuals meet on a gall containing a female. Some parasitic fig wasp males are highly modified for fighting, possessing sclerotized head capsules and large scythe-like mandibles (Hamilton 1979; Murray 1989, 1990). Sometimes the head is partly shielded by the sclerotized thorax. In another much-quoted passage, Hamilton (1979) vividly captures the life of these males in the interior of the fig "that can only be likened in human terms to a darkened room full of jostling people among whom, or else lurking in cupboards and recesses which open on all sides, are a dozen or so maniacal homicides armed with knives." Hamilton describes fighting in a species of *Idarnes* where even a small bite can result in death, possibly indicating the presence of a venom. Males vary in size, and small individuals attempt to avoid large males, burrowing among the fig galls to search for unmated females in concealed galls. Outmatched males sometimes hide in vacated galls and attempt to nip the legs of victorious males. Murray (1987) has observed similar behavior in *Philotrypesis pilosa*, a species where fighting is less vicious than in *Idarnes*.

Hamilton (1979) suggested fighting is least common when most wasps in a fig are siblings. Selection for fighting will be weaker if competitors are closely related and if fighting carries the risk that all males die so that sisters leave the fig unmated. Selection may even favor brothers allowing sisters to mate with unrelated individuals if there is a cost to inbreeding. Hamilton predicted that fighting will be found most frequently in species with a near-equal sex ratio, because female-biased sex ratios are found when most wasps are the progeny of just a few parents (see sec. 4.3.1). However, Murray (1987, 1989) failed to find a correlation between sex ratio and fighting. Instead, the number of other males in the fig seemed to be the most important correlate of fighting. In an intraspecific study of fighting in *Philotrypesis pilosa*, the greatest number of injuries were found when there were intermediate numbers of males in the fig (Murray 1987). Fighting was rare when the fig contained just a few males, although when fights did occur they tended to be severe. When many males inhabited the fig, fights were common although not particularly serious. These observations accorded with some predictions of a model of fighting in fig wasps (Murray and Gerrard 1984, 1985), although predicted correlations between fighting and female numbers, and between fighting and the ratio of females to males, were not found. In an interspecific survey, Murray (1989) also found that fighting was less common in species where many males developed in the same fig (although, as discussed by the author, phylogenetic correlates of fighting influence the result).

A number of factors predispose male parasitoids to search for mates at their emergence site. First, and most obviously, males and females must develop together at the same site. This mating system will also be favored if females are (1) hard to locate at other times in their life; (2) mate only once in their life; (3) are receptive to mating soon after emergence; and (4) when multiple matings occur, if the first males to mate enjoy sperm precedence. A consequence of males remaining at the oviposition site is that mating frequently takes place between siblings, leading to selection for a female-biased sex ratio. The likelihood that most siblings are females may reinforce any initial advantages of remaining at the emergence site. By remaining, the male may also increase its inclusive fitness by helping guarantee that its sisters are mated. A possible disadvantage of sibmating is a reduction in fitness due to inbreeding depression.

MALES SEARCH FOR EMERGENCE SITES

If females develop in predictable or easily detectable emergence sites, males may be selected to seek out these sites and wait for females to emerge. Such a strategy will also be favored if females mate only once, are immediately receptive after emerging, or if the first male to mate a female enjoys sperm precedence.

Good examples of this mating system are found in the Rhyssini, a group of ichneumonid parasitoids that attack siricoid wood wasps and possibly other hosts that live deep in wood. The female emerges in the feeding tunnel made by its host and burrows to the surface, a process that may take up to three days. Males congregate on the surface of the wood before the females emerge, attracted by the sound of the burrowing female (Heatwole et al. 1962; Eggleton 1990), and also by chemical cues associated with the host (Matthews et al. 1979; Davies and Madden 1985) or the host's symbiotic fungi (Madden 1968; Spradbery 1970b; the female uses similar cues in host location; see sec. 2.2.2). Quite frequently a number of males (up to thirty have been recorded) wait together for the same female. Different species adopt one of three mating strategies (Eggleton 1991): males of *Rhyssa persuasoria* wait until the female emerges and then scramble to be the successful mate (Myers 1928); species of *Megarhyssa* and *Rhyssella* also scramble for matings but in these species the female is mated before she actually leaves the emergence burrow (Yasumatsu 1937; Heatwole et al. 1962, 1964; Nuttall 1973; Matthews et al. 1979; Crankshaw and Matthews 1981; Davies and Madden 1985; Eggleton 1991). Finally, there is evidence that *Lytarmes maculipennis* defends a potential emergence site from conspecifics, mating with the female as she emerges from her burrow (Eggleton 1990). Eggleton (1991) has related the different mating systems to a phylogenetic hypothesis for the Rhyssini. *Rhyssa* species cannot bend their abdomen and are probably unable to mate with the female before she leaves the burrow. The other species, which form a taxonomic group derived from the *Rhyssa* clade, are able to bend the abdomen, which has a variety of morphological adaptations that facilitate its insertion into the female's tunnel. If abdominal morphology is used to infer mating system in the absence of behavioral observations, then the three mating systems appear to be localized in separate taxonomic groups. It is unclear exactly why some species defend female emergence sites from other males and others do not. Possibly phylogenetic constraints are responsible, but ecological factors may also be involved. In cases where many males congregate at the emergence site, the energy required for site defense, or the risks of injury, may outweigh any benefits (Thornhill and Alcock 1983; Eggleton 1991). In addition, where several species occur together, the male wasps may be unable to tell the species of the female until after her emergence, thus lessening the benefits of site defense.

Some species of parasitic fig wasp only have winged males (normally most or all males are wingless, see sec. 1.3.3). Murray (1989) notes that some winged males seek out and enter figs containing virgin females, or stand guard over the exit hole through which females are emerging, fighting off rival males. The males of yet another species searched the foliage of fig trees for recently emerged females.

Males of species that wait at the pupation site for females to emerge may also attempt to locate other emergence sites (e.g., *Nasonia vitripennis* (Pteromalidae), P. E. King et al. 1969; *Melittobia* spp. (Eulophidae), Buckell 1928, Matthews 1974; *Cotesia (=Apanteles) glomeratus* (Braconidae), Tagawa and Kitano 1981). Males may leave their emergence site after all females have hatched, may be ejected from the site by more powerful males, or the species may be dimorphic with dispersing and nondispersing males (see below).

MALES SEARCH FOR FEEDING SITES

Many adult parasitoids feed from flowers and honeydew, and males may search for females at suitable feeding sites. There are few studies of mating at feeding sites although it is easy to catch male ichneumonids and other wasps at flower heads. Males of the desert bee fly *Lordotus miscellus* (Bombyliidae) defend individual rabbit bush plants (*Chrysothamnus nauseosus*) where females come to feed (Toft 1984).

This mating system is most likely to occur in species where females feed predictably at well-defined feeding sites. Species that are not immediately receptive after mating, and which require food to mature their eggs, are also likely candidates.

MALES SEARCH FOR OVIPOSITION SITES

Male parasitoid wasps can frequently be collected in the same habitat as searching females. For example, the males and females of leaf-miner parasitoids can often be swept from the foodplant of their hosts, even if they emerged from the leaf litter. It seems likely in these cases that males are narrowing their search for females by congregating at sites where hosts are present. Males may make use of the same cues as females in searching for sites where hosts may be present.

McAuslane et al. (1990a) have studied the response of male *Campoletis sonorensis* (Ichneumonidae) to females and to their host's food plant in wind tunnel experiments. Although female *C. sonorensis* produce a volatile sex pheromone, males are inefficient at locating females unless the host plant is also present. McAuslane et al. demonstrate that sex pheromone, volatiles from the host plant, and the visual stimulus of the host plant act together to improve mate location.

This mating strategy is likely to occur when oviposition sites are relatively well defined so females are easy to locate. It will also be tend to occur in species where the female mates several times, is unreceptive to mating immediately after emergence, and where the last male to mate enjoys sperm precedence.

LEKS AND SWARMS

If males aggregate at sites that contain no resources, and if females visit the sites purely to obtain copulations, the mating system is termed a lek. In insects, the lek frequently takes the form of an aerial aggregation or swarm. Leks are currently the subject of intense study by bird and mammal behavioral ecologists (e.g., N. B. Davies 1991), but the little information available about lekking and swarming in parasitoids does not justify a lengthy discussion of this topic.

A few parasitoids are known to form all-male swarms that are visited by virgin females. This behavior is widespread in the braconid subfamily Blacinae (e.g., Southwood 1957; van Achterberg 1977) and found sporadically in other groups (e.g., Dryinidae, Jervis 1979; Chalcidoidea, Nadel 1987). Nadel found male swarms of three species of Encyrtidae and Pteromalidae around boulders at the summit of a hill. The desert bee fly (*Lordotus pulchrissimus*; Bombyliidae) forms small swarms in dry areas of California (Toft 1989a, 1989b). This species is almost certainly a parasitoid although its host is unknown. While the position of swarms appears arbitrary, they appear at the same sites each generation, suggesting either that the flies recognize landmarks that are not obvious to humans, or that they deposit long-lasting pheromones as has been described for some nonparasitoid hymenopterans (Toft 1989b; Thornhill and Alcock 1983). Females enter the swarm and are quickly mated, often by relatively large males. Although firm evidence is hard to obtain, the mating advantage of large males is probably due to intrasexual competition between males rather than female choice.

Antolin and Strand (1992) have recently described a related mating system in the braconid wasp *Bracon hebetor*. Today, this species is found most frequently parasitizing pyralid moths in grain stores and silos. The surface of the grain store has local undulations, and male wasps congregate at the top of molehill-sized mountains where females come to mate. This study is important because *Bracon hebetor* is a gregarious species with a female-biased sex ratio which has normally been interpreted as due to local mate competition. However, *B. hebetor* has single-locus sex determination (sec. 4.1) and thus would suffer a severe reduction in fecundity if mating occurred between siblings. This study partly explains the paradox by showing that *B. hebetor* is normally an outbred species; the explanation for the female-biased sex ratio remains to be discovered.

Why do males enter a lek or swarm and so expose themselves to strong competition for mates? If females are concentrated in only a few places in the environment, then males must congregate at these sites creating a de facto swarm. This explanation seems unsatisfactory as the position of swarms often appears arbitrary and may move over time. Possibly the swarm itself attracts females, either visually or olfactorily, so that all the males in the swarm gain an advantage by banding together to act as a beacon for females. Alternatively, females may preferentially choose to select their mates from swarms if it allows them to copulate with high-quality males. Females may themselves compare male quality and pick good mates, or there may be intrasexual competition among males for favored positions in the swarm.

7.2.2 MATING BEHAVIOR

ETHOLOGICAL STUDIES

There is a huge number of descriptions of mating behavior in different species of parasitoids. Gordh and DeBach (1978) list accounts of mating in 104 species of parasitoid wasp, and other reviews include Matthews (1974) and van den Assem (1986). Important studies include Barrass's (1976 and included references) work on the pteromalid *Nasonia vitripennis*, the work of Gordh and DeBach (1976, 1978) on *Aphytis* (Aphelinidae), and the long series of comparative studies of eulophid and pteromalid wasps by van den Assem and colleagues (van den Assem 1986 and included references). Here I give only a brief flavor of this work, largely based on the studies of van den Assem's group on mating in chalcidoids.

Experiments with dummies indicate that both morphological and chemical cues are required before a male will initiate courtship. Objects that are otherwise ignored by males can be made attractive by treating them with a solvent extract of female cuticle (van den Assem and Jachmann 1982; Obara and Kitano 1974). Males may mount dead females, but if the females' antennae or wings are removed, the male is apt to mount at the wrong end. In *Nasonia vitripennis*, female receptivity is induced by a complex series of male actions. Males nod and vibrate their wings, and a secretion from the mandibular gland is also important. In addition, and like most small chalcidoids, *N. vitripennis* uses its wing musculature to make noises that resonate within the thoracic cavity. If males are prevented from producing the mandibular gland secretion, or from making sounds, the female never becomes receptive. However, gluing the head so nodding is prevented, or removing the wings, slows down but does not prevent successful courtship (van den Assem et al. 1980b; van den Assem and Putters 1980). A female *N. vitripennis* signals her willingness to mate by lowering her antenna which is the cue for the male, until now riding high on the female's thorax, to back up and copulate. This behavior is easy to reproduce using a model wasp with movable antennae (van den Assem and Jachmann 1982). In many other species, females signal receptivity by raising the abdomen.

Females of most parasitoids are not receptive to further courtship after successful insemination, although there is much variation among species (van den Assem 1986). Females of some species irrevocably ignore male courtship after insemination while in other species females are less receptive but may succumb to prolonged courtship (van den Assem and Visser 1976). In *N. vitripennis*, loss of receptivity is triggered by successful courtship rather than by insemination itself: if males are removed immediately prior to copulation, females act as if they have been successfully mated (pseudovirginity, see below). In *Aphytis* spp. females become sexually unreceptive some time after insemination. Males guard females after copulation, presumably to prevent other males from supplanting their sperm (Gordh and DeBach 1978; see also Ridley 1983). Post-copulatory mate guarding has also been recorded from pteromalids in the genus *Asaphes* (van den Assem 1986), aphelinids in the genus *Encarsia* (Viggiani and Battaglia 1983), and in several fig wasps (Murray 1987, 1989). Male *Cecidostiba semifascia* (Pteromalidae) are able to recognize previously mated females and avoid courtship (van den Assem 1986). In other species, males attempt to court unreceptive females. The time a male should spend attempting to induce receptivity in a female will be influenced by the benefits of successful copulation (which may be lower in an unreceptive female if lack of response indicates she has already mated) and by the abundance of females. The decision can be modeled using a marginal-value model (Parker 1974; see sec. 2.5.2) which predicts more persistent courtship when females are rare. Van den Assem et al. (1984b) found that *N. vitripennis* males courted unreceptive females for longer periods of time when females are rare and discusses the proximate processes that may be responsible for this observation.

Courtship behavior tends to be similar in related species although useful taxonomic characters at the species level can sometimes be observed, for example in *Melittobia* (Eulophidae, van den Assem et al. 1982), *Muscidifurax* (Pteromalidae, van den Assem and Povel 1973), *Monodontomerus* (Torymidae, Goodpasture 1975), and *Aphytis* (Aphelinidae, Gordh and DeBach 1978). Trends can also be discerned in larger taxonomic groups. Van den Assem and Jachmann (1982) and van den Assem (1986) argue that in several chalcidoid families males originally courted the female from the copulatory position, but that over evolutionary time the male's position moved forward so that he rides on the female's thorax. This trend may be associated with a decrease in the size of the male relative to the female.

SEXUAL SELECTION

Higher fitness can be achieved in three ways: greater viability, higher fecundity, or more mates. The last route is confined to the sex, normally the male, which competes for access to the other sex. Traits that evolve to increase mating frequency are said to be sexually selected. Sexual selection may occur through competition among males for mates (intrasexual selection), or through intersexual sexual selection where females choose males with certain characteristics. Sexual selection is one of the most active areas in evolutionary and behavioral ecology with recent reviews by Harvey and Bradbury (1991) and Kirkpatrick and Ryan (1991).

Male adaptations for direct and indirect competition for mates at emergence and oviposition sites are obvious examples of traits favored by intrasexual selection (sec. 7.2.1). Are there any circumstances when females may be subject to intrasexual selection? In species with very female-biased sex ratios, females may compete for mates (or at least mates with sperm). Natural selection will, however, act on mothers to produce a sufficiently male-biased sex ratio that all her daughters normally get mated so shortages of males are likely to occur only exceptionally. Females that produce sex pheromones might compete with other females to attract males, but again intrasexual competition seems unlikely to be very important because after mating one female, the male is nearly always available for a second mating.

While the process of intrasexual selection is well understood, there is considerable debate and uncertainty about female choice and intersexual selection. In particular, while it is generally accepted that female choice can lead to the evolution of exaggerated male characters—the classic example being the peacock's tail—there is little consensus about the mechanisms involved.

A female should mate only with a male of her own species, and many of the elaborate courtship behaviors described in the last section are probably species isolation mechanisms. Although far harder to study, the most important species recognition cues are probably chemical. A second simple type of female choice is for males that provide useful resources for reproduction. In insects without paternal care, such resources are probably limited to nutrients passed to the female in the seminal fluid (Thornhill and Alcock 1983). However, I know of no evidence for nutrient transfer during mating in parasitoids.

Females may choose to mate with certain males solely on an assessment of their genetic quality. If inbreeding depression reduces progeny fitness, females may be selected to mate with unrelated males. One mechanism of inbreeding avoidance is to mate with rare males, that is, to choose males with an unusual genotype. There is some evidence for a rare-male effect in the pteromalid *Nasonia vitripennis* (Grant et al. 1974, 1980; White and Grant 1977). In laboratory experiments, males of a relatively rare genotype obtain more matings than those of a common genotype. The mating advantage of rare males is reduced if air that has previously passed over males of the common genotype is blown into the mating chamber. This experiment suggests an olfactory basis to mate choice. However, in a review of frequency-dependent mating, Partridge (1983) has pointed out both theoretical and methodological problems with the rare-male effect. Mating with rare males can carry disadvantages; the reason rare males are uncommon may not be because they are emigrants from other populations, but because they are of low fitness and hence at low frequency. Even if they are emigrants, they may be adapted to different microhabitats and hence are poor-quality mates. If the rare-male effect is studied using groups of males and females rather than individuals, a fixed female preference for certain male types can give rise to the misleading appearance of a rare-male effect. It is also necessary to exclude changes in male behavior in experiments designed to study the role of olfactory factors in male choice. The experiments with *N. vitripennis* need to be reexamined in the light of Partridge's criticisms.

Are there parasitoid equivalents of the peacock's tail that might be explained by sexual selection? Most parasitoid taxa contain examples of species with brightly colored or morphologically unusual males. For example, *Mesopolobus* is a genus of pteromalid wasps that chiefly attack cynipid gall wasp larvae; like many other pteromalids, females have metallic-green bodies but with rather dull-colored legs and antennae. The middle tibia of males, however, is frequently striped yellow and orange and may posses an unusual black knob (Graham 1969). The middle tibia is passed over the female's eyes during mating (Askew 1971), but whether species recognition or some other form of female choice accounts for the evolution of the tibial ornamentation is unknown.

MATING FREQUENCY, SPERM PRECEDENCE, AND TIMING OF MATING

There are at least four reasons why females may mate more than once. First, a single male ejaculate may not contain sufficient sperm for the female's future needs, the female may run of sperm, or sperm from a previous mating may have degraded (Ridley 1988b). The second reason is the female may encounter a superior male or may want to encourage sperm competition among the ejaculates of different males to ensure her eggs are fertilized by high-quality sperm (Thornhill and Alcock 1983). Third, if a female's offspring feed together in the same patch, and if patches vary in condition, there may be an advantage for a mother to produce genetically heterogeneous offspring (Ridley 1993): in a metaphor first used by G. C. Williams (1975), the production of genetically homogeneous offspring is like buying identically numbered tickets for the same lottery. Finally, the female may only accept a second mating to stop herself being pestered by a courting male. The costs of mating include wasting time, a possible increased risk of predation, and, more speculatively, an increased risk of contracting a sexually transmitted microorganism.

Mating frequency among parasitoids has been most closely studied in the Hymenoptera. In a review of mating behavior among parasitoid wasps, Gordh and DeBach (1978) found that multiple mating by males occurred in all forty-eight cases where mating behavior had been described, while multiple mating by females occurred in only five out of thirty-four cases. Recently, Ridley (1993) has collected information from the literature on the female mating behavior of ninety-nine species of parasitoid wasp of which nearly 80% were reported to mate only once. Applying a formal comparative analysis (Grafen

1989; Harvey and Pagel 1991), Ridley was able to show that there was a significant tendency for gregarious species to mate several times and for solitary species to be monandrous. Ridley argues that his results support a sibling competition explanation for multiple mating because the advantages of genetically heterogeneous offspring will be greatest in gregarious species where siblings compete together for host resources.

While I am convinced by Ridley's demonstration that multiple mating is associated with gregariousness, I believe there are alternative explanations to sibling competition. Gregariousness is normally associated with female-biased sex ratios. Quite often one male mates many females and may transfer little or occasionally no sperm. The variance in the insemination capacity of gregarious males may favor polyandry. In addition, it is more important for a gregarious species to be mated. As I described in section 4.5.3, an unmated female may suffer no or little fitness penalties by producing only sons if the population is at sex ratio equilibrium and if there is little sibling mating. However, species with a female-biased sex ratio due to local mate competition do suffer if they are unable to produce females because males in all-male broods may have little opportunity for mating. Thus gregarious species may face both an uncertain sperm supply and suffer a greater penalty if they run out of sperm. This double jeopardy favors multiple mating.

There have been few studies of sperm precedence in cases of multiple mating. Using genetic markers, van den Assem and Feuth-de Bruin (1977) found that a second male *Nasonia vitripennis* could successfully court and mate a female as long as some time had elapsed since the first mating. The fraction of daughters fathered by the second male remained constant over time, suggesting that the sperm in the two ejaculates had become mixed in the spermatheca. Van den Assem et al. (1989) also describe some evidence for sperm competition and a first-male advantage in the pteromalid *Lariophagus distinguendus*. Dyson (quoted in Antolin and Strand 1992) also used genetic markers to show that 11% of females of the braconid *Bracon hebetor* produced multifathered broods when large populations of males and females were maintained in the laboratory.

Some female parasitoids are not immediately receptive to males after emergence. There are at least two reasons why this behavior may evolve (Thornhill and Alcock 1983). First, in synovigenic species, eggs may not be matured until after the parasitoid has fed. Many tachinid flies have a lengthy incubation period before eggs are ready to be deposited. In such species early mating may not be favored as sperm may die and clog up the spermathecal ducts before they are required. In addition, resources may be required to keep sperm alive. A second reason might be the avoidance of inbreeding. If siblings emerge near to each other the first males a female encounters are likely to be brothers. A postemergence refractory period may lessen the risk of matings between siblings (Antolin and Strand 1992).

PROTANDRY

Protandry occurs when males emerge before females. In parasitoids, males may emerge early (1) relative to females in the same brood, or (2) relative to other females in the population. Where males remain at the pupation site to mate with females from the same brood, there is a clear advantage in emerging early to ensure females do not emerge and disperse before they can be inseminated. The advantages of the second type of protandry are less straightforward.

Consider a species with discrete generations where female emergence is more or less synchronized. If the risks of mortality over the female flight period is small, then all males will be selected to emerge before the first female to ensure that they have the greatest chance of mating with the most females. However, where the risk of male mortality is significant, early emergence may be penalized if the males risk dying while there are still females in their pupae. The optimal emergence time will also be influenced by competition with other males: late emergence may be advantageous if competition from other males is less severe at that time of the season. In the face of competition with other males, a mixed distribution of emergence times will be favored rather than all males emerging at the same time. The theoretical optimum emergence pattern can be thought of as a type of temporal ideal-free distribution (sec. 2.5.2): all males on average achieve the same number of matings, whenever they emerge. There have been no theoretical studies of protandry specifically addressed to parasitoids, although a series of models prompted by butterfly behavior are applicable (Wiklund and Fagerström 1977; Bulmer 1983; Iwasa et al. 1983; Parker and Courtney 1983). These models suggest that the most important factor influencing male emergence is the risk of mortality. When mortality is low, most males will emerge before females; as the risk of mortality increases, the emergence times of the two sexes become more similar. If mortality is very high, and if most females are mated soon after emergence, the pattern of male and female emergence becomes nearly identical.

If a species has overlapping generations and a constant population size, then there can be no advantage to the second type of protandry: males will achieve the same average mating success whenever they emerge. However, protandry could be favored in species with overlapping generations if the numbers of parasitoids emerging fluctuate, either seasonally or episodically.

Protandry might be achieved in a number of ways. First, individual ovipositing females might lay male eggs early in life and female eggs late in life. Second, in species which diapause prior to emergence as adults, the male may leave diapause before the female. Finally, in species without diapause, the development of males may be faster than that of females. I know of no evidence for the first two mechanisms, but faster male development is widespread in parasitoids (King 1988). Parasitoid development times are probably influenced by a trade-off between speed and final size because poorer assimilation of food is likely to be a cost of faster growth. Charnov et al. (1981) and others (see sec. 4.4.1) have argued that the benefits of large size may be less for male parasitoids than for females. If this is true, then males may be able to achieve faster development, and hence a mating advantage through protandry, relatively cheaply.

There is one further complication. The speed of development may depend on the size of the host. At least in some species, development proceeds faster in smaller hosts (e.g., Mackauer 1986; Sequeira and Mackauer 1992a). This suggests a second reason why females may tend to lay male eggs in smaller hosts: not only do males suffer less from being small than females (Charnov's size advantage hypothesis, see sec. 4.4.1), but small males may also gain an advantage through protandry.

CONSTRAINED SEX ALLOCATION

In the chapter on sex ratio (sec. 4.5.3), I argued that female hymenopteran parasitoids will sometimes be constrained to lay only male eggs and that this has interesting consequences for the optimal sex ratio of unconstrained females. I also argued that virginity is probably the most likely explanation for many cases of constrained sex allocation and asked when females might prefer to search for hosts rather than mates. In this section I discuss oviposition by virgin females and some other factors that may cause constrained sex allocation (see also Godfray 1990).

Virgin oviposition. In the laboratory, most hymenopteran parasitoids oviposit when unmated, producing male eggs. In some species the act of fertilization causes a noticeable pause during oviposition that allows an observer to detect if a female egg has been laid (sec. 4.5.1). Observations on virgin wasps reveal that they attempt to fertilize a proportion of their eggs, as if they were mated (Pallewatta 1986; Strand 1989b). This suggests that at least in some species oviposition behavior is unaffected by mating status. There are, however, reports in the literature of virgin females laying fewer eggs than mated females (McColloch and Yuasa 1915) or even laying eggs at a greater rate (Moratorio 1977; Tagawa 1987; Cronin and Strong 1990b). Eulophid wasps in the genus *Melittobia* lay large clutches of eggs with a very female-biased sex ratio on aculeate bees and wasps. An unmated female normally lays a very small clutch of males and then remains with her progeny until they emerge; she then mates with one of her sons and lays a much larger clutch, normally on the same host which is only partially consumed by the first generation (Balfour Browne 1922). Virgins of a number of bethylid wasps also remain with their brood and mate with their sons (van Emden 1931; de Toledo 1942; Rilett 1949), and similar behavior occurs in haplodiploid scolytid beetles (Hamilton 1967).

Sperm depletion. If parasitoids are supplied with unlimited numbers of hosts in the laboratory, the sex ratio often becomes increasingly male biased with time, presumably because of sperm depletion (e.g., van den Assem 1986; King 1987). However, parasitoids are unlikely to live as long in the field as in the laboratory, or to encounter as many hosts. Because sperm are relatively cheap to produce, one would normally expect males to transfer enough sperm to their mates to cover all but their most exceptionable requirements. However, it is possible that sperm cannot survive indefinitely in the female's spermatheca so that the male is selected to transfer a smaller ejaculate. A second possible exception to this generalization might occur in species with highly female-biased sex ratios where one male mates with a large number of his sisters. The male may be selected to withhold sperm in order to inseminate other individuals. It is interesting that the mother, the male, and the female all have subtly different optima. The female wants most sperm to reduce the probability of depletion while the male is selected to spread its sperm across the maximum number of individuals to maximize his number of granddaughters. The mother's optimum is probably intermediate: she wants to produce enough males that her daughters are all mated with a reasonably sized ejaculate, but not to waste the opportunity of laying daughters by overproducing sons. I am unaware of studies of sperm depletion in the field except for that of Antolin and Strand (1992) on the braconid *Bracon hebetor* in a corn silo in Wisconsin. In this species older females can be distinguished from younger by morphology and color pattern. Dissections of old females revealed that between 10% and 20% had empty spermatheca, probably because of sperm depletion.

Pseudovirginity. After insemination, females of species that mate just once in their lifetime become unreceptive to males. The change in behavior occurs late in courtship but prior to copulation (see above). Van den Assem (1969) found that if the courtship of the eucoilid *Leptopilina heterotoma* was disturbed prior to insemination, either by the experimenter or by conspecific jostling, the unmated female became unreceptive as if she had been mated. He found the same phenomenon (which he called pseudovirginity) in the pteromalid *Lariophagus distinguendus* (van den Assem 1970). Pseudovirginity also occurs if females that mate only once copulate with a male that has run out of sperm. It has often been recorded that males will continue to mate with females after they have run out of sperm (King 1987). For example, a male *Bracon lineatella* (Braconidae) mated thirty-one females although it ran out of sperm after the third one (Laing and Caltagirone 1969). Similarly, Simmonds (1953) found that the pteromalid *Spalangia cameroni* would mate eight times but inseminate only the first four females. The number of females that can be inseminated by *Nasonia vitripennis* (Pteromalidae) is influenced by the size of the male (Barrass 1961; van den Assem 1986). Unless males are able to monitor precisely their sperm reserves, sperm-exhausted males are likely to be selected to attempt to mate with most females just in case some successful sperm transfer occurs.

Limited time for mating. There is evidence from some parasitoids that if females fail to find a mate early in life, they are subsequently unreceptive to male courtship and thus produce only male eggs. Crandell (1939) found that female *Pachcrepoides dubius* (Pteromalidae) ignored males if unmated in the first five days of adult life. Drea et al. (1972) found that female *Microctonus stelleri* (Braconidae) were reluctant to mate after just one day, and similar behavior was found in the braconid *Aphidius smithi* (Wiackowski 1962). Stary (1970) has suggested that a limited window in which mating can occur is widespread in the braconid subfamily Aphidiinae. Other aphidiines appear to ignore males after the initiation of oviposition (e.g., *Trioxys indicus*, Subba Rao and Sharma 1962; *Aphidius matricariae*, Verai 1942). It is difficult to think of any positive benefits to the restriction of mating to the first few days of adult life, and more information is required about the reproductive biology of the species concerned.

Postmating constraint. If some female wasps are provided with hosts immediately after mating they invariably lay a male egg. This phenomenon has been recorded in the braconids *Bracon hebetor* (Genieys 1924) and *Aphidius smith* (Mackauer 1976), and in the pteromalid *Nasonia vitripennis*, in the latter both after a first and second mating (van den Assem 1977; van den Assem and Feuth-de Bruin 1977). Van den Assem (1977) has suggested that after mating, sperm are attracted to the spermatheca up a chemical gradient. Before sperm are able to leave the spermatheca, this chemical must dissipate, which takes a little time. Although the postmating constraint operates only for a short period, the effect is potentially significant in short-lived species. Mackauer (1976) calculates that the postmating constraint lasts for 15% of the adult life of *A. smithi*, which puts an upper bound of 85% on the maximum achievable female-biased sex ratio.

Too many matings. Both Flanders (1946), working with the braconid *Macrocentrus ancylivorus*, and Strand (pers. comm.), working with the encyrtid *Copidosoma floridanum*, have noticed that if a female is mated too many times she is unable to fertilize her eggs. What appears to happen is that the spermathecal ducts become blocked with excessive ejaculates.

POLYMORPHIC MATING STRATEGIES

Many examples of polymorphic male mating strategies have been studied by behavioral ecologists. In some cases males adopting each strategy have identical fitnesses, and the relative frequencies of the different strategies are maintained by frequency-dependent selection (mixed polymorphism). In other cases, males differ in some quality such as size that affects mating success, and poor competitors adopt a different mating strategy (conditional polymorphism). In this case small males are "making the best of a bad job" and have lower fitness than large males. Mating strategy polymorphisms may be purely behavioral or may have both a behavioral and a morphological basis.

The clearest example of polymorphic mating strategies among parasitoids occurs in a number of species of parasitic fig wasp. As described in (secs. 1.3.3 and 7.2.1), the males of most species are highly modified for competitive mate location in the specialized microhabitat of the interior of the fig. However, in

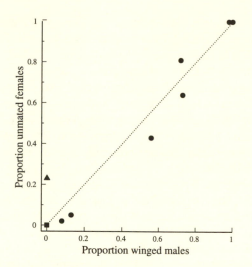

Figure 7.2 The relationship between the proportion of winged males in different species of fig wasp, and the proportion of females leaving figs that contain no males (and hence are unmated). (The circles are data from Hamilton 1979 and the triangle from Godfray 1988; the square represents four species, one from Hamilton 1979, the others from Godfray 1988.)

some species two types of male are found: the highly modified form with no wings, eyes, and pigmentation, and which never leaves the fig, and "normal-looking" males which resemble the female. The second type of male does not compete for mates in the fig but leaves the fig with the females to search for mates outside. Hamilton (1979) argued that dispersing males would be found only in species which produced all-female broods in at least some figs. All females in figs containing nondispersing males are likely to be mated before they disperse. However, if some females develop in figs without nondispersing males, dispersing males may be able to locate unmated females in the environment. A simple model suggests that the fraction of dispersing males should equal the fraction of females that leave their natal fig unmated. Hamilton (1979) tested this idea with data from Brazilian fig wasps and found a good match between theory and data (fig. 7.2). There are, however, some species of fig wasp in which a significant fraction of females develop in figs with no males, but in which all males are nondispersing and confined to the fig (Godfray 1988).

Winged and wingless males are found in some other species of parasitoid, perhaps associated with different mating strategies (sec. 7.5.1). A second form of polymorphic mating strategy occurs in those fig wasp species in which males fight for access to females. As mentioned above (sec. 7.2.1), small males tend to avoid large males and instead search for concealed females (Hamilton 1979). Murray (1990) noted that in species of *Sycorytes* large males possessed big heads and fighting mandibles while small males had little heads without fighting mandibles. It is not clear how the fitness of small males compares with large males in this system.

Males may adopt a "sneaky" strategy, loitering around courting males in the hope of achieving a surreptitious mating. In many chalcidoids, the male courts a female by standing on top of her thorax until she signals receptivity when he backs up and copulates. Males have been observed to wait in the vicinity of courting pairs and to dart forward and copulate after the female signals receptivity, and before the courting male has repositioned himself (van den Assem et al. 1980a, 1980b). In some species that mate in situations where many males are present, for example *Pteromalus puparum* (Pteromalidae) which mates on the surface of a parasitized butterfly pupae, the male backs up many times during courtship. Van den Assem et al. (1980a) have suggested that this behavior allows the courting male to detect potential competitors and so lessen the chance of another male achieving a sneak copulation.

Species of *Megarhyssa* (Ichneumonidae) parasitize hosts deep in tree trunks, and males compete to mate with females as they emerge from tunnels in the wood (sec. 7.2.1). Males insert their abdomens into the tunnel to mate with the emerging females, and small males with short abdomens stand little chance of mating in competition with larger males. There is some evidence that small males wait around the emergence site in case the female is still receptive after unsuccessful insemination before emergence (Yasumatsu 1937; Crankshaw and Matthews 1981; Eggleton 1990). It seems likely that the fitness of small males is less than that of large males (Eggleton 1991).

7.3 Resource Defense and Maternal Care

Some female parasitoids defend hosts, or sites containing hosts, from conspecific competitors. A few species continue to defend the host after oviposition, a form of maternal care.

If two searching females encounter each other they may engage in ritualistic display or actual fighting. Bark beetle parasitoids move over the surface of a tree or log searching for hosts and not infrequently come into contact with individuals of their own or another species of parasitoid. Such encounters are normally followed by aggressive displays of wing beating (Dix and Franklin 1974; Mills 1991). Mills suggests that some pteromalid and eurytomid wasps of the ash bark beetle (*Leperisinus varius*) deliberately approach ovipositing braconids which they displace from the host, and then lay their own eggs. The braconids (*Coeloides* sp.) appear to be better able to locate hosts and their superior skills are cleptoparasitized by the other species. Lawrence (1981b) observed that female *Diachasmimorpha (=Biosteres) longicaudatus* aggressively excluded other females from fruit while searching for their hosts, tephritid fruit flies. Larger females successfully dominated smaller individuals although the latter were sometimes able to rush in and make a "hit-and-run" oviposition.

The wasp family Scelionidae are exclusively egg parasitoids, and females often defend clumps of host eggs from other conspecifics (Wilson 1961). In a literature survey, Waage (1982a) found that twenty-nine out of thirty-three species that attacked small clumps of eggs (fewer than fifty) showed intra-specific aggression, while only one out of four species attacking larger clumps of eggs was recorded to fight. Waage suggests that fighting and territoriality may not be economic on large clutches of hosts eggs which are difficult to defend from competitors, and on which competition for hosts is less strong than on small clutches. The relationship between the size of the host egg mass and fighting among male scelionids after emergence has already been discussed (sec. 7.2.1).

Postoviposition maternal care is rare among parasitoids with the exception of the family Bethylidae. The anomalous braconid wasp *Cedria paradoxa* is a gregarious ectoparasitoid of forest Lepidoptera in India and China. The female remains with her brood after oviposition, apparently to protect them from chalcidoid hyperparasitoids (Beeson and Chatterjee 1935; Chu 1935). Many bethylids remain with their brood after oviposition (Griffiths and Godfray 1988). Some females prepare the host by removing hairs and bristles (Mertins 1980) or rupturing the host's skin (Bridwell 1919; Kearns 1934a) to allow easy access for the larvae. Hardy and Blackburn (1991) compared the survival of broods of *Goniozus nephantidis* with and without the female in attendance. At least in the laboratory, removing the mother had no effect on larval fitness, unless there was a risk of superparasitism or multiparasitism. The guarding female was able to protect her eggs from conspecific females which destroyed and replaced the eggs on unprotected hosts. The mother was also able to fend off a potential competitor (the braconid *Bracon* sp. nr. *hebetor*) whose larvae would have outcompeted her own.

Behavior approaching parental care is found in virgin wasps of the eulophid genus *Melittobia*. As was mentioned above (sec. 7.2.2), virgin females lay small clutches of eggs and remain with their young until they become adult; they then mate with their sons and lay normal clutches of eggs on the same host. While waiting for their sons to grow up, female wasps are recorded guarding and tending their offspring.

7.4 Host Defense against the Adult Parasitoid

In the last chapter I described how hosts defend themselves from larval parasitoids, and how parasitoids have evolved countermeasures to host defenses. In this section I explore how hosts defend themselves from the adult parasitoid.

In many cases, where parasitism results in death or permanent paralysis, this is the only line of defense available to the parasitoid. I distinguish between avoidance of the searching parasitoid, passive defense, and active defense.

7.4.1 AVOIDANCE

Parasitoids locate their hosts using a variety of chemical, visual, tactile, and other cues. Natural selection will act on the host to minimize, where possible, stimuli that can be used in host location by searching parasitoids. It is thus not surprising that most of the attractant and arrestant chemicals used by searching parasitoids are produced when the host eats or defecates, two activities that it cannot avoid. Although much harder to appreciate, hosts are likely to display as subtle an olfactory camouflage as the visual camouflage obvious to human eyes. Some hosts deliberately broadcast chemical or other signals to locate mates, or as aggregation or alarm pheromones. Where parasitoids have evolved to make use of these cues (sec. 2.2.3), the risk of parasitism will be a major factor influencing the timing and amplitude of the signal.

There are several ways in which hosts may reduce the cues they provide for searching parasitoids (Vet et al. 1991; Vet and Dicke 1992). Phytophagous hosts living inside plant tissue may conceal themselves so that they provide no external visual cues for the searching parasitoid. Leaf-mining insects, for example, often only eat part of the green leaf mesophyll, perhaps so that an obvious white mine is not produced. Kato (1984, 1985) has suggested that the tortuous linear mines of some agromyzids which double back and cross over themselves are adaptations that prevent the parasitoid from following the mine to the host.

The host can reduce the usefulness to the parasitoid of cues produced while eating, or by a host plant as a response to feeding, by "snacking": eating small amounts of food and then moving on to a new feeding site (Mauricio and Bowers 1990). Some lepidopteran caterpillars cut off partially eaten leaves which might otherwise have acted as cues for a variety of predators and parasitoids (Heinrich 1979). These behaviors are normally interpreted as removing clues for visually hunting predators such as birds, but parasitoid avoidance may also be involved. Host larvae often carefully ensure that frass falls off the host plant and does not contaminate the feeding area (Price 1981; Vinson 1984). The reason for this may be partly sanitary, but because parasitoids often use chemicals in frass as arrestant and attractant stimuli, this behavior may also reduce parasitism. Some hesperid butterflies can explosively eject frass over 1 m (Frohawk 1913). Oviposition sites can also be chosen to reduce the information provided to parasitoids. Faeth (1985) found that the cosmopterygid leaf miner *Stilbosis juvantis* avoided attacking damaged leaves as these attracted parasitoids. As has already been discussed (sec. 2.5.3), J. N. Thompson (1986) suggested that the pattern of oviposition across patches may have evolved to provide the parasitoid with the minimum information about the distribution of hosts.

Hosts that are located through movement or vibration cannot remain motionless but can reduce movement to a minimum. Sokolowski (1980; see also Sokolowski et al. 1986; de Belle et al. 1989) discovered a natural polymorphism in the behavior of larval *Drosophila melanogaster.* Some individuals, called rovers, moved actively through the medium while others, called sitters, were significantly less mobile. She suggested that the sitter morph might be at an advantage in the presence of parasitoids that located their prey through vibrotaxis (sec. 2.2.3). In a laboratory experiment, Sokolowski and Turlings (1987) compared parasitism of a rover strain with a temperature-sensitive paralytic mutant: below 29°C the mutant behaves as a rover, but at or above 29° it neither moves nor feeds (in this regard they are more immobile than sitters, which move a little and also feed, thereby producing some vibration). The eucoilid parasitoid *Leptopilina heterotoma* did not distinguish between the two strains, while the braconid *Asobara tabida* was unable to locate the temperature-sensitive mutant above 29°. *L. heterotoma* locates its hosts by stabbing the medium with its ovipositor, while *A. tabida* uses vibrotaxis. Carton and Sokolowski (1992) compared the rates of parasitization of sitters and rovers by two eucoilids: *Leptopilina boulardi*, which locates its host by stabbing with its ovipositor; and *Ganaspis xanthopoda*, which uses vibrotaxis. Rovers were parasitized significantly more often by *G. xanthopoda* and sitters significantly more often by *L. boulardi*. They also obtained preliminary results suggesting that in population cage experiments the presence of *L. boulardi* could lead to an increase in the frequency of the sitter morph, and that in the field the distribution of the two forms was correlated with the types of parasitoids present in the environment. Carton and David (1985) have also studied two *Drosophila* behavioral morphs, digger and non-digger, which tend to feed at different depths. They found that *Leptopilina boulardi* attacked more individuals of the non-digger morphs, which remain nearer the surface.

Roitberg and Lalonde (1991) have discussed another host polymorphism that may be maintained by parasitoid pressure. Many herbivorous insects mark oviposition sites which are subsequently avoided by other conspecific females (Roitberg and Prokopy 1987). Both the marking female and females that subsequently encounter the oviposition site benefit from a reduction in larval competition (in a manner exactly analogous to host marking by parasitoids, see sec. 3.3.3). However, the oviposition mark may attract parasitoids or cause them to remain longer at the oviposition site (for examples see sec. 2.2.2). Roitberg and Lalonde constructed a model that demonstrated how parasitoid attack could result in the persistence of both marking and nonmarking morphs in the host population. In the absence of the parasitoid, the host population evolved to consist solely of the marking morph. The pteromalid wasp *Halticoptera rosae* responds to the oviposition site marker of its host, the tephritid *Rhago-*

letis basiola. Roitberg and Lalonde note that nonmarking flies are found at low frequencies in the field and speculate that parasitoid attack may be maintaining a polymorphism in the host population.

Over long periods of evolutionary time, hosts may evolve to avoid parasitism by changing their feeding pattern or phenology. For example, parasitism is often invoked as the selection pressure responsible for the evolution of gall formation. I shall return to the long-term influence of parasitism on the shape of the host niche in the next chapter (sec. 8.2.4).

7.4.2 Passive Defense

Many lepidopteran hosts live gregariously in large webs which afford at least some protection from searching parasitoids. Larvae may retreat into the interior of the web when parasitoids are present. Species that feed partially concealed in plants also may retreat into the safest part of their feeding site when they detect a parasitoid. Another common response is for hosts to remain absolutely motionless in the presence of parasitoids so that movement does not reveal their position (e.g., Richerson and DeLoach 1972).

The thickness of the host's cuticle (or chorion) may be sufficient to deter attack by some species of parasitoid. Some tropical butterflies in the nymphalid subfamily Charaxinae hide in rolled leaves and plug the entrance to the roll with their head capsule. The head capsule appears to be too hard to be pierced by the ovipositor of a parasitoid wasp and also affords some protection against the larvae of tachinid flies which lay their eggs on the exposed part of the caterpillar (DeVries 1987). The waxy covering of mealy bugs and the hard covering of scale insects may also deter parasitoid attack. The gregarious braconid parasitoid *Cotesia (=Apanteles) congregatus* is unable to oviposit through the tough final instar cuticle of its host, *Manduca sexta*, the tobacco hornworm (Beckage and Riddiford 1978). Trichogrammatid egg parasitoids (*Trichogramma* spp.) typically have a wide host range but are unable to attack species with thick chorion (Flanders 1935; Salt 1938).

Many species of parasitoid have specialized behaviors that allow them to oviposit through less armored parts of the host's epidermis such as intersegmental membranes. Species of euphorine braconids have particular problems as they attack heavily sclerotized adult beetles; oviposition often occurs between the abdominal sclerites, or in the mouth, base of the antenna, or anus (Shaw and Huddleston 1991).

7.4.3 Active Defense

One of the most common responses of hosts to the presence of parasitoids is violent wriggling (Askew 1971; Matthews 1974). The movement may be sufficient to throw the parasitoid off the host, or to prevent it from laying an egg.

The very smooth surface of some butterfly pupae may be an adaptation to prevent parasitoids gaining a firm footing on the wriggling host (Cole 1959a). The intersegmental grooves of the pupae of some tenebrionid beetles have sclerotized and toothed margins which act as gin traps, trapping the tarsi of parasitoids walking over the host (Hinton 1955; Askew 1971). The cocoons of a number of genera of ichneumonid jump if disturbed. Jumping is caused by the sudden straightening of the larva inside the cocoon, and this ability is lost after pupation. A jump, often of several centimeters, may land the cocoon in a safer place for pupation, or it may be a means of escaping from hyperparasitism. Aphids kick violently when attacked, and large individuals can prevent a parasitoid from getting to within striking distance (e.g., Kouamé and Mackauer 1991). Gregarious caterpillars such as first instar white butterflies (*Pieris brassicae*) simultaneously jerk their bodies in response to the presence of parasitoids. Such behavior may frighten a generalist species but is ignored by their specialist parasitoids.

Some insects deliberately fall off the plant if they detect a nearby parasitoid. Many lepidopterans throw themselves off the plant but remain attached by a silk thread. After dangling below the plant for some while, they climb back up and resume feeding. An example of this behavior is provided by the larvae of the green cloverworm (*Plathypena scabra*, a noctuid moth) which rests between feeding bouts suspended on a thread beneath the leaf. However, one of its parasitoids, the braconid (*Diolcogaster facetosa*), is able to slide down the silken thread and oviposit on the hanging larva (Yeargan and Braman 1986). To do this, the wasp grasps the thread with one fore tarsus and one hind tarsus, using legs on the same side of the body. Oviposition may be clean and quick although in other cases host and parasitoid drop to the ground. It is not necessary for the wasp to encounter a host on the leaf to detect the thread, and the parasitoid is thus able to locate hosts resting between feeding bouts. *P. scabra* is also attacked by the braconid *Cotesia marginiventris* which in its turn is parasitized by the ichneumonid hyperparasitoid *Mesochorus discitergus*. On encountering a caterpillar hanging from the thread, the ichneumonid reels the thread up and examines the caterpillar, laying an egg if it already contains the primary parasitoid (Yeargan and Braman 1989).

Another common defense strategy is to exude drops of sticky liquid from the mouth. The liquid may contain toxins or it may coagulate and gum up the parasitoid. Contamination by the oral exudate of *Heliothis virescens* elicits prolonged grooming in the braconid wasp *Cardiochiles nigriceps* (Hays and Vinson 1971). This defense strategy is often less effective when performed by young larvae (Johansson 1950) and may select for early attack by the parasitoid (Slansky 1986). Aphids secrete a waxy substance from their siphunculi which may deter parasitoids. Some species bleed copiously when wounded, and it is possible that the coagulating hemolymph deters the parasitoid from laying an egg (Vinson 1990a).

The most important countermeasure against host movement is paralysis. Many parasitoids dart in and sting the host before retreating until the venom has caused paralysis (Vinson and Iwantsch 1980b). Only then do they return to the host and lay their eggs at leisure. The host may be permanently paralyzed, or in the case of koinobionts, may recover and continue feeding. Another adaptation is the possession of large chelae or mandibles that can be used to grasp the host firmly. Female wasps in the family Dryinidae have large raptorial tarsi with an opposable claw which they use to grab their homopteran prey. Several genera of euphorine braconids attack adult beetles and have large mandibles which they use to grasp the host while they carry out the difficult task of ovipositing into a heavily armored insect. Other euphorines grasp their hosts with their legs while one aberrant genus, *Streblocera*, appears to possess raptorial antennae (Shaw and Huddleston 1991).

A number of host species turn on their attackers and try and kill them. As might be expected, the species of host that use this strategy most effectively are predators rather than herbivores. In Hawaii, some geometrid moths have carnivorous larvae which are quite capable of attacking and eating their bethylid parasitoids (Bridwell 1919, 1920). There is limited evidence that some large moth larvae which cannot be subdued by one parasitoid are successfully attacked by two which go on to share the host (Bridwell 1920). Ichneumonids that attack spiders (e.g., *Polysphincta*) can also be killed by their hosts (Askew 1971).

Some parasitoids deliberately allow themselves to be attacked by a predacious host. The chalcidid *Lasiochalcidia igiliensis* incites ant lions (*Myrmeleon* sp.) to emerge from their burrows and seize the wasp's heavily armored legs. From this position, the wasp is able to lay an egg through the membrane between the host's head and thorax (Steffan 1961).

If the host shows parental care, the parasitoid must avoid the adult before it can attack the larvae. Female bees (*Halictus* spp.) engage in combat with mutillid wasps (*Mutilla canadensis*) that attempt to enter their nests, although the males just block the nest entrance with their abdomens (Melander and Brues 1903). Eberhard (1975) has carefully described the way in which a pentatomid (*Antiteuchus tripterusi*) defends its egg masses from two scelionid wasps (*Trissolcus bodkini* and *Phanuropsis semiflaviventris*). The bug uses its legs and antennae to ward off wasps which approach the egg mass. This strategy is only partly successful as wasps are able to rush in and lay an egg before being noticed or while the bug is distracted. Wasps can sometimes crawl under the bug and remain there, ovipositing undetected. Eberhard describes subtle differences in the strategies adopted by the two species of parasitoid.

Parasitoids that attack the larvae of well-guarded social insects face particularly severe problems. Two main strategies seem to have evolved. Some species are well armored and immune to the stings and mandibles of the defending insects. For example, the ichneumonid *Ichneumon eumerus* attacks Large Blue

Butterfly larvae (*Maculinea rebeli*) in ants' nests where they are attacked by numerous worker ants (Thomas and Elmes 1993). The wasps survive attack through the possession of a tough cuticle and also through the production of a fight-inducing pheromone that causes ants to attack each other. Other species of parasitoid are ignored and presumably chemically mimic their hosts. The planidial larvae of the Eucharitidae which parasitize ant larvae and pupae are apparently ignored by adult ants. Michener (1969) has described the morphology and behavior of perilampid wasp larvae (*Echthrodape africana*) which are ectoparasitoids of the larvae of anthophorid bees (*Braunsapis* sp.). The bee nests in hollow stems, and adults actively rearrange the position of larvae as the grow. Unlike all other perilampids, *E. africana* larvae have long hairs and pseudopods which cause it to resemble the host and which presumably allow the wasp to escape detection.

Finally, some hosts are guarded from parasitoids. Members of a specialized caste of leaf-cutting ants protect normal workers from attack by a phorid fly (sec. 2.2.3 Eibl-Eibesfeldt and Eibl-Eibesfeldt 1968). Lycaenid butterflies tended by ants enjoy protection from at least some parasitoids (Atsatt 1981; Pierce and Mead 1981). Stary (1966, 1970) reviews the evidence that ant tending influences aphid parasitism by the braconid subfamily Aphidiinae. In many cases, the parasitoids are ignored by the ants and in some cases parasitism of tended colonies can be very high. Ants often tap parasitoids in the act of oviposition with their antennae without disturbing them. There are a few cases where ants appear to be effective in defending aphids. Ants in the genus *Formica* drive away *Xenostigmus bifasciatus* (Braconidae) from aphids (*Cinara* spp.) and may even form a protective screen between the aphids and the parasitoid (Stary 1970).

7.5 Dispersal

It is perhaps not surprising that we know relatively little about parasitoid dispersal because measuring the movement of small insects presents formidable technical problems. Some indirect evidence suggests that parasitoids regularly move quite large distances. For example, Askew (1968b) and Copland and Askew (1977) have noted that even quite small parasitoids may be collected at some distance from the nearest habitats where their hosts are found.

Dispersal can be studied either by marking individual parasitoids, or by releasing a species into a habitat where it does not occur and plotting the distribution of recaptures. The first technique has been used to study movement between artificial patches in the eucoilid parasitoid *Leptopilina heterotoma* which attacks *Drosophila* species (sec. 2.3.2; Papaj and Vet 1990), although here the chief intention was to study orientation to different odor cues rather than dispersal as such. We have marked two pteromalid (*Habrocytus* spp.) and two torymid (*Torymus* spp.) parasitoids of a tephritid fly (*Terellia ruficauda*) that lives in thistle seed heads. The average daily movement of all four parasitoid species exceeded that of their host (T. H. Jones, M. P. Hassell, and H.C.J. Godfray, unpublished). Another promising technique for marking adult parasitoids is with trace elements administered to the parasitoids' hosts (Payne and Wood 1984; C. G. Jackson et al. 1988; Hopper and Woolson 1991). Hopper and Woolson experimented with marking the braconid *Microplitus croceipes* with cesium (Cs), dysprosium (Dy), rubidium (Rb), and strontium (Sr). The elements could be detected in the adult insect, and in the concentrations used they had little effect on viability and fecundity.

The release of parasitoids during biological control programs offers an unparalleled opportunity for studying dispersal over geographical scales. Unfortunately, most programs are undertaken in response to an agricultural emergency, leaving little time for monitoring spread beyond checking that establishment has occurred. DeBach (1974) states that a tachinid fly that parasitizes larval eucalyptus snout beetles (*Gonipterus scutellatus*) was recorded over 100 miles from the nearest release site after only one year. The small aphelinid *Aphytis lepidosaphes* which attacks purple scale (*Lepidosaphes beckii*) on citrus took "a few years" to spread 100 miles from its release site in California to Baja California in Mexico. DeBach and Argyriou (1967) estimated that *Aphytis melinus* spread at a rate of approximately 100 km per year after its release in Greece to combat the scale insect *Chrysomphalus dictyospermi*. Onillon (1990) reviews the rates of spread of different whitefly parasitoids (Aphelinidae); for example, *Cales noacki* colonizes an area of 80 km^2 in France in eighteen months, while *Encarsia lahorensis* moved only 1,600 m in ten years in Florida. There are other similar examples in the biological control literature but, as far as I am aware, no attempts to survey rates of spread of different species or to relate range extension to the movement of individual insects (see Kareiva 1987b, 1990 and Kareiva and Odell 1987 for a discussion of the relevant theory and applications to insect predators).

7.5.1 APTERY

The primary means of dispersal in insects is flight and thus loss of wings may severely reduce an insect's capacity to move long distances. Loss of wings in either one or both sexes occurs sporadically throughout the parasitoid Hymenoptera and is also found in a few other groups. In some parasitoids, both winged and wingless forms of the same sex occur.

LOSS OF WINGS IN BOTH SEXES

The loss of wings in both sexes often occurs in parasitoids that attack hosts in grass and low vegetation. Examples include the braconid *Chasmodon apterus* which attacks stem-boring Diptera (Moore 1983), and many species in the ichneumonid genus *Gelis*. As one might expect, aptery seems to be less common in related species that attack hosts on trees. Wings and flight muscles require energy and resources that might otherwise be used for reproduction. However, flight is important in host and mate location and also in dispersal to find new habitats. Species that live in dense vegetation may not use flight when searching for mates and hosts; indeed, wings may be a positive disadvantage when moving through this habitat. If the environment is relatively stable so that long-distance dispersal is not of great importance, natural selection may favor redirection of resources away from wings and flight muscles.

Wings may also have been abandoned by some other parasitoids that search for hosts in environments where flight is not useful. Gärdenfors (1990) has recently described a new species of wingless aphidiine braconid (*Trioxys apterus*) from a collection of insects made at high altitude (4100 m) in Ecuador. He notes that the same collection contained an unusually high proportion of wingless parasitoids (Ichneumonoidea, Proctotrupoidea) and speculated that flight might be positively dangerous is a very windy, montane habitat.

WING LOSS IN ONE SEX

Wings are lost when the advantages of redirecting resources outweigh the disadvantages of not being able to fly. This delicate balance may tip in different directions for the two sexes. For example, eggs are more expensive to produce than sperm which, other things being equal, may favor female aptery. On the other hand, a male may not require flight to locate a female, for example when most matings take place at the emergence site. Furthermore, if early emergence is important in competition for mates (protandry: sec. 7.2.2), males may be selected to develop as fast as possible, even at the expense of not forming wings.

Female aptery is found in many species of Bethylidae, Dryinidae, and in the Tiphiidae. The costs of aptery in the inability to disperse are mitigated in some species by phoretic copulation: the female is transported *in copula* by the winged male (Clausen 1940a).

The most common correlate of male aptery is mating at the emergence site. If nearly all females are mated immediately after emergence, then there is little point in a male being able to fly. Examples of flightless males include the very well studied pteromalid *Nasonia vitripennis* and eulophids in the genus *Melittobia*. As has already been discussed (sec. 7.2.1), the most extreme adaptations to mating at the emergence site occur in fig wasps, many of which have wingless males that never leave the fig. Some species have normal wings or reduced wings but seem to be unable to fly. Perhaps the main cost of flight is the production of wing muscles and, these having been lost, it is possible that the wings remain as vestigial organs because the necessary genetic variation to allow their elimination has not occurred.

If wings are an encumbrance when searching for hosts, yet long-distance dispersal to locate microhabitats containing hosts is still important, then the best policy may be to retain wings but bite them off after a suitable habitat is located. Some diapriids wasps that mimic army ants adopt this strategy (see below, sec. 7.6), as does the curious pteromalid *Bairamlia fuscipes (=nidicola)*, which has been reared from the pupae of fleas collected in birds' nests (Graham 1969). Some scelionids that ride phoretically on adult female insects and parasitize their eggs lose their wings after locating the adult host (see sec. 2.2.5).

WING POLYMORPHISMS

Wing polymorphisms occur in both sexes, in males normally associated with mating strategies, and in females with different dispersal tactics. Like other polymorphisms, wing polymorphisms may be mixed or conditional. In a mixed polymorphism, each morph has on average the same fitness, at least when the numbers of the two morphs are at an evolutionary equilibrium determined by frequency-dependent selection. In a conditional polymorphism, one morph is adopted by disadvantaged individuals; for example, small individuals may lose their wings. Here the two morphs do not have equal fitnesses but disadvantaged individuals are doing as well as they can given their handicap. In a mixed polymorphism, all individuals may share the ability to develop as any morph, or the polymorphism may be genetically determined. Little is known of the developmental basis of wing polymorphism although Kearns (1934b) provides compelling evidence that winglessness among male *Cephalonomia gallicola* (Bethylidae) is controlled by two alleles segregating at a single locus. Apterous males are quite common in *C. gallicola* (van Emden 1931; Kearns 1934a). Simple wingless mutations are known in other parasitoids (e.g., *Bracon hebetor (=juglandis)*, Petters et al. 1978) although in these cases aptery is not found in the wild.

Some of the best examples of wing polymorphism among male parasitoids are found among fig wasps. Here, wingless individuals specialize on mating females within the fig while winged individuals search for unmated females in the environment (sec. 7.2.2). A similar polymorphism may occur in some scelionid wasps (Hamilton 1979). The egg parasitoid *Telenomus polymorphus*, which attacks the large egg masses of the reduviid bug *Heza insignis*, has both winged and wingless males (da Costa Lima 1944). The wingless forms have large heads, probably an adaptation for fighting other males for the possession of the host egg mass.

Wing polymorphisms are common in the family Bethylidae, especially in the genera *Cephalonomia* and *Scleroderma*. Many species of *Sceleroderma* are dimorphic in both sexes (Bridwell 1920; Evans 1963) although at least in some cases wingless individuals are runts (Hamilton 1979, see below). The situation in *Cephalonomia* is more complex. In *C. gallicola*, males are both winged and wingless (Kearns 1934b), while in *C. formiciformis* only the females are polymorphic (Richards 1939; Hamilton 1979). *C. perpusilla* has both winged and wingless males and, in addition, Evans (1963) distinguishes four female forms: macropterous, micropterous, subapterous, and apterous. *Cephalonomia* are very small wasps that attack beetles in bracket fungi and similar substrates. Evans suggests that wingless forms remain in the bracket fungus in which they developed, or crawl to nearby fungi on the same tree, while the winged forms disperse to locate other trees. It is not clear whether the four female forms represent different dispersal strategies or whether morphology is influenced by host quality (Hamilton 1979).

A curious example of male wing dimorphism is found in the trichogrammatid *Trichogramma semblidis*. Males reared from moth eggs are winged while those reared from the eggs of the alder fly *Sialis lutaria* (Neuroptera) are wingless (Salt 1937b, 1939). The two morphs differ in other more minor characters and are quite distinct. Salt reared only nineteen wasps from the "wrong" host among 1847 males from moth eggs and 1740 males from alder fly eggs. Rearing parasitoids for many generations on moth eggs did not destroy the ability of the strain to produce wingless morphs on alder fly eggs. Wasps reared from the two hosts are of similar size, and there is no evidence that wingless males in alder flies are of low fitness. It is possible that the explanation of the polymorphism is related to the mating system of wasps attacking different host species. Alder flies lay their eggs in large clusters of 500–700 eggs (Salt 1939); the natural alternative hosts of *T. semblidis* are unknown but are likely to be the solitary eggs of Lepidoptera, or eggs laid in relatively small clutches. Males emerging from *Sialis lutaria* egg masses may not need wings if they have ample mating opportunities at the emergence site and if all other females in the environment are mated soon after leaving the pupa. A large cluster of alder fly eggs is likely to contain many potential females. In these circumstances, the wasp may be selected to redirect investment from wing formation to sperm maturation or some other more useful function. A male emerging from a solitary host egg will require wings to locate females; even if it develops with females in a gregarious clutch of host eggs, wings may be advantageous if there are unmated females in the local environment. If this suggestion is correct, the advantage of having wings depends on the type of host in which the male develops. The conditional strategy observed by Salt appears to be a solution to the conflicting selection pressures.

Winged and wingless males also occur in an unnamed species of *Trichogramma* which parasitizes eggs of the lymantriid moth *Ivela auripes* (Kurosu 1985). On average nine wasps develop in each host egg, one or two of which are normally males. Mating takes place both inside and outside the host egg; in the latter case winged males appear to have a distinct advantage although wingless males attempt to achieve "sneaky" copulations. Kurosu (1985) suggests that winged and wingless males are both able to inseminate all the females developing on the same egg which raises the question of why males are not monomorphic. Kurosu argues that if a male develops alone with a clutch of females, he may be selected to use fewer resources, and forgo wings, so that his sisters have more to feed on—a form of sibling altruism. However, if a second female superparasitizes the host, laying a single male egg, the first male is selected to pre-empt resources, grow larger, and become a winged male. Kurosu supports his argument with data on the relative sizes of the two sexes in clutches containing one or two males. I am not convinced that the data provides more than circumstantial support for this idea and also think that the role of superparasitism in the production of winged males needs better demonstration. Nevertheless, this is a fascinating hypothesis that deserves further study.

Females of some species of the eulophid genus *Melittobia* are dimorphic or even trimorphic (Schmeider 1933; Freeman and Ittyeipe 1976, 1982). In a study of an unnamed species in the *hawaiiensis* complex, Freeman and Ittyeipe (1982) distinguished three types of female: (1) "crawlers" with no wings and no phototactic response; (2) "jumpers" with wings and a phototactic response which readily jumped but never flew; and (3) "fliers" which both possessed and used wings and were positively phototactic. A typical female of any morph might lay about 450 eggs of which 400 would be fliers, and 25 each of jumpers and crawlers. This species attacks prepupae of the sphecid wasp *Sceliphron assimile* in nests containing up to about forty cells, and Freeman and Ittyeipe (1982) suggest that the three morphs may be adapted to different dispersal ranges. The crawlers remain in the cell laying a further clutch of eggs on the same partially eaten host; the jumpers attack other cells in the same nest, while the fliers search for new nests. The frequency of different morphs appears to be environmentally determined and strongly affected by the number of larvae developing in the host—fewer crawlers are produced when resources are limiting. This makes good sense, as when competition is severe few resources will be left for a second generation on the same host.

A related form of wing polymorphism is found in the ichneumonid *Sphecophaga vesparum*, a parasitoid of social wasps, *Vespula* spp. The first mention of the natural history of this species was by Kirby (1835) in one of the *Bridgewater Treatises*, but a number of aspects of its biology are still obscure. The recent introduction of *S. vesparum* into New Zealand to control social wasps has led to renewed interest in this ichneumonid, and my account of its biology is based on Donovan (1991), who provides a full bibliography of earlier studies. Winged females enter *Vespula* nests in spring and early summer and lay eggs through recently capped cells onto the host larvae or pupae. The parasitoids feed externally and pupate in cocoons within the host cell. Three different types of cocoon can be distinguished: (1) white cocoons that give rise to brachypterous females; (2) thin yellow cocoons that give rise to winged females (and possibly winged males) in the same summer; and (3) thick yellow cocoons that give rise to winged ichneumonids of both sexes the following or subsequent summers. Brachypterous females remain in the nest attacking unparasitized larvae and pupae; the winged females from thin yellow cocoons quit the nest (although it is not clear if they lay any eggs before leaving) and presumably search for other wasp nests. Finally, wasps from thick yellow co-coons emerge from the abandoned remains of the nest either the next summer, or after two or three years. There may be many generations of brachypterous females within the same nest over one summer. Each type of female appears to be capable of laying eggs that develop into any other type of female. It seems that brachypterous females are adapted to capitalizing on the normally abundant resources in the same nest, while winged females are adapted to finding new nests and for overwintering.

The most obscure aspect of the biology of *Sphecophaga vesparum* is the role of sex. Donovan (1991) and his colleagues have maintained this species in culture for many generations without allowing females to mate. In the field, parthenogenetic reproduction appears to be the rule for brachypterous females and is possible for winged females. Yet males are commonly produced from thick yellow cocoons and, at least in the laboratory, males and females appear to copulate successfully. Superficially, the reproductive strategy of this species resembles the type of cyclical parthenogenesis found in aphids, cladocerans, and rotifers: there are a number of parthenogenetic generations during the summer when resources are abundant and conditions benign, followed by a sexual generation coinciding with dispersal and overwintering. If true, this would be the first example of cyclical parthenogenesis from a parasitoid, al-though many gall-forming Cynipidae, a group derived from parasitoids, prac-tice this form of reproduction. However, the importance of sex to overwinter-ing *S. vesparum* is not clear, nor indeed whether the female uses the males' sperm to fertilize her eggs. Another possibility is that some cytoplasmic factor similar to those that cause thelytoky in chalcidoids (sec. 5.2) is influencing reproduction in this species. There is an urgent need for genetic studies of *S. vesparum* to resolve these questions: whatever the answer, the reproductive biology of *S. vesparum* is unique among known parasitoids.

Gelis corrupter is a hyperparasitoid of a number of hymenopterous para-sitoids and can be reared in the laboratory on ant larvae. The female is apterous but the male exists in two very distinct forms: one winged, the other wing-less. Aptery in males is not genetic but depends on the size of the host; small hosts give rise to wingless wasps and large hosts to winged individuals. It appears that males switch to a more female-like development path in small hosts (Salt 1952).

As host size decreases, the size of the parasitoid emerging from the host normally decreases until a threshold is reached when the host is too small to support parasitoid development. Near the threshold, malformed individuals, called "runts" by Salt (1941), are often observed. Runts are commonly wing-less, which may be a simple consequence of the developmental trauma associ-ated with lack of food, or it may be a last-ditch attempt by the parasitoid to salvage some adult fitness by redirecting resources away from investment in dispersal to more important functions. Runts are particularly common in trichogrammatid egg parasitoids but are recorded from many other species of parasitoid, both Hymenoptera and Diptera (Salt 1941).

7.6 Defense from Predators

Like other insects, adult parasitoids are attacked by a wide spectrum of predators and have evolved a variety of antipredator countermeasures. The head and thorax of many hymenopterous parasitoids is covered by a thick, sculptured cuticle which probably provides protection against predators. The abdomen of most species is less well protected, although in many groups there is a tendency for at least the dorsal surface of the abdomen to become sclerotized, sometimes forming a dorsal protective carapace. The Chrysididae are beautifully colored and very well armored wasps that can roll up into a ball when attacked by predators. Many species of the family, which contains numerous nonparasitoids, attack the larvae of solitary wasps, and the thick cuticle may be required as a protection against female wasps guarding their young.

Ichneumonids and braconids use their sting to defend themselves against predators (Askew 1971; Quicke 1984; Quicke et al. 1992). The sting can be felt by humans but is not normally very painful. The sting of the Mutillidae is decidedly painful: Clausen (1940a) quotes an account by Bouwman[1] of a belief in Cyprus that the sting was fatal and caused a large number of deaths among the British garrison during the occupation of 1878. Many ichneumonids and braconids produce a repugnant odor when disturbed (Townes 1939; Quicke 1988). Other ichneumonids have large claws (Townes 1939) or long bristles that project beyond their tarsal claws, and which are thought to break off and become embedded in the throats of predators; yet other species have sharp spinelike protuberances (Gauld 1987).

Many parasitoids, particularly species from the tropics, are brightly colored or possess distinctive patterns. A number of authors, beginning with Marshall (1902) and Shelford (1902), have suggested that parasitoids form parts of Mullerian or Batesian mimicry rings. More recently, Quicke (1986) has surveyed the color patterns found in one group of parasitoids, the Afrotropical braconids in the subfamily Braconinae. He argues that a number of distinct color patterns can be distinguished and describes the twelve most common examples. Wasps with very similar markings and coloration are found in taxonomically unrelated genera suggesting convergent evolution. Moreover, the same patterns are found in other braconid subfamilies, in various species of Reduviidae (predatory bugs, Hemiptera), Syrphidae (hoverflies, Diptera), Mantispidae (Neurop-

---

[1] I have failed to trace the origin of this anecdote; it is not in Bouwman (1909), the reference given by Clausen (1940a).

tera), Amatidae (moths, Lepidoptera), and, most commonly, in wood-boring beetles of the cerambycid subfamily Lamiinae. The beetles are apparently edible but little is known about the palatability of the braconids. Of the twelve color patterns distinguished by Quicke (1986), redder variants tend to be more common in dryer areas (where red, lateritic soils are common) and blacker variants in rainforest. Mimicry rings seem to occur in other ichneumonoid faunas but have been less thoroughly investigated (Mason 1964; Harris 1978; Quicke 1986; Quicke et al. 1992). Some species of leucospid wasp that mimic bees and social wasps show distinct geographical races. Boucek (1974) suggests that different populations are Batesian mimics of different aculeate models, a form of geographical polymorphism common in butterflies.

There are some clear cases of mimicry in temperate parasitoids. Conopid flies which attack and parasitize aculeate bees and wasps are normally striped black and yellow and resemble bees and wasps. There is seldom a close match between a parasitoid and its specific host (Askew 1971) and the resemblance is almost certainly Batesian mimicry. The bee flies (Bombyliidae) obtain their name from the dense hair that clothes adults of *Bombylius* and related genera and which to humans gives them a beelike appearance. However, it is doubtful whether these species are true bee mimics. Tropical bee flies in the genus *Systropus* have a curious elongate appearance and are probably solitary wasp mimics. Eggleton (1991) suggests that tropical and temperate rhyssine ichneumonids mimic vespids, sphecids, and evil-smelling braconids.

A particularly interesting type of mimicry occurs when hosts and parasitoids evolve similar coloration. Quicke et al. (1992) have recently reviewed examples of pattern convergence between both hosts and parasitoids, and predator and prey. A few leucospid wasps resemble their hosts (aculeate Hymenoptera) although most are Batesian mimics of unrelated bees and wasps (Boucek 1974). Some cerambycids resemble their ichneumonid parasitoids, but here the wasp is probably the model rather than the mimic (Marshall 1902; Harris 1978). This is difficult to explain as the parasitoid is normally rarer than the host and thus potential predators would not learn to associate the color pattern with distastefulness (Quicke et al. 1992). Possibly the parasitoid may be part of a wider mimicry ring. A very few ground beetles (Carabidae) are parasitoids of chrysomelid beetles. Some species of *Lebia* closely resemble their hosts, species of flea beetles (Halticinae) (Lindroth 1971). The hosts are not distasteful but are named for their ability to jump when confronted by a predator. Lindroth suggests that predators may avoid flea beetles as uncatchable and that their parasitoids have evolved to be Batesian mimics. Some tropical chrysomelids are highly toxic, so much so that they are used by Bushmen as an arrow poison. These beetles are also parasitized by carabids (*Lebistina*) which have become Batesian mimics of their host (Lindroth 1971). Another example of convergence between host and parasitoid is described by Quicke et al. (1992) and involves cerambycid beetles in the genus *Phoracantha*—important pests of eucalyptus in Australia. The beetles have light and dark bands on their thorax and elytra, as do members of no less than four genera of their braconid parasitoids. Each genus appears to have evolved the characteristic coloration independently; in two genera the light bands are achieved through pigmentation, while in two other unrelated genera the light bands are produced by a dense mat of pubescence, the hairs highly flattened and reflective. It is unclear whether the beetles and wasps are part of a Mullerian or a Batesian mimicry ring and, if the latter is the case, which species are mimics and which are models. Aggressive mimicry occurs where the parasitoid or predator resembles the host or prey to camouflage its approach. There are a number of examples of aggressive mimicry by predators, but Quicke et al. (1992) found no cases among parasitoids. Although not an example of visual mimicry, some conopid flies appear to mimic the flight of their hosts, which they attack in mid-air (sec. 2.2.3; Raw 1968). Diapriids that mimic ants (see below) may also be aggressive mimics.

Many species of parasitoids, especially ichneumonids, have characteristic white markings on their antennae. Either the tip may be white, or there may be a subdistal white band. The ovipositor sheaths of a number of tropical ichneumonoids are also distally white. The function of these markings is unknown, but one hypothesis is that they deflect the attack of a predator from a more vital part of the body (Quicke 1984). Tropical parasitoids with long ovipositors may be most at risk from predation while in the act of ovipositing, at which time the expendable ovipositor sheaths (but not the important ovipositor) stick out behind the insect. Alternatively, the white markings on the antenna might be involved in estimating host size, while the sheath markings may allow the wasp to see the position of its ovipositor (Quicke 1984).

The wingless females of many Dryinidae have a striking superficial resemblance to ants (Donisthorpe 1927; Clausen 1940a; Olmi 1993) although their gait is not antlike (Richards 1939). It is unclear whether this is an example of mimicry or morphological convergence under similar selection pressures (Askew 1971). Clearer ant mimicry is found among diapriid wasps that are associated with ants, although the life history of these insects is not clearly resolved. Wing (1951) describes the morphology and behavior of *Solenopsia imitatrix*, a species normally found associated with the ant *Solenopsis fugax*. The wasp is "visually not too dissimilar" to the ant and is either ignored by the ants or actively fed. Although initially winged, wasps associated with ants loose their wings. The host of *S. imitatrix* is not known and may be the ant or an ant symbiont. In the New World a number of diapriid genera have evolved remarkable morphological and behavioral mimicry of army ants (*Eciton*) (Masner 1959, 1976, 1977). The wasps run with the ants and are able to perform a wide repertoire of ant behaviors such as trail following, attraction, and appeasement, and the elicitation of trophallaxis (feeding). There is some evidence that they provide secretions used by the ant. Although initially winged, on finding an ant column they lose their wings. These wasps are thought to be parasitoids of ant larvae or pupae. Parasitoids constitute only a small fraction of the different ant guests and symbionts that have found a place in what Wing (1951) calls this "fascinating Alice-in-Wonderland society" (E. O. Wilson 1971).

Some parasitoids search for hosts at night, which may be an adaptation to reduce predation (Gauld and Huddleston 1976; Gauld and Mitchell 1977; Huddleston and Gauld 1988; Quicke 1992). The ichneumonid subfamily Ophionae contains many nocturnal species and are often caught at ultraviolet moth traps. Nocturnal wasps in this subfamily are normally reddish brown, with large eyes, ocelli, and antennae. Similar sets of characters (the ophionoid facies) have evolved independently in other nocturnal ichneumonoid groups.

7.7 Host Synchronization

Parasitoids should obviously time their emergence as adults to coincide with the presence of susceptible host stages. This is particularly important when the host has one or a few generations a year. To achieve synchronization, parasitoids often diapause or enter resting stages when hosts are unavailable. In the first part of this section I review very briefly some physiological and behavioral aspects of parasitoid synchronization. In the second section I discuss some of the evolutionary and ecological issues of host-parasitoid synchronization in species with discrete generations. Host synchronization may also be important in species with overlapping generations if the populations have cyclic dynamics, and I finish by speculating about the selection pressures involved in such cases.

7.7.1 DIAPAUSE AND QUIESCENCE

This account of parasitoid diapause is based largely on two recent reviews by Tauber et al. (1983, 1986; see also Doutt 1959, Fisher 1971, and Saunders 1982). Diapause can occur during any stage of the parasitoid's life cycle. Environmental stimuli that promote diapause may be perceived directly by the parasitoid or indirectly, mediated through a change in host physiology. If diapause occurs during the pupal stage, the structure of diapausing and nondiapausing pupae and cocoons often varies. Diapausing cocoons tend to be tougher as they have to survive for a much longer period of time (Stary 1970; Parrish and Davis 1978; Shaw and Huddleston 1991; Donovan 1991).

The larvae of many endoparasitoids that delay killing their hosts (koinobiont) remain in the first instar until the host is full grown. There is some debate about whether this slow-down in development should be classified as diapause. Diapause is normally associated with low metabolic activity and reduced locomotion; because first instar koinobiont larvae often move actively around the host searching and fighting with competitors, I believe it is best not to use diapause to describe this developmental pause. The advantages of delaying development are that the parasitoid has more resources when it finally moults into the second instar, and that the ovipositing parasitoid can attack hosts that otherwise would be too small to support the development of a larva. In addition, the first parasitoid in a host may be at a competitive advantage should super- or multiparasitism occur. Yet another possible factor selecting for a delay in development is the necessity of emerging in synchrony with the host.

If a host containing a first instar koinobiont larva enters diapause, the parasitoid normally remains as a first instar until growth resumes. I know of no study which has investigated whether the parasitoid larva enters a true diapause with reduced metabolic and behavioral activity. There are many examples of parasitoids that attack several species of host and have development times determined by the host (Tauber et al. 1983, 1986). For example, the European Cornborer (*Ostrinia (=Pyrausta) nubilalis*) has both single-brooded and multibrooded strains, and the development time of its braconid parasitoid *Chelonus annulipes* depends on whether the host enters diapause (Bradley and Arbuthnot 1938). Polgár et al. (1991) have recently shown that the larvae of the braconid parasitoid of aphids, *Aphidius matricariae*, normally enter diapause when placed in oviparae but not in other aphid morphs. Oviparae are produced late in the season. In these cases the timing of parasitoid development is determined by host physiology; possibly the parasitoid responds to changes in the level of host hormones associated with diapause (Beckage 1985; Lawrence 1986, 1990). The dependence of parasitoid phenology on host species may have important implications as an isolating mechanism in sympatric speciation (sec. 8.3.1). Not all parasitoids respond passively to the host: some species accelerate or decelerate host development to ensure they emerge at the correct time (Fisher 1971; see sec. 6.4.1). Diapause in idiobiont parasitoids may also be influenced by the diapause status of their host. For example, the hyperparasitoid *Catolaccus aeneoviridis* (Pteromalidae) is more likely to diapause if it attacks a diapausing primary parasitoid (*Cotesia (=Apanteles) congregatus,* Braconidae) (McNeil and Rabb 1973). Similarly, the egg parasitoid *Trichogramma cacaeciae* does not enter diapause if it attacks nondiapausing eggs of its normal host *Archips (=Cacoecia) rosana* (Tortricidae), and other moth eggs in the laboratory, but will diapause if it attacks overwintering *A. rosanus* eggs (Marchal 1936; Strand 1986).

Parasitoids that enter diapause after they have killed the host, typically as pupae or adults, normally respond directly to environmental stimuli. The factors that initiate and (although less well studied) terminate diapause are as varied in parasitoids as in other insects and include, in approximate order of importance, photoperiod, temperature, thermoperiod, and humidity (Tauber et al. 1983, 1986). Frequently, diapause is influenced by combinations of factors, for example photoperiod and temperature. The ichneumonid *Pimpla instigator* is extremely sensitive to red light, which is rare among insects; only red light penetrates the thick walls of the host cuticle (Claret 1982). There is also strong evidence implicating maternal effects in diapause control. To induce diapause in the pteromalid *Nasonia vitripennis*, the mother must be kept at the correct photoperiod (Schneiderman and Horwitz 1958; Saunders 1965, 1966). The mother also influences diapause in the trichogrammatid egg parasitoid *Trichogramma evanescens* (Zaslavsky and Umarova 1981).

Different geographical strains of the same species often show different diapause strategies. This can lead to problems if an inappropriate strain of parasitoid is released during a biological control campaign and climatic matching of target and collection sites is often undertaken. Problems can also arise if a species from the Southern Hemisphere is required to be released in the Northern Hemisphere: diapause has to be broken and the species' phenological response reset.

Temporal fluctuations in the abiotic environment can select for variable diapause strategies (see, for example, Leon 1985 for a discussion in the context of seed dormancy, and Hanski 1988 for insect diapause). This may explain why collections of parasitized hosts made during the summer often contain both individuals that hatch the same year and that overwinter, and why some parasitoids may diapause for two winters (e.g., Turnock 1973; Sunose 1978). Diapause strategies may be influenced not only by temporal fluctuations in the abiotic environment, but also by fluctuations due to deterministic population dynamics, and by coevolution with the diapause strategy of the host. As yet, this subject has received very little investigation.

7.7.2 Host Synchronization with Discrete Generations

Consider a host with just a single generation a year attacked by a specialist parasitoid. What is the optimal emergence pattern for the parasitoid? If the parasitoids suffer virtually no mortality as adults (at least until after all hosts have disappeared), then the obvious strategy is to emerge early and wait for the host to hatch. However, mortality can seldom be ignored and early emergence carries the risk of dying before the main peak of hosts. The picture is further complicated by competition among parasitoids: it may be advantageous to emerge when hosts are relatively scarce if there are also few competitors.

This problem can be studied using models of protandry (sec. 7.2.2). Instead of males emerging and competing for mates, we now have females emerging and competing for hosts. The optimal pattern of emergence is a temporal ideal free distribution where no individual can increase its fitness by emerging either earlier or later. Any model should also take into account other factors that affect fitness and vary with time: for example, individuals that emerge early might suffer greater mortality due to poor weather, while late-emerging individuals may encounter many previously parasitized hosts. There may also be costs to emerging early if this requires faster development, or emerging later if pupal mortality is significant.

Yet other considerations may complicate the evolution of emergence times. Females emerging at certain times may have difficulty in finding mates. This is likely to be of relatively minor importance as male emergence patterns will evolve in response to the temporal distribution of females. A more important factor is the evolution of the host emergence patterns. If parasitoids are a major source of host mortality, then their emergence times may evolve to minimize parasitoid attack. I am not aware of any formal treatment of this complicated reciprocal interaction. It is reasonable to suppose that while the main selection pressures influencing female parasitoid emergence is the availability of hosts, the emergence pattern of the host will be less strongly influenced by parasitoid activity (especially when parasitism is not a major source of mortality) because of the importance of other factors such as host plant phenology and intraspecific competition (Iwasa 1991). Parasitoids may thus respond faster than hosts in any coevolutionary interaction involving emergence patterns. The pattern of emergence of hosts and parasitoids is of interest to population dynamicists because the presence of temporal refuges for the host from parasitoid attack can contribute to the dynamic stability of host-parasitoid interactions (Godfray et al. 1993).

7.7.3 HOST SYNCHRONIZATION WITH OVERLAPPING GENERATIONS

If host and parasitoid generations completely overlap, and if populations are constant or fluctuate randomly, there will be no selection pressures on parasitoid emergence times. However, if hosts fluctuate predictably, increasing in abundance at certain times of year or, for example, after prolonged rain, then parasitoids may be selected to delay emergence to coincide with their hosts. The costs and benefits of different emergence times could be studied using methods similar to those discussed in the last section.

A different type of problem involving synchronization can arise in species with overlapping generations. Models of host parasitoid interactions in continuous time often predict that both the host and the parasitoid populations cycle with a period of approximately one host generation (Godfray and Hassell 1987, 1989; Godfray and Chan 1990; Gordon et al. 1991). The two main conditions for generation cycles are that the parasitoid development period should be roughly 0.5 or 1.5 times that of the host, and that the juvenile period of the host and parasitoid should be relatively long compared with their adult lifespan. There are a number of examples of generation cycles from tropical insects, especially plantation pests, that have been interpreted as due to parasitoids (Godfray and Hassell 1989). Cycles are maintained because most parasitoid eggs are laid at times of host abundance. If the generation time of the parasitoid differs from that of the host, then eggs laid at times of host abundance will become searching adults, not at the next peak of host abundance (one host generation later), but either before or after (fig. 7.3). Individual hosts out of

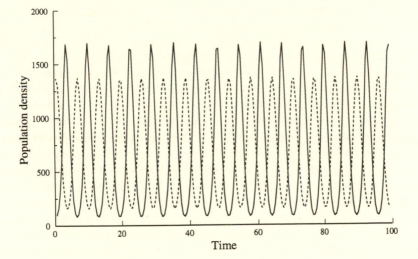

Figure 7.3 Generation cycles predicted by an age-structured host-parasitoid model. The numbers of susceptible hosts (solid lines) and searching parasitoids (dashed lines) are plotted against time. Both hosts and parasitoid densities cycle with a period of approximately one host generation (in this example six time units; a parasitoid generation takes three time units), giving the appearance of discrete generations. The peak of searching parasitoids coincides with the time when susceptible hosts are least abundant. (The particular model used to generate this figure is in the appendix of Godfray and Chan 1990.)

phase with the majority of the population are thus penalized and the host generation cycles maintained. A relative short adult lifespan is important in maintaining cycles because if adults lay eggs over a long period of time, host cohorts become smeared and parasitoids will always find hosts to parasitize.

Selection pressures act in very different ways on hosts and parasitoids in populations with generation cycles. Selection on hosts tends to support the status quo, as individuals emerging earlier or later are penalized. Host development periods should thus remain constant. Selection on the parasitoid favors either longer or short development periods because both result in the emergence of the parasitoid when hosts are more abundant. It is difficult to predict the result of these conflicting selection pressures: the speed with which each population can respond in evolutionary time, the type of genetic variance in development period, and the constraints and other selection pressures experienced by host and parasitoid are all important. Two possible outcomes are (1) equal (or nearly equal) host and parasitoid generations times leading to the loss of generation cycles and the removal of selection on both host and parasitoid, and (2) persistent cycles where the parasitoid is selected to reduce its generation rate but is unable to do so because of a physiological constraint. In general we know rather little about the evolution of life history traits that affect demographic parameters in populations with cyclic or chaotic dynamics.

7.8 Conclusions

Many of the topics covered in this chapter have received considerably less attention from behavioral ecologists in comparison with host searching, oviposition, and sex ratio. Part of the explanation is that the most interesting questions concern the behavior of adult parasitoids in the field where observing and experimenting with small mobile insects is very difficult. Nevertheless, more workers are turning to field observations and experiments, a trend that will hopefully continue.

In my view, the single factor that poses the greatest problem for the development of a quantitative parasitoid behavioral ecology is the lack of a full understanding of the consequences to an adult parasitoid of being large or small. Nearly every aspect of host selection, clutch size, superparasitism, and sex ratio has implications for adult size. Although the effect of size on some components of adult fitness are relatively easy to measure, for example maximum fecundity, the most important consequences are probably felt through reduced longevity and can be measured only in the field. Obtaining these measurements, and using them in quantitative tests of theories about parasitoid reproductive strategies, is a major challenge to experimentalists.

Our knowledge of parasitoid mating systems is biased toward species where mating takes place at the pupation site. These species usually have female-biased sex ratios and, not infrequently, other interesting behaviors such as male aggression. The suite of adaptations associated with this mating system, Hamilton's (1967) "biofacies of extreme inbreeding," are now rather well understood and provide some of the most satisfying examples of intraspecific sexual selection among males. Work on fig wasps in particular has moved beyond the descriptive stage to quantitative comparisons of adaptations by different species. However, it would be wrong to conclude that we have a good understanding of parasitoid mating systems. The large majority of species do not mate at the pupation site but elsewhere in the environment. Our knowledge of these other types of mating system is poor and lags behind studies of other groups of insects. Again, the small size of most parasitoids leads to difficulties in studying sexual behavior in the field.

Although host synchronization and diapause are subjects that have received a lot of attention from applied entomologists, they have been somewhat ignored by evolutionary biologists. There are, however, many interesting problems, both theoretical and empirical: How widespread is prolonged diapause (i.e., diapause through more than one adverse season) in parasitoids? Do host and parasitoid diapause strategies coevolve, and are they influenced by demographic variability? And what is the optimal emergence pattern for males and females that maximizes their chances of finding mates and hosts, respectively?

8 Life Histories and Community Patterns

Ecologists attempt to understand the population dynamics and community structure of interacting populations and species. Their raw material, the different animals and plants they study, are the products of biological evolution. Insights into population dynamics can be gained from studying the evolution of parasitoid life histories which determine demographic parameters. Similarly, an understanding of the interplay of evolutionary and ecological processes helps make sense of patterns in community structure.

In this chapter I discuss the evolution of parasitoid life histories, and how the interaction of ecological and evolutionary processes influences host breadth, parasitoid species load, and parasitoid diversity. The first section of the chapter concentrates on life history evolution including explanations of interspecific differences in parasitoid fecundity, and the suggestion that the life histories of parasitoids (especially those attacking host larvae) fall into two main categories. In the second section I discuss why some hosts are attacked by many more species of parasitoids than others, what determines parasitoid host breadth, and how parasitism may mold the shape of the host niche. The final section is about parasitoid diversity: why there are so many species of parasitoids, the question of whether parasitoid diversity is unusually low in the tropics, and explanations for radical evolutionary innovations by parasitoids.

8.1 Life History Evolution

The informal study of parasitoid life histories dates back to the first naturalists who unraveled the complexities of parasitoid life cycles. As more data became available, the broad distinctions among parasitoid life histories described in the introductory chapter became established. A few workers methodically surveyed the morphological and biological variation among individual groups of parasitoids, adopting an almost taxonomic approach to the description and nomenclature of different life histories. The dipteran family Tachinidae was particularly well served by this approach, and several very full life history classifications were developed (Townsend 1908, 1934–1939; Pantel 1910–1912). However, the modern study of parasitoid life histories begins in the 1960s with a series of studies by R. R. Askew of the comparative biologies of the parasitoids of gall-forming and leaf-mining insects, and the work of

P. W. Price who in the early 1970s sought to explain the differences in fecundity among a number of related parasitoids attacking the same species of host. In this section I first discuss the evolution of parasitoid fecundity and then the more general hypothesis that there is a broad dichotomy in parasitoid life histories, at least among larval parasitoids. The concept of r- and K- selection has often been applied to different parasitoid species; I argue that this is a useful concept but that it should be used in a manner more in keeping with its origins in population and ecological genetics. Finally, I describe the application of modern comparative biology to the study of parasitoid life histories, a subject still in its infancy.

8.1.1 THE EVOLUTION OF PARASITOID FECUNDITY

P. W. Price (1972b) studied a guild of ten ichneumonid wasps attacking the Swaine jack pine sawfly (*Neodiprion swainei*). Females of different species of wasp varied in the number of ovarioles per ovary, a morphological feature that correlates with the number of eggs available for oviposition (Price 1975). The wasps also differed in the host stage they attacked although all species emerged from cocoons. Price (1972b, 1974) discovered that wasps attacking early host stages had many more ovarioles than those attacking later stages (fig. 8.1). He also found, using data from a large survey of the reproductive morphology of the Ichneumonidae (Iwata 1960), that the mean ovariole number in an ichneumonid subfamily was correlated with the stage of the host life cycle that members of the subfamily typically attacked (fig. 8.2) (Price 1973a). Species with many ovarioles also tended to have a greater capacity to store mature oocytes in the ovariole and oviduct. Finally, the same pattern was found yet again in a comparative study of fifty-nine species of ichneumonids with particularly well-known biologies (Price 1975).

Why should species attacking young host stages have greater fecundities? Price (1972b, 1973b, 1974, 1975) argued that the key to explaining this pattern was the schedule of immature mortality experienced by the host. Before proceeding it is useful to make a distinction between potential fecundity, as measured by ovariole counts and similar techniques, and realized fecundity, the number of eggs actually deposited by a parasitoid. Now, unparasitized hosts and hosts containing the eggs or first instar larvae of parasitoids are likely to experience similar rates of mortality. Thus parasitoids that lay their eggs in very young hosts, but whose larvae delay development until the host matures, suffer far greater levels of immature mortality than species that attack a pupa or cocoon and immediately kill the host. Many hosts suffer very severe mortality in their youngest instars, and parasitoids attacking the smallest host stages are particularly severely affected. Now, if an average is taken over many generations, realized fecundity must balance immature mortality: this is a truism

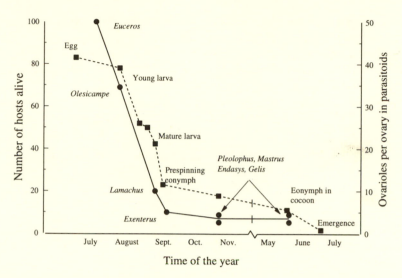

Figure 8.1 Cohort survival of juvenile Swaine jack pine sawfly (*Neodiprion swainei*) (dashed line) and the number of ovarioles per ovary in ten of its ichneumonid parasitoids that attack the host at different times during the year (solid line). The host suffers heavy mortality in its larval stage but less mortality as a prespinning eonymph, and as an eonymph or pupa in the cocoon. The earliest attacking parasitoid is *Euceros frigidus*, which lays its eggs in the vicinity of sawfly egg masses. *Olesicampe lophyri, Lamachus lophyri, Exenterus amictorius*, and *E. diprionis* attack successively later larval stages, while the five species *Pleolophus basizonus, P. indistinctus, Endasys subclavatus, Mastrus aciculatus*, and *Gelis urbanus* attack eonymphs and pupae within cocoons, both in the autumn and in spring. (From Price 1974.)

for any persistent species that in the long run neither increases nor decreases in abundance. Thus the amount of juvenile mortality experienced by a parasitoid species can be used to predict its average realized fecundity. With some caveats and exceptions that I discuss below, potential fecundity is likely to be highly correlated with average realized fecundity. Parasitoids that can expect to lay relatively few eggs throughout their life are unlikely to evolve large potential fecundities: instead of using resources to make more eggs, they will increase the size of individual eggs or use the resources to increase searching efficiency. In conclusion, the stage of host attacked influences juvenile parasitoid mortality and hence average realized fecundity, which is correlated to measures of potential fecundity such as ovariole number.

Price's argument is called the "balanced mortality hypothesis" and has a long pedigree in population biology (see Price 1974). It has sometimes been used to argue that species facing severe mortality evolve larger fecundity *in order to* maintain population stability. Such an argument requires group selection, and this is emphatically not the interpretation adopted by Price. One of

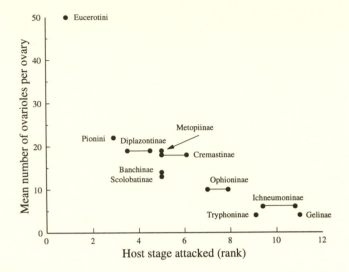

Figure 8.2 The relationship between the average number of ovarioles per ovary in different ichneumonid subfamilies and tribes, and the stage of host they attack. The ranking used is: 0–2 oviposition on foliage near host; 2–4 eggs; 4–6 young larvae; 6–8 mature larvae; 8–10 old larvae; 10–12 pupae and puparia. Lines represent taxa that attack a range of host stages. (From Price 1973a.)

the most elegant features of the balanced mortality hypothesis is that its predictions are independent of the population dynamics of the interaction, or of the main details of the species biology.

The balanced mortality hypothesis predicts that average realized fecundity, and thus measures of potential fecundity such as ovariole number, should be the reciprocal of juvenile mortality. Figure 8.3 shows this relationship for the ten parasitoids of *Neodiprion swainei* (Price 1974). There is an impressive correlation accounting for 93% of the variance in the data.

Parasitoid fecundity should also be correlated with other factors that influence juvenile mortality in addition to the host stage attacked. Thus the average realized fecundity of a parasitoid attacking a host that suffers severe predation (which simultaneously affects both host and parasitoid) must be higher than that of a parasitoid attacking a host with little predation. Other things being equal, the higher realized fecundity of the first parasitoid will select for greater potential fecundity. Price (1973a, 1975) suggests that parasitoids of hosts in relatively inaccessible locations such as burrows, mines, leaf rolls, and webs should be protected from extrinsic mortality and in consequence have a relatively low potential fecundity. In a survey of ninety genera of Ichneumonidae, ovariole number was inversely correlated with ovipositor length (Price 1973a). The hosts of parasitoids with long ovipositors are likely to be well concealed in substrate or plant tissue and protected from much extrinsic mortality. There

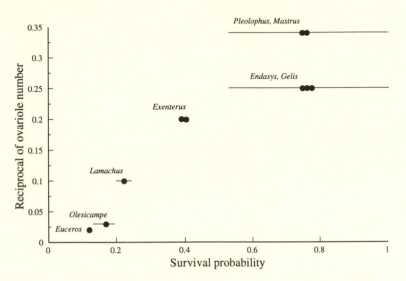

Figure 8.3 A graph of the reciprocal of ovariole number per ovary and the probability of juvenile survival for the ten ichneumonid parasitoids of Swaine jack pine sawfly (*Neodiprion swainei*) (the same species as in fig. 8.1). The horizontal lines represent the range of juvenile mortalities experienced by different species of parasitoids. Note that there are two species of *Pleolophus* and *Exenterus*. (Adapted from Price 1974.)

is evidence that the mortality schedules of herbivores with different feeding niches vary: for example, leaf miners suffer relatively constant levels of mortality throughout development, while the mortality experienced by juvenile gall makers tends to be concentrated in early instars (Cornell 1990). Such observations can be used to make predictions about the fecundity of parasitoids attacking different instars of these herbivores.

A special cause of juvenile parasitoid mortality is death caused by competition due to superparasitism or multiparasitism. Species that suffer from this type of mortality, either because they are competitively inferior to other species, or because individual selection favors superparasitism, will have greater realized fecundity and hence are likely to have greater potential fecundity. In the case of superparasitism, there is an added complication that more eggs will be required for placement on previously parasitized hosts. How the balanced mortality hypothesis is affected by superparasitism requires further study.

The balanced mortality hypothesis highlights the fact that average realized fecundity must balance juvenile mortality, and thus mortality is a good indicator of selection on potential fecundity. However, variation in realized fecundity will also influence selection on ovariole number, egg storage capacity, and the other components of potential fecundity. Consider first between-generation variation in realized fecundity. Imagine two species that suffer similar levels of juvenile mortality when attacking two different species of host. One species has boom-and-bust population dynamics: parasitoids lay large numbers of eggs in most generations leading to rapid growth in population density. The number of parasitoids cannot increase indefinitely and there are intermittent catastrophes that decimate the population. In contrast, the second species has very stable population densities. Although the long-term, average realized fecundity is the same in both species and predicted by the balanced mortality hypothesis, the first species experiences higher realized fecundity in most generations and thus selection for greater fecundity. This is the classic r- and K-selection argument of life history theory. Populations of the first species are increasing in density most of the time and high fecundity is at a premium. Populations of the second species are relatively stable, and natural selection favors a redirection of resources away from egg number to traits that promote competitive ability in a crowded environment. Price (1974) comes to different conclusions about how r- and K-selection relates to the balanced mortality hypothesis. The discrepancy is due to different interpretations of the concept which I shall return to below (sec. 8.1.3).

Realized fecundity may also vary within a generation. Imagine two species, one that attacks a solitary host, the other a clumped host. In the second case, an individual parasitoid that locates a large clump of hosts is able to lay many eggs. In consequence the variance in realized fecundity across individuals is likely to be much greater than in the case of the species attacking the solitary host. The response of natural selection to greater variance in realized fecundity will normally be to redirect resources to increasing potential fecundity to avoid the risk of egg depletion.

Will host accessibility per se influence realized fecundity and selection on potential fecundity? The answer, I think, depends on the population dynamics of the host-parasitoid interaction. If a host and a specialist parasitoid are at population dynamic equilibrium, standard theory (Hassell 1978) states that the ratio of host and parasitoids at equilibrium is independent of the rate of host discovery by the parasitoid. Parasitoids depress the densities of accessible hosts more than the densities of inaccessible hosts and, as a result, realized fecundity is independent of accessibility. This need not be the case for more complicated equilibrium models, and certainly not for nonequilibrium populations. In the case of parasitoids with boom-and-bust dynamics, fluctuations in parasitoid density will probably be accentuated by high rates of host discovery leading to a correlation between host accessibility and parasitoid potential fecundity.

Thus to conclude, in relatively stable populations the mortality schedule experienced by the immature parasitoid is an important indicator of parasitoid realized fecundity and hence of selection on ovariole numbers and other aspects of potential fecundity. However, higher fecundities are predicted as the temporal or spatial variance in realized fecundity increases, and in species where superparasitism is frequent or which suffer heavy immature mortality due to competition with other parasitoids. The balanced mortality hypothesis in effect uses a population-dynamic constraint to solve for an optimal life history. This is a different approach from much of modern life history theory, though one that is increasingly being adopted (Charnov 1990; Charnov and Berrigan 1991). I turn now briefly to parasitoids that oviposit away from their hosts. These species demand a slightly different treatment as there is no longer a simple relationship between the number of hosts located and the number of eggs laid.

OVIPOSITION AWAY FROM THE HOST

Many species of the large dipteran family Tachinidae, and all members of the small hymenopteran family Trigonalyidae, lay eggs on the food of their host and require that the eggs are ingested for parasitism to occur. These insects typically lay a very large number of eggs; tachinid fecundities may be as high as 13,000, and trigonalyid fecundities of over 10,000 have been recorded (Clausen 1940a; Askew 1971).

Most solitary parasitoids that lay eggs directly onto their hosts evolve a fecundity roughly the same as the maximum number of unparasitized hosts they are likely to meet in their life. There is no point in producing more eggs as they will go unused (except for those used in superparasitism). Trigonalyids, and those tachinids which lay their eggs away from the host, differ in that the more eggs placed in the environment, the greater the number of hosts likely to be parasitized. The eggs of ordinary parasitoids are like limpet mines—your stock of these should approximate the number of targets; the eggs of parasitoids that lay their eggs away from the host are like land mines—the more you lay, the greater the chance of hitting the target. Species that lay their eggs away from the host are thus selected to lay many eggs, even at the cost of reducing the amount of investment per egg. In fact, the trend toward many small (microtype) eggs may be accentuated by positive advantages of small size: small eggs may be more likely to be eaten by the host and less likely to be damaged by chewing. It may also be unnecessary to supply the embryo with large amounts of resource if it can obtain nourishment from the host as soon as it hatches.

Species that lay their eggs away from the host, but which have planidial or triungulin larvae that seek out or are carried to the host (sec. 2.2.5), are also likely to suffer heavy juvenile mortality. The number of hosts parasitized will increase with the number of eggs deposited, so these species will be subject to some of the same selection pressures for high fecundity experienced by the tachinids and trigonalyids discussed in the last paragraph. However, there are now positive disadvantages to the production of very small eggs that prevent the evolution of the massive fecundities seen in species whose eggs are ingested. Eggs must carry sufficient resources to produce larvae large enough to locate hosts efficiently. It is a reasonable guess that large larvae are able to survive longer and have a greater probability of finding a host.

Considering the difficulties first instar parasitoid larvae must experience in locating hosts, I doubt whether laying eggs randomly away from the host is often a viable strategy. The deposition of eggs on the substrate is most likely to evolve when the parent can place eggs near the host, or where adaptations such as phoresy are possible (sec. 2.2.5). Its prevalence in tachinids and other nonhymenopteran parasitoids may be due to the fact that these groups normally lack a well-developed piercing ovipositor. However, an active larva may be valuable for burrowing through plant tissue and entering the host in cases where the host is completely inaccessible to the adult parasitoid (Askew 1971). If the parasitoid egg stage is particularly vulnerable to predation, parasitism or abiotic mortality, a parasitoid may be selected to incubate its eggs and deposit either mature eggs or larvae (larviposition) (Price 1975). These strategies are found in many tachinid flies where the posterior regions of the oviduct is enlarged and highly vascularized to form an organ for egg maturation, the uterus. Some tachinids in which the adults locate the hosts also larviposit. The chief disadvantage to internal egg maturation is that the fly must avoid death for often a week or two weeks while the eggs ripen before it can begin to reproduce.

Price (1975) compared fecundity in tachinid flies with different life histories (fig. 8.4). He ranked the species in order of likely increasing juvenile survival and found a broad inverse correlation with the number of ovarioles and hence fecundity.

8.1.2 THE DICHOTOMOUS HYPOTHESIS

So far the discussion has focused largely on one component of parasitoid life history: fecundity. An alternative approach is to study suites of interrelated characters. Askew (1975) suggested that most parasitoid life histories could be classified into "one of two broad strategies" and similar arguments are present in papers written by Price and Force between 1970 and 1975. The origin of the dichotomy can be related to a major division in parasitoid natural history. Many parasitoids feed externally (ectoparasitoids) and kill or permanently parasitize their hosts at the time of attack (idiobionts). A second large assemblage consists of species that feed internally (endoparasitoids) and which allow their hosts to continue to feed and grow after parasitism (koinobionts). As is discussed below, not all ectoparasitoids are idiobionts, nor all endoparasitoids koinobionts, but in the majority of cases the two traits are correlated, especially among larval parasitoids. What might be called the "dichotomous hypothesis" states that natural selection operates on the life history strategies of the two sorts of parasitoids to magnify the initial differences.

Figures 8.5 and 8.6 describe some of the selection pressures that may operate on the two classes of parasitoid (Askew 1975; Askew and Shaw 1986; Force 1972, 1974, 1975; Price 1972b, 1973a, 1973b, 1974, 1975; Blackburn

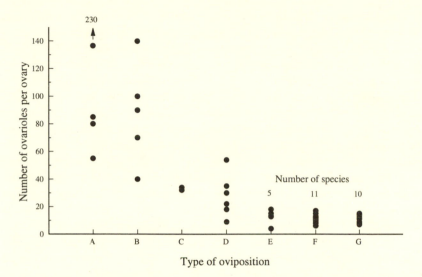

Figure 8.4 The relationship between tachinid fecundity and life history. The graph plots the number of ovarioles per ovary for different tachinid life histories ranked in order of likely decreasing juvenile mortality. *A*, oviposition on food, eggs eaten by host; *B*, oviposition on food or soil, planidial larvae; *C*, oviposition in vicinity of host, planidial larvae; *D*, oviposition on host larvae, nonpiercing ovipositor; *E*, oviposition on host larvae, piercing ovipositor; *F*, oviposition on host adult, nonpiercing ovipositor; *G*, oviposition on host adult, piercing ovipositor. Category *F* contains three Conopidae; all other species are Tachinidae. (This graph is taken from Price 1975, where the original studies from which the data were obtained are listed.)

1991a, 1991b). Consider koinobiont endoparasitoids first. As the hosts of koinobionts continue to grow after parasitism, these species are able to attack relatively small insects that are not large enough to support the developing larvae at the time of parasitism. There may be other advantages to the parasitism of small hosts. Endoparasitoids must cope with the internal defenses of the host and, as was discussed in section 6.3.1, small hosts often have less efficient defenses. Further, in competition with other endoparasitoids (both in superparasitism and multiparasitism), the oldest larva in a host normally has an advantage (Askew 1975), which again favors the parasitism of young hosts. As the balanced mortality argument shows, parasitoids that attack young hosts have higher realized fecundities and thus experience selection for greater reproductive capacity: koinobionts require more eggs. Other factors may also influence their optimal reproductive capacity. The ability of koinobionts to delay their development until the host has matured leads to relatively large adult parasitoids that may be able to locate more hosts. On the other hand, the specialist adaptations required by exposure to host defenses in endoparasitism result in a narrow host range and tend to reduce the number of hosts encountered. This trend may be offset by the greater host synchronization and specialism that a

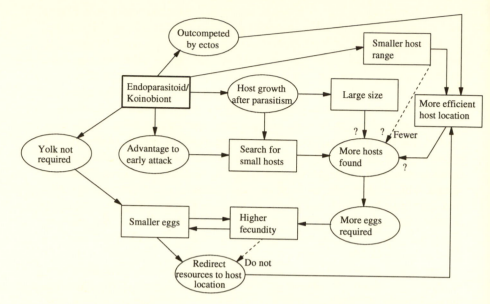

Figure 8.5 The web of selection pressures influencing the life history of koinobiont endoparasitoids (see text).

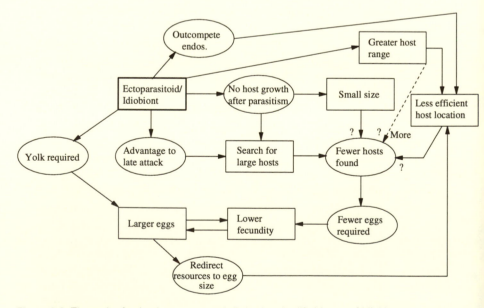

Figure 8.6 The web of selection pressures influencing the life history of idiobiont ectoparasitoids (see text).

narrow host range allows. The resources available for reproduction are finite and therefore a cost of higher fecundity may be a reduction in egg size. However, endoparasitoid eggs are bathed in hemolymph and so require less yolk, thus lessening selection for egg size. Endoparasitoids are normally poor competitors in comparison with ectoparasitoids, and in consequence there may be selection to redirect resources away from provisioning eggs to searching for hosts in marginal areas where exploitation by ectoparasitoids is less severe.

Essentially the opposite selection pressures operate on ectoparasitoid idiobionts. The fact that hosts do not grow after parasitism, and that the youngest ectoparasitoid on a host is often at a competitive advantage (Askew 1975), leads to selection for oviposition on relatively mature hosts. Fewer hosts are found, which reduces the pressure for large potential fecundities. Ectoparasitoid eggs must have yolk and thus need to be larger than endoparasitoid eggs, a factor that again tends to reduce fecundity. The competitive advantage of ectoparasitoids over endoparasitoids allows resources to be diverted from host location to provisioning eggs.

The schema depicted in figures 8.5 and 8.6 should not be treated too seriously but are meant broadly to indicate the type of selection pressures that operate on the different classes of parasitoids. Not all factors will apply to every parasitoid and many exceptions can be expected. In addition, the magnitude of the different selection pressures will differ within each class of parasitoid. For example, among endoparasitoids, egg-larval parasitoids attack the youngest hosts and may be expected to show some of the strongest adaptations involving increased fecundity.

The life history dichotomy described in this section is likely to be most marked among larval parasitoids. Although some egg and pupal parasitoids delay the death of their host and so might be called koinobionts, the differences between the koinobiont and idiobiont parasitoids of nongrowing host stages is small compared with the differences among larval parasitoids. Not all larval parasitoids are koinobiont endoparasitoids or idiobiont ectoparasitoids. Some endoparasitoids immediately kill their host while some ectoparasitoids allow their host to continue growing and have special adaptations to ensure they can survive moulting by the host. The life histories of these species are likely to involve a compromise between the two syndromes described in this section.

8.1.3 r- AND K-SELECTION

How useful is the application of r- and K-selection to parasitoid biology? Mueller (1988) points out that the concept is used in at least two ways in evolutionary biology. Originally it was developed as a mathematical population genetic tool to contrast natural selection operating in species near population dynamic equilibrium (K-selection) and in species perturbed away from equilibrium (r-selection). This is the sense in which I use the terms here. How-ever, r- and K-selection is frequently used much less formally to contrast suites of life history characters associated with high (r) and low (K) fecundity. I think this is the sense in which the concept has been employed in discussions of parasitoid biology (Force 1972, 1974, 1975; Price 1973b; Askew 1975; Askew and Shaw 1986; Barbosa 1977). Thus, parasitoids with comparatively high fecundity are by definition r-strategists and subject to r-selection: for example, "*Tetrastichus* is clearly the most r-selected [of a guild of parasitoids] because of its high r and inferior competitive characteristics" (Force 1975).

The problem with the second usage of the concept is that r- and K-selection cease to have any explanatory power but become merely labels for sets of life history characters. I believe there are positive advantages to restricting r- and K-selection to their original meanings, that is, to contrast selection pressures in populations that spend most of their time growing rapidly, and selection pres-sures in relatively stable populations. r-selection will certainly tend to increase fecundity, but not all differences in fecundity will be due to r-selection. To see this, consider two specialist parasitoids, each existing at a stable equilibrium with their host. One species attacks very early in its host's life cycle and suffers high juvenile mortality, while the second species attacks late in its host life cycle and suffers little juvenile mortality. The balanced mortality argument predicts that at equilibrium the first species will have higher realized fecundity than the second, and so experience selection for higher egg capacity. I do not think these differences in fecundity should be attributed to r-selection because, at least in this hypothetical case, both populations remain at their equilibrium population densities.

Parasitoids whose population densities fluctuate in numbers between gener-ations, either unpredictably or due to the influence of seasonality, will be sub-ject to the strongest r-selection. For example, species such as some scale insect parasitoids which attack large aggregations of hosts that support a number of parasitoid generations will experience boom-and-bust population dynamics and strong r-selection. Similarly, species that have several generations a year and that build up rapidly in numbers before being decimated during the winter will also be r-selected, for example many temperate aphid parasitoids. These species will normally, although not always, evolve relatively high fecundities: there will also be strong selection for reduced generation time (Lewontin 1965), and it is conceivable that if a trade-off exists between generation time and fecundity, r-selection can actually lead to a reduction in fecundity. (This is not as perverse as it may appear because the Malthusian parameter r in life history theory combines the effects of fecundity and generation time).

Several workers have suggested that parasitoids with high fecundity are found in areas supporting relatively low populations of the host, while parasit-oids with low fecundity occupy the center of the host's range. There is evidence for this from studies of the ichneumonid complex attacking *Neodiprion swainei* (Price 1973b) and the chalcidoid community attacking the gall midge *Rhopalomyia californica* (Force 1972, 1974, 1975). Populations of the sawfly in early successional habitats contain fewer species than those in later successional habitats, and these species tend to be larval parasitoids with high fecundity. Larval parasitoids also tend to predominate in low-density, peripheral sawfly populations, while pupal parasitoids, with lower fecundity, are found in high-density, core populations. When a stand of coyote bush (*Baccharis pilularis*, the host plant of *R. californica*) is cut back, the first parasitoid to attack the colonizing hosts is the eulophid *Tetrastichus* sp., the species with the highest fecundity.

There are several reasons why species with high fecundity may often be found in marginal habitats. First, species that attack early in the life cycle often tend to lose in competition with later parasitoids in cases of multiparasitism (and also hyperparasitism). This might lead to selection to search less productive habitats where competition is not so severe. This explanation generates the observed pattern without invoking *r*- and *K*-selection in the sense I use the terms. It is also possible that species preadapted to search marginal populations have more fluctuating, boom-and-bust populations, and that *r*-selection is partially responsible for their increased fecundity. However, the very strong correlation between fecundity and juvenile mortality revealed by Price (1974) suggests that the fecundity of the ichneumonids he studied may be explained by the balanced mortality hypothesis alone.

8.1.4 COMPARATIVE STUDIES

The ideas discussed in this section have been supported largely by the informal comparison of parasitoid natural histories. However, there have been a few quantitative tests of theoretical predictions. As discussed above, Price (1972b, 1973a, 1974, 1975) demonstrated strong correlations between ichneumonid fecundity and juvenile mortality and obtained suggestive evidence for the same pattern in Tachinidae (Price 1975). Price (1973a) also showed that among ichneumonids, egg length is negatively correlated with the number of ovarioles, the latter strongly correlated with fecundity. A trade-off between the number and size of eggs is predicted in most life history models (Smith and Fretwell 1974).

Although cross-species comparisons are valuable, they can also be misleading (Harvey and Pagel 1991). The chief problem is the lack of statistical independence among data points caused by the phylogenetic relationship among species and higher taxa. A variety of statistical techniques has been proposed to avoid these problems, and a consensus is emerging that methods should be based on the analysis of statistically independent contrasts calculated for a known or partially known phylogeny (Felsenstein 1985; Harvey and Pagel 1991). The techniques work best with a well-resolved phylogeny but can be applied to less well understood groups such as parasitoids by using the standard taxonomic classification.

Blackburn (1991a, 1991b) amassed data from the literature on a variety of life history parameters and ecological variables for 474 species of parasitoid wasps. Using the methods of Harvey and Pagel (1991), he searched for correlations among life history parameters, and between life history and ecological variables. In most animals, especially vertebrates, body size is a major correlate of nearly all life history variables. However, Blackburn found that body size failed to exert its normal pervasive influence. Different components of preadult life span tended to be correlated with each other (e.g., species with a long pupal period also tended to have a long larval period). Controlling for body size, the length of the preadult period was correlated in a fairly obvious fashion with a number of ecological and life history variables: koinobionts, parasitoids that attack eggs, and temperate species had longer preadult life spans than idiobionts, pupal parasitoids, and tropical species. There was some suggestion that species attacking exposed hosts had shorter preadult life spans than those attacking concealed hosts, although this result may have arisen because of correlations between host concealment and other factors. The most interesting results related to fecundity. Wasps with large fecundities (the maximum number of eggs reported as having been laid by a wasp of that species) tended to have smaller eggs, and to be capable of laying eggs at a faster rate. Further, species with large fecundities required a shorter period of time after eclosion before being able to begin oviposition. Curiously, there was an unexpected negative correlation between fecundity and the window of parasitism, the length of time the host is susceptible to parasitism.

Blackburn's analysis supports the hypothesis that there is a trade-off between fecundity and egg size. Small eggs require less yolk and enable the wasp to adopt a pro-ovigenic strategy, that is, to eclose from the pupae with mature eggs rather than to mature eggs sequentially throughout the adult life. Pro-ovigeny allows a greater rate of oviposition and obviates the necessity of an adult prereproductive period in which to mature eggs. The lack of importance of body size is also interesting. Harvey et al. (1989) have suggested that body size is an important predictor of mammal life histories because it is highly correlated with mortality rates. Blackburn argues that body size is less important in parasitoids because, as Price stresses, the major factors influencing mortality are associated with host ecology rather than parasitoid morphology. Although an important first step, the data set used in these analyses is a very sparse representation of a very diverse group of wasps, with a rather poorly resolved taxonomy. In addition, information about particular variables was often missing from the literature accounts of the different parasitoid species.

The shortcomings of the data set may explain the rather few significant correlations to emerge. Better phylogenies and more comprehensive biological information are required before comparative analyses can be used as other than a very blunt instrument.

8.2 Host Range and Parasitoid Species Load

Individual parasitoid species often interact with more than one host species and with other species of parasitoid. This section is about the evolutionary processes influencing both the host range of parasitoids, and interactions between different parasitoid species. I begin by discussing patterns in parasitoid species loads, which I define as the number of parasitoid species attacking a species of host. This is an area that has received considerable attention by parasitoid community ecologists. I then review the evolutionary determinants of parasitoid host range. Finally, I discuss whether competitive interactions among parasitoids, and among hosts mediated by parasitoids, are important in structuring insect communities.

8.2.1 PARASITOID SPECIES LOAD

There is enormous variation in the number of parasitoid species that attack different host species, and much effort has been devoted to the search for statistical correlations between parasitoid species load and different aspects of host ecology. Unlike other areas of parasitoid community ecology, there is a considerable amount of data that can be used to address this question. Large collections of hosts and parasitoids are often reared as part of applied entomological or ecological studies. Several workers, Hawkins in particular, have amassed large data sets of parasitoid species loads and used them to test a variety of hypotheses.

Statistical analysis of host-rearing records is the only feasible way to study parasitoid species load, but this technique is not without problems. First, the number of parasitoid species reared from a host is strongly influenced by sample size (fig. 8.7). Sample size is likely to be correlated with many ecological factors: agricultural pests generally receive more attention from entomologists than economically unimportant species, while groups that are easy to rear (e.g., leaf miners, gall formers) have a larger number of rearing records than more taxing groups (e.g., external folivores). It is essential to control for sample size to avoid spurious correlation. There are two ways this may be done; perhaps the best way is to fit sample size as an explanatory variable in a statistical model and to use the residuals from the model in tests of hypotheses about the importance of other explanatory variables. A simpler procedure is to work

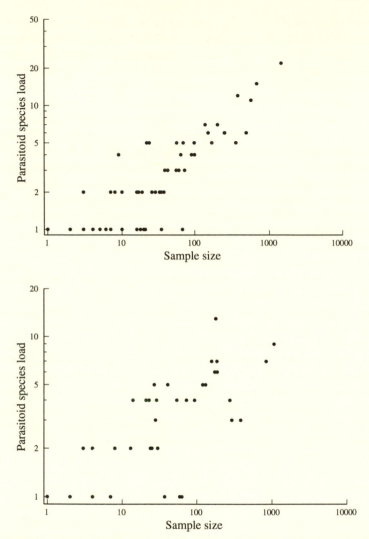

Figure 8.7 The relationship between parasitoid species load and sample size for two leaf-miner communities. *Top*: Parasitoids of agromyzid flies reared in 1981 from one site in southern England (C. V. Towner and H.C.J. Godfray, unpublished). *Bottom*: Parasitoids reared in 1989 and 1990 from all species of leaf miner growing in a 100 × 200 m area of tropical regrowth forest in Costa Rica (J. Memmott, H.C.J. Godfray, and I. D. Gauld, unpublished).

only with host species with large numbers of rearing records. The assumption behind the second method is that parasitoid load asymptotes with increasing sample size. This assumption is almost certainly valid for specialist parasitoids, but the asymptote may be reached very slowly in the case of generalist ectoparasitoids which show low host specificity and often occur at very low frequencies on strange hosts. A related problem in using literature data is that different host species may be sampled over different geographical ranges. Geographical patterns in parasitoid species load is itself an interesting question, although hidden correlations between the geographical range over which samples were collected and other ecological factors may give rise to spurious results. Finally, Askew and Shaw (1986) discuss a variety of other difficulties: samples taken (1) from the edge (geographical or ecological) of a host species range, (2) from one host generation or during only part of the year, or (3) that include only part of the host life cycle may all underestimate parasitoid species numbers. All these shortcomings of the data base may be annoying yet benign if their sole effect is to increase the error term in any statistical analysis. However, more insidious problems arise if, as is often the case, the sampling bias is correlated with potential explanatory variables.

The second drawback is a general problem of all attempts to search for statistical correlations in community data: a significant correlation does not indicate causality. Statistics cannot help with this problem, and biological intuition is required to judge the plausibility of any proposed causal link. A final problem may, at least in part, have a statistical solution. Most analyses of parasitoid species load assume each host species to be a statistically independent data point. As discussed above (sec. 8.1.4) there are problems with this assumption because host species are phylogenetically related. For example, in the present context, two very similar host species, which cannot be distinguished by their parasitoids, will have the same parasitoid species load. It makes no sense to treat them as separate data points. Extension of comparative techniques (Harvey and Pagel 1991) to the analysis of community data such as parasitoid species load would be very valuable.

HOST FEEDING NICHE

The clearest pattern to have emerged from the analysis of parasitoid species loads is the pervasive importance of the feeding niche of the host. Nearly all host species available for analysis are herbivores that can be classified by their feeding site. Hawkins and colleagues have usually distinguished six categories of feeding niche: (1) external folivores; (2) species that feed in spun leaves, webs, etc.; (3) leaf-mining insects that feed inside the leaf lamina; (4) gall-forming insects; (5) species that feed inside stems, in wood, or within buds, fruits, flowers, and seeds; and (6) root-feeding insects. In a recent paper (Hawkins 1993), a further two categories are examined: case-bearing insects and insects that spend part of their juvenile period feeding internally and part externally. In some analyses the two categories (1–2) of external feeding hosts (exophytes) and the four categories (3–6) of internal feeders (endophytes) are lumped together. The categories (1) to (6) are ordered roughly in accordance with the concealment of the host within the host plant.

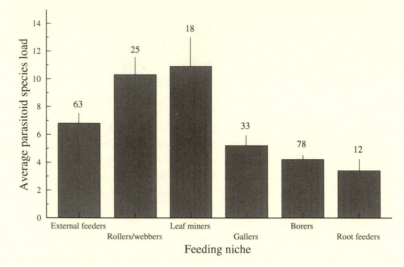

Figure 8.8 The average parasitoid species load on hosts with different feeding niches. Only data from species where more than one thousand hosts have been sampled is included. The vertical lines represent standard error; the number of host species sampled is given above the bar. (From Hochberg and Hawkins 1992.)

Hawkins and Lawton (1987) and Hawkins (1988a) compiled a data base of the number of parasitoid species attacking a variety of phytophagous hosts in the British Isles. They found that leaf-mining insects possessed the highest parasitoid species load, with a trend to lower loads on both less and more concealed hosts. Since these first studies, Hawkins has steadily increased the size of his data base, which now includes records of 12,079 parasitoid-host associations on 2188 species of hosts in 110 families (Hawkins 1993). The same association of parasitoid species load and host feeding niche has been found using data from different zoogeographic zones, and from temperate and tropical regions (Hawkins 1990, 1993; Hawkins et al. 1990, 1992; fig. 8.8). The pattern is also found if analysis is restricted to host species with over one thousand rearing records, and hence is very unlikely to be an artefact of differential rearing effort (Hawkins 1988a).

Why should feeding niche affect parasitoid species load? This question is complex because it involves processes operating in both evolutionary and ecological time. In the case of specialist parasitoids, the enlargement of a species parasitoid load must involve evolutionary change, presumably over a long period of time. Ecological processes may play a part if recruitment of a new species is influenced by the number of parasitoids already attacking a host. The inclusion of generalist parasitoids will also involve a mixture of evolutionary and ecological processes. One can envisage a pool of potential parasitoids of a particular host species whose size is set by evolutionary processes. The fraction of the pool that is actually recorded from a host will be determined by ecological processes, and also by the statistics of sampling. In addition, if data on parasitoid loads are collected from part of the host range, for example in studies restricted to the British Isles, biogeographic processes may influence both the numbers of generalists and specialists attacking a host. A final complication is strong evidence, reviewed below, that the ratio of specialists to generalists may vary with feeding site.

Hawkins and Lawton (1987) and Hawkins (1988a, 1990) did not present a detailed hypothesis of how feeding niche influences the ecological and evolutionary processes determining parasitoid species load but argued that two major facets of parasitoid biology were affected by feeding niche. First, the ease of host discovery is influenced by feeding niche. External feeders are relatively mobile and are able to move away from areas where their feeding damage or frass might attract parasitoids (sec. 2.2.2). Species feeding in webs and spinnings are less able to escape cues that may attract parasitoids, while leaf miners are effectively imprisoned between the leaf lamina. Gall formers are better off than leaf miners because, although the gall may be easier to detect, the host within it is better concealed. Species that burrow in stems or wood, or feed within buds or flowers, are even better hidden from parasitoids, while root feeders enjoy protection from the soil as well as from plant tissue. The second factor that may covary with feeding niche is parasitoid juvenile mortality. Species feeding concealed in plant tissue will suffer less risk of death from predation than species feeding exposed on the plant.

To see how these factors may influence parasitoid species load, assume first that the host populations are not regulated by their parasitoids. Assume also that the recruitment of new species of parasitoids is unaffected by percentage parasitism and by the number of species already attacking the host. I shall call this the noninteractive model. Parasitoid species load is determined by rates of gain and loss of parasitoids in both evolutionary and ecological time. Host detectability is likely to influence the recruitment of parasitoids. Consider a rare mutant of a specialist parasitoid that is able to survive on a novel host; the chances of this individual achieving a host shift is improved if the novel host is easily found. Similarly, a generalist parasitoid that is able to attack a host is more likely to realize this potential if the host is easy to locate. Once a parasitoid has colonized a host, protection from other sources of mortality may reduce the probability of stochastic extinction.

Can a noninteractive model be defended as a basis for studying parasitoid species load? A number of authors have argued that parasitoids exert little or no influence on the distribution and abundance of their host, that hosts are regulated by competition or predation, and that parasitoid abundance and distribution tracks that of the host without influencing it (e.g., Hawkins 1992). The main evidence for this view is the frequent failure to detect the type of host and parasitoid density dependence necessary for regulation (Dempster 1983).

Against these arguments must be set the outstanding success of many biological control attempts where parasitoids have patently influenced host numbers, and also the great difficulty in detecting the regulating influence of parasitoid populations in field data: there is a real danger of confusing absence of evidence with evidence of absence (Hassell 1986; Hassell et al. 1989). Of course, the true answer is almost certainly that in some systems parasitoids are crucial in determining host abundance while in other cases they are not. For example, Price and colleagues have put forward convincing arguments that populations of several shoot-galling sawflies (*Euura*) and insects with related life histories are limited by the availability of oviposition sites. Although these species are attacked by parasitoids, they are probably of secondary importance in population regulation compared with the role of the host plant. However, other sawflies with less specialized oviposition requirements are unlikely to be limited by oviposition sites, and a role for parasitoids is strongly indicated by the success of classical biological control programs directed against these species (see Price 1990 and Price et al. 1990 for a review of extensive experimental work on these systems). Even if host populations are not regulated by parasitoids, interactions between species may influence the recruitment of parasitoids to new hosts. There is evidence that parasitoid load and percentage parasitism are correlated (Hawkins and Gagné 1989; Hawkins 1993), while evidence for competition among parasitoids is discussed below (sec. 8.2.3). A straightforward noninteractive model is thus almost certainly too simplistic in the majority of cases.

If parasitoids influence host density, the role of host detectability in determining parasitoid load is more complicated. Consider two specialist parasitoids that regulate their different hosts at a stable equilibrium. Suppose that the only difference in the natural history of the two interactions is that one host is far easier to discover than the other. At population dynamic equilibrium, by definition, each parasitoid exactly replaces itself. If one host species is easier to find, its equilibrium population density must be lower than the other. Thus in this admittedly artificial example, the population dynamics lead to the neutralization of differences in detectability.

Recently, Hochberg and Hawkins (1992) have made an important start at trying to develop an interactive model of parasitoid species load. They argue that the major effect of host feeding niche is to render different fractions of the host population immune from parasitoid attack—in other words, to be in a refuge. This is an important variation on earlier arguments because it assumes that hosts in a refuge can never be found by parasitoids, however many are searching in the environment. Hochberg and Hawkins simulate a host population that is initially attacked by fifty species of generalist and fifty species of specialist parasitoids. This assemblage of species is iterated over one thousand generations, during which time many species fall out of the simulation, either becoming extinct (specialists) or so rare on the host that they are deemed undetectable (generalists). At the end of the computer experiment, the number of hosts and parasitoids are related to the size of the host refuge. Of course, many details of the biology of the interactions have to be assumed—for example, the competitive hierarchy of species, and the details of numerical and functional responses. Hochberg and Hawkins chose standard forms of these functions from the theoretical literature on host parasitoid population dynamics (Hassell 1978; Hassell and May 1986; Pacala et al. 1990). Although the detailed results with different parameter assumptions vary, in most simulations parasitoid species loads peak at immediate levels of host refuges. Most of the variation in parasitoid load is due to changes in the numbers of generalists, while specialist diversity is comparatively constant. Hochberg and Hawkins go on to equate the feeding niches (1) to (6) described above with increasing percentage refuges and argue that their modeling results are consistent with the patterns shown in the data (fig. 8.8).

As Hochberg and Hawkins (1992) stress, their results are provisional but the sensitivity analysis they perform suggests that the predictions are robust. However, I believe their equation of the spectrum of feeding niches (1–6) with increasing refuges needs stronger justification. For example, it does not necessarily follow that a greater proportion of leaf miners than semiconcealed feeders are in structural refuges because leaf miners are completely concealed: all leaf mines may be equally accessible to parasitoids.

The predictions of the model can best be understood by considering the relationship between the size of the refuge and the number of hosts available for parasitism. When the refuge is large (and the host population stabilized by direct host density dependence), only a few parasitoid species are able to maintain themselves on the small number of hosts available for parasitism. In these populations only one or a few specialists can coexist and the numerical response by generalists will be very weak. As the refuge decreases in size, the number of hosts available for parasitism increases, and more of both generalists and specialist can be supported. An important parameter determining the number of specialists is the amount of spatial heterogeneity in the system that prevents competitive exclusion (see sec. 8.2.3). At some point, the number of hosts available for parasitism becomes sufficiently large that parasitoids are able to depress and regulate the host population a considerable way below its carrying capacity. In consequence, the number of hosts available for parasitism again decreases, leading to a fall in parasitoid diversity. Hochberg and Hawkins's (1992) explanation for parasitoid species load is purely ecological and assumes that parasitoids have a major influence on the abundance of host species with few refuges. It also accords a small role to evolutionary processes and implicitly assumes parasitoid diversity is not controlled by the rate at which parasitoids adapt to hosts. In many ways it is the antithesis of the noninteractive model sketched above.

I finish this section by discussing another factor that may affect the relationship between parasitoid species load and host feeding niche, but which has received little attention (but see Askew 1980 and below for a similar argument in a slightly different context). This might be called host homogeneity. Consider two species, one taxonomically or ecologically isolated, with few or no close relatives feeding in a similar niche, the other part of a swarm of species all feeding in very similar ways and perhaps taxonomically related. Several arguments indicate that the second species will have a larger parasitoid species load. First, the swarm of species will be a larger evolutionary target than the isolated species. More generalists will recruit to the swarm, and be able to attack each of its members, than will become adapted to the isolated species. Once able to attack one member of the swarm, generalists may speciate in isolated geographical areas and then reinvade, amplifying the number of parasitoid species. In addition, the possibility of feeding on many host species lessens the risk of the extinction of a generalist. The number of specialists is also likely to be higher on a member of the species swarm as there are many closely related hosts supporting parasitoids partially preadapted to colonize the focal species. I have highlighted these processes by a stark comparison, but more generally there is likely to be a correlation between the parasitoid species load of a host, and the number of other similar species in the environment. The possible importance of host homogeneity is supported by studies of herbivore species loads on plants. The diversity of specialized herbivores such as leaf miners and gall formers that have an intimate relationship with their host is strongly influenced by host plant taxonomy (Askew 1961; Claridge and Wilson 1982; Godfray 1984).

Could there be a relationship between host similarity and feeding niche? I suggest that the strong selection pressures exerted by feeding within a leaf lamina lead to great ecological homogeneity among leaf miners and that this may explain their high parasitoid species load. In addition, relatively few taxa have evolved the leaf-mining habit and the ecological homogeneity is reinforced by taxonomic homogeneity—most leaf miners are members of large groups of related species. Externally feeding species are taxonomically more diverse, and the idiosyncratic selection pressures involved in feeding exophytically on different plant species result in greater ecological heterogeneity. Increasing heterogeneity is also found in the sequence of feeding niches beginning with leaf miners and moving toward more concealed hosts. Galls makers are normally members of large taxonomically homogeneous groups, yet variable gall morphology leads to ecological heterogeneity. The greatest ecological heterogeneity is probably found among borers in stems, wood and flowers, and also in root feeders. Adaptations that allow the location of one species of borer or root feeder may be of little help in locating other species. Thus host heterogeneity offers a possible explanation for the observed trends in parasitoid species load and host feeding niche. I have outlined these ideas as a noninteractive hypothesis, but it would also be interesting to study them in a population-dynamic context.

To conclude, Hawkins and colleagues have demonstrated a robust pattern in parasitoid species load across different feeding niches. There are a number of explanations that need not be mutually exclusive.

1. Parasitoid diversity is determined by the relative rates of recruitment and loss of parasitoids on different host species. Recruitment is proportional to the detectability of hosts in different feeding niches, and the rate of loss is influenced by host mortality in different niches.

2. Parasitoid diversity is determined by the size of the host population available for parasitoid attack. Diversity is low when most hosts are protected from parasitism in proportional refuges, and also low when most hosts are susceptible to parasitism, as in these circumstances population regulation by the parasitoids themselves decreases host abundance. Diversity peaks at intermediate levels of susceptibility to parasitism.

3. The hosts feeding in some feeding niches are ecologically and/or taxonomically more uniform than those in others. Greater uniformity leads to greater opportunities for the transfer or sharing of parasitoid species among hosts and thus greater diversity.

GALL-FORMING INSECTS

The parasitoids of gall-forming insects have received particular attention by community ecologists and some of the best examples of parasitoid food webs are studies of gall-forming cynipids (Hymenoptera; Askew 1961) and cecidomyiids (Diptera; Force 1974; Hawkins and Goeden 1984). Askew (1961) suggested that gall formation by the herbivore may have evolved as a response to parasitoid attack. Further, surveying the extraordinary structural diversity of cynipid galls on oaks, he suggested that parasitoid pressure may lead to the evolution of dissimilar gall morphologies. Parasitoid pressure is not the only hypothesis that can explain the evolution of galling (Price et al. 1987) although it has received considerable attention. There are two main approaches to studying this hypothesis. First, the parasitoid species load and the percentage parasitism experienced by gall-forming and non-gall-forming insects can be compared to see if gall formers enjoy protection from their enemies. Second, the role of parasitoids in selecting for larger galls can be studied in individual host-parasitoid interactions. I return to the second approach below (sec. 8.2.4) but in this section discuss the determinants of the parasitoid species loads of gall formers and ask whether this can tell us anything about gall evolution.

Hawkins and Gagné (1989) collected information from the literature on the parasitoid species loads of 191 species of phytophagous Cecidomyiidae and statistically explored correlations with a variety of ecological variables. The most important single determinant of parasitoid species load was, yet again, feeding habit. Cecidomyiids were classified (in order of increasing parasitoid species load) into (1) endophytic nongalling species that leave no external evidence of their activities; (2) species that live in rolled leaves; (3) endophytic nongalling species that do leave external evidence of their activities; (4) species that form galls that are similar externally in texture and color to the surrounding plant tissue; and finally (5) species that form distinct and easily recognizable galls. Gall midges that are easy to locate appear to have greater parasitoid species loads. In addition to feeding habit, both bivoltine cecidomyiids (as opposed to those with one or many generations a year) and species that pupate in plant tissue rather than in the soil tend to be attacked by more species of parasitoid.

The results of this study are generally consistent with the wider survey of herbivorous insects and can be explained by the same mechanisms (see also Tscharntke 1992). Species that are easier to locate might accumulate larger parasitoid loads through noninteractive processes operating in ecological or evolutionary time, or there may be subtle population dynamic interactions between the proportion of hosts protected from parasitoids and the number of wasps that can coexist on a single host. In addition, host homogeneity may play a part. The majority of cecidomyiids form galls and may together form a pool of potential hosts that are shared by generalists and between which there is easy cross-colonization by specialists.

Price and Pschorn-Walcher (1988) compared the average parasitoid species loads on gall-forming and external-feeding tenthredinid sawflies (Hymenoptera) and found that while the former were on average attacked by four parasitoid species, the latter were attacked by an average of sixteen. They argued that this was evidence in favor of the importance of parasitoids in the evolution of galling. However, leaf-mining sawflies are also attacked by relatively few species (Askew and Shaw 1986) so the major factor effecting sawfly parasitoid load may be endophagy versus exophagy.[1] Hawkins (1988b; see also Hawkins and Gagné 1989) compared gall-forming and non-gall-forming members of two families of Diptera (Cecidomyiidae and Tephritidae) and found in both cases that gall-forming species had larger parasitoid loads and also suffered higher levels of percentage parasitism. He concluded that this was evidence against the role of parasitoids in the evolution of galling.

There are, however, major problems in using data on contemporary parasitoid species load and on percentage parasitism to test explanations of the origin of the gall formation (Price et al. 1987; Price and Pschorn-Walcher 1988). The evolution of gall formation has occurred on relatively few occasions, and the specific selection pressures responsible for its appearance in the Cretaceous or early Tertiary in ancestral gall flies or gall midges may be impossible to reconstruct. Once galls have evolved, parasitoid radiation and adaptation may completely obscure any initial correlations between galling and escape from parasitism. To illustrate this point consider the following sequence of events: (1) large galls are selected as they provide protection from

[1] Sawflies are an exception to the normal pattern of greater parasitoid loads on endophagous species. The majority of sawflies feed externally and I believe this exception is consistent with the importance of taxonomic and ecological isolation in determining parasitoid species load.

parasitism; (2) parasitoid ovipositor length increases in response to larger gall size; (3) the coevolutionary dance continues until other selection pressures or constraints prevent further gall enlargement; (4) the parasitoid ovipositor length stabilizes at a size that allows all hosts to be attacked. In this case, parasitism was responsible for the evolution of large galls; yet after gall structure has stabilized, rates of parasitism are the same or even higher than before. After galls have evolved, parasitoid radiation among members of the gall clade may subsequently increase parasitoid species load. A less troublesome question, which I shall return to below (sec. 8.2.4), is whether parasitism in contemporary time maintains gall size and structure.

PLANT ARCHITECTURE AND SUCCESSION

It has often been suggested that hosts attacking early successional plants such as grasses, herbs, and shrubs have fewer parasitoids than those attacking plants characteristic of the later stages of succession such as trees. This is undoubtedly true for a number of host taxa. Cynipid wasps that form galls on oak (*Quercus*) have many more parasitoids than species that form galls on low-growing plants, and gracillariid moths (*Phyllonorycter*) that mine the leaves of trees have more parasitoids than species attacking herbs and shrubs (Askew 1980). However, the evidence from large surveys of parasitoid species loads is more equivocal. For example, in his most recent analysis of British herbivores, Hawkins (1992a) found that only 4% of the variance in parasitoid species load was attributable to host plant type.

There is a strong correlation between herbivore diversity and plant growth form (Lawton and Schröder 1977). Askew (1980) first suggested that parasitoid species loads were correlated with the diversity of taxonomically and ecologically similar hosts on plants of the same growth form. The taxonomic distribution of the two groups studied by Askew (1980) is consistent with this explanation: cynipid wasps have undergone a remarkable radiation on oak while moths in the genus *Phyllonorycter* are largely confined to trees. Exceptions to the rule that herbivore diversity is higher on later successional plants also supports Askew's suggestion. Agromyzid flies are predominantly leaf miners that attack all types of plants but are very much more common on herbs than on trees or shrubs. There is evidence that the parasitoid species loads of flies attacking herbaceous plants is higher than that of species mining trees (Hawkins et al. 1990) (although the parasitoid fauna of the latter group is rather poorly known). The parasitoid community of agromyzids on herbs is similar in size to that of *Phyllonorycter* on trees, although there is greater variance in agromyzid parasitoid numbers, perhaps reflecting the greater morphological and taxonomic diversity of herbaceous plants (C. V. Towner and H.C.J. Godfray, unpublished). Cecidomyiid midges are another group that have radiated on low-growing plants. Hawkins and Gagné (1989) found no significant effect of plant growth form in their analysis of cecidomyiid parasitoid loads.

A number of other explanations have been proposed to explain the relationship between parasitoid species load and plant structure. Askew and Shaw (1986) suggest that differences in apparency may be important. The concept of apparency was introduced by Feeny (1976) to explain differences in the chemical defense strategies of plants against insect herbivores. Feeny argued that the distribution of early successional plants is spatially and temporally unpredictable so that they are largely able to escape from specialized herbivores. In consequence, early successional plants invest in relative cheap qualitative chemical defenses that, while effective against generalists, are comparatively easy to detoxify by adapted herbivores. Late successional plants are always located by their herbivores and invest in quantitative chemical defenses such as tannins that act to reduce the digestibility of the plant. Today, plant chemical defenses are most frequently studied using plant resource budgets, which place defenses against insects in the context of other factors vying for limited plant resources (Coley et al. 1985). It has also become clear that many late successional plants possess toxins in addition to quantitative chemical defenses.

Plant apparency may affect parasitoid species loads in several ways (Southwood 1977, 1988; Price 1991). First, herbivores on early successional plants may be harder to find because of the temporal and spatial unpredictability of their host plants. In discussing the correlation between parasitoid species loads and host feeding niche, I described the argument that the ease of host discovery influences parasitoid species loads. Exactly the same process may operate here although it is again important to distinguish between interactive and noninteractive models of parasitoid species loads. The second explanation is that the hosts themselves differ in apparency to parasitoids, either independently of plant apparency, or compounding the latter's effects. Early successional hosts find it difficult to locate their food plants and so themselves may have an even less predictable distribution. Finally, early successional hosts may sequester the toxins in their food plants, reducing the rate at which they are attacked by generalist parasitoids, and reducing the likelihood of host shifts by specialist parasitoids. As yet, there is no evidence that efficiency of host discovery is correlated with the growth form of host plant, nor that hosts that sequester toxins are attacked by fewer parasitoids.

The population densities of insects feeding on herbs is probably less than species on trees: herb biomass is much less than that of trees, and individual species of herb are often relatively rare in the environment compared with trees. The resource fragmentation hypothesis, as applied to this problem (see sec. 8.3.2. for its original context), states that few parasitoids can be maintained on the relatively small populations of hosts attacking early successional plants. Hawkins et al. (1990) tested this idea by comparing the numbers of generalist and specialist parasitoids attacking early and late successional hosts (generalists and specialists were equated with idiobionts and koinobionts respectively; see sec. 8.3.3 below). Specialist parasitoids will be most affected by declining hosts densities and should thus be relatively rare on early successional hosts. The results depended on feeding niche and were contradictory: fewer specialists attacked early successional, externally feeding herbivores, but the reverse trend was found for endophytic hosts. They concluded that their study provided little if any support for resource fragmentation.

Finally, differences in the number of herbivore species per plant are often related to the architectural complexity of the plant (Lawton 1983). Trees are larger and have many different architectural features (bark, wood, etc.) that are absent in herbs but which may support herbivores. Studies of parasitoid species loads have concentrated on the number of species attacking individual hosts of similar biologies on trees and herbs, and hence the gross effects of plant architecture are unlikely to be important. However, trees provide many different microhabitats which allow a wider range of suitable places for parasitoids to rest or hibernate, and there may thus be an indirect affect of plant architecture.

In conclusion, the parasitoid species loads of some taxa are higher on species feeding on late successional plants, while the reverse trend is found in other groups. In these cases parasitoid species load can be most easily explained by the much greater diversity of the particular taxa on early or late successional plants. Beyond this, the overall association between parasitoid species load and succession is unclear. In concluding this I differ markedly from Price (1991), who views changes in parasitoid species load along the successional gradient as one of the few clear patterns in the community ecology of herbivores and their parasitoids.

In addition to broad differences in plant structure, individual plant species vary in morphology and ecology in ways that affect the parasitoid species load of their herbivores. For example, parasitoids may find difficulty searching for hosts on glandular plants or species with very hairy leaves. As well as absolute barriers to parasitoid activity such as hairiness, other plants may just have atypical morphology or ecology and so be avoided by more generalist parasitoids.

A good example of the influence of plant structure is provided by Gross and Price's (1988) study of two congeneric leaf-mining moths (*Tildenia inconspicuella* and *T. georgei*, Gelechiidae), which attack horse nettle (*Solanum carolinense*) and ground cherry (*Physalis heterophylla*), respectively (both plants are in the tomato and nightshade family [Solanaceae] and are unrelated to either nettles or cherry!). The ground-cherry miner has a typical gelechiid life history, forming a small mine but also folding the leaf and leaving the mine to feed in a silken tunnel. The species on horse nettle mines throughout its life, a less typical behavior for this family. Two species of parasitoid were reared from the ground-cherry miner but no less than nine species were reared from the horse-nettle species. Why should there be such a large difference? A direct effect of host plant was ruled out because ground-cherry miners that had been transferred to horse nettle, or had naturally colonized this species, were not attacked by the extra parasitoids reared from the horse-nettle miner: the parasitoids appeared to be species-specific rather than plant-specific. Instead, Gross and Price concluded that the completely endophytic habits of the horse-nettle miner, and its less mobile behavior, rendered it susceptible to a number of generalist parasitoids, particularly ectoparasitoids, that were unable to attack the vagile and semiendophytic ground-cherry miner. They also argued that the difference in host ecology was caused by differences in plant structure. Horse nettle has firm stellate hairs which makes movement outside the mine difficult, while ground cherry has thin flexible hairs. Ground-cherry leaves tend to be relatively small and often wilt when attacked by a gelechiid, making movement between leaves essential for the host. In contrast, horse-nettle leaves are larger and do not wilt when attacked, so that the ability to move between leaves is unnecessary. Thus plant structure has an indirect effect on parasitoid species load by determining the feeding habits of the host.

OTHER FACTORS

Studies of the number of herbivore species on different food plants predate those of parasitoid species load by about twenty-five years (Southwood 1961). Can the more extensive investigations of the former provide us with any clues about the latter? One of the primary correlates of herbivore species load is host plant range (Opler 1974; Strong 1974; Lawton and Schröder 1977; Cornell and Washburn 1979). The influence of host plant range probably operates in two ways: first, widespread hosts are larger evolutionary targets; and second, widespread hosts are more likely to occur in different microhabitats and occupy greater latitudinal and altitudinal ranges, thus providing a greater range of niches for potential colonists (Strong 1979; Strong et al. 1984). Unfortunately, the data do not exist to carrying out equivalent studies at the next trophic level: the ranges of few herbivores with known parasitoid species loads are understood with sufficient precision. Finally, the length of time a plant has been exposed to colonization may influence its herbivore species load (Birks 1980). Equivalent processes may operate in parasitoids as introduced species typically have fewer parasitoids than their endemic relatives (Greathead 1986; Cornell and Hawkins 1993).

8.2.2 HOST RANGE AND FOOD-WEB STRUCTURE

The observed host range of a parasitoid will be influenced by factors acting on an evolutionary time scale that determine the set of hosts that can at least potentially support a parasitoid, and behavioral factors that determine whether a particular host is accepted for oviposition. I have already discussed the behavioral ecology of host acceptance in chapter 3 (sec. 3.1) and in this section will concentrate on the evolutionary aspects of this question. The evolution of ecological specialization has been reviewed by Futuyma and Moreno (1988).

Host range data is far harder to collect than data on parasitoid species load because it involves rearing many different species of host. Existing data are of two main types: (1) large catalogs of known host associations, frequently, although not always, concerned with limited taxonomic groups of parasitoids and very seldom including any quantitative information; (2) food webs containing information on all parasitoids attacking a restricted range of hosts, often in a single geographical area. Parasitoid webs may contain no quantitative information (connectance webs), or may be semiquantitative (giving the relative abundance of parasitoids per hosts) or fully quantitative (giving, in addition, the relative abundances of different hosts) (Memmott and Godfray 1993). Table 8.1 summarizes some of the major studies of parasitoid webs.

Information from host catalogs must be treated with extreme caution (Askew and Shaw 1986). The difficulties of parasitoid taxonomy, plus the risk of associating the parasitoid with the wrong host, render many large catalogs almost useless for ecological studies. Important exceptions are catalogs compiled by experts who have either personally examined reared voucher specimens, or have at least carefully weeded host records in the literature (e.g., Boucek and Askew 1968; Griffiths 1964–1968).

Most discussions of specialization assume a trade-off between the efficiency of resource utilization and the number of resources that can be used for food. In the context of parasitoid biology, the advantages of increased specialization might be more efficient host location, or a greater ability to overcome host defenses. The exact nature of the trade-off will depend on the biology of the parasitoid. In particular, those species with the most intimate connection with their hosts—endoparasitoid koinobionts—will suffer greater exposure to the physiological defenses of the host and in consequence are more likely to be specialists. I am aware of no experimental studies of host range trade-offs in parasitoids, and only a few studies using other organisms (Futuyma and Moreno 1988).

No parasitoid successfully parasitizes all hosts in the environment, and species that are attacked by the same parasitoid share certain characteristics. The two most important determinants of host range are host taxonomy and shared ecology (Askew and Shaw 1986; Shaw 1988).

The correlation between host taxonomy and parasitoid range has been demonstrated on numerous occasions. To choose but two examples, Griffiths (1964–1968) demonstrated in a survey of the dacnusine (Braconidae) parasitoids of agromyzids (primarily leaf-mining flies) a variety of levels of taxonomic specialization. Some species attack nearly all leaf-mining hosts while others are restricted to certain host genera, host species groups, or are completely restricted to a single host species. Among the parasitoids of cynipid gall wasps on oak (chiefly chalcidoids in the families Pteromalidae, Torymi-

Table 8.1

Some studies of parasitoid webs with a summary of their properties.

System	Feeding Niche	Trophic Levels	Number of Species	Quantification	Temperate/ Tropical	Spatial Summation	Temporal Summation	Reference
Oak galls	Gall wasp	8	61(62)	Semi	Temperate	Yes	Yes	Askew 1961
Leaf-miners on trees	Leaf miner	4	24(32)	Semi	Temperate	No	No	Askew & Shaw 1979
Aphids on trees	Sap sucker	3	21(38)	Connectance	Temperate	Yes	Yes	Rejmanek & Stary 1979
Gall midges on *Atriplex*	Gall fly	6	26(51)	Semi	Temperate	No	No	Hawkins & Goeden 1984
Gall midges on creosote	Gall fly	4	7(8)	Semi	Temperate	No	No	Force 1974
Weevils on dock	Stem miner	6	8(24)	Connectance	Temperate	Yes	Yes	Hopkins 1984
Mistletoe herbivores	Mixture of herbivores	5	3(23)	Connectance	Temperate	Yes	No	Whittaker 1984
Dry forest leaf miners	Leaf miners	4	86(220)	Fully	Tropical	No	No	Memmott, Gauld, & Godfray in prep.

Source: Memmott & Godfray 1993.

Key:

Feeding niche: Mode of feeding of herbivores in web.

Trophic levels: Maximum number of trophic levels in web.

Numbers of species: Number of parasitoids species in web and, in parentheses, total number of species.

Quantification: As described in the text, parasitoid webs can be classified as connectance, semi-quantitative, or fully quantitative.

Temperate/Tropical: Study conducted in temperate or tropical regions.

Spatial summation: Web contains data from one or more than one locality.

Temporal summation: Web contains data from one or more than one year.

dae, Eulophidae, and Eurytomidae), Askew (1961) found examples of species restricted to hosts of one species, one genus, and also of wasps that attacked most cynipid gall makers. Parasitoids also differed in their willingness to attack nongalling inquilines and to develop as facultative hyperparasitoids.

Correlations between host taxonomy and parasitoid host range can arise for at least two reasons. First, parasitoids may attack closely related hosts because they share similar physiological properties and defense mechanisms. Second, closely related parasitoids are likely to have similar biologies—for example, they are more likely to feed on the same host plant or to have similar feeding niches.

The importance of shared ecology is best illustrated by examples of unrelated hosts that share host plants or feeding niches and are attacked by the same parasitoid. As has already been discussed in chapter 2 (sec. 2.2.1), hosts that feed on the same food plant not infrequently share the same parasitoids (e.g., Picard and Rabaud 1914; Vinson 1981, 1985). Plant chemistry may influence parasitoid host range if hosts sequester toxins from their food plants. Chemical similarity is known to influence polyphagy at the herbivore trophic level, and chemical diversity has been linked with host range (e.g. Strong et al. 1984). Similar considerations may influence parasitoid host range. Feeding niche is also important. For example, some parasitoids such as the braconid *Colastes braconius* attack only leaf-mining insects but are able to develop on leaf-mining flies, beetles, moths, or sawflies, representatives of four insect orders. The ichneumonid *Scambus sagax* attacks moths, beetles, and sawflies associated with resinous galls, shoots, and cones of conifers, while the ichneumonid *Endromopoda detrita* is a parasitoid of moths, flies, and phytophagous chalcidoids in grass stems (Fitton et al. 1988). Stary (1970) records that the braconid *Ephedrus persicae* attacks most aphids feeding in galls and curled leaves but only attacks externally feeding species when they are sufficiently abundant to cause the leaf of the host plant to curl. Townes and Townes (1960) and Gauld (1984) describe how many pimpline ichneumonids have specialized ecological preferences but wide host ranges. For example, one species of *Dolichomitus* attacks only timber borers in certain tree species, but parasitizes both beetle and hymenopteran hosts. In a recent study, Hoffmeister (1992) surveyed the parasitoids attacking seven races or species of tephritid fly feeding in the fleshy seeds of a variety of trees, shrubs, and climbers in Europe. He found that host ecology—broadly defined as phenology, feeding habit, and host plant taxonomy—was more important than host taxonomy in determining the makeup of the parasitoid complex. Host phenology has frequently been shown to influence host range. Several parasitoids of cynipid gall wasps attack most hosts available at one time of year (Askew 1961). Finally, an experimental demonstration of the importance of host ecology is provided by Zwölfer and Kraus (1957), who transplanted pupae of a tortricid moth (*Choristoneura murinana*) that feeds on silver fir (*Abies alba*) into artificial leafrolls on oaks.

Even though oak and fir grow together, *C. murinana* is normally never attacked by a generalist ichneumonid (*Apechthis rufatus*), which is a common parasitoid of oak tortricids. However, *A. rufatus* was reared from about 10% of the *C. murinana* transferred to oak.

The importance of shared ecology must not be overemphasized. There are also many examples of parasitoids that attack one or a few closely related hosts in a wide variety of habitats. Price (1981) illustrates this point with the ichneumonid *Hoplismenus morulus* which attacks a variety of closely related butterflies (in the family Nymphalidae) throughout most of subarctic North America and in habitats as different as deciduous woodland and chapparal. The braconid parasitoid *Chorebus nana* attacks a closely related group of agromyzid leaf miners (the *Phytomyza obscura* group) on labiate plants both in marshes and on chalk downs (Griffiths 1964–1968, who also gives a number of similar examples).

There are other possible determinants of parasitoid host range. Specialization is promoted by environmental constancy, and thus narrower host ranges might be expected among the parasitoids of predictable hosts. Hosts feeding on early successional plants may be less predictable to parasitoids than those on late successional plants, and hence more specialists may be found on the second group (Askew and Shaw 1986; Hawkins et al. 1990; see also sec. 8.2.1). Holt and Lawton (1993) have recently argued that one parasitoid attacking two hosts may drive one species to local extinction (see also sec. 8.2.4), giving rise to apparent specialism which they term "dynamic monophagy."

Some predictions can be made about the relative host ranges of species with an intimate biochemical and physiological connection with their hosts (larval koinobiont endoparasitoids), and species that do not have to contend with active host defenses (larval idiobiont ectoparasitoids and species attacking nongrowing host stages). The former group should be relatively specialized and their host range will be strongly influenced by host taxonomy. The latter group should be less specialized and their host range will be influenced by both host taxonomy and host ecology. Thus (1) koinobionts should have fewer hosts than idiobionts; (2) pupal and egg parasitoids should be less specialized than larval parasitoids (Strand 1986); (3) the koionobiont parasitoids of taxonomically isolated hosts should attack few other species; and (4) the idiobiont parasitoids of ecologically isolated hosts should attack few other species. Idiobiont larval parasitoids more often attack hosts in concealed feeding niches where death or permanent paralysis is less likely to increase the risk of predation (Hawkins 1990). Hosts in concealed feeding niches will thus share more parasitoids than externally feeding species. There will be numerous exceptions to these broad generalizations. For example, many tachinid flies are koinobiont endoparasitoids yet can subvert the host immune system of a wide variety of species (sec. 6.3.1) and thus enjoy a remarkably broad host range.

The suggestion that koinobionts have broader host ranges than idiobionts has some empirical support. In informal surveys of parasitoids of lepidopteran and hymenopteran leaf miners, Askew and Shaw (1986), Pschorn-Walcher and Altenhofer (1989) and Sato (1990) all observed greater host ranges among idiobionts than koinobionts. Sheehan and Hawkins (1991) used data from the Canadian Forest Insect and Disease Survey (Bradley 1975) to compare the host range of two subfamilies of Ichneumonidae, the idiobiont Pimplinae and the koinobiont Metopiinae. Controlling for the data available on each species, they found wider host ranges among the Pimplinae. This study is valuable in providing quantitative data on host range, although it is also limited because it includes only two taxa and one evolutionary transition between the two types of biology. The host ranges of species within the two families are unlikely to be statistically independent data points because they may be influenced by other phylogenetically correlated aspects of the wasps' biology (Harvey and Pagel 1991).

The evolution of host range will affect the overall structure of parasitoid food webs. A broad host range will increase the connectance of the web, the fraction of all possible host-parasitoid interactions that are in fact observed, while the pattern of host specialization will determine the amount of compartmentalization, the degree to which the web can be deconstructed into isolated or semi-isolated subwebs (Pimm and Lawton 1980). Food web structure will also be influenced by competitive interactions among parasitoids and hyperparasitoids.

Finally, there are a variety of macroevolutionary aspects of parasitoid host range (Futuyma and Moreno 1988). In some host taxa, particular specializations appear to be taxonomically conserved: all Eucharitidae parasitize ants; the complete Opiine+Alysiine clade (Braconidae) are restricted to cyclorraphous Diptera, and the ichneumonid subfamily Ichneumoninae and the braconid subfamily Microgasterinae parasitize only Lepidoptera. In other groups, for example the Eulophidae and Pteromalidae, nearly all species parasitize a restricted set of hosts, yet the clade is not committed to any particular host group. Generalism also may be phylogenetically conserved. The braconid genus *Dacnusa* is comprised of many species specialized on particular agromyzid leaf miners, but the few species with wide host ranges are closely related. The study of these patterns among parasitoids, as well as the exploration of old questions such as the degree to which specialism is an evolutionary cul de sac, is in its infancy. It is unlikely that much progress will be made until good phylogenies are available for parasitoid taxa.

8.2.3 COMPETITION AND COEXISTENCE AMONG PARASITOIDS

Two species of competing animals cannot exist on identical resources in a constant environment (Volterra 1926; Gause 1934), yet most species of host are attacked by more than one species of parasitoid, often including several specialists. What are the ecological factors that allow the coexistence of competing parasitoids, and will natural selection promote or reduce the likelihood of coexistence? Where coexistence is not possible, which species ultimately triumphs? The analysis of competition between parasitoids is particularly complicated when host population densities are regulated by parasitoids. These questions also have a practical importance: there is a long-standing debate in the biological-control literature about the wisdom of introducing several parasitoids that might compete with one another (e.g., Turnbull and Chant 1961).

For coexistence to be possible, each species must increase in numbers when rare or, in other words, interspecific competition must be weaker than intraspecific competition. This is achieved most simply when the niches of two competitors only partially overlap. There are many ways in which parasitoid niches may differ. Generalist species may attack different spectra of hosts while both specialist and generalist parasitoids may search for hosts in different microhabitats or at different times of year. If the niches of two species are sufficiently similar so that coexistence is impossible, the winner is the species that can survive on the lowest equilibrium host density. The dominant species reduces the population of the host to so low a level that the subordinate species is unable to replace itself and so goes extinct.

Spatial or temporal variability may also promote species coexistence. Consider first spatial variability. If new patches containing hosts are continually appearing in the environment, and if the poorer competitor is better at locating these patches, then both species may be able to coexist. In these circumstances the poorer competitor is a fugitive species that maintains itself by discovering and exploiting new patches before they are located by the superior competitor (e.g., Skellam 1951; Hutchinson 1951). A different mechanism has been analyzed by Atkinson and Shorrocks (1981) and Hanski (1981). Two species of competitor can coexist in an environment made up of ephemeral patches if both species have a clumped distribution across patches, and if the two distributions are uncorrelated (or show little covariance; Ives 1988). Because members of the same species tend to occur in the same patch, intraspecific competition is stronger than interspecific competition, and coexistence is possible. Temporal variability may also promote species coexistence although I am not aware of this idea being applied to parasitoids. Chesson (1986) has shown how differential responses by two competitors to temporal environmental change may allow coexistence, while Armstrong and McGehee (1980) have studied a predator-prey model where the coexistence of two predators is allowed by their differential responses to temporal fluctuations in population density generated by predator-prey limit cycles.

Parasitoid biologists have often argued that the superior dispersal capabilities of poorer competitors are important in explaining coexistence (the fugitive species hypothesis). For example, Pschorn-Walcher and Zwölfer (1968) and Zwölfer (1971) call this mechanism "counter-balanced competition" and distinguish between "intrinsic competition" (i.e., ordinary competition) and "extrinsic competition" (the superior dispersal and host-finding prowess of the fugitive species). Zwölfer (1971), Schröder (1974), and colleagues have argued that the coexistence of parasitoids on a number of forest pests can be explained by counterbalanced competition and have classified the parasitoids of these insects as either intrinsically or extrinsically superior competitors. However, while it is relatively straightforward to determine intrinsic competitive superiority by examining the outcome of multiparasitism, it is far harder to demonstrate extrinsic competitive superiority. Experimental studies of comparative host location are needed to assess objectively extrinsic competitive superiority.

The coexistence of competing parasitoids is complicated when the host population is regulated by its parasitoids. A number of population dynamic models have examined the requirements for coexistence in these circumstances. Models by Hassell and Varley (1969), May and Hassell (1981), Hogarth and Diamond (1984), Kakehashi et al. (1984), and Godfray and Waage (1991) all assume some form of density dependence in the parasitoid populations that cause intraspecific competition to exceed interspecific competition. Hassell and Varley assumed mutual interference between conspecific parasitoids while May and Hassell, Hogarth and Diamond, and Kakehashi et al. either explicitly or implicitly assumed some form of heterogeneity in parasitoid attack: either the two parasitoids have clumped but independent distributions, or the poorer competitor is more efficient at locating hosts (as in the counterbalanced competition hypothesis). Kakehashi et al. also explicitly study a model where the two parasitoids have only partially overlapping niches. All these studies have modeled hosts and parasitoids with discrete and perfectly synchronized generations. Godfray and Waage (1991) briefly studied coexistence in a species with overlapping generations but without specifying the source of density dependence in the parasitoid population. Briggs (1993) also studied species with overlapping generations and modeled two parasitoids attacking successive host stages. She found that coexistence was possible only when the second parasitoid could successfully develop on hosts previously parasitized by the other parasitoid. Without this capability, the second species suffers very severe interspecific competition and can be driven extinct by the earlier-acting parasitoid. The aim of these studies was primarily to explore the problem of whether to introduce one or several species of parasitoids during a biological control problem. Their results suggest that the strategy which gives the greatest depression in host equilibrium abundance may depend quite critically on the biological details of the interaction.

So far I have assumed fixed differences in the ecologies of potential competitors and asked whether coexistence is possible. Another question is whether natural selection will tend to cause niche divergence or in other ways promote coexistence. This question has been widely studied in other areas of evolutionary ecology (e.g., Slatkin 1980; Milligan 1985), although as far as I am aware there are no studies that have specifically modeled parasitoid competition.

I turn now to empirical studies of competition in parasitoids. Competition among parasitoids can be investigated by experimental manipulation of parasitoid densities, or by looking for evidence of the influence of competition on community structure.

EVIDENCE

Very few studies have experimentally manipulated two or more parasitoids attacking the same host in controlled population experiments. Force (1970, 1974) inoculated greenhouse populations of gall midge (*Rhopalomyia californica*) with different combinations of its six natural parasitoids, two of which show extensive facultative hyperparasitism. As the experiments did not run long enough for an equilibrium to be obtained, the results are difficult to interpret. However, in experiments excluding the facultative hyperparasitoids, a combination of intrinsic rate of increase and competitive superiority appeared to determine the species that achieved numerical dominance. The inclusion of the facultative hyperparasitoids changed the pattern of species dominance as they attacked one primary host in preference to another, and even attacked parasitized hosts in preference to unparasitized ones.

Although not designed as ecological experiments, and thus unfortunately seldom replicated, interesting information can be obtained from biological-control releases. There are a few examples where one parasitoid has appeared to replace another species through competitive exclusion. A number of braconid wasps in the subfamily Opiinae were introduced to control the oriental fruit fly (*Dacus dorsalis*, Tephritidae) in Hawaii. The first species to become established was *Diachasmimorpha longicaudatus* in 1948. However, during 1949 this species was replaced by *Biosteres vandenboschi*, which was itself replaced in 1950 by *Biosteres arisanus*.[2] By 1951 the first two species had largely disappeared. The successive displacements by *B. vandenboschi* and *B. arisanus* were interpreted as being due to competition (Bess and Haramoto 1958; Clausen et al. 1965). DeBach (1974) notes that "in spite of the competition between these three species, each replacement of one by another was accompanied by a higher total parasitization and a greater reduction in the fruit fly infestation." This observation is in accord with the theoretical prediction that the competitively superior species reduces the resource to the lowest level. *B. arisenus* is an egg-pupal parasitoid, while *D. longicaudatus* and *B. vandenboschi* are larval-pupal parasitoids. The competitive superiority of the earlier-acting species is predicted by Briggs's (1993) age-structured models of parasitoid competition.

---

[2] There have been many taxonomic problems with the opiine parasitoids of Tephritidae, and this group is currently under intensive study by R. A. Wharton (1988 and included references). In most ecological texts, *longicaudatus, vandenboschi,* and *arisanus* are placed in the genus *Opius;* in addition, *arisanus* is called *oophilus.* The species *vandenboschi, arisanus,* and their relatives are not closely related to "true" *Biosteres* (parasitoids of anthomyiid leaf miners in temperate regions) and when more is known of their taxonomy will be placed in a separate genus (Wharton 1988).

The second classic example of competitive displacement among introduced parasitoids concerns the parasitoids of the citrus red scale (*Aonidiella aurantii*, Diaspididae), California's worst citrus pest until its successful control in the 1950s. The most important parasitoids are aphelinid wasps in the genus *Aphytis*. Before the Second World War red scale was attacked by *A. chrysomphali*, which failed to give adequate control. A second species, *A. lingnanensis*, was introduced in 1948 and led to the virtual extinction of *A. chrysomphali* from most areas. In 1957 a third species, *A. melinus*, was introduced. This species was discovered in the arid regions of Pakistan and northern India and was able to replace *A. lingnanensis* in the dry regions of central California although not in moist coastal areas (DeBach and Sisojevic 1960; DeBach and Sundby 1963; DeBach 1966). DeBach and colleagues suggested that *A. melinus* outcompeted *A. lingnanensis* because it was less severely affected by bad winter weather. More recently, the competitive interactions between these two species have been dissected by Luck and coworkers (Luck et al. 1982; Luck and Podoler 1985; Opp and Luck 1986; Luck 1990), who have put forward a different explanation. Both species show host-size specific sex allocation, laying male eggs on small hosts and female eggs on large hosts (see sec. 4.4). *A. melinus* oviposits female eggs on smaller hosts than *A. lingnanensis*, and because smaller hosts are always more abundant than larger hosts, it may drive the host to a density so low that *A. lingnanensis* is unable to replace itself, at least in areas that are climatically favorable to *A. melinus*.

The second way to assess the importance of competition is to look for evidence of its influence on community structure. There is a massive literature on this subject, especially concerning birds and mammals. The types of patterns that have been identified include (1) character displacement: niche divergence in regions of sympatry; (2) the regular distribution of niches along resource axes; (3) checkerboard distributions: the occurrence of two species individually in many localities but never or only occasionally together; and (4) community convergence: parallelism in community structure in different localities. The potential importance of interspecific competition was highlighted by the "MacArthur school" of population ecologists in the 1960s and 1970s, and it is probably true to say that the influence of this school led to a rather uncritical acceptance of many cross-species patterns as evidence for competition. A backlash came around 1980 with the arguments of the "Florida school" that evidence for competition should be tested against rigorously defined null models. Applying their methods to some of the most famous examples of patterns attributed to competition, members of the Florida school often rejected competition in favor of a null hypothesis containing purely statistical processes. More recently still, there has been a growing acceptance that many null models are too rigorous, and may lead to the erroneous rejection of competition as an organizing force. These issues are vigorously discussed in chapters of the book edited by Strong et al. (1985).

There is some evidence for the nonrandom distribution of parasitoid niches. For example, Askew and Shaw (1974) surveyed the parasitoids attacking leaf-mining moths of the genus *Phyllonorycter* (Gracillariidae) on British trees. The majority of parasitoids attack miners on more than one tree species, but one genus, *Achrysocharoides* (Eulophidae), consists of species largely confined to single-host plant genera. Askew and Shaw point out that there is a strong tendency for each host genus to have its own species of *Achrysocharoides*, a pattern that suggests the genus has divided up niche space under the influence of interspecific competition. However, a formal statistical test has not been performed and there are exceptions to the rule; for example, two species are found on miners of the oak genus, *Quercus*. In a larger survey of parasitoid communities on cynipid gall wasps and leaf-mining moths, Askew (1980) writes that "the host ranges of specific parasites, particularly of congeneric species, tend to be mutually exclusive and most communities studied support one species of specific chalcid or ichneumonid parasite in some abundance." Similarly, in reviewing the host associations of the many braconids in the *Chorebus ovalis/lateralis* complex, Griffiths (1964–1968) remarks that "all its species show a high degree of host specificity, but they are remarkably evenly spread over *Agromyza* and almost all Phytomyzinae [the major groups of leaf-mining agromyzid flies] . . . yet there are few hosts from which more than one species of the group can be bred." Finally, although very little is known about the host range of parasitoids in the tropics, there is one example of possible mutually exclusive specialist parasitoids. Gauld (1988b) discusses work of D. H. Janzen on ichneumonids in the *Enicospilus americanus* complex; Janzen found that "each of the reared species has only ever been bred from a single species of saturniid [moth larva], and no saturniid is known to be attacked by more than a single species in this complex."

While it is clear that many experts in the field believe they can observe community patterns that are probably attributable to competition, the evidence is really too meager to say anything definite about its role as a major selection pressure influencing parasitoid host range and community organization. However, I think it is useful to draw a distinction between two positions on this issue that might be called the weak and strong belief in the role of competition. Adherents of the weak position believe that interspecific competition is of frequent occurrence among parasitoids and is responsible for some patterns in community organization. Adherents of the strong position believe that it is the primary structuring force in parasitoid communities, largely responsible for their size and composition. Belief in the strong position allows a number of testable predictions to be made. First, communities of parasitoids on similar hosts in different geographic regions should be of the same size and made up of similar numbers of specialists and generalists, egg, larval, and pupal parasitoids. In other words, there should be strong community convergence. Second, there should be no vacant niches (Lawton 1983; Price 1984). If one species of host can support two specialist parasitoids, then hosts with similar biologies should also be able to support a pair of specialists.

Not surprisingly, there is good evidence to reject the strong position. Price (1980) discusses the distribution of alysiine parasitoids (Braconidae) of agromyzid flies using the data set collected by Griffith (1964–1968). There is enormous variation in the number of parasitoid species attacking different species of fly and very little evidence of convergent parasitoid community structure on this morphologically very similar group of hosts. One might quibble about whether it is fair to draw such conclusions from a data set that represents rearings from throughout Europe of only part of a parasitoid complex, and with many host associations represented by a single or very few records. However, my own rearings of 6000 agromyzid parasitoids from a single locality do not contradict Price's conclusions. Hawkins (1990) compared parasitoid species loads on similar hosts in different geographic regions. He found that parasitoid species load varied considerably on exophytic hosts, but was relatively constant on endophytic species. He tentatively suggested that this may indicate competition to be relatively more important on endophytic species but at best this is weak evidence for the strong position since other aspects of community structure on endophytic insects (for example, the ratio of koinobionts to idiobionts) varied geographically.

Another prediction of the strong position is that in cross-species comparisons, the abundance of a parasitoid should be inversely related to the number of other parasitoid species attacking the same host or hosts. Dean and Ricklefs (1979) attempted to test this idea using a large data set on the larval parasitoids of Lepidoptera collected by the Canadian Forest Insect Survey. They used two measures of parasitoid abundance, the number of reared individuals and the percentage parasitism, and controlled for a variety of extraneous variables such as the number of food plants used by the host, host abundance, and host aggregation. The hypothesis was tested by examining the partial correlation between parasitoid abundance and the number or abundance of other parasitoids on the same host. In nearly all cases positive rather than negative correlations were found. Their conclusions are summarized in the title of their paper: "Do parasites of Lepidoptera larvae compete for hosts? No!"

This conclusion was fiercely attacked by Force (1980) and Bouton et al. (1980). Force argued that competition between parasitoids might be important in many individual cases and yet still not show the communitywide patterns sought by Dean and Ricklefs. He also pointed out that interference competition would not generate the predicted patterns. Bouton et al. concentrated on the suitability of the data set and the appropriateness of the statistical analysis. Variation in sample size and inconsistencies in the treatment of gregarious and solitary hosts might have obscured any pattern in the data. The omission of egg and pupal parasitoid removes two important guilds of potential competitors. More seriously, treating host species as statistically independent data points cannot be justified. Species will vary in the susceptibility to parasitoid attack, which will tend to generate the positive correlations found by Dean and Ricklefs (1979). Dean and Ricklefs (1980) moderated their claims in a brief reply to their critics entitled: "Do parasites of Lepidoptera larvae compete for hosts? No evidence"—in my view, "no evidence" should be interpreted as a lack of suitable data, rather than positive evidence for the absence of competition.

In conclusion, few if any would defend the strong position that parasitoid community structure is dominated by competitive effects. Yet the strong position must not be used as an Aunt Sally to discredit the more moderate weak position which accords a lesser but still significant role to interspecific competition at the parasitoid trophic level. The true role of competition in parasitoid communities is most likely to be revealed by experimental studies of pairs and small groups of species, and by the detection of patterns in quantitative food webs describing small guilds of co-occurring parasitoids and hosts.

8.2.4 PARASITISM AND THE HOST NICHE

Herbivorous insects frequently appear to partition resources nonrandomly, and the same is true of other insects that serve as hosts for parasitoids. For example, herbivores are frequently restricted to one or a few host plants, and often feed only on restricted parts of the plant. Such specializations may simply reflect natural selection for efficient resource use; alternatively, competition between host species may restrict their realized niches. Yet another alternative is that natural enemies in general, and parasitoids in particular, influence the shape of the host niche.

How a parasitoid affects the evolution of its host's niche depends on how many other host species it attacks. It is perhaps useful to distinguish three cases: (1) attack by specialist parasitoids; (2) attack by very generalist parasitoids whose abundance is largely unaffected by the abundance of the focal host or of other hosts sharing the same microhabitat; and (3) attack by relatively specialized parasitoids that also attack other hosts in the same environment and whose abundance is determined by the sum of potential hosts in one microhabitat. The last case has received considerable attention because it can lead to the evolutionary divergence of host niches in a manner exactly parallel to competition.

In order for parasitoids to influence the evolution of a host's niche, the risk of parasitism must vary in niche space. There is abundant evidence to support this assumption (Lawton 1986). Herbivores feeding on different parts of the same plant very often suffer different levels of parasitoid attack (Vinson 1981, 1985; Price et al. 1980). Similarly, it has often been shown that parasitoid attack can depend on the host's food plant or, in the case of nonherbivorous insects, the microhabitat in which it feeds (Gilbert and Singer 1975; DeBach 1966; Weseloh 1976b; Price et al. 1980; Price 1981; Vinson 1981, 1985; Lawton 1986; van Alphen and Vet 1986). Host susceptibility may even depend on host plant clone (Price and Clancy 1986; Craig et al. 1990). Finally, parasitoid attack often varies at different times of the year (e.g., Myers 1981; West 1985; Weis and Abrahamson 1985; Clancy and Price 1986). There is thus little room to doubt that most hosts experience nonuniform levels of parasitism throughout their ecological range.

The ubiquity of the nonuniform risk of parasitism raises a number of evolutionary questions. First, given the status quo, how should hosts distribute themselves across the environment? Second, in the long term, how may the fundamental niche of a host be shaped by parasitism? Further problems are raised by the second question: how frequently do hosts and parasitoids coevolve, and can shared parasitism lead to the divergence of host niches?

HOST DISTRIBUTIONS MOLDED BY PARASITISM

If the risks of parasitism vary, why do not all hosts change their phenology, feeding habits, or location to minimize parasitoid attack? The obvious answer is that host fitness depends on more than parasitoid attack. In particular, resource quality, susceptibility to predation, and abiotic mortality may counteract any benefit of reduced parasitism. A number of studies have demonstrated trade-offs between parasitoid attack and other components of fitness. Myers (1981) has argued that the lasiocampid moth *Malacosoma californicum* feeds early in the year to avoid parasitism, even though food quality is poorer at the beginning of the season. The leaf-mining gracillariid moth *Phyllonorycter harrisella* normally mines the leaves of oaks during the summer months and is attacked by a large suite of parasitoids. West (1985) created an artificial generation early in the year and found the moth larvae suffered less parasitism and also grew better on the young succulent foliage. However, oak is attacked by many external folivores early in the season that happily eat mined leaves, causing a mortality that probably counters the other advantages to premature emergence. Gibson and Mani (1984) have shown that while Tiger butterflies (*Danaus chrysippus*) obtain protection from birds by feeding on alkaloid-rich milkweeds (Asclepiadaceae), they also suffer high levels of parasitism on these plants. Finally, Clancy and Price (1986) compared the phenology of the gall-making sawfly *Pontania* sp. on the leaves of arroyo willow (*Salix lasiolepis*) in two sites in Northern Arizona. At one site, parasitoids were rare and oviposition occurred early in the season; at the other site parasitoids were common and oviposition occurred late in the season. Clancy and Price (1986) suggested that late oviposition at the second site was an adaptive response to heavy parasitoid attack, a hypothesis supported by the observation that at this site early sawfly larvae were especially prone to parasitism. Late oviposition was not without costs as it may be difficult for the ovipositing sawfly to find leaves in the correct developmental stage, and there is also the risk that the leaves abscise before larval development is completed.

Although these arguments explain why hosts do not necessarily minimize parasitism, they still predict an optimum trade-off between the benefits of reduced parasitism and costs due to other processes. The question remains why variable levels of parasitism should be observed so frequently. Of course, some variability will arise by chance or because the herbivore cannot predict parasitoid distributions or is poorly adapted to a novel environment. However, the risk of parasitism, and also the risk of predation and the severity of resource competition, will be influenced by host density. A host may tolerate an increased risk of parasitism in an area with few competitors where it can monopolize limiting resources. Hosts are predicted to distribute themselves across the environment in an ideal free distribution—where no host can increase its fitness by changing position (Fretwell and Lucas 1970; Whitham 1980; Valladares and Lawton 1991). In reality, a population of animals can only be expected to approximate an ideal free distribution because an individual's assessment of the distribution of competitors and parasitoids is obviously limited.

In addition to its application to hosts, the ideal free distribution has also been used to describe the distribution of parasitoids in a patchy environment (sec. 3.1.1). If the scale of movement of host and parasitoid individuals is roughly the same, one can envisage a joint ideal free distribution where both hosts and parasitoids are distributed in the environment such that no individual of either species can increase its fitness by movement. If parasitoids can disperse further than hosts, or vice versa, the distributions will be more complicated, and the fitnesses of individuals of the less mobile participant may not be constant.

THE EVOLUTION OF THE HOST NICHE

Many host populations suffer severe mortality from parasitoid attack. It is reasonable to suppose that there will be strong selection in favor of any mutation that lessens the risk of parasitoid attack and that the avoidance of parasitism is an important determinant of the shape of host niches (Lawton 1986). This raises a variety of interesting questions: Can hosts evolve complete immunity from parasitism? What are the differences between the selection pressures exerted by specialist and generalist parasitoids? Can host evolution lead to the extinction of specialist parasitoids? How often do hosts and parasitoids coevolve? In discussing these issues, it is frustratingly difficult to move beyond speculation. The time scale of the processes involved normally precludes experimentation and the reconstruction of past selection pressures by comparative studies is fraught with problems: the "ghost of parasitism past" (Price 1990) is just as spectral an entity as the "ghost of competition past" (Connell 1980).

It is often suggested that parasitoids are involved in the evolution of concealed feeding niches by herbivores. In particular, parasitism is frequently invoked to explain the evolution of gall making (Askew 1961). Washburn and Cornell (1979) found that the toughness of the galls of a cynipid wasp affected the probability of parasitism, as did Weis (1982a, 1982b) working with a cecidomyiid fly. In the latter case, the hardening is partially caused by a symbiotic fungus. Weis and Abrahamson (1985) found that hosts (the tephretid *Eurosta solidaginis*) in thick galls on *Solidago altissima* were more likely to escape parasitism by the eurytomid *Eurytoma gigantea*. What is more, there appears to be genetic variability in gall size within populations of *Eurosta solidaginis*. In this species, the benefits of large gall size may be offset by increased bird predation (Weis and Abrahamson 1985; Weis et al. 1985). Price and Clancy (1986), working with the sawfly *Euura lasiolepis* on arroyo willow (*Salix lasiolepis*), also found evidence that parasitism by a wasp in the genus *Pteromalus* (Pteromalidae) led to selection for large galls. Understanding why larger galls do not evolve in these systems is an interesting challenge.

Another aspect of host biology that may be influenced by parasitoid attack is the length of time spent in developmental stages during which attack occurs. Cornell (1983, 1990) and Craig et al. (1990) argue that many gall-forming insects are only susceptible to parasitism during a relative short "window of vulnerability," while only certain instars of many species of hosts can be attacked by parasitoids. Standard life history theory predicts that natural selection will tend to minimize the time spent in these stages. Pimentel et al. (1978) found evidence that housefly populations exposed in the laboratory for a number of generations to high levels of parasitism by the pteromalid *Nasonia vitripennis* evolved to spend less time in the vulnerable pupal stage.

COEVOLUTION

In the past, coevolution has been used in a rather loose way to describe adaptations by one class of organism to competition or predation by a second class. Thus the swiftness of both antelope and cheetah is explained as due to the coevolution of predators and prey. Today, coevolution is normally used in a more narrow sense to refer to reciprocal adaptations between individual pairs of species. Under this definition, coevolution appears to be rather rare among predators and prey (Endler 1986): adaptations such as the hard shells of many mollusks and the speed of escaping antelope have arisen as defenses against broad guilds of predators. However, among hosts and true parasites or pathogens, coevolution appears to be much more common, and specific gene-for-gene interactions have been established in a number of cases. I have already discussed the possibility of physiological coevolution between hosts and parasitoids (sec. 6.3.3); is there any evidence for ecological coevolution?

It is hard to demonstrate coevolution among parasitoids and hosts, but the most convincing (although anecdotal) evidence comes from hosts that have adaptations that appear to have evolved under parasitoid selection pressure, but where a parasitoid has managed to keep up with the host. To give one example, the torymid wasp *Apocryptaphagus* sp. makes a large gall inside the figs of *Ficus hispidioides*. This species has probably evolved from a parasitoid or inquiline of pollinating fig wasps to become a true gall former and is now approximately five times the size of the pollinating agaonid. A possible reason why it increased in size and became a gall former was to avoid parasitism: the gall juts out into the interior of the fig, placing the torymid larva beyond the range of the ovipositors of the normal parasitoids of the pollinating agaonid. However, if this was the true reason for the change in feeding habit, it has been ultimately unsuccessful because the wasp is heavily parasitized by another torymid, *Apocrypta mega*, which is specific to *Apocryptaphagus* (the generic name of the host is unfortunate). This species too is much larger than others in the genus and appears to have increased in size and evolved a longer ovipositor in concert with its host (Godfray 1988).

Weis et al. (1989) make the interesting point that there may be reciprocal changes in host and parasitoid phenotypes without genetic change in both parties. As was mentioned above, the eurytomid *Eurytoma gigantea* attacks galls of the tephritid *Eurostoma solidaginis* and appears to exert a selection pressure favoring an increase in gall size. However, a larger gall results in a larger fly and, if parasitism occurs, a larger wasp. The length of the ovipositor in female wasps is allometrically related to body size, and hence the female *Eurytoma* emerging from large galls are able to attack hosts in larger galls. Thus selection for larger galls causes a response by the wasp, leading to selection for yet larger galls. However, the change in the wasp is nongenetic and purely an effect of the environment: an "arms race without coevolution."

APPARENT COMPETITION

When species compete for a limiting resource, an increase in the abundance of one competitor has a negative impact on the growth rate of other competitors. However, the same pattern may be found if two species share a common parasitoid. An increase in the abundance of one species will lead to more parasitoids, and as a consequence the second species may suffer higher rates of parasitism. Holt (1977) described interactions of this type between two species mediated by a shared natural enemy as apparent competition. Like traditional competition, apparent competition can result in the exclusion of one of the two interacting species (Holt 1977, 1984). Apparent competition can also lead to selection for niche divergence in the two host insects.

Parasitoids can cause host niche divergence in two ways. First, hosts attacked by different parasitoid species may experience different selection pressures and so may evolve in different directions. In this case divergence is an accidental consequence of adaptation to parasitism. Second, when hosts share parasitoids, there may be direct selection for niche divergence to reduce apparent competition. Lawton and colleagues (Lawton 1978, 1986; Lawton and Strong 1981; Jeffries and Lawton 1984; see also Askew 1961, Gilbert and

Singer 1975, Zwölfer 1975, Price et al. 1980, Freeland 1983, and Gilbert 1984 for related ideas involving parasitoids, and Jeffries and Lawton 1984 for a history of the concept of apparent competition) have stressed the potential importance of apparent competition in structuring host communities. They argue that host species compete for "enemy-free space"—niche space where they obtain respite from shared natural enemies. Apparent competition provides an explanation for nonrandom patterns in the structure of communities of herbivores that appear not to compete directly.

There are several ways that ideas about apparent competition may be tested. First, it is important to demonstrate that the increased abundance of one species does affect the growth rate of another species when the only interaction between the species is through shared parasitoids. Second, it may be possible to demonstrate that patterns in herbivore community structure are caused by apparent competition. For example, an observation that herbivore guilds that are heavily attacked by natural enemies tend to show less niche overlap than those attacked by few parasitoids would be consistent with apparent competition. Gilbert (1984) suggests that the community structure of tropical *Heliconius* butterflies may be influenced by apparent competition, although more work is needed to exclude other explanations. The unambiguous association of patterns of niche differentiation with natural enemies is a very difficult task.

I know of only one experimental study that provides firm evidence of apparent competition mediated by parasitoids. The variegated leafhopper (*Erythroneura variabilis*), which feeds on the leaves of cultivated grapes, invaded the San Joaquin Valley, California, in 1980. Its spread was associated with a marked decline of the grape leafhopper (*E. elegantula*), a species with which it shares a common egg parasitoid, the mymarid *Anagrus epos*. Settle and Wilson (1990) conducted a series of experiments to determine whether the decline of the grape leafhopper was due to direct competition or to apparent competition via *A. epos*. They found that the two species did compete, but that intra- and interspecific competition was very similar and unlikely to explain the marked drop in numbers of the grape leafhopper. However, while the invading variegated leafhopper lays its eggs deep in the leaf mesophyll, the grape leafhopper places its eggs just below the leaf's epidermis. As a result, the grape leafhopper is far more susceptible to *A. epos* than the variegated leafhopper. The reason grape leafhopper numbers are depressed is that partial protection from parasitism allows relatively high population densities of variegated leafhoppers which support *A. epos* populations that preferentially attack the grape leafhopper. A secondary affect of the parasitoid is that by reducing grape leafhopper numbers below their carrying capacity, invasion by the variegated leafhopper is not hampered by resource competition (Settle and Wilson 1990).

Apparent competition can also prevent the establishment of a species (Goeden and Louda 1976). Zwölfer (1979, and quoted by Lawton 1986) was unable to set up experimental colonies of the thistle-head tephritid *Urophora cardui* in several localities outside its normal range because of severe attack by the generalist eurytomid *Eurytoma robusta*. Gilbert (1984) found that glasshouse cultures of rare *Heliconius* butterflies cultured with common congeners could be destroyed by shared egg parasitoids (*Trichogramma* sp.). There are several examples in the biological control literature where the successful control of one host species depends on the presence of alternative hosts in the same environment (Lawton 1986). It is possible that apparent competition may be involved in these interactions, although the more likely (and traditional) explanation is that an alternative host must be present to prevent the parasitoid dying out through lack of hosts at certain times of the year.

The larvae of many butterflies in the family Lycaenidae are tended by ants and enjoy protection from parasitoids and predators (Atsatt 1981; Pierce and Mead 1981). In a comparative survey, Pierce and Elgar (1985) found that ant-tended species had a greater host range than species not tended by ants. There are many possible explanations for this observation, one of which is that freedom from apparent competition allows the colonization of a greater variety of host plants (Pierce and Elgar 1985; Lawton 1986).

8.2.5 TRI-TROPHIC INTERACTIONS

The influence of plants on the interactions between parasitoids and herbivorous insects has been repeatedly discussed throughout this book and has been reviewed by Price et al. (1980) and Price (1981). Parasitoids use stimuli emanating from the plant as aids to host location (sec. 2.2.2), and plant species and plant structure frequently influence the risk of parasitism (sec. 8.2.4). The growth form and antiherbivore defenses of the plant have major effects on the insects that feed on it, which in turn may influence the parasitoid species load and percentage parasitism of the herbivore (sec. 8.2.1). The quality of a host for a parasitoid will also be affected by the nutritional status of the plant and whether the plant contains toxins that have been sequestered by the host (sec. 6.1.3). Finally, a parasitoid may find it easier to search some types of plants than others. In particular, host location may be difficult on plants covered in hairs or trichomes, or that have sticky glandular secretions. The walking speed of the aphelinid whitefly parasitoid *Encarsia formosa* is dramatically reduced on hairy as opposed to glabrous varieties of cucumber, and biological control is unsuccessful on these plants (Hulspas-Jordaan and van Lenteren 1978; see also Obrycki 1986). Variation in plant characteristics may also strongly influence the population dynamics of host-parasitoid interactions (van Emden 1966; Lawton and McNeill 1979). For example, poor plant quality may reduce the intrinsic rate of increase of the host and allow or improve biological control. Finally, plants other than the food plant of the host may influence parasitism, either by masking volatiles produced by the food plant or host, or by providing food for parasitoids such as honeydew or nectar (Atsatt and O'Dowd 1976; Price et al. 1980; Price 1981).

There is thus no doubt that the plant has a huge influence on the evolution-ary and behavioral ecology of host-parasitoid interactions. A less well under-stood problem is to what extent selection acts directly on the plant to increase the success of its herbivores' parasitoids. The benefits of increased parasitoid efficiency may be either (1) an immediate decrease in herbivore attack, or (2) a reduction in the future population density of herbivores. Parasitism does not always reduce the damage done by a herbivore. Many koinobiont parasitoids allow their hosts to complete feeding before they are killed, while some gre-garious species cause their hosts to feed through one or more supernumerary instars (sec. 6.4.1; Price et al. 1980). Long-term reductions in herbivore popu-lation density may benefit the individual plant if colonies of herbivores persist for many generations on a single food plant (for example, many aphids, mealy bugs, scale insects, and other homopterans), or if the plant population is com-posed of closely related or clonal individuals.

There are various ways a plant may attract parasitoids and increase their efficiency. Many parasitoids feed as adults and it is known that they are at-tracted to flowers and to extrafloral nectaries (e.g., van Emden 1963; Leius 1967; Syme 1977). Shahjahan (1974) showed that the higher parasitism of tarnished plant bugs (*Lygus lineolaris*) on *Erigeron* spp. compared with other plants occurred because the braconid *Leiophron pallipes (=Peristenus pseudo-pallipes)* was attracted to their flowers. The main function of flowers is of course to attract pollinators, and I know of no suggestion that the attraction of parasitoids has played a role in floral evolution. The evolution of extrafloral nectaries is a better candidate to be influenced by parasitoids (Gilbert 1975; Atsatt and O'Dowd 1976; Smiley 1978; Price et al. 1980; Koptur 1985). How-ever, the prime purpose of these structures in many plants is to attract ants that protect the plant from harmful insects. Indeed, herbivores that are tolerated by ants such as honey-dew-producing Homoptera and ant-associated lycaenid butterflies appear often to enjoy protection from their parasitoids (Atsatt 1981; Pierce and Mead 1981). In a study over an elevational gradient in Costa Rica, Koptur (1985) found more parasitoids attracted to nectar sources at higher altitudes where ants were less common.

It is clear that parasitoids frequently use volatile chemicals emanating from the plant to locate sites where hosts may be found. Might the plant produce volatile chemicals deliberately to assist the parasitoid? While certainly possi-ble, the drawback of this strategy to the plant is that the same chemicals might be used by its herbivores. The advantage of this strategy thus depends on a delicate balance of costs and benefits. A better strategy might be to produce volatile chemicals only in response to herbivore feeding. Recently, it has been shown that plants produce specific volatile chemicals (terpenoids and indole) when fed on by herbivores and that the herbivores' parasitoids uses these chemicals in host location (Turlings et al. 1989, 1990b, 1991a, 1991b; Tur-lings and Tumlinson 1991, 1992; sec. 2.2.2). Similar tritrophic interactions have been discovered between plants, predatory mites, and phytophagous mites (Dicke et al. 1990a, 1990b). However, terpenoids can act as a feeding deterrent for herbivores and may also be involved in wound healing and as a protection against pathogen infection. Turlings and Tumlinson (1991) suggest that the primary function of these induced chemicals is defense against herbivores and pathogens, although the possibility of a secondary benefit for the plant of attracting parasitoids is not ruled out.

New research will almost certainly demonstrate more examples of possible chemical communication between the first and third trophic levels and suggest new areas of evolutionary inquiry (Vet and Dicke 1992). For example, Sabelis and de Jong (1988) and Dicke and Sabelis (1989) have modeled the release of volatile chemicals that attract predators and parasitoids by plants. If there are costs to the production of the chemical, selection may favor a plant to rely on its conspecific neighbors to attract herbivore natural enemies. Obviously the benefits of this strategy are frequency dependent and the ESS is often a mixture of individuals that do or do not release volatiles.

Many trees and shrubs contain tannins, resins, and silica that reduce the digestibility of their leaves. Feeny (1976), Rhoades and Cates (1976) and others have suggested these chemicals are a defense against herbivores. However, decreasing the nutritive value of the leaf may actually cause a herbivore to increase its consumption of plant material so the value of these compounds to the plant is not immediately obvious. One idea is that the main role of substances that reduce digestibility is to slow development and thus increase the probability that the herbivore is killed by its natural enemies (Feeny 1976; Moran and Hamilton 1980).

Atsatt and O'Dowd (1976) suggest that the availability of parasitoids may influence plant phenology. Wild grape and wild blackberry grow together and are attacked by leafhoppers that share the same egg parasitoid, the mymarid *Anagrus epos* (Doutt and Nakata 1973). The leafhoppers feed on young leaves and Atsatt and O'Dowd suggest that the main leaf flush of wild grape has evolved to occur at a time when many parasitoids are hatching from the leafhoppers on wild blackberry whose leaf flush occurs earlier in the year.

8.3 Parasitoid Diversity

I began this book by commenting on the large number of parasitoid species that have been described and the still larger number suspected to remain undescribed. It seems clear that the ancestral hymenopteran and dipteran parasitoids found themselves in a relatively unexploited adaptive zone. The resultant adaptive radiation onto different host species, possibly occurring simultaneously with the adaptive radiation of the modern insect orders, is responsible for the huge number of species we observe today. In this section I ask whether there is anything special about the parasitoid way of life, or the natural history of parasitoid taxa, that may have led to particularly high parasitoid diversity. For example, are speciation rates higher among parasitoids, and are certain mechanisms of speciation found more frequently among parasitoids? I also review arguments suggesting that unlike most other major groups of insects, parasitoid diversity does not increase toward the tropics. Finally, I discuss studies that have sought to explain major evolutionary changes within parasitoid lineages, for example radical host shifts and the transition between endoparasitism and ectoparasitism.

8.3.1 WHY ARE THERE SO MANY SPECIES OF PARASITOIDS?

Speciation leads to an increase in parasitoid diversity while extinction leads to a decline. An important issue is whether the rates of speciation and extinction are influenced by parasitoid diversity. If this is so, parasitoid diversity may achieve an equilibrium where, although new species may arise and others go extinct, overall diversity is constant. If this is not so, diversity will rise or fall determined solely by the relative rates of speciation and extinction. Competition for hosts among parasitoids is the most likely process to cause extinction or speciation rates to be functions of parasitoid diversity. However, as reviewed above (sec. 8.2.3) there is no consensus about the importance of interspecific competition for parasitoid population and community dynamics, let alone macroevolutionary dynamics. The question of whether parasitoid diversity is or is not at equilibrium is thus probably unanswerable, at least at present. Perhaps the most we can ask is whether parasitoid taxa are likely to display high rates of speciation or low rates of extinction, two factors that will lead to high parasitoid diversity under both equilibrium and nonequilibrium models. Even these questions are difficult: speciation is "more thoroughly awash in unfounded and often contradictory speculation than any other single topic in evolutionary theory" (Futuyma 1983)!

There are probably three main forms of speciation in parasitoid taxa. First, a population may be divided by a geographical barrier and evolution then take independent courses in each subpopulation. If the barrier is removed and the two populations re-meet, speciation will have occurred if the populations are reproductively isolated (allopatric speciation). Second, speciation may occur if gene exchange is in some way prevented between populations attacking different hosts in the same geographical locality (sympatric speciation). Finally, and really a special case of the other two mechanisms, parasitoids may speciate at the same time and in the same way as their host, a process called co-cladogenesis.

The parallel evolution of hosts and parasites is sometimes called Fahrenholtz's rule, and there are good examples from chewing lice (Mallophaga: Trichodectidae) attacking gophers and pinworms (*Enterobius*) attacking primates (Hafner and Nadler 1988; Brooks and Glen 1982; Brooks 1988; Harvey and Keymer 1991). There is some evidence for co-cladogenesis in ichneumonid parasitoids of larval sawflies (Pschorn-Walcher 1957, 1965) and braconid parasitoids of aphids (Mackauer 1967). However, the general opinion among parasitoid systematists is that co-cladogenesis is of relatively minor importance in parasitoid phylogeny.[3]

Are parasitoids particularly susceptible to the geographical division of their range and consequent allopatric speciation? Most parasitoids are relatively specialized animals with narrow ecological tolerances. The resources they use, their hosts, are likely to have a fragmented distribution rendering them prone to allopatric speciation (e.g., Futuyma and Moreno 1988). Against this argument is the fact that many parasitoids, despite their small size, disperse widely and are often found a great distance from their normal habitats (Askew 1968b). I am aware of no evidence from parasitoids on whether genetic divergence following geographic isolation is most frequently due to natural selection or genetic drift.

One of the most contentious issues in evolutionary biology is the possibility of sympatric speciation: the genetic divergence of two populations without geographical barriers (Bush 1975; Futuyma and Mayer 1980; Feder et al. 1988; Tauber and Tauber 1989; Coyne 1992). Most work in this area concerns phytophagous insects: if some individuals begin feeding on a novel host plant and mate with other individuals with the same feeding habits, the new subpopulation may become genetically isolated from the parent population. It is quite possible that similar processes happen in parasitoids. One aspect of the biology of many parasitoids that increases the likelihood of sympatric speciation is the dependence of parasitoid phenology on host phenology (sec. 7.7.1). If part of a parasitoid population colonizes a new host species, and if as a consequence they tend to emerge at a different time of year and mate with individuals from the same host, then genetic divergence is more likely to occur. Until recently, it was widely thought that partial genetic isolation might be converted to total reproductive isolation by natural selection, reducing the number of less fit interstrain matings, a process called reinforcement. However, recent theoretical studies have cast doubt on the importance of reinforcement in nature. One unsatisfactory aspect of the controversy about sympatric speciation is that it is always possible to invent a scenario involving allopatric isolation to explain any putative case of speciation without geographic barriers. This has led Coyne (1992) to remark that it may be "impossible to estimate the importance of sympatric speciation in nature."

Some specific aspects of hymenopteran biology may influence the rate of speciation of hymenopteran parasitoids. First, their haplodiploid genetics can

---

[3] This statement is based on the interrogation of three senior parasitoid systematists at the Natural History Museum, London.

influence the speed of evolution and thus the length of time two populations need be separated for reproductive isolation to occur. The rate at which favorable mutations are fixed is probably higher in haplodiploid species than diplo-diploid species (Hartl 1972; Crozier 1985). Second, Askew (1968b) has suggested that the frequency of sibling mating among hymenopteran parasitoids, especially chalcidoids, may influence speciation rates. The ecology of parasitoids increases the likelihood that members of one family mate among themselves and this "will have the effect of restricting gene flow in the population, . . . and of promoting speciation in populations exposed to disruptive selection pressures." Another way that sibling mating (and the female-biased sex ratios that often accompany it) can influence speciation is by reducing the effective population size and by increasing the rate at which random genetic drift leads to reproductive isolation. A third consideration is the possibility that the incompatibility microorganisms found in *Nasonia vitripennis* (sec. 5.1) are widespread in parasitoids. Infection of two populations of one species by different strains of the microorganism can cause instantaneous genetic isolation and speciation. Incompatibility microorganisms are now known from five orders of insects (Werren 1991; O'Neill et al. 1992), and more information about their distribution in parasitoids is urgently required.

I finish this section by briefly considering extinction, the process that destroys diversity. Low extinction rates might contribute to high parasitoid diversity. Parasitoid extinction has received little attention, although one factor that may increase the rate of allopatric speciation, the fragmentation of parasitoid resources, might also act to increase the rate of extinction. Askew (1968b) has argued that the frequency of sibling mating among hymenopteran parasitoids may decrease the likelihood of extinction at low population densities due to failure to find a mate. In conclusion, we do not know why there are so many species of parasitoids. A partial answer is probably because there are so many species of host but exactly how parasitoid speciation and extinction interact with host diversity to produce the million or more species of parasitoids on earth today is shrouded in uncertainty.

8.3.2 THE DIVERSITY OF TROPICAL PARASITOIDS

There had been a tacit assumption that parasitoid diversity, like the diversity of their hosts, increased toward the equator until Owen and Owen (1974) published the results of comparative sweep-net sampling in England and tropical West Africa. They found that the diversity of Ichneumonidae was approximately the same in Africa and in England. This observation was confirmed by Janzen (1973), Janzen and Pond (1975), and Janzen (1981), also using sweep nets, and later by Gauld (1986b, 1987), Gauld et al. (1992), Noyes (1989), and Askew (1990) in comparisons using other techniques. Janzen and Pond (1975) and Janzen et al. (1976) found other parasitoid taxa, in addition to ichneumo-nids, to be less diverse in their tropical sweep-net samples. The Owens' and Janzen's observations have been criticized on several grounds (Morrison et al. 1979; Hespenheide 1979). First, the Ichneumonidae may be unrepresentative of other parasitoid taxa. It is quite common for some insect herbivore taxa to decrease in abundance toward the equator, for example sawflies and the pre-dominantly leaf-mining flies in the family Agromyzidae, even though overall herbivore diversity is much higher in the tropics. Perhaps the Ichneumonidae are an equivalent exception among the parasitoids. Second, sweep sampling is likely to underestimate the abundance of smaller species of parasitoids that may compensate for the rarity of ichneumonids (Hespenheide 1979). Recent surveys of parasitoids on the tropical island of Sulawesi (Noyes 1989; Askew 1990; see also table 1.1) also suggest high diversities in the tropics of groups of small parasitoids such as the chalcidoid families Encyrtidae and Apheli-nidae. To conclude, there is a clear trend in the Ichneumonidae for diversity to decrease toward the tropics. The diversity of some groups of parasitoids in-creases toward the tropics but overall the weight of evidence points to either a decrease in parasitoid diversity toward the equator, or at the very least to a much lower increase in parasitoid diversity in comparison to that of their hosts.

Putting on one side the empirical problem of demonstrating the presence and magnitude of the trend, a number of explanations have been put forward to explain a dearth of parasitoids in the tropics. Janzen and Pond (1975) and Janzen (1981) proposed that as the diversity of hosts rises toward the equator, the density of each host population drops until a point is reached when they are too rare to support specialist parasitoids. This idea, known as the resource fragmentation hypothesis, predicts that specialist parasitoids should be less common in the tropics. Morrison et al. (1979) suggest that resource fragmenta-tion may lead to selection in favor of egg parasitism because eggs are the most abundant of host stages. Janzen (1973) and Gauld et al. (1992) have also ar-gued that the reduced seasonality of hosts in the tropics offer fewer and less intense seasonal peaks during which parasitoids can find otherwise rare hosts.

A different explanation has been given by Rathcke and Price (1976). They pointed out that predation on herbivores in the tropics is typically greater than in temperate regions, in large part because of the abundance of ants. They argued that parasitized herbivores are predated more often than healthy indi-viduals, and thus tropical parasitoids may suffer particularly high levels of juvenile mortality. Exactly how increased parasitoid mortality decreases spe-cies diversity was not specified, although presumably there is an increase in the likelihood of parasitoid extinction or a decrease in the probability of successful speciation. Hawkins et al. (1992) have suggested that this hypothesis applies more to exophytic than endophytic herbivores because the latter are frequently protected from predation. There are many variants of the predation hypothesis. Gauld (1987) has argued that the chief predation pressures may be experienced by adult rather than immature parasitoids. In support of this argument, tropical ichneumonids display a variety of antipredator adaptations that are seldom seen in temperate species, and searching for hosts at night is also relatively more common. Alternatively, parasitoids may be affected indirectly by predators through competition for hosts. Ants in particular are likely to be important competitors in the tropics. In both temperate and tropical regions, most parasitoids have to pass periods of time when their hosts are not available. In temperate regions, this period normally coincides with the cool winter when predators and hyperparasitoids are also inactive. In the tropics, predators and hyperparasitoids may be active all year round, and the more intense predation might influence parasitoid diversity (Janzen 1973; Gauld et al. 1992).

Gauld (1986b) attempted to distinguish among the hypotheses by examining in detail the Australian ichneumonid fauna which declines slightly in diversity toward the tropics. Part of the reason is the absence of sawfly parasitoids in tropical regions where their hosts are scarce or missing. Gauld found that specialist parasitoids decline in frequency toward the tropics as predicted by the resource fragmentation hypothesis. However, parasitoids of concealed hosts are also more common in tropical areas, a prediction of the predation hypothesis since these hosts are protected from ants and other predators. There is some evidence that the few specialist parasitoids of concealed pupae are less abundant in the tropical parts of Australia, a trend predicted by the resource fragmentation hypothesis but not the predation hypothesis. Hawkins et al. (1992) have also tried to distinguish between these ideas by comparing the parasitoid species loads of hosts in the southern and northern regions of North America. Although not a true temperate/tropical comparison, their study covered a wide latitudinal band from the subarctic to the subtropical. They argued that a drop in the number of specialists (equated with koinobionts) per host toward the south would be evidence for the resource fragmentation hypothesis. A disproportionate drop in the number of specialist parasitoids on exophytic hosts would support Rathcke and Price's predation hypothesis. However, their results (see also Hawkins 1990) gave strong support to neither hypothesis, although an increase in the number of tachinids toward the south is consistent with the resource fragmentation hypothesis because this family is unusual among koinobionts in typically having large host ranges.

Gauld et al. (1992) have recently suggested another reason why parasitoids may be less diverse in the tropics. They point out that tropical woody plants tend to have more chemical toxins than their temperate counterparts. In consequence, their herbivores also tend to contain more toxins, requiring their parasitoids to have special adaptations to cope with these potentially harmful chemicals. The difficulty of adapting to feed on such hosts limits the numbers of generalist parasitoids and reduces the probability of host transfer by specialists, the two processes acting together to reduce parasitoid diversity in the tropics. Gauld et al. (1992) call this the "nasty host hypothesis." They predict that tropical parasitoid diversity will be most impoverished in species attacking the best-defended hosts (external folivores) and least impoverished in species attacking poorly defended hosts such as insect eggs and herbivores which feed on parts of the plant that lack toxins.

8.3.3 EVOLUTIONARY INNOVATIONS

The history of a parasitoid lineage often involves radical biological innovations. The innovation might be the switch to a taxonomically unrelated host, or a shift in the type of habitat used for host location. Alternatively, there may be changes in life history, for example between endoparasitism and ectoparasitism, between being a koinobiont or an idiobiont, and between primary parasitism and hyperparasitism. Such radical innovations occur infrequently in the evolutionary history of a clade, and it is often difficult or impossible to reconstruct the specific selection pressures (if any) responsible for the rare transitions. Parasitoid systematists frequently suggest plausible evolutionary pathways that link species with diverse biologies or that attack disparate host taxa. Although sometimes informative, the problem with this approach is the difficulty of falsifying often informally expressed hypotheses. The adoption of phylogenetic methods by many parasitoid systematists has led to a more scientific approach to the study of major evolutionary innovations. At the moment, the lack of good phylogenies hinders the widespread application of these methods, and much progress is to be expected in the coming decade or so. My aim in this section is to give a flavor of some of this work, but with no pretense at comprehensiveness.

TRANSITIONS BETWEEN HOST TAXA AND HOSTS WITH
DIFFERENT ECOLOGIES

Hymenopteran parasitoids probably evolved from a symphytan ancestor that switched from feeding on plant tissue and any insect larvae it happened across, to the obligate consumption of another insect (sec. 1.5.1). Thus the first hymenopterous parasitoids were probably ectoparasitoid idiobionts (killing or permanently parasitizing their hosts) that attacked insects concealed in plant tissue. This assumption is supported by a study of the Ichneumonoidea (Ichneumonidae + Braconidae) where the most plesiomorphic (primitive) groups tend to have this biology. Using a mixture of formal phylogenetic methods and traditional evolutionary systematics, Gauld (1988a) has identified several evolutionary pathways originating with this ancestral life history which have occurred on a number of separate occasions in the Ichneumonoidea:

1. From parasitoids of immature boring insects to parasitoids of the larvae of aculeate wasps nesting in abandoned tunnels in plant tissue; to parasitoids of the larvae of aculeate wasps nesting in other situations such as burrows in the ground or in nests constructed by the adult.

2. From parasitoids of immature boring insects to parasitoids of the cocoons of species pupating in plant tissue; to the cocoons of species pupating in more exposed sites on the plant; to cocoons or cocoonlike structures in a variety of situations. The latter type of host includes the cocoonlike egg sacs of spiders which are exploited by at least three unrelated groups of ichneumonid (Fitton et al. 1987).

Both evolutionary pathways illustrate how shared taxonomy and shared ecology interact to influence evolutionary innovations. Other patterns can be identified within koinobiont endoparasitoids. A number of ichneumonid clades are restricted to certain taxa such as the sawflies and the cyclorraphous Diptera. The most primitive members of these groups attack phytophagous hosts such as leaf miners (Gauld 1988a) in feeding niches where taxonomically disparate hosts often have highly convergent ecologies. In addition, leaf miners and other endophagous hosts with confined feeding sites are particularly susceptible to parasitism by relatively unspecialized parasitoids. The homogeneous biology of groups such as leaf miners offer a gateway through which parasitoid lineages can transfer from one taxon of host to another.

Relatively few parasitoids attack adult insects. A major exception is the braconid subfamily Euphorinae which parasitizes a variety of adult insects, especially beetles. Shaw (1988) constructed a phylogenetic hypothesis for the subfamily using cladistic techniques and based on morphological characters. He then used the phylogeny to examine host relations within the group. The basal group attacks adult chrysomelid beetles which feed on plants; in many species adults and larvae feed together, and this close association may have allowed the transition from larval to adult parasitism. Shaw surveyed the host relationships within the family and suggested that from attacking adult phytophagous beetles feeding externally on plants, there had been host shifts to predatory beetles and phytophagous hemipterans. Several lineages have switched from attacking externally feeding beetles to parasitizing concealed species. One strange group attacks adult hymenopterans such as bumblebees, and Shaw speculates that a host shift may have occurred from adult beetles feeding on plants to adult bees in the same habitat.

TRANSITIONS BETWEEN ECTOPARASITISM AND ENDOPARASITISM, AND BETWEEN BEING AN IDIOBIONT AND A KOINOBIONT

Idiobiont ectoparasitism is almost certainly the plesiomorphic state in most if not all parasitoid lineages. This life history makes the least demands on any insect adapting to parasitism. How did more specialized life histories such as endoparasitism, and especially koinobiont endoparasitism, evolve from these simpler progenitors?

The distinction between koinobionts and idiobionts is less clear-cut in egg and pupal parasitoids than in larval parasitoids, and the transition between the two states is probably easier in parasitoids attacking nongrowing host stages.

Pupal endoparasitoids are common in several parasitoid families, and at least in the Ichneumonidae probably evolved from pupal ectoparasitoids. Gauld (1988a) argues that the switch from pupal ectoparasitism to endoparasitism is associated with parasitism of hosts in flimsy cocoons that provide poor protection from predators, hyperparasitoids, and the environment. Once pupal endoparasitism is established, radiation onto hosts with naked pupae is possible. A few pupal endoparasitoids probably evolved from larval-pupal endoparasitoids that attacked the host at successively later stages of its life cycle (for example, some eulophid parasitoids of leaf-miner pupae).

Koinobiont endoparasitism is arguably the most specialized form of parasitism. Did the evolution of endoparasitism precede the adoption of a koinobiont life history or vice versa? There is phylogenetic evidence for both evolutionary pathways. In the Ichneumonidae, the most primitive members of several koinobiont endoparasitoid clades (for example, the subfamily Ichneumoninae) are pupal endoparasitoids. Gauld (1988a) suggests that larval koinobiont endoparasitism arose through selection to attack progressively earlier stages in the host's life cycle. Adaptations by the larvae to cope with the active defenses of a nonparalyzed host would have to evolve concomitantly with the switch to earlier attack.

Shaw (1983) has hypothesized an evolutionary pathway in rogadine braconids from larval idiobiont ectoparasitoids to koinobiont ectoparasitoids, and then to koinobiont endoparasitoids. The hosts of some rogadine braconids feed in relatively exposed sites (leaf mines) and then move to a safe concealed site for pupation (outside the mine). There is thus an advantage in being a koinobiont and delaying the immobilization of the host until it has reached the secure haven. This strategy is most easily achieved by koinobiont ectoparasitism, which does not require the immature parasitoid to face the host's immune system. However, koinobiont ectoparasitism is a risky strategy because it is relatively easy for the parasitoid egg to be dislodged from an active larva. Selection will thus favor oviposition of eggs just below the skin or in host organs where they are protected from hemocytes. Once internal oviposition has evolved, natural selection can act to improve the parasitoid's response to the host's defenses and so ultimately to produce full koinobiont endoparasitism. Koinobiont endoparasitism may have arisen in a similar way in other ichneumonoid lineages (Gauld 1988a). One interesting group of koinobiont ectoparasitoids are the Polysphinctini (a tribe of the ichneumonid subfamily Pimplinae). These wasps attach themselves externally to the abdomen of spiders and probably evolved from species attacking spider egg sacs. The transition from egg sac to adult spider probably occurred in a species that attacked a nest-making spider which guards its young. The ovipositing wasp would need to sting the adult spider to gain access to the egg sac. From feeding on the egg sac and paralyzed spider, the Polysphinctini evolved to become koinobiont ectoparasitoids of the adult spider (I. D. Gauld and M. R. Shaw, pers. comm.).

The family Dryinidae is exclusively koinobiont but shows a complicated mixture of endoparasitism and ectoparasitism. Eggs may be laid externally or internally on different species of Homoptera but the first instar larva is always an endoparasitoid. In the vast majority of species, the second instar larva pushes its way through an intersegmental membrane so that only its head protrudes into the host while the rest of its body lies in a sac formed from discarded exuviae. It seems likely that this specialized life history has evolved as a compromise between the protection afforded to the egg and first larva by endoparasitism and the benefits of ectoparasitism in avoiding host defenses. In this light, the anomalous species *Crovettia theliae* is particularly interesting: it alone among related dryinids attacks treehoppers (Membracidae), practices polyembryony, and is wholly endoparasitic (Olmi 1994).

HYPERPARASITISM

Hyperparasitism is known in about seventeen families of hymenopteran parasitoids and in a few species of dipteran and coleopteran parasitoids (Gordh 1981; Sullivan 1987; Gauld and Bolton 1988). The evolutionary origins of facultative hyperparasitism seem easy to explain. A facultative hyperparasitoid is able to develop either as a primary parasitoid, or as a parasitoid of other parasitoid species attacking the same host. This form of hyperparasitism is a particularly efficient way of competing with another parasitoid for the host. If the facultative hyperparasitoid feeds externally, no special adaptations may be needed to attack a primary parasitoid in addition to the host. However, some facultative hyperparasitoids do develop as koinobiont endoparasitoids and thus need to be able to survive in two taxonomically unrelated hosts. Possibly the fact that the primary parasitoid obtains its nutrition from the host leads to a certain degree of physiological similarity between host and primary parasitoid.

Obligate hyperparasitoids have a wide taxonomic distribution although it is unusual for large taxa to be exclusively composed of species with this biology—two exceptions are the charipid subfamily Alloxystinae, which is composed exclusively of obligate hyperparasitoids of aphids, and the Mesochorinae, a subfamily of Ichneumonidae that attack other ichneumonids, as well as braconids and tachinids (Gauld and Bolton 1988). Obligate hyperparasitism might evolve in at least two ways: (1) via facultative hyperparasitism, or (2) by a host shift from a primary parasitoid of one host to a secondary parasitoid of another species.

Although it might seem desirable for a parasitoid to retain the ability to attack both host and primary parasitoid, obligate hyperparasitism could evolve from facultative hyperparasitism if there is a trade-off between the efficiency of utilization of primary and secondary hosts, and if the hyperparasitic species very frequently encounters parasitized hosts. The latter might occur if host behavior changes after parasitism to increase the likelihood of encounter by the hyperparasitoid, or if the hyperparasitoid uses a chemical cue associated with the primary parasitoid in host selection (see sec. 2.2.2). Gauld and Bolton (1988) give two possible examples of this evolutionary pathway: many encyrtids are primary parasitoids or facultative hyperparasitoids of coccoid homopterans, but some in the genus *Prochilonerus* are obligate hyperparasitoids of coccoids through other encyrtids (Rosen 1981); most aphelinids attack aphids and whiteflies but a few genera such as *Marietta* are obligate hyperparasitoids (Viggiani 1984).

The evolution of obligate hyperparasitism by host transfer is most likely if the old primary and new secondary hosts share physiological or ecological attributes. The closest phylogenetic relatives to most hymenopteran parasitoids are sawflies with which they conceivably share some physiological features. Cooper (1954) suggested that the obligate hyperparasitoid Trigonalyidae evolved from parasitoids of sawflies, and there is some evidence that obligate hyperparasitoids in the Perilampidae and the ichneumonid subfamily Mesochorinae also evolved from primary parasitoids of sawflies (Gauld and Bolton 1988). Shared ecology is also likely to be important. Many ichneumonid groups have diversified on cocooned hosts and include species which have become obligate hyperparasitoids attacking the primary parasitoid after it has formed a cocoon (Gauld 1988a). Shaw and Askew (1976) call these species "pseudohyperparasitoids" because they attack the primary parasitoid after it has separated from the original host). Once a species has evolved to parasitize a primary parasitoid in its cocoon, natural selection might favor earlier attack before pupation, when the primary parasitoid is still associated with the host. Possibly the host may be easier to locate before it secretes itself for pupation. Another parasitoid that might be classed as a pseudohyperparasitoid is the braconid *Syntretus lyctae* which develops in the abdomen of an adult ichneumonid (*Phaeogenes invisor*) (Cole 1959b). Related species attack the adults of nonparasitic Hymenoptera (Shaw and Huddleston 1991). Finally, Gauld and Bolton (1988) argue that some obligate hyperparasitoids have evolved from parasitoids of predators of their primary hosts. The charipid subfamily Alloxystinae are hyperparasitoids of aphids through aphidiine braconids; their nearest relatives appear to be parasitoids of aphid predators such as hover flies and ladybird beetles (Syrphidae and Coccinellidae).

To conclude, a number of plausible hypotheses can be put forward to explain the evolution of different forms of hyperparasitism. Further progress will be slow until good phylogenies are available for parasitoid taxa that include hyperparasitoids.

8.4 Conclusions

The balanced mortality hypothesis uses a population dynamic constraint to predict the average number of unparasitized hosts encountered by a parasitoid and hence its requirement for eggs. This technique is so successful at predicting parasitoid fecundity because the availability of unparasitized hosts is probably the major source of density dependence in parasitoid life cycles. It is curious that after Price's work in the early 1970s the study of parasitoid fecundity has received so little attention. Perhaps one reason is that the use of population dynamic assumptions in life history theory has not been popular and is frequently misunderstood as group selection. Such clear confirmation of theoretical predictions are unusual in life history studies and deserve to be known to a wider audience. New work needs to focus on quantitative tests of theory, using modern comparative methods. It will be interesting to see whether deviations from the predictions of the balanced mortality hypothesis can be explained by boom-and-bust population dynamics or by temporal and spatial variance in egg requirements. Modern comparative techniques also need to be applied to other problems in parasitoid life history theory where I believe there is a need for more formally phrased hypotheses: many ideas in this field are vague and speculative and often ignore possible population dynamic interactions between the host and parasitoid. I also argue strongly for a restriction of the terms r- and K-selection to their original meaning.

Studies of parasitoid species load have revealed a tantalizing glimpse of the structure of host-parasitoid communities. It is only a glimpse because by itself information on the number of parasitoids per host can tell us little about community organization when most parasitoids attack more than one species of host. There is an urgent need for studies that try and reconstruct the web of interactions between guilds of hosts and their parasitoids, ideally using fully quantitative methods. Compared with predator-prey interactions, it is comparatively easy to study host-parasitoid webs and it is perhaps surprising there are so few examples in the literature. The availability of more webs will be a major step toward understanding the determinants of parasitoid species loads and parasitoid host breadths.

There are numerous demonstrations that the risks of parasitism vary across the ecological range of their host. This naturally suggests that parasitoids may mold the shape of host niches, although measurements of the selective pressures exerted by parasitoids are rare. The demonstration that parasitoids may select for larger gall sizes, but that other selection pressures may favor smaller sizes, is an important advance, and similar studies need to be carried out on other systems. The role of apparent competition in structuring host communities is also poorly understood and, considering how influential this idea has been, the paucity of experimental studies using parasitoids is perhaps surprising. The sophistication of the tritrophic chemical interactions between parasitoid, host, and host plant are only just beginning to be revealed. This is an exciting area that is likely to generate a variety of new evolutionary problems.

We know there are at a lot of parasitoid species on earth, but not how many, and in particular how many in the tropics. More basic data on parasitoid biodiversity is required, although I hope this will be collected as part of programs designed to study the biological basis of biodiversity rather than simply to collect raw numbers that are interpretable in many different ways. The biology of parasitoids is relatively simple and understanding the determinants of parasitoid diversity may be easier than explaining the diversity of other trophic groups.

The study of major evolutionary innovations by parasitoids is only just beginning to advance beyond speculation and description. Progress requires good phylogenies which are largely unavailable at the moment. Molecular techniques now make it feasible routinely to reconstruct the phylogenies of relatively large groups of organism and the next decade is likely to see an explosion in the phylogenetic information available for parasitoids. It is an exciting prospect that many of the ideas discussed in this chapter, often in a frankly speculative and unsatisfactory manner, are likely to receive hard answers in the not too distant future.

References

Abdelrahman, I. 1974. Studies in ovipositional behaviour and control of sex in *Aphytis melinus* De Bach, a parasite of California red scale, *Aonidiella aurantii* (Mask.). *Australian Journal of Zoology* **22**, 231–247.

van Achterberg, C. 1977. The function of swarming in *Blacus* species (Hymenoptera, Braconidae, Helconinae). *Entomologische Berichten* **37**, 151–152.

Alcock, J. 1981. Notes on the reproductive behavior of some Australian thynnine wasps (Hymenoptera: Tiphiidae). *Journal of the Kansas Entomological Society* **54**, 681–693.

Allen, H. W. 1925. Biology of the red-tailed tachina-fly, *Winthemia quadripustulata* Fabr. *Mississippi Agricultural Experimental Station Technical Bulletin* **12**, 1–32.

van Alphen, J.J.M. 1980. Aspects of the foraging behaviour of *Tetrastichus asparagi* Crawford and *Tetrastichus* spec. (Eulophidae), gregarious egg parasitoids of the asparagus beetles *Crioceris asparagi* L. and *C. duodecimpunctata* L. (Chrysomelidae). I. Host species selection, host stage selection and host discrimination. *Netherlands Journal of Zoology* **30**, 307–325.

van Alphen, J.J.M. 1988. Patch time allocation by insect parasitoids: Superparasitism and aggregation. In G. de Jong, ed., *Population Genetics and Evolution,* pp. 215–221. Springer-Verlag, Berlin.

van Alphen, J.J.M., and Drijver, R.A.B. 1982. Host selection by *Asobara tabida* Nees (Braconidae; Alysiinae), a larval parasitoid of fruit inhabiting *Drosophila* species. I. Host stage selection with *Drosophila melanogaster* as host. *Netherlands Journal of Zoology* **32**, 215–231.

van Alphen, J.J.M., and Galis, F. 1983. Patch time allocation and parasitization efficiency of *Asobara tabida* Nees, a larval parasitoid of *Drosophila. Journal of Animal Ecology* **52**, 937–952.

van Alphen, J.J.M., and van Harsel, H. H. 1982. Host selection by *Asobara tabida* Nees (Braconidae: Alysiinae), a larval parasitoid of fruit inhabiting *Drosophila* species. III. Host species selection and functional response. In J.J.M. van Alphen, "Foraging Behaviour of *Asobara tabida*, a Larval Parasitoid of Drosophilidae," pp. 61–93. Ph.D. diss., University of Leiden.

van Alphen, J.J.M., and Janssen, A.R.M. 1982. Host selection by *Asobara tabida* Nees (Braconidae: Alysiinae), a larval parasitoid of fruit inhabiting *Drosophila* species. II. Host species selection. *Netherlands Journal of Zoology* **32**, 215–231.

van Alphen, J.J.M., and Nell, H. W. 1982. Superparasitism and host discrimination by *Asobara tabida* Nees (Braconidae: Alysiinae), larval parasitoid of Drosophilidae. *Netherlands Journal of Zoology* **32**, 232–260.

van Alphen, J.J.M., and Thunnissen, I. 1983. Host selection and sex allocation by *Pachycrepoides vindemiae* Rondani (Pteromalidae) as a facultative hyperparasitoid of *Asobara tabida* Nees (Braconidae: Alysiinae) and *Leptopilina heterotoma* (Cynipoidea: Eucoilidae). *Netherlands Journal of Zoology* **33**, 497–514.

van Alphen, J.J.M., and Vet, L.E.M. 1986. An evolutionary approach to host finding and selection. In J. K. Waage and D. Greathead, eds., *Insect Parasitoids,* pp. 23–61. Academic Press, London.

van Alphen, J.J.M., and Visser, M. E. 1990. Superparasitism as an adaptive strategy for insect parasitoids. *Annual Review of Entomology* **35**, 59–79.

van Alphen, J.J.M.; van Dijken, M. J.; and Waage, J. K. 1987. A functional approach to superparasitism, host discrimination need not be learnt. *Netherlands Journal of Zoology* **37**, 167–179.

van Alphen, J.J.M.; Nordlander, G.; and Eijs, I. 1991. Host habitat finding and host selection of the *Drosophila* parasitoid *Leptopilina australis* (Hymenoptera, Eucoilidae), with a comparison of the niches of European *Leptopilina* species. *Oecologia* **87**, 324–329.

van Alphen, J.J.M.; Visser, M. E.; and Nell, H. W. 1992. Adaptive superparasitism and patch time allocation in solitary parasitoids: Searching in groups versus sequential patch visits. *Functional Ecology* **6**, 528–535.

Antolin, M. F., and Strand, M.R. 1992. Mating system of *Bracon hebetor* (Hymenoptera: Braconidae). *Ecological Entomology* **17**, 1–7.

Antolin, M. F., and Strong, D. R. 1987. Long-distance dispersal by a parasitoid (*Anagrus delicatus*, Mymaridae) and its host. *Oecologia* **73**, 288–292.

Arakawa, R. 1987. Attack on the parasitized host by a primary solitary parasitoid, *Encarsia formosa* (Hymenoptera: Aphelinidae): The second female pierces with her ovipositor the egg laid by the first one. *Applied Entomology and Zoology* **22**, 644–645.

Armstrong, R. A., and McGehee, R. 1980. Competitive exclusion. *American Naturalist* **115**, 151–170.

Arthur, A. P. 1962. Influence of host tree on abundance of *Itoplectis conquisitor* (Say), a polyphagous parasite of the European pine shoot moth *Rhyacionia buoliana* (Schiff). *Canadian Entomologist* **94**, 337–347.

Arthur, A. P. 1966. Associative learning in *Itoplectis conquisitor* (Say) (Hymenoptera: Ichneumonidae). *Canadian Entomologist* **98**, 213–223.

Arthur, A. P. 1967. Influence of position and size of host on host searching by *Itoplectis conquisitor*. *Canadian Entomologist* **99**, 877–886.

Arthur, A. P. 1971. Associative learning by *Nemeritis canescens* (Hymenoptera: Ichneumonidae). *Canadian Entomologist* **103**, 1137–1141.

Arthur, A. P. 1981. Host acceptance. In D. A. Nordlund, R. L. Jones, and W. J. Lewis, eds., *Semiochemicals, Their Role in Pest Control,* pp. 97–120. John Wiley, New York.

Arthur, A. P., and Ewen, A. B. 1975. Cuticular encystment: A unique and effective defense reaction by cabbage looper larvae against parasitism by *Banchus flavescens* (Hymenoptera: Ichneumonidae). *Annals of the Entomological Society of America* **68**, 1091–1094.

Arthur, A. P.; Stainer, J.E.R.; and Turnball, A. L. 1964. The interaction between *Orgilus obscurator* (Nees) (Hymenoptera: Braconidae) and *Temelucha interruptor* (Grav.) (Hymenoptera: Ichneumonidae), parasites of the pine shoot moth *Rhyacionia buoliana* (Schiff.) (Lepidoptera: Olethreutidae). *Canadian Entomologist* **96**, 1030–1034.

Arthur, A. P.; Hegdekar, B. M.; and Rollins, L. 1969. Component of the host haemolymph that induces oviposition in a parasitic insect. *Nature* **223**, 966–967.

Arthur, A. P.; Hegdekar, B. M.; and Batsch, W. W. 1972. A chemically defined synthetic medium that induces oviposition in the parasite *Itoplectis conquisitor* (Hymenoptera: Ichneumonidae). *Canadian Entomologist* **104**, 1251–1258.

Askew, R. R. 1961. On the biology of the inhabitants of oak galls of Cynipidae (Hymenoptera) in Britain. *Transactions of the Society for British Entomology* **14**, 237–268.

Askew, R. R. 1968a. A survey of leafminers and their parasites on laburnum. *Transactions of the Royal Entomological Society London* **120**, 1–37.

Askew, R. R. 1968b. Considerations on speciation in Chalcidoidea (Hymenoptera). *Evolution* **22**, 642–645.

Askew, R. R. 1971. *Parasitic Insects*. Heinemann, London.

Askew, R. R. 1975. The organisation of chalcid-dominated parasitoid communities centred upon endophytic hosts. In P. W. Price, ed., *Evolutionary Strategies of Parasitoids*, pp. 130–153. Plenum, New York.

Askew, R. R. 1980. The diversity of insect communities in leaf-mines and plant galls. *Journal of Animal Ecology* **49**, 817–829.

Askew, R. R. 1990. Species diversity of hymenopteran taxa in Sulawesi. In W. J. Knight and J. D. Holloway, eds., *Insects and Rain Forests of South East Asia*, pp. 255–260. Royal Entomological Society, London.

Askew, R. R., and Ruse, J. M. 1974. Biology and taxonomy of species of the genus *Enaysma* Delucchi (Hym., Eulophidae, Entedonitinae) with sprecial reference to the British fauna. *Transactions of the Royal Entomological Society of London* **125**, 257–294.

Askew, R. R., and Shaw, M. R. 1974. An account of the Chalcidoidea (Hymenoptera) parasitising leafmining insects of deciduous trees in Britain. *Biological Journal of the Linnean Society* **6**, 289–335.

Askew, R. R., and Shaw, M. R. 1979. Mortality factors affecting the leaf-mining stages of *Phyllonorycter* (Lepidoptera: Gracilariidae) on oak and birch. 2. Biology of the parasite species. *Zoological Journal of the Linnean Society* **67** 51–64.

Askew, R. R., and Shaw, M. R. 1986. Parasitoid communities: Their size, structure and development. In J. K. Waage and D. Greathead, eds., *Insect Parasitoids,* pp. 225–264. Academic Press, London.

van den Assem, J. 1969. Reproductive behaviour of *Pseudeucoila bochei* (Hymenoptera, Cynipoidea). I. A description of courtship behaviour. *Netherlands Journal of Zoology* **19**, 641–648.

van den Assem, J. 1970. Courtship and mating in *Lariophagus distinguendus* (Forst.) Kurdj. (Hymenoptera, Pteromalidae). *Netherlands Journal of Zoology* **20**, 329–352.

van den Assem, J. 1971. Some experiments on sex ratio and sex regulation in the Pteromalid *Lariophagus distinguendus*. *Netherlands Journal of Zoology* **21**, 373–402.

van den Assem, J. 1977. A note on the ability to fertilize following insemination. *Netherlands Journal of Zoology* **27**, 230–235.

van den Assem, J. 1986. Mating behaviour in parasitic wasps. In J. K. Waage and D. Greathead, eds., *Insect Parasitoids,* pp. 137–167. Academic Press, London.

van den Assem, J., and Feuth-de Bruin, E. 1977. Second matings and their effect on the sex ratio of the offspring in *Nasonia vitripennis* (Hym. Pteromalidae). *Entomologia Experimentalis et Applicata* **21**, 23–28.

van den Assem, J., and Jachmann, F. 1982. The coevolution of receptivity signalling and body size in the Chalcidoidea. *Behaviour* **80**, 96–105.

van den Assem, J., and Kuenen, D. J. 1958. Host finding of *Chaetospila elegans* Westw. a parasite of *Sitophilus granarius* L. *Entomologia Experimentalis et Applicata* **1**, 174–180.

van den Assem, J., and Povel, G.D.E. 1973. Courtship behaviour of some *Muscidifurax* species (Hym., Pteromalidae): A possible example of a recently evolved ethological isolating mechanism. *Netherlands Journal of Zoology* **23**, 465–487.

van den Assem, J., and Putters, F. A. 1980. Patterns of sound produced by courting chalcidoid males and its biological significance. *Entomologia Experimentalis et Applicata* **27**, 293–302.

van den Assem, J., and Visser, J. 1976. Aspects of sexual receptivity in female *Nasonia vitripennis. Biologie du Comportement* **1**, 37–56.

van den Assem, J.; Gijswijt, M. J.; and Nübel, B. K. 1980a. Observations on courtship and mating strategies in a few species of parasitic wasps (Chalcidoidea). *Netherlands Journal of Zoology* **30**, 208–227.

van den Assem, J.; Jachmann, F.; and Simbolotti, P. 1980b. Courtship behaviour of *Nasonia vitripennis* (Hym. Pteromalidae): Some qualitative, experimental evidence for the role of pheromones. *Behaviour* **75**, 301–307.

van den Assem, J.; in den Bosch, H.A.J.; and Prooy, E. 1982. *Melittobia* courtship behaviour, a comparative study of the evolution of a display. *Netherlands Journal of Zoology* **32**, 427–471.

van den Assem, J.; Putters, F. A.; and Prins, T. C. 1984a. Host quality effects on sex ratio of the parasitic wasp *Anisopteramalus calandrae* (Chalcidoidea, Pteromalidae). *Netherlands Journal of Zoology* **34**, 33–62

van den Assem, J.; Putters, F. A.; and van der Voort-Vinkestijn, M. J. 1984b. Effects of exposure to an extremely low temperature on recovery of courtship behaviour after waning in the parasitic wasp *Nasonia vitripennis. Journal of Comparative Physiology* **155**, 233–237.

van den Assem, J.; van Iersal, J. A.; and Los-den Hartogh. R. L. 1989. Is being large more important for female than male parasitic wasps? *Behaviour* **108**, 160–195.

Atkinson, W. D., and Shorrocks, B. 1981. Competition on a divided and ephemeral resource: A simulation model. *Journal of Animal Ecology* **50**, 461–471.

Atsatt, P. R. 1981. Lycaenid butterflies and ants: Selection for enemy-free space. *American Naturalist* **118**, 638–654.

Atsatt, P. R., and O'Dowd, D. J. 1976. Plant defence guilds. *Science* **193**, 24–29.

Aubert, J. F. 1961. L'expérience de la bourre de coton démontre que le volume de l'hôte intervient en tant que facteurs essentials dans la détermination du sex chez les Ichneumonides Pimplines (Hym.). *Bulletin de la Société Entomologique de France* **66**, 89–93.

Austin, A. D. 1984. The fecundity, development and host relationships of *Ceratobaeus* spp. (Hymenoptera: Scelionidae), parasites of spider eggs. *Ecological Entomology* **9**, 125–138.

Austin, A. D. 1985. The function of spider egg sacs in relation to parasitoids and predators with special reference to the Australian fauna. *Journal of Natural History* **19**, 359–376.

Avilla, J.; Anadón, J.; Sarasúa, M. J.; and Albajes, R. 1991. Egg allocation of the autoparasitoid *Encarsia tricolor* at different relative densities of primary host (*Trialeurodes vaporariorum*) and two secondary hosts (*Encarsia formosa* and *E. tricolor*). *Entomologia Experimentalis et Applicata* **59**, 219–227.

Ayal, Y. 1987. The foraging strategy of *Diaeretiella rapae* I. The concept of the elementary unit of foraging. *Journal of Animal Ecology* **56**, 1057–1068.

Bai, B., and Mackauer, M. 1990. Self and conspecific host discrimination by the aphid parasitoid *Aphelinus asychis* Walker (Hymenoptera: Aphelinidae). *Canadian Entomologist* **122**, 363–372.

Bai, B., and Mackauer, M. 1991. Recognition of heterospecific parasitism: Competition between aphidiid (*Aphidius ervi*) and aphelinid (*Aphelinus asychis*) parasitoids of aphids (Hymenoptera: Aphidiidae; Aphelinidae). *Journal of Insect Behaviour* **4**, 333–345.

Baker, J. L. 1976. Determinants of host selection for species of *Aphytis* (Hymenoptera: Aphelinidae), parasites of diaspine scales. *Hilgardia* **44**, 1–25.

Bakker, K.; Bagchee, S. N.; van Zwet, W. R.; and Meelis, E. 1967. Host discrimination in *Pseudeucoila bochei* (Hymenoptera, Cynipidae). *Entomologia Experimentalis et Applicata* **10**, 295–311.

Bakker, K.; Eijsackers, H.J.P.; van Lenteren, J. C.; and Meelis, E. 1972. Some models describing the distribution of eggs of the parasite *Pseueucoila bochei* (Hym., Cynip.) over its host, larvae of *Drosophila melanogaster*. *Oecologia* **10**, 29–57.

Bakker, K.; van Alphen, J.J.M.; van Batenberg, F.H.D.; van der Hoeven, N.; Nell, H.W.; van Strien-van Liempt, W.T.F.H.; and Turlings, T. C. 1985. The function of host discrimination and superparasitism in parasitoids. *Oecologia* **67**, 572–576.

Bakker, K.; Peulet, P.; and Visser, M. E. 1990. The ability to distinguish between hosts containing different numbers of parasitoid eggs by the solitary parasitoid *Leptopilina heterotoma* (Thomson) (Hym., Cynip.). *Netherlands Journal of Zoology* **40**, 514–520.

Balfour Browne, F. 1922. On the life-history of *Melittobia acasta* Walker, a chalcid parasite of bees and wasps. *Parasitology* **14**, 349–370.

Baltensweiler, W., and Moreau, J. P. 1957. Ein Beitrag biologisch-systematischer Art zur Kenntnis der Gattung *Phytodietus* (Hymenoptera). *Zeitschrift für Angewandte Entomologie* **41**, 272–276.

Barbosa, P. 1977. *r* and *K* strategies in some larval and pupal parasitoids of the Gypsy Moth. *Oecologia* **29**, 311–327.

Barbosa, P. 1988. Natural enemies and herbivore plant interactions: Influence of plant allelochemicals and host specificity. In P. Barbosa and D. Letourneau, eds., *Novel Aspects of Insect-Plant Allelochemicals and Host Specificity,* pp. 201–229. John Wiley, New York.

Barbosa, P.; Saunders, J. A.; Kemper, J.; Trumbule, R.; Olechno, J.; and Martinat, P. 1986. Plant allelochemicals and insect parasitoids: Effects of nicotine on *Cotesia congregata* and *Hyposoter annulipes*. *Journal of Chemical Ecology* **12**, 1319–1328.

Baronio, P., and Sehnal, F. 1980. Dependence of the parasitoid *Gonia cinerescens* on the hormones of its lepidopterous host. *Journal of Insect Physiology* **26**, 619–626.

Barrass, R. 1961. A quantitative study of the behaviour of the male *Mormoniella vitripennis* towards two constant stimulus situations. *Behaviour* **18**, 288–312.

Barrass, R. 1976. Inhibitory effects of courtship in the wasp *Nasonia vitripennis* and a new interpretation of the biological significance of courtship in insects. *Physiological Entomology* **1**, 229–234.

Bartlett, B. R. 1964. Patterns in the host feeding habit of adult parasitic Hymenoptera. *Annals of the Entomological Society of America* **57**, 344–350.

Beckage, N. E. 1985. Endocrine interactions between endoparasitic insects and their hosts. *Annual Review of Entomology* **30**, 371–413.

Beckage, N. E., and Riddiford, L. M. 1978. Developmental interactions between the tobacco hornworm *Manduca sexta* and its braconid parasite *Apanteles congregatus*. *Entomologia Experimentalis et Applicata* **23**, 139–151.

Beckage, N. E., and Riddiford, L. M. 1982. Effects of parasitism by *Apanteles congregatus* on the endocrine physiology of the tobacco hornworm *Manduca sexta*. *General Comparative Endocrinology* **47**, 308–322.

Beckage, N. E., and Riddiford, L. M. 1983a. Growth and development of the endoparasitic wasp *Apanteles congregatus*, dependence on host nutritional status and parasitoid load. *Physiological Entomology* **8**, 231–241.

Beckage, N. E., and Riddiford, L. M. 1983b. Lepidopteran anti-juvenile hormones: Effects on development of *Apanteles congregatus* in *Manduca sexta*. *Journal of Insect Physiology* **29**, 633–637.

Beckage, N. E., and Templeton, T. J. 1986. Physiological-effects of parasitism by *Apanteles congregatus* in terminal-stage tobacco hornworm larvae. *Journal of Insect Physiology* **32**, 299.

Beeson, C.F.C., and Chatterjee, S. N. 1935. Biology of the Braconidae. *Indian Forest Record, New Series, Entomology* **1**, 105–138.

Beling, I. 1932. Zur Biologie von *Nemeritis canescens* Grav. (Hym., Ophion.). I. Züchtungserfahrungen und ökologische Beobachtungen. *Zeitschrift für Angewandte Entomologie* **19**, 223–249.

Bell, G. 1982. *The Masterpiece of Nature*. Croom Helm, London.

de Belle, J. S.; Hilliker, A. J.; and Sokolowski, M. B. 1989. Genetic localization of foraging (*for:*), a major gene for larval behaviour in *Drosophila melanogaster*. *Genetics* **123**, 157–163.

Benn, M.; Degrave, J.; Gnanasunderam, C.; and Hutchins, R. 1979. Hostplant pyrrolizidine alkaloids in *Nyctemera annulata* Boisduval: Their persistence through the life-cycle and transfer to a parasite. *Experientia* **35**, 731–732.

Benson, J. F. 1973. Intraspecific competition in the population dynamics of *Bracon hebetor* Say (Hymenoptera: Braconidae). *Journal of Animal Ecology* **42**, 105–124.

Berberet, R. C.; Willson, L. J.; and Odejar, M. 1987. Probabilities for encapsulation of eggs of *Bathyplectes curculionis* (Hymenoptera: Ichneumonidae) by larvae of *Hypera postica* (Coleoptera: Curculionidae) and resulting reduction in effective parasitism. *Annals of the Entomological Society of America* **80**, 483–485.

Bernstein, C.; Kacelnik, A.; and Krebs, J. R. 1988. Individual decisions and the distribution of predators in a patchy environment. *Journal of Animal Ecology* **57**, 1007–1026.

Bernstein, C.; Kacelnik, A.; and Krebs, J. R. 1991. Individual decisions and the distribution of predators in a patchy environment. II: The influence of travel costs and the structure of the environment. *Journal of Animal Ecology* **60**, 205–226.

Bess, H. A., and Haramoto, F. H. 1958. Biological control of the oriental fruit fly in Hawaii. *Proceedings of the 10th International Congress of Entomology (1956)* **4**, 835–840.

Beukeboom, L. W. and Werren, J. H. 1992. Population genetics of a parasitic chromosome—experimental analysis of *psr* in subdivided populations. *Evolution* **46**, 1257–1268.

Beukeboom, L., and Werren, J. H. 1993. Transmission and expression of the parasitic paternal sex ratio (*psr*) chromosome. *Heredity* **70**, 437–443.

Beukeboom, L.; Reed, K. M.; and Werren, J. H. 1993. Effects of deletions on mitotic stability of the paternal-sex-ratio (*psr*) chromosome. Manuscript.

Birks, H.J.B. 1980. British trees and insects: A test of the time hypothesis over the last 13,000 years. *American Naturalist* **115**, 600–605.

Blackburn, T. M. 1991a. A comparative examination of life-span and fecundity in parasitoid Hymenoptera. *Journal of Animal Ecology* **60**, 151–164.

Blackburn, T. M. 1991b. Evidence for a "fast-slow" continuum of life-history traits among parasitoid Hymenoptera. *Functional Ecology* **5**, 65–74.

Blissard, G. W.; Fleming, J.G.W.; Vinson, S. B.; and Summers, M. D. 1986. *Camploletis sonorensis* virus: Expression in *Heliothis virescens* and identification of expressed sequences. *Journal of Insect Physiology* **32**, 351–359.

Blumberg, D., and DeBach, P. 1981. Effects of temperature and host age upon the encapsulation of *Metaphycus stanleyi* and *Metaphycus helvolus* eggs by brown soft scale *Coccus hesperidium*. *Journal of Invertebrate Pathology* **37**, 73–79.

Boldt, P. E. 1974. Temperature, humidity and host: Effect on rate of search of *Trichogramma evanescens* and *T. minutum* auctt. (not Riley 1871). *Annals of the Entomological Society of America* **67**, 706–708.

Boronio, P.; and Sehnal, F. 1980. Dependence of the parasitoid *Gonia cinerescens* on the hormones of its lepidopterous host. *Journal of Insect Physiology* **26**, 619–626.

von Borstel, R. C. [and 20 junior authors]. 1968. Mutational response of *Habrobracon* in the Biosatellite II experiment. *Bioscience* **18**, 598–601.

van den Bosch, R. 1964. Encapsulation of the eggs of *Bathyplectes curculionis* (Thompson) (Hymenoptera: Ichneumonidae) in larvae of *Hypera brunneipennis* (Boheman) and *Hypera postica* (Gyllenhal) (Coleoptera: Curculionidae). *Journal of Invertebrate Pathology* **6**, 343–367.

Boucek, Z. 1974. A revision of Leucospidae (Hymenoptera: Chalcidoidea) of the world. *Bulletin of the British Museum (Natural History), Entomology Supplement* **23**, 1–247.

Boucek, Z. 1988. *Australian Chalcidoidea (Hymenoptera)*. CAB International Institute of Entomology, Wallingford, U.K.

Boucek, Z., and Askew, R. R. 1968. Index of Palearctic Eulophidae (excluding Tetrastichinae). *Index of Entomophagous Insects* **3**. Le François, Paris.

Boucek, Z.; Watsham, A.; and Wiebes, J. T. 1981. The fig wasp fauna of the receptacles of *Ficus thonningii* (Hymenoptera, Chalcidoidea). *Tijdschrift voor Entomologie* **124**, 149–233.

Bouchard, Y., and Cloutier, C. 1984. Honeydew as a source of host-searching kairomones for the aphid parasitoid *Aphidius nigripes* (Hymenoptera: Aphidiidae). *Canadian Journal of Zoology* **62**, 1513–1520.

Bouchard, Y., and Cloutier, C. 1985. Role of olfaction in host finding by the aphid parasitoid, *Aphidius nigripes* (Hymenoptera: Aphidiidae). *Journal of Chemical Ecology* **11**, 801–808.

Boulétreau, M. 1971. Crossance larvaire et utilisation de l'hote chez *Pteromalus puparum*: Influence de la densité de population. *Annales de Zoologie Ecologique Animale* **3**, 305–318.

Boulétreau, M. 1976. Effect of photoperiod on sex ratio of the progeny in a parasitoid wasp, *Pteromalus puparum*. *Entomologia Experimentalis et Applicata* **19**, 197–204.

Boulétreau, M. 1986. Coevolution between parasitoids and hosts. In J. K. Waage and D. Greathead, eds., *Insect Parasitoids,* pp. 169–200. Academic Press, London.

Boulétreau, M., and David, J. R. 1981. Sexually dimorphic response to host habitat toxicity in *Drosophila* parasitic wasps. *Evolution* **35**, 395–399.

Boulétreau, M., and Fouillet, P. 1982. Variabilité génétique intrapopulation de l'aptitude de *Drosophila melanogaster* à permettre le développement d'un Hyménoptère parasite. *Compte rendu de l'Académie des Sciences* **295**, 775–778.

Bouton, C. E.; McPheron, B. A.; and Weis, A. E. 1980. Parasitoids and competition. *American Naturalist* **116**, 876–881.

Bouwman, B. E. 1909. Über die lebensweise von *Methoca ichneumonides* Latr. *Tijdschrift voor Entomologie* **52**, 284–294.

Boyce, M. S.; and Perrins, C. M. 1988. Optimizing great tit clutch size in a fluctuating environment. *Ecology* **68**, 142–153.

Bradley, G. A. 1975. *Parasites of Forest Lepidoptera in Canada.* Part 1. *Subfamilies Metopiinae and Pimplinae (Hymenoptera: Ichneumonidae).* Publication 1336, Canadian Forest Service, Ottawa.

Bradley, W. G., and Arbuthnot K. D. 1938. Relationship of host physiology to development of the braconid parasite *Chelonus annulipes. Annals of the Entomological Society of America* **31**, 359–365.

Bragg, D. E. 1974. Ecological and behavioural studies of *Phaeogenes cyriarae*: Ecology, host specificity; searching and oviposition; and avoidance of superparasitism. *Annals of the Entomological Society of America* **67**, 931–936.

Braman, S. K., and Yeargan, K. V. 1989. Reproductive strategy of *Trissolcus euschisti* (Hymenoptera: Scelionidae) under conditions of partially used host resources. *Annals of the Entomological Society of America* **82**, 172–176.

Breeuwer, J.A.J., and Werren, J. H. 1990. Microorganisms associated with chromosome destruction and reproductive isolation between two insect species. *Nature* **346**, 558–560.

Breeuwer, J.A.J.; Stouthamer, R.; Werren, J. H.; and Weisburg, W. G. 1992. Phylogeny of cytoplasmic incompatibility microorganisms in the parasitoid was genus *Nasonia* (Hymenoptera: Pteromalidae) based on 16S ribosomal DNA sequences. *Insect Molecular Biology* **1**, 25–36.

Brehélin, M., and Zachery, D. 1986. Insect haemocytes: A new classification to rule out the controversy. In M. Brehélin, ed., *Immunity in Invertebrates,* pp. 36–48. Springer-Verlag, Berlin.

Bridwell, J. C. 1919. Some notes on Hawaiian and other Bethylidae (Hymenoptera) with descriptions of new species. *Proceedings of the Hawaii Entomological Society* **4**, 21–38.

Bridwell, J. C. 1920. Some notes on Hawaiian and other Bethylidae (Hymenoptera) with the description of a new genus and species. *Proceedings of the Hawaii Entomological Society* **4**, 291–314.

Bridwell, J. C. 1929. Thelytoky or arrhenotoky in *Sclerodermus immigrans. Psyche* **36**, 119–120.

Briggs, C. J. 1993. Competition among parasitoid species on a stage-structured host, and its effect on host suppression. *American Naturalist* **141**, 372–396.

Brodeur, J., and McNeil, J. N. 1989. Seasonal microhabitat selection by an endoparasitoid through adaptive modification of host behavior. *Science* **244**, 226–228.

Brodeur, J., and McNeil, J. N. 1990. Overwintering microhabitat selection by a diapausing endoparasitoid: Induced phototactic and thigmotactic responses in dying hosts. *Journal of Insect Behavior* **3**, 751–763.

Brodeur, J., and McNeil, J. N. 1992. Host behaviour modification by the endoparasitoid *Aphidius nigripes*: A strategy to reduce hyperparasitism. *Ecological Entomology* **17**, 97–104.

Bronstein, J. L. 1988a. Limits to fruit production in a monoecious fig: Consequences of an obligate mutualism. *Ecology* **69**, 207–214.

Bronstein, J. L. 1988b. Mutualism, antagonism and the fig-pollinator interaction. *Ecology* **69**, 1298–1302.

Brooks, D. R. 1988. Macroevolutionary comparisons of host and parasite phylogenies. *Annual Review of Ecology and Systematics* **19**, 235–259.

Brooks, D. R., and Glen, D. R. 1982. Pinworms and primates: A case study in evolution. *Proceedings of the Helminothological Society of Washington* **49**, 76–85.

Brunson, M. H. 1937. The influence of the instars of host larvae on the sex of progeny of *Tiphia popilliavora*. *Science* **86**, 197.

Bryan, G. 1983. Seasonal biological variation in some leaf-miner parasites in the genus *Achrysocharoides* (Hymenoptera, Eulophidae). *Ecological Entomology* **8**, 259–270.

Buckell, E. R. 1928. Notes on the life history and habits of *Melittobia chalybii* Ashmead (Chalcidoidea: Elachertidae). *Pan-Pacific Entomologist* **5**, 14–22.

Buijs, M. J.; Pirovana, I.; van Lenteren, J. C. 1981. *Encarsia pergandiella*, a possible biological control agent for the greenhouse whitefly, *Trialeurodes vaporariorum*: A study on intra- and interspecific host selection. *Mededelingen van de Faculteit Landbouwwetenschappen Rijksuniversiteit Gent* **46**, 465–475.

Bull, J. J. 1981. Sex ratio evolution when fitness varies. *Heredity* **46**, 9–26.

Bull, J. J. 1983. *Evolution of Sex Determining Mechanisms.* Benjamin/Cummings, Menlo Park, California.

Bull, J. J., and Charnov, E. L. 1988. How fundamental are Fisherian sex ratios? *Oxford Surveys in Evolutionary Biology* **5**, 96–135.

Bulmer, M. G. 1983. Models for the evlution of protandry in insects. *Theoretical Population Biology* **23**, 314–322.

Bulmer, M. G. 1986. Sex ratios in geographically structured populations. *Trends in Ecology and Evolution* **1**, 35–38.

Bush, G. L. 1975. Sympatric speciation in phytophagous parasitic insects. In P. W. Price, ed., *Evolutionary Strategies of Parasitoids*, pp. 187–206. Plenum, New York.

Cade, W. 1975. Acoustically orienting parasitoids: Fly phonotaxis to cricket song. *Science* **190**: 1312–1313.

Cade, W. 1981. Field cricket spacing, and the phonotaxis of crickets and parasitoid flies to clumped and isolated cricket songs. *Zeitschrift für Tierpsychologie* **55**, 365–375.

Cade, W. 1984. Effects of fly parasitoids on nightly calling duration in field crickets. *Canadian Journal of Zoology* **62**, 226–228.

Camors, F. B., and Payne, T. L. 1972. Response of *Heydenia unica* to *Dendroctonus frontalisi pheromones* and a host-tree terpene. *Annals of the Entomological Society of America* **65**, 31–33.

Campbell, B. C., and Duffey, S. F. 1979. Tomatine and parasitic wasps: Potential incompatibility of plant antibiosis with biological control. *Science* **205**, 700–702.

Campbell, B. C., and Duffey, S. F. 1981. Alleviation of α-tomatine induced toxicity to the parasitoid, *Hyposoter exiguae*, by phytosterols in the diet of the host, *Heliothis zea*. *Journal of Chemical Ecology* **7**, 927–946.

Cappuccino, N. 1992. Adjacent trophic-level effects on spatial density dependence in a herbivore-predator-parasitoid system. *Ecological Entomology* **17**, 105–108.

Carton, Y., and Boulétreau, M. 1985. Encapsulation ability of *Drosophila melanogaster*: A genetic analysis. *Developmental and Comparative Immunology* **9**, 211–219.

Carton, Y., and David, J. R. 1983. Reduction of fitness in *Drosophila* adults surviving parasitization by a cynipid wasp. *Experientia* **39**, 231–233.

Carton, Y., and David, J. R. 1985. Relation between the genetic variability of digging behaviour of *Drosophila* larvae and their susceptibility to a parasitic wasp. *Behavioural Genetics* **15**, 403–408.

Carton, Y., and Kitano, H. 1981. Evolutionary relationships to parasitism by seven species of the *Drosophila melanogaster* subgroup. *Biological Journal of the Linnean Society* **16**, 227–241.

Carton, Y., and Nappi, A. 1991. The *Drosophila* immune reaction and the parasitoid capacity to evade it: Genetic and coevolutionary aspects. *Acta Œecologia* **12**, 89–104.

Carton, Y., and Sokolowski, M. B. 1992. Interactions between searching strategies of *Drosophila* parasitoids and the polymorphic behavior of their hosts. *Journal of Insect Behavior* **5**, 161–175.

Casas, J. 1988. Analysis of searching movements of a leafminer parasitoid in a structured environment. *Physiological Entomology* **13**, 373–380.

Casas, J. 1989. Foraging behaviour of a leafminer parasitoid in the field. *Ecological Entomology* **14**, 257–265.

Caspari, E., and Watson, G. S. 1959. On the evolutionary importance of cytoplasmic sterility in mosquitos. *Evolution* **13**, 568–570.

Chabora, P. C. 1970a. Studies of parasite-host interactions. II. Reproductive and developmental response of the parasite *Nasonia vitripennis* to strains of the housefly host, *Musca domestica*. *Annals of the Entomological Society of America* **63**, 1632–1636.

Chabora, P. C. 1970b. Studies of parasite-host interactions. III. Host race effect on the lifetable and population growth statistics of the parasite *Nasonia vitripennis*. *Annals of the Entomological Society of America* **63**, 1637–1642.

Chabora, P. C., and Chabora, A. J. 1971. Effects of an interpopulation hybrid host on parasite population dynamics. *Annals of the Entomological Society of America* **64**, 558–562.

Chabora, P. C., and Pimentel, D. 1977. Effect of host (*Musca domestica* Linnaeus) age on the pteromalid parasite *Nasonia vitripennis*. *Canadian Entomologist* **98**, 1226–1231.

Chacko, M. J. 1969. The phenomenon of superparasitism in *Trichogramma evanescens minutum* Riley. I. *Beiträge zur Entomologie* **19**, 617–635.

Chan, M. S., and Godfray, H.C.J. 1993. Host-feeding strategies of parasitoid wasps. *Evolutionary Ecology* (in press).

Charnov, E. L. 1976. Optimal foraging: The marginal value theorem. *Theoretical Population Biology* **9**, 129–136.

Charnov, E. L. 1979. The genetical evolution of patterns of sexuality: Darwinian fitness. *American Naturalist* **113**, 465–480.

Charnov, E. L. 1982. *The Theory of Sex Allocation.* Princeton University Press, Princeton.

Charnov, E. L. 1990. On evolution of age of maturity and the adult lifespan. *Journal of Evolutionary Biology* **3**, 139–144.

Charnov, E. L. 1993. *Life History Invariants.* Oxford University Press, Oxford.

Charnov, E. L., and Berrigan, D. 1990. Dimensionless numbers and life history evolution: Age of maturity versus the adult lifespan. *Philosophical Transactions of the Royal Society of London B* **332**, 41–48.

Charnov, E. L., and Krebs, J. R. 1974. On clutch size and fitness. *Ibis* **116**, 217–219.

Charnov, E. L., and Skinner, S. W. 1984. Evolution of host selection and clutch size in parasitoid wasps. *Florida Entomologist* **67**, 5–21.

Charnov, E. L., and Skinner, S. W. 1985. Complementary approaches to the understanding of parasitoid oviposition decisions. *Environmental Entomology* **14**, 383–391.

Charnov, E. L., and Skinner, S. W. 1988. Clutch size in parasitoids: The egg production rate as a constraint. *Evolutionary Ecology* **2**, 167–174.

Charnov, E. L., and Stephens, D. W. 1988. On the evolution of host selection in solitary parasitoids. *American Naturalist* **132**, 707–722.

Charnov, E. L.; Los-den Hartogh, R. L.; Jones, W. T.; and van den Assem, J. 1981. Sex ratio evolution in a variable environment. *Nature* **289**, 27–33.

Chesson, P. L. 1986. Environmental variation and the coexistence of species. In J. Diamond and T. J. Case, eds., *Community Ecology,* pp. 240–256. Harper and Row, New York.

Chesson, P. L., and Murdoch, W. W. 1986. Aggregation of risk: Relationships among host-parasitoid models. *American Naturalist* **127**, 696–715.

Chewyreuv, I. 1913. Le rôle des femelles dans la détermination du sexe de leur descendance dans le groupe des Ichneumonides. *Compte Rendu des Séances de la Société de Biologie* **74**, 695–699.

Chopard, L. 1923. Les parasites de la mante religieuse. *Annales de Societe Entomologique de France* **91**, 249–274.

Chow, A., and Mackauer, M. 1991. Patterns of host selection by four species of aphidiid (Hymenoptera) parasitoids: Influence of host switching. *Ecological Entomology* **16**, 403–410.

Chow, A., and Mackauer, M. 1992. The influence of prior ovipositional experience on host selection in four species of aphidiid wasps (Hymenoptera: Aphidiidae). *Journal of Insect Behaviour* **5**, 99–108.

Chow, F. J., and Mackauer, M. 1984. Inter- and intraspecific larval competition in *Aphidius smithi* and *Praon pequodorum* (Hymenoptera: Aphidiidae). *Canadian Entomologist* **116**, 1097–1107.

Chow, F. J., and Mackauer, M. 1985. Multiple parasitism of the pea aphid: Stage of development of parasite determines survival of *Aphidius smithi* and *Praon pequodorum* (Hymenoptera: Aphidiidae). *Canadian Entomologist* **117**, 133–134.

Chow, F. J., and Mackauer, M. 1986. Host discrimination and larval competition in the aphid parasite *Ephedrus californicus. Entomologia Experimentalis et Applicata* **41**, 243–254.

Chu, J. 1935. Notes on the biology of *Cedria paradoxa* Wilkinson, a hymenopterous parasite of the mulberry pyralid (*Mararonia pyloalis* Walker). *Chekiang Province Bureau of Entomology Yearbook* **1933**, 193–202.

Chumakova, B. M., and Goryunova, Z. S. 1963. Development of males of *Prospaltella perniciosa* Tow. (Hymenoptera: Aphelinidae) parasite of San Jose scale (Homoptera: Coccoidea). *Entomological Review* **42**, 178–181.

Clancy, K. M., and Price, P. W. 1986. Temporal variation in 3-trophic-level interactions among willows, sawflies, and parasites. *Ecology* **67**, 1601–1607.

Claret, J. 1982. Modification du signal photopériodique par la cuticle de l'hôte pour un endoparasite. *Compte Rendu des Séances de la Société de Biologie* **176**, 834–838.

Claridge, M. F. 1959. Notes on the genus *Systole* Walker, including a previously undescribed species (Hym.; Eurytomidae). *Entomologist's Monthly Magazine* **95**, 38–43.

Claridge, M. F. 1961. A contribution to the biology and taxonomy of some Palaearctic species of *Tetramesa* Walker (=*Isosoma* Walk.; =*Harmolita* Motsch.)(Hymenoptera: Eurytomidae), with particular reference to the British fauna. *Transactions of the Royal Entomological Society of London* **113**, 175–216.

Claridge, M. F., and Wilson, M. R. 1982. Insect herbivore guilds and species-area relationships: leafminers on British trees. *Ecological Entomology* **7**, 19–30.

Clark, A. B. 1978. Sex ratio and local resource competition in a prosimian primate. *Science* **201**, 163–165.

Clark, A. M.; Bertrand, H. A.; and Smith, R. E. 1963. Lifespan differences between haploid and diploid males in *Habrobracon serinopae*. *American Naturalist* **97**, 203–208.

Clausen, C. P. 1939. The effect of host size upon the sex ratio of Hymenopterous parasites and its relation to methods of rearing and colonization. *Journal of the New York Entomological Society* **47**, 1–9.

Clausen, C. P. 1940a. *Entomophagous Insects.* McGraw-Hill, New York.

Clausen, C. P. 1940b. The oviposition habits of the Eucharidae (Hymenoptera). *Journal of the Washington Academy of Science* **30**, 504–516.

Clausen, C. P. 1940c. The immature stages of the Eucharidae. *Proceedings of the Entomological Society of Washington* **42**, 161–170.

Clausen, C. P. 1941. The habits of the Eucharidae. *Psyche* **48**, 57–69.

Clausen, C. P. 1976. Phoresy among entomophagous insects. *Annual Review of Entomology* **21**, 343–368.

Clausen, C. P.; Clancy, D. W.; and Chock, Q. C. 1965. Biological control of the oriental fly (*Dacus sorsalis* Hendel) and other fruit flies in Hawaii. *United States Department of Agriculture Technical Bulletin* **1322**, 1–102.

Cloutier, C. 1984. The effect of host density on egg distribution by the solitary parasitoid, *Aphidius nigripes* (Hymenoptera, Aphelinidae). *Canadian Entomologist* **116**, 805–811.

Cloutier, C.; Dohse, L. A.; and Bauduin, F. 1984. Host discrimination in the aphid parasitoid *Aphidius nigripes*. *Canadian Journal of Zoology* **62**, 1367–1372.

Cole, L. R. 1959a. On the defences of lepidopterous pupae in relation to the oviposition behaviour of certain Ichneumonidae. *Journal of the Lepidopterist's Society* **13**, 1–10.

Cole, L. R. 1959b. On a new species of *Syntretus* Förster (Hym.: Braconidae) parasitic on an adult ichneumonid, with a description of the larva and notes on its life history and that of its host, *Phaeogenes invisor* (Thunberg). *Entomologist's Monthly Magazine* **95**, 18–21.

Cole, L. R. 1981. A visable sign of a fertilization act during oviposition by an ichneumonid wasp, *Itoplectis maculator. Animal Behaviour* **29**, 299–300.

Coley, P. D.; Bryant, J. P.; and Chapin, F. S. 1985. Resource availability and plant
anti-herbivore defense. *Science* **230**, 895–899.

Colgan, P., and Taylor, P. 1981. Sex-ratio in autoparasitic Hymenoptera. *American
Naturalist* **117**, 564–566.

Collins, M. D., and Dixon, A.F.G. 1986. The effect of egg depletion on the foraging
behaviour of an aphid parasitoid. *Journal of Applied Entomology* **102**, 342–352.

Collins, M. D.; Ward, S. A.; and Dixon, A.F.G. 1981. Handling time and the functional
response of *Aphelinus thomsoni*, a predator and parasite of the aphid *Drepanosi-
phum platanoides. Journal of Animal Ecology* **50**, 479–487.

Colwell, R. K. 1981. Group selection is implicated in the evolution of female-biased
sex ratios. *Nature* **290**, 401–404.

Comins, H. N., and Hassell, M. P. 1979. The dynamics of optimally foraging predators
and parasites. *Journal of Animal Ecology* **48**, 335–351.

Connell, J. H. 1980. Diversity and the coevolution of competitors, or the ghost of
competition past. *Oikos* **35**, 131–138.

Conner, G. W., and Saul, G. B. 1986. Acquisition of incompatibility by inbred wild-
type stocks of *Mormoniella. Journal of Heredity* **77**, 211–213.

Cook, J. M. 1991. Sex Determination and Sex Ratios in Parasitoid Wasps. Ph.D. diss.,
University of London.

Cook, J. M.; Rivero Lynch, A. P.; and Godfray, H.C.J. 1994. Sex ratio and foundress
number in the parasitoid wasp *Bracon hebetor. Animal Behaviour* (in press.

Cook, R. M., and Hubbard, S. F. 1977. Adaptive searching strategies in insect parasitoids. *Journal of Animal Ecology* **46**, 115–125.

Cooper, K. W. 1954. Biology of Eumenine wasps, IV. A Trigonalid wasp parasitic on
Rygchium rugosum (Saussure). *Proceedings of the Entomological Society of
Washington* **56**, 280–288.

Copland, M.J.W., and Askew, R. R. 1977. An analysis of the chalcidoid (Hymenoptera)
fauna of a sand-dune system. *Ecological Entomology* **2**, 27–46.

Corbet, S. A. 1971. Mandibular gland secretion of larvae of the flour moth, *Anagasta
kuehniella*, contains an epideictic pheromone and elicits oviposition movement in
a hymenopteran parasite. *Nature* **232**, 481.

Corbet, S. A. 1973. Concentration effects and the response of *Nemeritis canescens* to a
secretion of its host. *Journal of Insect Physiology* **19**, 2119–2128.

Corbet, S. A. 1985. Insect chemosensory response: A chemical legacy hypothesis. *Ecological Entomology* **10**, 143–153.

Cornell, H. V. 1976. Search strategies and the adaptive significance of switching in
some general predators. *American Naturalist* **110**, 317–320.

Cornell, H. V. 1983. The secondary chemistry and complex morphology of galls
formed by the Cynipidae (Hymenoptera), why and how? *American Midland Naturalist* **110**, 225–234.

Cornell, H. V. 1988. Solitary and gregarious brooding, sex ratios and the incidence of
thelytoky in the parasitic Hymenoptera. *American Midlands Naturalist* **119**, 63–
70.

Cornell, H. V. 1990. Survivorship, life history, and concealment: A comparison of leaf
miners and gall formers. *American Naturalist* **136**, 581–597.

Cornell, H. V., and Hawkins, B. A. 1993. Accumulation of native parasitoids on introduced hosts: A comparison of "hosts-as-natives" and "hosts-as-invaders." *American Naturalist* **141**, 847–865

Cornell, H. V., and Pimentel, D. 1978. Switching in the parasitoid *Nasonia vitripennis* and its effect on host competition. *Ecology* **59**, 297–308.

Cornell, H. V., and Washburn, J. O. 1979. Evolution of the richness-area correlation for cynipid gall wasps on oak trees: A comparison of two geographical areas. *Evolution* **33**, 257–274.

Corrigan, J. E., and Lashomb, J. H. 1990. Host influences on the bionomics of *Edovum puttleri* (Hymenoptera: Eulophidae): Effects on size and reproduction. *Environmental Entomology* **19**, 1496–1502.

Cosmides, L. M., and Tooby, J. 1981. Cytoplasmic inheritance and intragenomic conflict. *Journal of Theoretical Biology* **89**, 83–129.

da Costa Lima, A. 1928. Notas sobre a biologia de *Telenomus fariai* Lima parasito dos ovos de *Triatoma*. *Memorias do Institutio Oswaldo Cruz* **21**, 201–218 (with English translation).

da Costa Lima, A. 1944. Quarta contribuicao ao conhecimento da biologia do *Telenomus polymorphus* n. sp. (Hymenoptera: Scelionidae). *Anais da Academia Brasileira de Ciencias* **15**, 211–227.

Coudron, T. A., and Puttler, B. 1988. Response of natural and factitious hosts to the ectoparasite *Euplectrus plathypenae* (Hymenoptera: Eulophidae). *Annals of the Entomological Society of America* **81**, 931–937.

Coudron, T. A.; Kelly, T. J.; and Puttler, N. 1990. Developmental responses of *Trichoplusia ni* (Lepidoptera: Noctuidae) to parasitism by the ectoparasite *Euplectrus plathypenae* (Hymenoptera: Eulophidae). *Archives of Insect Biochemistry and Physiology* **13**, 83–94.

Cox, D. R., and Oakes, D. 1984. *Analysis of Survival Data*. Chapman and Hall, London.

Coyne, J. A. 1992. Genetics and speciation. *Nature* **355**, 511–515.

Craig, T. P.; Itami, J. K.; and Price, P. W. 1990. The window of vulnerability of a shoot-galling sawfly to attack by a parasitoid. *Ecology* **71**, 1471–1482.

Crandell, H. A. 1939. The biology of *Pachycrepoides dubius* Ashm. (Hymenoptera), a pteromalid parasite of *Piophila casei* Linne (Diptera). *Annals of the Entomological Society of America* **32**, 632–654.

Crankshaw, O. S., and Matthews, R. W. 1981. Sexual behaviour amongst parasitic *Megarhyssa* wasps (Hymenoptera: Ichneumonidae). *Behavioural Ecology and Sociobiology* **9**, 1–7.

Cronin, J. T., and Strong, D. R. 1990a. Density-independent parasitism among host patches by *Anagrus delicatus* (Hymenoptera: Mymaridae): Experimental manipulation of hosts. *Journal of Animal Ecology* **59**, 1019–1026.

Cronin, J. T., and Strong, D. R. 1990b. Biology of *Anagrus delicatus* (Hymenoptera: Mymaridae), an egg parasitoid of *Prokelesia marginata* (Hymenoptera: Delphacidae). *Annals of the Entomological Society of America* **83**, 846–854.

Crosskey, R. W. 1980. Family Tachnidae. In R. W. Crosskey, ed., *Catalogue of the Diptera of the Afrotropical Region,* pp. 822–882. British Museum (Natural History), London.

Crowson, R. A. 1981. *The Biology of the Coleoptera*. Academic Press, London.

Crozier, R. H. 1971. Heterozygosity and sex determination in haplodiploidy. *American Naturalist* **105**, 399–412.

Crozier, R. H. 1977. Evolutionary genetics of the Hymenoptera. *Annual Review of Entomology* **22**, 263–288.

Crozier, R. H. 1985. Adaptive consequences of male-haploidy. In W. Helle and M. W. Sabelis, eds., *Spider Mites: Their Biology, Natural Enemies and Control,* vol. IA, pp. 201–222. Elsevier, Amsterdam.

Cruz, Y. P. 1981. A sterile defender morph in a polyembryonic hymenopterous parasite. *Nature* **294**, 446–447.

da Cunha, A. B., and Kerr, W. E. 1957. A genetical theory to explain sex determination by arrhenotokous parthenogenesis. *Forma et Functio* **1**, 33–36.

Cuthill, I. C.; Kacelnik, A.; Krebs, J.R.; Haccou, P.; and Iwasa, Y. 1990. Starlings exploiting patches: The effect of recent experiences on foraging decisions. *Animal Behaviour* **40**, 625–640.

Dahlman, D. L. 1990. Evaluation of teratocyte functions: An overview. *Archives of Insect Biochemistry and Physiology* **13**, 159–166.

Daley, D. J., and Maindonald, J. H. 1989. A unified view of models describing the avoidance of superparasitism. *IMA Journal of Mathematics Applied in Medicine and Biology* **6**, 161–178.

Daly, H. V. 1983. Taxonomy and ecology of Ceratinini of North Africa and the Iberian Peninsula (Hymenoptera: apoidea). *Systematic Entomology* **8**, 29–62.

Daniel, D. M. 1932. *Macrocentrus ancylivorus* Rohwer, a polyembryonic braconid parasite of the oriental fruit moth. *New York State Agricultural Experimental Station Technical Bulletin* **187**.

Darling, D. C., and Werren, J. H. 1990. Biosystematics of two new species of *Nasonia* (Hymenoptera: Pteromalidae) reared from birds' nests in North America. *Annals of the Entomological Society of America* **83**, 352–369.

David, J., and van Herrewege, J. 1983. Adaptation to alcoholic fermentation in *Drosophila* species: Relationships between alcohol tolerance and larval habitats. *Comparative Biochemical Physiology* **74**, 283–288.

Davies, D. H.; Strand, M. R.; and Vinson, S. B. 1987. Changes in differential haemocyte count and *in vitro* behavior of plasmatocytes from host *Heliothis virescens* caused by *Campoletis sonorensis* polydnavirus. *Journal of Insect Physiology* **33**, 143–153.

Davies, N. B. 1991. Mating strategies. In J. R. Krebs and N. B. Davies, eds., *Behavioural Ecology: An Evolutionary Approach,* 3d ed., pp. 263–294. Blackwell Scientific, Oxford.

Davies, N. W., and Madden, J. L. 1985. Mandibular gland secretions of two parasitoid wasps (Hymenoptera: Ichneumonidae). *Journal of Chemical Ecology* **11**, 1115–1127.

Dawkins, R. 1976. *The Selfish Gene.* Oxford University Press, Oxford.

Dean, J. M., and Ricklefs, R. E. 1979. Do parasites of Lepidoptera larvae compete for hosts? No! *American Naturalist* **113**, 302–306.

Dean, J. M., and Ricklefs, R. E. 1980. Do parasites of Lepidoptera larvae compete for hosts? No evidence. *American Naturalist* **116**, 882–884.

DeBach, P. 1943. The importance of host-feeding by adult parasites in the reduction of host populations. *Journal of Economic Entomology* **36**, 647–658.

DeBach, P. 1966. The competitive displacement and coexistence principles. *Annual Review of Entomology* **11**, 183–212.

DeBach, P. 1974. *Biological Control by Natural Enemies*. Cambridge University Press, Cambridge, U.K.

DeBach, P., and Argyriou, L. C. 1967. The colonization and success in Greece of some important *Aphytis* spp. (Hym. Aphelinidae) parasitic on citrus scale insects (Hom. Diaspidae). *Entomophaga* **12**, 325–342.

DeBach, P., and Sisojevic, P. 1960. Some effects of temperature and competition on the distribution and relative abundance of *Aphytis lingnanensis* and *A. chrysomphali* (Hymenoptera: Aphelinidae). *Ecology* **41**, 153–160.

DeBach, P., and Sundby, R. A. 1963. Competitive displacement between ecological homologues. *Hilgardia* **46**, 1–35.

DeLoach, C. J., and Rabb, R. L. 1972. Seasonal abundance and natural mortality of *Winthemia manducae* (Diptera, Tachinidae) and degree of parasitization of its host, the tobacco hornworm. *Annals of the Entomological Society of America* **65**, 779–790.

Dempster, J. P. 1983. The natural control of populations of butterflies and moths. *Biological Reviews* **58**, 461–481.

Dethier, V. G.; Browne, L. B.; and Smith, C.N. 1960. The designation of chemicals in terms of the responses they elicit from insects. *Journal of Economic Entomology* **53**, 134–136.

DeVries, P. J. 1984. Butterflies and Tachinidae: Does the parasite always kill its host? *Journal of Natural History* **18**, 323–326.

DeVries, P. J. 1987. *The Butterflies of Costa Rica and Their Natural History*. Princeton University Press, Princeton.

Dicke, M. 1988. Microbial allelochemicals affecting the behaviour of insects, mites, nematodes and protozoa in different trophic levels. In P. Barbosa and D. Letourneau, eds., *Novel Aspects of Insect-Plant Allelochemicals and Host Specificity*, pp. 125–163. John Wiley, New York.

Dicke, M., and Sabelis, M. W. 1988. Infochemical terminology: Based on cost-benefit analysis rather than origin of compounds. *Functional Ecology* **2**, 131–139.

Dicke, M., and Sabelis, M. W. 1989. Does it pay plants to advertise for bodyguards? Towards a cost-benefit analysis of induced synomone production. In H. Lambers, M. L. Cambridge, H. Konings, and T. L. Pons, eds., *Causes and Consequences of Variation in Growth Rate in Higher Plants*, pp. 341–358. SPB Academic, The Hague.

Dicke, M.; van Lenteren, J. C.; Boskamp, G.J.F.; and van Dongen-van Leeuwen, E. 1984. Chemical stimuli in host-habitat location by *Leptopilina heterotoma* (Thomson) (Hymenoptera: Eucoilidae), a parasite of *Drosophila melanogaster*. *Journal of Chemical Ecology* **10**, 695–712.

Dicke, M.; Sabelis, M. W.; Takabayashi, J.; Briun, J.; and Posthumus, M. A. 1990a. Plant strategies of manipulating predator-prey interactions through allelochemicals: Prospects for application in pest control. *Journal of Chemical Ecology* **16**, 3091–3118.

Dicke, M.; van Beek, T. A.; Posthumus, M.A.; Ben Dom, N.; van Bokhoven, H.; and de Groot, A. 1990b. Isolation and identification of volatile kairomone that affects acarine predator-prey interactions. Involvement of host plant in its production. *Journal of Chemical Ecology* **16**, 381–396.

van Dijken, M. J. 1991. A cytological method to determine primary sex ratio in haplo-diploid species. *Entomologia Experimentalis et Applicata* **60**, 301–304.

van Dijken, M. J., and Waage, J. K. 1987. Self and conspecific superparasitism by the egg parasitoid *Trichogramma evanescens*. *Entomologia Experimentalis et Applicata* **43**, 183–192.

van Dijken, M. J.; van Alphen, J.J.M.; van Stratum, P. 1989. Sex allocation in *Epidinocarsis lopezi*: Local mate competition. *Entomologia Experimentalis et Applicata* **52**, 249–255.

van Dijken, M. J.; Neuenschwander, P.; van Alphen, J.J.M.; and Hammond, W.N.O. 1991. Sex ratios in field populations of *Epidinocarsis lopezi*, an exotic parasitoid of the cassava mealybug in Africa. *Ecological Entomology* **16**, 233–240.

van Dijken, M. J.; van Stratum, P.; and van Alphen, J.J.M. 1992. Recognition of individual-specific marked parasitized hosts by the solitary parasitoid *Epidinocarsis lopezi*. *Behavioural Ecology and Sociobiology* **30**, 77–82.

van Dijken, M. J.; van Stratum, P.; and van Alphen, J.J.M. 1993. Superparasitism and sex ratio of the solitary parasitoid *Epidinocarsis lopezi*. *Entomologia Experimentalis et Applicata* . **68**, 51–58.

Dijkstra, L. J. 1986. Optimal selection and exploitation of hosts in the parasitic wasp *Colpoclypeus florus* (Hym.: Eulophidae). *Netherlands Journal of Zoology* **36**, 177–301.

Ding, D.; Swedenborg, P. D.; and Jones, R. 1989. Plant odor preferences and learning in *Macrocentrus grandii* (Hymenoptera: Braconidae), a larval parasitoid of the European Corn Borer, *Ostrinia nubialis* (Lepidoptera: Pyralidae). *Journal of the Kansas Entomological Society* **62**, 164–176.

Dix, M. E., and Franklin, R. T. 1974. Inter- and intraspecific encounters of southern pine beetle parasitoids under field conditions. *Environmental Entomology* **3**, 131–134.

Donaldson, J. S., and Walter, G. H. 1984. Sex ratios of *Spalangia endius* (Hymenoptera: Pteromalidae), in relation to current theory. *Ecological Entomology* **9**, 395–402.

Donaldson, J. S., and Walter, G. H. 1991a. Brood sex ratios of the solitary parasitoid wasp, *Coccophagus atratus*. *Ecological Entomology* **16**, 25–33.

Donaldson, J. S., and Walter, G. H. 1991b. Host population structure affects field sex ratios of the heteronomous hyperparasitoid, *Coccophagus atratus*. *Ecological Entomology* **16**, 35–44.

Donaldson, J. S.; Clark, M. M.; and Walter, G. H. 1986. Biology of the heteronomous hyperparasitoid *Coccophagus atratus* Compère (Hymenoptera: Aphelinidae): Adult behaviour and larval development. *Journal of the Entomological Society of Southern Africa* **49**, 349–357.

Donisthrope, H.St.J.K. 1927. *British Ants: Their Life History and Classification*. 2d ed. George Routledge and Sons, London.

Donovan, B. J. 1991. Life cycle of *Sphecophaga vesparum* (Curtis) (Hymenoptera: Ichneumonidae), a parasitoid of some vespid wasps. *New Zealand Journal of Zoology* **18**, 181–192.

Doolittle, W. F., and Sapienza, C. 1980. Selfish genes, the phenotype paradigm and genome evolution. *Nature* **284**, 601–603.

Doutt, R. L. 1959. The biology of parasitic Hymenoptera. *Annual Review of Entomology* **4**, 161–182.

Doutt, R. L. 1964. Biological characteristics of entomophagous adults. In P. DeBach, ed., *Biological Control of Insect Pests and Weeds,* pp. 145–167. Reinhold, New York.

Doutt, R. L., and Nakata, J. 1973. The *Rubus* leafhopper and its egg parasitoid: An endemic biotic system useful in grape-pest management. *Environmental Entomology* **2**, 381–386.

Dover, B. A.; Davies, D. H.; and Vinson, S. B. 1988. Degeneration of last instar *Heliothis virescens* prothoracic glands by *Camploletis sonorensis* polydnavirus. *Journal of Invertebrate Pathology* **51**, 80–91.

Dowden, P. B. 1961. The gypsy moth egg parasite, *Ooencyrtus kuwanai*, in Southern Connecticut in 1960. *Journal of Economic Entomology* **54**, 876–878.

Dransfield, R. D. 1979. Aspects of host-parasitoid interactions of two aphid parasitoids, *Aphidius urticae* (Haliday) and *Aphidius uzbeckistanicus* (Luzhetski) (Hymenoptera, Aphidiidae). *Ecological Entomology* **4**, 307–316.

Drea, J. J.; Dysart, R. J.; Coles, L. W.; and Loan, C. C. 1972. *Microctonus stelleri* (Hymenoptera, Braconidae, Euphorinae), a new parasite of the alfalfa weevil introduced into the United States. *Canadian Entomologist* **104**, 1445–1456.

Dreyfus, A., and Breuer, M. E. 1944. Chromosomes and sex determination in the parasitic hymenopteran *Telenomus fariai* Lima. *Genetics* **29**, 75–82.

Driessen, G., and Hemerik, L. 1992. The time and egg budget of *Leptopilina clavipes*, a parasitoid of larval *Drosophila. Ecological Entomology* **17**, 17–27.

Driessen, G.; Hemerik, L.; and Boonstra, B. 1991. Host selection behaviour in relation to survival in hosts of *Leptopilina clavipes*, a parasitoid of larval *Drosophila. Netherlands Journal of Zoology* **41**, 99–111.

Drost, Y. C.; Lewis, W. J.; Zanen, P. O.; and Keller, M. A. 1986. Beneficial arthropod behavior mediated by airborne semiochemicals. I. Flight behavior and influence of preflight handling of *Microplitus croceipes* (Cresson). *Journal of Chemical Ecology* **12**, 1247–1262.

Drost, Y. C.; Lewis, W. J.; and Tumlinson, J. H. 1988. Beneficial arthropod behavior mediated by airborne semiochemicals. V. Influence of rearing method, host plant and adult experience on host searching behavior of *Microplitus croceipes* (Cresson), a larval parasitoid of *Heliothis. Journal of Chemical Ecology* **14**, 1607–1616.

Eberhard, W. G. 1975. The ecology and behavior of a subsocial pentatomid bug and two scelionid wasps: Strategy and counter strategy in a host and its parasite. *Smithsonian Contributions to Zoology* **205**, 1–39.

Edson, K. M., and Vinson, S. B. 1976. The function of the anal vesicle in respiration and excretion in the braconid wasp, *Microplitus croceipes. Journal of Insect Physiology* **22**, 1037–1043.

Edson, K. M.; Stoltz, D. B.; and Summers, M. D. 1981. Virus in a parasitoid wasp: Suppression of the cellular immune response in the parasitoid's host. *Science* **211**, 582–583.

Eggleton, P. 1990. The male reproductive behaviour of *Lytarmes maculipennis* (Smith) (Hymenoptera: Ichneumonidae). *Ecological Entomology* **15**, 357–360.

Eggleton, P. 1991. Patterns in male mating strategies of the Rhyssini: A holophyletic group of parasitoid wasps (Hymenoptera: Ichneumonidae). *Animal Behaviour* **41**, 829–838.

Eggleton, P., and Belshaw, R. 1992. Insect parasitoids: An evolutionary overview. *Proceedings of the Royal Society of London B* **337**, 1–20.

Eggleton, P., and Gaston, K. J. 1990. "Parasitoid" species and assemblages: Convenient definitions or misleading compromises? *Oikos* **59**, 417–421.

Eibl-Eibesfeldt, I., and Eibl-Eibesfeldt, E. 1968. The workers' bodyguard. *Animal's Magazine* **11**, 16–17.

Eickbush, D. G.; Eickbush, T. H.; and Werren, J. H. 1992. Molecular characterization of repetitive DNA sequences from a B chromosome. *Chromosoma.* **101**, 575–583.

Eller, F. J.; Bartelt, R. J.; Jones, R. L.; and Kulman, H. M. 1984. Ethyl(Z)-9-hexadecenoate, a sex pheromone of *Syndipnus rubiginosus*, a sawfly parasitoid. *Journal of Chemical Ecology* **10**, 291–300.

Eller, F. J.; Tumlinson, J. H.; and Lewis, W.J. 1992. Effect of host diet and preflight experience on the flight response of *Microplitus croceipes* (Cresson). *Physiological Entomology* **17**, 235–240.

Elzen, G. W.; Williams, H. J.; and Vinson, S. B. 1983. Responses by the parasitoid *Campoletis sonorensis* to chemicals (Synonomes) in plants: Implications for habitat location. *Environmental Entomology* **12**, 1872–1876.

Elzen, G. W.; Williams, H. J.; and Vinson, S. B. 1984a. Isolation and identification of cotton synomones mediating searching behaviour of parasitoid *Campoletis sonorensis*. *Journal of Chemical Ecology* **10**, 1251–1264.

Elzen, G. W.; Williams, H. J.; and Vinson, S. B. 1984b. Role of diet in host selection of *Heliothis virescens* by parasitoid *Campoletis sonorensis* (Hymenoptera: Ichneumonidae). *Journal of Chemical Ecology* **10**, 1535–1541.

Elzen, G. W.; Williams, H. J. and Vinson, S. B. 1986. Wind tunnel flight responses by hymenopterous parasitoid *Campoletis sonorensis* to cotton cultivars and lines. *Entomologia Experimentalis et Applicata* **42**, 285–289.

Elzen, G. W.; Williams, H. J.; Vinson, S. B.; and Powell, J. E. 1987. Comparative behaviour of parasitoids *Campoletis sonorensis* and *Microplitus croceipes*. *Entomologia Experimentalis et Applicata* **45**, 175–180.

van Emden, F. 1931. Zur Kenntnis der Morphologie und Oekologie des Brotkäfer-Parasiten *Cephalonomia quadridentata* Duchaussoy. *Zeitschrift für Morphologie und Ökologie der Tiere* **23**, 425–574.

van Emden, H. F. 1963. Observations on the effect of flowers on the activity of parasitic Hymenoptera. *Entomologist's Monthly Magazine* **98**, 265–270.

van Emden, H. F. 1966. Plant insect relationships and pest control. *World Review of Pest Control* **5**, 115–123.

Endler, J. A. 1986. *Natural Selection in the Wild*. Princeton University Press, Princeton.

English-Loeb, G. M.; Karban, R.; and Brody, A. K. 1990. Arctiid larvae survive attack by a tachinid parasitoid and produce viable offspring. *Ecological Entomology* **15**, 361–362.

Evans, H. E. 1963. A new species of *Cephalonomia* exhibiting an unusually complex polymorphism (Hymenoptera, Bethylidae). *Psyche* **70**, 151–163.

Ewen, A. B., and Arthur, A. P. 1976. Cuticular encystment in three noctuid species (Lepidoptera): Induction by acid gland secretion from an ichneumonid parasite

(*Banchus flavescens*). *Annals of the Entomological Society of America* **69**, 1087–1090.

Fabre, J. H. 1857. Mémoirs sur l'hypermetamorphose et les moeurs des Meloides. *Annales des Sciences Naturelles* **7**, 299–365.

Faeth, S. H. 1985. Host leaf selection by leaf miners: Interactions among three trophic levels. *Ecology* **66**, 870–875.

Feder, J. L.; Chilcote, C. A.; and Bush, G. L. 1988. Genetic differentiation between sympatric host races of apple maggot fly *Rhagoletis pomenella*. *Nature* **336**, 61–64.

Feeny, P. P. 1976. Plant apparency and chemical defense. In J. W. Wallace and R. L. Mansell, eds., *Biochemical Interactions between Insects and Plants*, pp. 1–40. Plenum, New York.

Felsenstein, J. 1985. Phylogenies and the comparative method. *American Naturalist* **125**, 1–15.

Ferrière, C. 1965. *Hymenoptera Aphelinidae d'Europe et du bassin Méditerranéen*. Masson, Paris.

Finlayson, L. H. 1951. The biology of *Cephalonomia waterstoni* Gahan (Hymenoptera, Bethylidae), a parasite of *Laemophloeus* (Col.; Cucujidae). *Bulletin of Entomological Research* **41**, 79–97.

Fisher, R. A. 1930. *The Genetical Theory of Natural Selection*. Oxford University Press, Oxford.

Fisher, R. C. 1961. A study in insect multiparasitism. II. The mechanism and control of competition for the host. *Journal of Experimental Biology* **38**, 605–628.

Fisher, R. C. 1963. Oxygen requirements and the physiological suppression of supernumerary insect parasitoids. *Journal of Experimental Biology* **40**, 531–540.

Fisher, R. C. 1971. Aspects of the physiology of endoparasitic Hymenoptera. *Biological Reviews* **46**, 243–278.

Fiske, W. F. 1910. Superparasitism: An important factor in the natural control of insects. *Journal of Economic Entomology* **3**, 88–97.

Fiske, W. F., and Thompson, W. R. 1909. Notes on the parasites of the Saturniidae. *Journal of Economic Entomology* **2**, 450–460.

Fitton, M. G.; Graham, M.W.R. de V.; Boucek, K.K.J.; Ferguson, N.D.M.; Huddleston, T.; Quinlan, J.; and Richards, O. W. 1978. Kloet and Hincks, *A Checklist of British Insects*, 2d ed. (completely revised), part 4, Hymenoptera. *Handbooks for the Identification of British Insects* **11(4)**, 1–159.

Fitton, M. G.; Shaw, M. R.; and Austin, A. A. 1987. The Hymenoptera associated with spiders in Europe. *Zoological Journal of the Linnean Society* **90**, 65–93.

Fitton, M. G.; Shaw, M. R.; and Gauld, I. D. 1988. Pimpline ichneumon-flies. Hymenoptera, Ichneumonidae (Pimplinae). *Handbooks for the Identification of British Insects* **7(1)**, 1–110.

Flanders, S. E. 1935. Host influence on the profligacy and size of *Trichogramma*. *Pan Pacific Entomologist* **11**, 175–177.

Flanders, S. E. 1936. A reproduction phenomenon. *Science* **83**, 499.

Flanders, S. E. 1942. Abortive development in parasitic Hymenoptera, induced by the foodplant of the insect host. *Journal of Economic Entomology* **35**, 834–835.

Flanders, S. E. 1946. The role of the spermatophore in the mass propagation of *Macrocentrus ancylivorus*. *Journal of Economic Entomology* **38**, 323–327.

Flanders, S. E. 1953. Variations in susceptibility of citrus-infesting coccids to parasitization. *Journal of Economic Entomology* **46**, 266–269.

Flanders, S. E. 1959. Differential host relations of the sexes in parasitic Hymenoptera. *Entomologia Experimentalis et Applicata* **2**, 125–142.

Flanders, S. E. 1967. Deviate-ontogenies in the aphelinid male (Hymenoptera) associated with the ovipositional behaviour of the parental female. *Entomophaga* **12**, 415–427.

Flanders, S. E. 1973. Particularities of diverse egg deposition phenomena characterizing carniveroid Hymenoptera. *Canadian Entomologist* **105**, 1175–1187.

Fleming, J.G.W. 1992. Polydnaviruses: Mutualists and pathogens. *Annual Review of Entomology* **37**, 401–425.

Fleming, J.G.W., and Summers, M.D. 1986. *Campoletis sonorensis* endoparasitic wasps contain forms of *C. sonorensis* virus DNA suggestive of integrated and extrachromosomal polydnavirus DNAs. *Journal of Virology* **57**, 552–562.

Fleming, J.G.W.; Blissard, G. W.; Summers, M. D.; and Vinson, S. B. 1983. Expression of *Campoletis sonorensis* virus in the parasitised host, *Heliothis virescens*. *Journal of Virology* **48**, 74–78.

Force, D. C. 1970. Competition among four hymenopterous parasites of an endemic host. *Annals of the Entomological Society of America* **63**, 1675–1688.

Force, D. C. 1972. r- and K-strategists in endemic host-parasitoid communities. *Bulletin of the Entomological Society of America* **18**, 135–137.

Force, D. C. 1974. Ecology of host-parasitoid communities. *Science* **184**, 624–632.

Force, D. C. 1975. Succession of *r* and *K* strategists in parasitoids. In P. W. Price, ed., *Evolutionary Strategies of Parasitoids*, pp. 112–129. Plenum, New York.

Force, D. C. 1980. Do parasites of Lepidoptera compete for hosts? Probably! *American Naturalist* **116**, 873–875.

Frank, S. A. 1983. A hierarchical view of sex-ratio patterns. *Florida Entomologist* **66**, 42–75.

Frank, S. A. 1984. The behaviour and morphology of the fig wasps *Pegoscapus assuetus* and *P. jimeneza*. Descriptions and suggested behavioural characters for phylogenetic studies. *Psyche* **91**, 289–308.

Frank, S. A. 1985. Hierarchical selection theory and sex ratios. II. On applying the theory, and a test with fig wasps. *Evolution* **39**, 949–964.

Frank, S. A. 1986. Hierarchical selection theory and sex ratios. I. General solution for structured populations. *Theoretical Population Biology* **29**, 312–342.

Frank, S. A. 1990. Sex allocation theory for birds and mammals. *Annual Review of Ecology and Systematics* **21**, 13–55.

Frank, S. A., and Swingland, I. R. 1988. Sex ratio under conditional sex expression. *Journal of Theoretical Biology* **135**, 415–418.

Freeland, W. J. 1983. Parasites and the coexistence of animal host species. *The American Naturalist* **121**, 223–236.

Freeman, B. E., and Ittyeipe, K. 1976. Field studies on the facultative response of *Melittobia* sp. (*Hawaiiensis* complex) (Eulophidae) to varying host densities. *Journal of Animal Ecology* **45**, 415–423.

Freeman, B. E., and Ittyeipe, K. 1982. Morph determination in *Melittobia*, a eulophid wasp. *Ecological Entomology* **7**, 355–363.

Fretwell, S. D., and Lucas, H. J. 1970. On territorial behavior and other factors influencing habitat distribution in birds. *Acta Biotheoretica* **19**, 16–36.

Fritz, R. S. 1982. Selection for host modification by insect parasitoids. *Evolution* **36**, 283–288.

Frohawk, F. W. 1913. Fecal ejection in hesperids. *Entomologist* **49**, 201–202.

Führer, E., and Willers, D. 1986. The anal secretion of the endoparasitic larva *Pimple turionellae*: Sites of production and effects. *Journal of Insect Physiology* **32**, 361–367.

Fulton, B. B. 1933. Notes on *Habrocystus cerealellae*. *Annals of the Entomological Society of America* **26**, 536–553.

Fulton, B. B. 1940. The hornworm parasite, *Apanteles congregatus* Say, and the hyper-parasite *Hypopteromalus tatracum* (Fitch). *Annals of the Entomological Society of America* **33**, 231–244.

Futuyma, D. J. 1983. Mechanisms of speciation. *Science* **219**, 1059–1060.

Futuyma, D. J., and Mayer, G. C. 1980. Non-allopatric speciation in animals. *Systematic Zoology* **29**, 254–271.

Futuyma, D. J., and Moreno, G. 1988. The evolution of ecological specialization. *Annual Review of Ecology and Systematics* **19**, 207–233.

Galis, F., and van Alphen, J.J.M. 1981. Patch time allocation and search intensity of *Asobara tabida* Nees (Hym.: Braconidae). *Netherlands Journal of Zoology* **31**, 701–712.

Galloway, K. S., and Grant, B. 1989. Reverse sex-ratio adjustment in an apparently outbreeding wasp, *Bracon hebetor*. *Evolution* **43**, 465–468.

Gärdenfors, U. 1990. *Trioxys apterus* sp. n. from Ecuador, a new wingless species of Aphidiinae (Hymenoptera: Braconidae). *Entomologica Scandinavica* **21**, 67–69.

Gardner, S. M., and Dixon, A.F.G. 1985. Plant structure and foraging success of *Aphidius rhopalosiphi* (Hymenoptera, Aphidiidae). *Ecological Entomology* **10**, 171–179.

Gardner, S. M., and van Lenteren, J. C. 1986. Characterisation of the arrestment responses of *Trichogramma evanescens*. *Oecologia* **68**, 265–270.

Gardner, S. M.; Ward, S. A; and Dixon, A.F.G. 1984. Limitation of superparasitism by *Aphidius rhopalosiphi*: A consequence of aphid defensive behaviour. *Ecological Entomology* **9**, 149–155.

Gaston, K. J. 1991. The magnitude of global insect species richness. *Conservation Biology* **5**, 283–296.

Gauld, I. D. 1984. The Pimplinae, Xoridinae, Acaenitinae and Lycorininae (Hymenoptera: Ichneumonidae) of Australia. *Bulletin of the British Museum (Natural History), Entomology Series* **49**, 235–339.

Gauld, I. D. 1986a. Taxonomy, its limitations and its role in understanding parasitoid biology. In J. K. Waage and D. Greathead, eds., *Insect Parasitoids*, pp. 1–22. Academic Press, London.

Gauld, I. D. 1986b. Latitudinal gradients in ichneumonid species-richness in Australia. *Ecological Entomology* **11**, 155–161.

Gauld, I. D. 1987. Some factors affecting the composition of tropical ichneumonid faunas. *Biological Journal of the Linnean Society* **30**, 299–312.

Gauld, I. D. 1988a. Evolutionary patterns of host utilization by ichneumonid parasitoids (Hymenoptera: Ichneumonidae and Braconidae). *Biological Journal of the Linnean Society* **57**, 137–162.

Gauld, I. D. 1988b. The species of the *Enicospilus americanus* complex (Hymenoptera: Ichneumonidae) in eastern North America. *Systematic Entomology* **13**, 31–53.

Gauld, I., and Bolton, B. 1988. *The Hymenoptera*. Oxford University Press, Oxford.

Gauld, I. D., and Huddleston, T. 1976. The nocturnal Ichneumonoidea of the British Isles, including a key to genera. *Entomologist's Gazette* **27**, 35–49.

Gauld, I. D., and Mitchell, P. A. 1977. Nocturnal Ichneumonidea of the British Isles: The genus *Alexeter* Foerster. *Entomologist's Gazette* **28**, 51–55.

Gauld, I.; Gaston, K. J.; and Janzen, D. H. 1992. Plant allelochemicals, tritrophic interactions and the anomalous diversity of tropical parasitoids: The "nasty" host hypothesis. *Oikos.* **65**, 353–357.

Gause, G. F. 1934. *The Struggle for Existence.* Williams and Wilkins, Baltimore.

Genieys, P. 1924. *Habrobracon brevicornis* Wesm.; the effects of the environment and the variation which it produces. *Annals of the Entomological Society of America* **18**, 143–202.

Gerber, H. S., and Klostermeyer, E. C. 1970. Sex control by bees: A voluntary act of egg fertilization during oviposition. *Science* **167**, 82–84.

Gerling, D., and Limon, S. 1976. A biological review of the genus *Euplectrus* (Hym.: Eulophidae) with special emphasis on *E. laphygmae* as a parasite of *Spodoptera littoralis* (Lep.: Noctuidae). *Entomophaga* **21**, 179–187.

Gerling, D., and Rotary, N. 1973. Hypersensitivity, resulting from host-unsuitability, as exemplified by two parasite species attacking *Spodoptera littoralis*. *Entomophaga* **18**, 391–396.

Gerling, D.; Orion, T.; and Delarea, Y. 1990. *Eretmoceros* penetration and immature development: a novel approach to overcome host immunity. *Archives of Insect Biochemistry and Physiology* **13**, 247–253.

Gherna, R. L.; Werren, J. H.; Weisburg, W.; Cote, R.; Woese, C. R.; Mandelco, L.; and Brenner, D. J. 1991. *Arsenophonus nasoniae* gen. nov.; sp. nov.; the causative agent of the son-killer trait in the parasitic wasp *Nasonia vitripennis*. *International Journal of Systematic Bacteriology* **41**, 563–565

Gibson, D. O., and Mani, G. S. 1984. An experimental investigation on the effects of selective predation by birds and parasitoid attack on the butterfly *Danaus chrysippus* (L.). *Proceedings of the Royal Society London* **221**, 31–51.

Gilbert, L. E. 1975. Ecological consequences of a coevolved mutualism between butterflies and plants. In L. E. Gilbert and P. H. Raven, eds., *Coevolution of Animals and Plants*, pp. 210–240. University of Texas Press, Austin.

Gilbert, L. E. 1984. The biology of butterfly communities. In R. I. Vane Wright and P. R. Ackery, eds., *The Biology of Butterflies*, pp. 41–54. Academic Press, London.

Gilbert, L. E., and Singer, M. C. 1975. Butterfly ecology. *Annual Review of Ecology and Systematics* **6**, 365–397.

Glas, P.C.G., and Vet, L.E.M. 1983. Host-habitat location and host location by *Diachasma alloeum* Muesebeck (Hym.: Braconidae), a parasitoid of *Rhagoletis pomonella* Walsh (Dipt. Tephritidae). *Netherlands Journal of Zoology* **33**, 41–54.

Glas, P.C.G.; Smits, P. H.; Vlaming, P.; and van Lenteren, J. C. 1981. Biological control of lepidopteran pests in cabbage crops by means of inundative releases of *Trichogramma* species (*T. evanescens* Westwood and *T. cacoeciae* March): A combination of field and laboratory experiments. *Mededelingen van de Faculteit Landbouwwetenschappen Rijksuniversiteit Gent* **46**, 487–497.

Godfray, H.C.J. 1984. Patterns in the distribution of leaf miners on British trees. *Ecological Entomology* **9**, 163–168.

Godfray, H.C.J. 1986a. Clutch size in a leaf-mining fly (*Pegomya nigritarsis*: Anthomyiidae). *Ecological Entomology* **11**, 75–81.

Godfray, H.C.J. 1986b. Models for clutch size and sex ratio with sibling interaction. *Theoretical Population Biology* **30**, 215–231.

Godfray, H.C.J. 1987a. The evolution of clutch size in invertbrates. *Oxford Surveys in Evolutionary Biology* **4**, 117–154.

Godfray, H.C.J. 1987b. The evolution of clutch size in parasitic wasps. *American Naturalist* **129**, 221–233.

Godfray, H.C.J. 1988. Virginity in haplodiploid populations: A study on fig wasps. *Ecological Entomology* **13**, 283–291.

Godfray, H.C.J. 1990. The causes and consequences of constrained sex allocation in haplodiploid animals. *Journal of Evolutionary Biology* **3**, 3–17.

Godfray, H.C.J. 1991. The signalling of need by offspring to their parents. *Nature* **352**, 328–330.

Godfray, H.C.J., and Chan, M. S. 1990. How insecticides trigger single-stage outbreaks in tropical pests. *Functional Ecology* **4**, 329–337.

Godfray, H.C.J., and Grafen, A. 1988. Unmatedness and the evolution of eusociality. *American Naturalist* **131**, 303–305.

Godfray, H.C.J., and Hardy, I.C.W. 1993. Virginity in haplodiploid animals. In S. Wrensch and D. Krainacker, eds., *Insect Sex Ratios*, pp. 404–417. Chapman & Hall, New York.

Godfray, H.C.J., and Hassell, M. P. 1987. Natural enemies may be a cause of discrete generations in tropical insects. *Nature* **327**, 144–147.

Godfray, H.C.J., and Hassell, M. P. 1989. Discrete and continuous insect populations in tropical environments. *Journal of Animal Ecology* **58**, 153–174.

Godfray, H.C.J., and Hassell, M. P. 1991. Encapsulation and host-population dynamics. In C. A. Toft, A. Aeschlimann, and L. Bolis, eds., *Parasite-Host Associations, Coexistence or Conflict*, pp. 131–147. Oxford University Press, Oxford.

Godfray, H.C.J., and Hunter, M. S. 1991. Sex ratios of heteronomous hyperparasitoids—adaptive or non-adaptive. *Ecological Entomology* **17**, 89–90.

Godfray, H.C.J., and Ives, A. R. 1988. Stochasticity in invertebrate clutch-size models. *Theoretical Population Biology* **33**, 79–101.

Godfray, H.C.J., and Pacala, S. W. 1992. Aggregation and the population dynamics of parasitoids and predators. *American Naturalist* **140**, 30–40.

Godfray, H.C.J., and Parker, G. A. 1991. Clutch size, fecundity and parent-offspring conflict. *Philosophical Transactions of the Royal Society of London B* **332**, 67–79.

Godfray, H.C.J., and Parker, G. A. 1992. Sibling competition, parent-offspring conflict and clutch size. *Animal Behaviour* **43**, 473–490.

Godfray, H.C.J., and Shaw, M. R. 1987. Seasonal variation in the reproductive strategy of the parasitic wasp *Eulophus larvarum* (Hymenoptera: Chalcidoidea: Eulophidae). *Ecological Entomology* **12**, 251–256.

Godfray, H.C.J., and Waage, J. K. 1988. Learning in parasitic wasps (News and Views), *Nature* **331**, 211.

Godfray, H.C.J., and Waage, J. K. 1990. The evolution of highly skewed sex ratios in aphelinid wasps. *American Naturalist* **136**, 715–721.

Godfray, H.C.J., and Waage, J. K. 1991. Predictive modelling in biological control: The mango mealy bug (*Rastrococcus invadens*) and its parasitoids. *Journal of Applied Ecology* **28**, 434–453.

Godfray, H.C.J.; Partridge, L.; and Harvey, P. H. 1991. Clutch size. *Annual Review of Ecology and Systematics* **22**, 409–429.

Godfray, H.C.J.; Hassell, M. P., and Holt, R. D. 1993. The population dynamic consequences of phenological asynchrony between parasitoids and their hosts. *Journal of Animal Ecology* (in press).

Goeden, R. D., and Louda, S. M. 1976. Biotic interference with insects imported for weed control. *Annual Review of Entomology* **21**, 325–342.

Goeden, R. D., and Lok, H. T. 1986. Comments on a proposed "new" approach for selecting for the biological control of weeds. *Canadian Entomologist* **118**, 51–58.

Goertzen, R., and Doutt, R. L. 1975. The ovicidal propensity of *Goniozus. Annals of the Entomological Society of America* **68**, 869–870.

Goodpasture, C. 1975. Comparative courtship behavior and karyology in *Monodontomerus* (Hymenoptera: Torymidae). *Annals of the Entomological Society of America* **68**, 391–397.

Gordh, G. 1976. *Goniozus gallicola* Fouts, a parasite of moth larvae, with notes on other bethylids (Hymenoptera: Bethylidae; Lepidoptera: Gelechiidae). *United States Department of Agriculture, Agricultural Research Service, Technical Bulletin* **1524**, 1–27.

Gordh, G. 1981. The phenomenon of insect hyperparasitism and its taxonomic occurrence in the Insecta. In D. Rosen, ed., *The Role of Hyperparasitism in Biological Control: A Symposium*, pp. 10–18. Division of Agricultural Science, University of California, Berkeley.

Gordh, G., and DeBach, P. 1976. Male inseminative potential in *Aphytis lingnanensis* (Hym.; Aphelinidae). *Canadian Entomologist* **108**, 583–589.

Gordh, G., and DeBach, P. 1978. Courtship behavior in the *Aphytis lingnanensis* group, its potential usefulness in taxonomy, and a review of sexual behavior in the parasitic Hymenoptera (Chalc.; Aphelinidae). *Hilgardia* **46**, 37–75.

Gordon, D. M.; Nisbet, R. M.; de Roos, A.; Gurney, W.S.C.; and Stewart, R. K. 1991. Discrete generations in host-parasitoid models with contrasting life cycles. *Journal of Animal Ecology* **60**, 295–308.

Götz, P. 1986. Encapsulation in arthropods. In M. Brehélin, ed., *Immunity in Invertebrates,* pp. 153–170. Springer-Verlag, Berlin.

Götz, P., and Boman, H. 1985. Insect immunity. In G. A. Kerkut and L. I. Gilbert, eds., *Comprehensive Insect Physiology, Biochemistry, Physiology and Pharmacology,* vol. 3, pp. 453–485. Pergamon Press, Oxford.

Grafen, A. 1984. Natural selection, kin selection and group selection. In J. R. Krebs and N. B. Davies, eds., *Behavioural Ecology, an Evolutionary Approach,* 2d ed., pp. 62–86. Blackwell Scientific, Oxford.

Grafen, A. 1986. Split sex ratios and the evolutionary origins of eusociality. *Journal of Theoretical Biology* **122**, 95–121.

Grafen, A. 1989. The phylogenetic regression. *Philosophical Transactions of the Royal Society of London* **326**, 119–156.

Grafen, A., and Godfray, H.C.J. 1991. Vicarious selection explains some paradoxes in dioecious fig/pollinator systems. *Proceedings of the Royal Society B* **245**, 73–76.

Graham, M.W.R. de V. 1969. The Pteromalidae of north-western Europe. *Bulletin of the British Museum (Natural History), Entomology Supplement* **16.**

Grant, B.; Snyder, G. A.; and Glesser, S. F. 1974. Frequency-dependent mate selection in *Mormoniella vitripennis. Evolution* **28**, 259–264.

Grant, B.; Burton, S.; Contoreggi, C.; and Rothstein, M. 1980. Outbreeding via frequency-dependent mate selection in the parasitoid wasp, *Nasonia (=Mormoniella) vitripennis. Evolution* **34**, 983–992.

Grbic, M.; Ode, P. J.; and Strand, M. R. 1992. Sibling rivalry and brood sex ratios in polyembryonic wasps. *Nature* **312**, 234–456.

Greany, P. D., and Oatman, E. R. 1972. Analysis of host discrimination in the parasite *Orgilus lepidus* (Hyemnoptera: Braconidae). *Annals of the Entomological Society of America* **65**, 377–383.

Greany, P. D.; Tumlinson, D. L.; Chambers, D. L.; and Bousch, G. M. 1977. Chemically mediated host finding by *Biosteres (Opius) longicaudatus*, a parasitoid of tephritid fruit fly larvae. *Journal of Chemical Ecology* **3**, 189–195.

Greathead, D. J. 1986. Parasitoids in classical biological control. In J. K. Waage and D. Greathead, eds., *Insect Parasitoids*. pp. 290–318. Academic Press, London.

Green, R. F. 1980. Bayesian birds: A simple example of Oaten's stochastic model of optimal foraging. *Theoretical Population Biology* **18**, 244–256.

Green, R. F. 1982. Optimal foraging and sex ratio in parasitic wasps. *Journal of Theoretical Biology* **95**, 43–48.

Green, R. F. 1984. Stopping rules for optimal foragers. *American Naturalist* **123**, 30–40.

Green, R. F.; Gordh, G.; and Hawkins, B. A. 1982. Precise sex ratios in highly inbred parasitic wasps. *American Natiralist* **120**, 653–665.

Greenblatt, J. A., and Barbosa, P. 1981. Effects of host's diet on two pupal parasitoids of the gypsy moth: *Brachymeria intermedia* (Nees) and *Coccygomimus turionellae* (L.). *Journal of Applied Ecology* **18**, 1–10.

Griffiths, D. 1977a. Models for avoidance of superparasitism. *Journal of Animal Ecology* **46**, 59–62.

Griffiths, D. 1977b. Avoidance of modified generalised distributions and their application to studies of superparasitism. *Biometrics* **33**, 103–112.

Griffiths, G.C.D. 1964–1968. The Alysiinae (Hymenoptera, Braconidae) parasites of the Agromyzidae (Diptera). I–VI. *Beiträge zur Entomologie* **14**, 823–914 (1964); **16**, 551–605 (1966); **16**, 775–951 (1966); **17**, 653–696 (1967); **18**, 5–62 (1968); **18**, 63–152 (1968). (A supplement was published in the same journal in 1984: **34**, 343–362.)

Griffiths, K. J., and Holling, C. S. 1969. A competition submodel for parasites and predators. *Canadian Entomologist* **101**, 785–818.

Griffiths, N., and Godfray, H.C.J. 1988. Local mate competition, sex ratio and clutch size in bethylid wasps. *Behavioural Ecology and Sociobiology* **22**, 211–217.

Grissell, E. E., and Goodpasture, C. E. 1981. A review of Nearctic Podagrionini with description of sexual behavior of *Podagrion mantis* (Hymenoptera: Torymidae). *Annals of the Entomological Society of America* **74**, 226–241.

Grosch, D. S. 1948. Dwarfism and differential mortality in *Habrobracon. Journal of Experimental Zoology* **107**, 289–313.

Gross, H. R. 1981. Employment of kairomones in the management of parasitoids. In D. A. Nordlund, R. L. Jones, and W. J. Lewis, eds., *Semiochemicals, Their Role in Pest Control,* pp. 137–152. John Wiley, New York.

Gross, P., and Price, P. W. 1988. Plant influences on parasitism of two leafminers: A test of enemy-free space. *Ecology* **69**, 1506–1516.

Guillot, F. S., and Vinson, S. B. 1972. Sources of substances which elicit a behavioural response from the insect parasitoid *Campoletis perdistinctus. Nature* **235**, 169–170.

Gunasena, G. H.; Vinson, S. B.; and Williams H. J. 1989. Interrelationships between growth of *Heliothis virecens* (Lepidoptera, Noctuidae) and that of its parasitoid *Campoletis sonorensis* (Hymenoptera, Ichneumonidae). *Annals of the Entomological Society of America* **82**, 187–191.

Gupta, A. P. 1985. Cellular elements in the haemolymph. In G. A. Kerkut and L. I. Gilbert, eds., *Comprehensive Insect Physiology, Biochemistry, Physiology and Pharmacology*, vol. 3, pp. 401–451. Pergamon Press, Oxford.

Gupta, A. P., ed., 1986. *Hemocytic and Humoral Immunity in Arthropods*. John Wiley, New York.

Haccou, P., and Hemerik, L. 1985. The influence of larval dispersal in the cinnabar moth (*Tyria jacobaeae*) on predation by the red wood ant (*Formica polyctena*): An analysis based on the proportional hazards model. *Journal of Animal Ecology* **54**, 755–769.

Haccou, P.; de Vlas, S. J.; van Alphen, J.J.M.; and Visser, M. E. 1991. Information processing by foragers: Effects of intra-patch experience on the leaving tendency of *Leptopilina heterotoma. Journal of Animal Ecology* **60**, 93–106.

Hadorn, E., and Walker, I. 1960. *Drosophila* und *Pseudeucoila*. I. Selektionsversuche zur Steigerung der Abwehrreaktion des Wirtes gegen den Parasiten. *Revue Suisse de Zoologie* **67**, 216–225.

Haeselbarth, E. 1979. Zur Parasitierung der Puppen von Forleule (*Panolis flammea* [Schiff.]), Kiefernspanner (*Bupalus piniarius* [L.]) und Heidelbeerspanner (*Boarmia bistortana* [Goeze]) in bayerischen Keifernwäldern. *Zeitschrift für Angewandte Entomologie* **87**, 186–202.

Hafner, M. S., and Nadler, S. A. 1988. Phylogenetic trees support the coevolution of parasites and their hosts. *Nature* **332**, 258–259.

Hagen, K. S. 1964. Developmental stages of parasites. In P. DeBach, ed., *Biological Control of Insect Pests and Weeds*, pp. 168–246. Reinhold, New York.

Hågvar, E. B. 1988. Multiparasitism of the green peach aphid, *Myzus persicae*: Competition in the egg stage between *Aphidius matricariae* and *Ephedrus cerasicola. Entomologia Experimentalis et Applicata* **47**, 275–282.

Hails, R. S. 1989. Host size and sex allocation of parasitoids in a gall forming community. *Oecologia* **81**, 28–32.

Hamilton, W. D. 1961. Geometry for the selfish herd. *Journal of Theoretical Biology* **31**, 295–311.

Hamilton, W. D. 1967. Extraordinary sex ratios. *Science* **156**, 477–488.

Hamilton, W. D. 1979. Wingless and fighting males in fig wasps and other insects. In M. S. Blum and N. A. Blum, eds., *Sexual Selection and Reproductive Competition in Insects*, pp. 167–220. Academic Press, London.

Hamilton, W. D. 1975. Innate social aptitudes of man: An approach from evolutionary biology. In R. Fox, ed., *Biosocial Anthropology*, pp. 133–155. John Wiley, New York.

Hamm, J. J.; Styer, E. L.; and Lewis, W. J. 1988. A baculovirus pathogenic to the parasitoid *Microplites croceipes* (Hymenoptera, Braconidae). *Journal of Invertebrate Pathology* **52**, 189–191.

Hamm, J. J.; Styer, E. L. and Lewis, W. J. 1990. Comparative virogenesis of filamentous virus and polydnavirus in the female reproductive tract of *Cotesia marginiventris* (Hymenoptera: Braconidae). *Journal of Invertebrate Pathology* **52**, 357–374.

Hammond, W.N.O.; van Alphen, J.J.M.; Neuenschwander, P.; and van Dijken, M. J. 1993. Aggregative foraging by field populations of *Epidinocarsis lopezi* (De Santis) (Hym: Encyrtidae), a parasitoid of the cassava mealybug (*Phenacoccus manihoti* Mat.-Ferr.) (Hom: Pseudococcidae). *Oecologia* (in press).

Hanski, I. 1981. Coexistence of competitors in patchy environment with and without predation. *Oikos* **37**, 306–312.

Hanski, I. 1988. Four kinds of extra long diapause in insects: A review of theory and observations. *Annales Zoologici Fennici* **25**, 37–53.

Hardie, J.; Nottingham, S. F.; Powell, W.; and Wadhams, L. J. 1991. Synthetic aphid sex pheromone lures female parasitoids. *Entomologia Experimentalis et Applicata* **61**, 97–99.

Hardy, I.C.W. 1992. Non-binomial sex allocation in the parasitoid Hymenoptera. *Oikos* **65**, 143–150.

Hardy, I.C.W., and Blackburn, T. M. 1991. Brood guarding in a bethylid wasp. *Ecological Entomology* **16**, 55–62.

Hardy, I.C.W., and Godfray, H.C.J. 1990. Estimating the frequency of constrained sex allocation in field populations of Hymenoptera. *Behaviour* **114**, 137–147.

Hardy, I.C.W.; Griffiths, N. T.; and Godfray, H.C.J. 1992. Clutch size in a parasitoid wasp: A manipulation experiment. *Journal of Animal Ecology* **61**, 121–129.

Harris, A. C. 1978. Mimicry by a longhorn beetle *Neocalliprason elegans* (Coleoptera: Cerambycidae) of its parasitoid *Xanthocryptus novozealandicus* (Hymenoptera: Ichneumonidae). *New Zealand Entomologist* **6**, 406–408.

Harris, V. E., and Todd, J. W. 1980. Male-mediated aggregation of male, female, and 5th-instar southern green stink bugs and concomitant attraction of a tachinid parasite, *Tricopoda pennipes*. *Entomologia Experimentalis et Applicata* **27**, 117–126.

Harrison, E. 1985. Oviposition Behaviour of *Venturia canescens*: A Study of the Effect of a Pheromone. Ph.D., diss., University of London.

Harrison, E.; Fisher, R. C.; and Ross, K. M. 1985. The temporal effects of Dufour's gland secretion in host discrimination by *Nemeritis canescens*. *Entomologia Experimentalis et Applicata* **38**, 215–220.

Hartl, D. L. 1971. Some aspects of natural selection in arrhenotokous populations. *American Zoologist* **11**, 309–325.

Hartl, D. L. 1972. A fundamental theorem of natural selection for sex linkage or arrhenotoky. *American Naturalist* **106**, 516–524.

Harvey, I.; Marris, G.; and Hubbard, S. 1987. Adaptive patterns in the avoidance of superparasitism by solitary parasitic wasps. *Les Colloques de l'INRA* **48**, 137–142.

Harvey, P. H., and Bradbury, J. W. 1991. Sexual selection. In J. R. Krebs and N. B. Davies, eds., *Behavioural Ecology, an Evolutionary Approach*, 3d ed., pp. 203–233. Blackwell Scientific, Oxford.

Harvey, P. H., and Keymer, A. E. 1991. Comparing life histories using phylogenies. *Philosophical Transactions of the Royal Society of London B* **332**, 31–39.

Harvey, P. H., and Pagel, M. 1991. *The Comparative Method in Evolutionary Biology*. Oxford University Press, Oxford.

Harvey, P. H., and Partridge, L. 1987. Murderous mandibles and black holes in hymenopteran wasps (News and Views). *Nature* **326**, 128–129.

Harvey, P. H.; Read, A. F.; and Promislow, D.E.L. 1989. Life history variation in placental mammals: unifying the data with theory. *Oxford Surveys in Evolutionary Biology* **4**, 117–154.

Hassell, M. P. 1968. The behavioural responses of a tachinid fly (*Cyzenis albicans* (Fall.)) to its host, the winter moth (*Operophtera brumata* (L.)). *Journal of Animal Ecology* **37**, 627–639.

Hassell, M. P. 1978. *The Dynamics of Arthropod Predator-Prey Systems*. Princeton University Press, Princeton.

Hassell, M. P. 1982. Patterns of parasitism by insect parasitoids in patchy environments. *Ecological Entomology* **7**, 365–377.

Hassell, M. P. 1984. Parasitism in patchy environments: Inverse density dependence can be stabilizing. *IMA Journal of Mathematics Applied to Medicine and Biology* **1**, 123–133.

Hassell, M. P. 1986. Parasitoids and population regulation. In J. K. Waage and D. Greathead, eds., *Insect Parasitoids,* pp. 201–224. Academic Press, London.

Hassell, M. P., and May, R. M. 1973. Stability in insect host-parasite models. *Journal of Animal Ecology* **42**, 693–726.

Hassell, M. P., and May, R. M. 1974. Aggregation in predators and insect parasites and its effect on stability. *Journal of Animal Ecology* **43**, 567–594.

Hassell, M. P., and May, R. M. 1986. Generalist and specialist natural enemies in insect predator-prey interactions. *Journal of Animal Ecology* **55**, 923–940.

Hassell, M. P., and May, R. M. 1988. Spatial heterogeneity and the dynamics of parasitoid-host systems. *Annales Zoologici Fennici* **25**, 55–61.

Hassell, M. P., and Pacala, S. W. 1990. Heterogeneity and the dynamics of host-parasitoid interactions. *Philosophical Transactions of the Royal Society of London B* **330**, 203–220.

Hassell, M. P., and Varley, G. C. 1969. New inductive population model for insect parasites and its bearing on biological control. *Nature* **223**, 1133–1136.

Hassell, M. P.; Latto, J.; and May, R. M. 1989. Seeing the wood for the trees: Detecting density dependence from existing life table studies. *Journal of Animal Ecology* **58**, 883–892.

Hassell, M. P.; May, R. M.; Pacala, S.; and Chesson, P. L. 1991. The persistence of host-parasitoid associations in patchy environments. I. A general criterion. *American Naturalist* **138**, 568–583.

Hawkins, B. A. 1988a. Species diversity in the third and fourth trophic levels: Patterns and mechanisms. *Journal of Animal Ecology* **57**, 137–162.

Hawkins, B. A. 1988b. Do galls protect endophytic herbivores from parasitoids? A comparison of galling and non-galling Diptera. *Ecological Entomology* **13**, 473–477.

Hawkins, B. A. 1990. Global patterns of parasitoid assemblage size. *Journal of Animal Ecology* **59**, 57–72.

Hawkins, B. A. 1992. Parasitoid-host food webs and donor control. *Oikos* **65**, 159–162.

Hawkins, B. A. 1993. Refuges, host population dynamics, and the genesis of parasitoid diversity. In J. LaSalle and I. D. Gauld, eds., *Hymenoptera and Biodiversity*, pp. 235–256. CAB International Press, Wallingford, U.K.

Hawkins, B. A., and Gagné, R. J. 1989. Determinants of assemblage size for the parasitoids of Cecidomyiidae. *Oecologia* **81**, 75–88.

Hawkins, B. A., and Goeden, R. D. 1984. Organization of a parasitoid community associated with a complex of galls on *Atriplex* spp. in southern California. *Ecological Entomology* **9**, 271–292.

Hawkins, B. A., and Lawton, J. H. 1987. Species richness for parasitoids of British phytophagous insects. *Nature* **326**, 788–790.

Hawkins, B. A.; Askew, R. R.; and Shaw, M. R. 1990. Influences of host feeding-niche and foodplant type on generalist and specialist parasitoids. *Ecological Entomology* **15**, 275–280.

Hawkins, B. A.; Shaw, M. R.; and Askew, R. R. 1992. Relationships among assemblage size, host specialization and climatic variability in North American parasitoid communities. *American Naturalist* **139**, 58–79.

Hays, D. B., and Vinson, S. B. 1971. Acceptance of *Heliothis virescens* (F.) as a host by the parasite *Cardiochiles nigriceps* Viereck. *Animal Behaviour* **19**, 344–352.

He, J., and Chen, X. 1991. *Xiphyropronia* gen. nov., a new genus of Roproniidae (Hymenoptera: Proctotrupoidea) from China. *Canadian Journal of Zoology* **69**, 1717–1719.

Heatwole, H.; Davis, D. M.; and Wenner, A. M. 1962. The behaviour of *Megarhyssa*, a genus of parasitic hymenopterans (Ichneumonidae; Ephialtinae). *Zeitschrift für Tierpsychologie* **19**, 652–664.

Heatwole, H.; Davis, D. M.; and Wenner, A. M. 1964. Detection of mates and hosts by parasitic insects of the genus *Megarhyssa* (Hymenoptera: Ichneumonidae). *American Midland Naturalist* **71**, 374–381.

Hederwick, M. P.; El Agose, M.; Garaud, P.; and Periquet, G. 1985. Mise en evidence de males heterozygotes chez l'hymenoptère *Diadromus pulchellus*. *Genetique Selection Evolution* **17**, 303–310.

Hefetz, A. 1987. The role of Dufours gland secretions in bees. *Physiological Entomology* **12**, 243–253.

Hefetz, A. 1990. Individual badges and specific messages in multicomponent pheromones of bees (Hymenoptera, Apidae). *Entomologia Generalis* **15**, 103–113.

Hegdekar, B. M., and Arthur, A. P. 1973. Host haemolymph chemicals that induce oviposition in the parasite *Itoplectis conquisitor* (Hymenoptera: Ichneumonidae). *Canadian Entomologist* **105**, 787–793.

Heinrich, B. 1979. Foraging strategies of caterpillars. *Oecologia* **42**, 325–337.

Heinz, K. M. 1991. Sex-specifc reproductive consequences of body size in the solitary ectoparasitoid, *Diglyphus begini*. *Evolution* **45**, 1511–1515.

Heinz, K. M., and Parrella, M. P. 1990. The influence of host size on sex ratios in the parasitoid *Diglyphus begini* (Hymenoptera: Eulophidae). *Ecological Entomology* **15**, 391–399.

Heitmans, W.R.B.; Haccou, P., and van Alphen, J.J.M. 1992. Egg supply, clutch size and survival probability in *Aprostocetus hagenowii* (Ratz.) (Hymenoptera: Eulophidae), a gregarious parasitoid of cockroach oothecae. *Proceedings of the Section of Experimental and Applied Entomology of the Netherlands Entomological Society* **3**, 62–69.

Hemerik, L.; Driessen, G.; and Haccou, P. 1993. Effects of intra-patch experiences on patch time, search time and searching efficiency of the parasitoid *Leptopilina clavipes* (Hartig). *Journal of Animal Ecology* **62**, 33–44.

Heong, K. L. 1981. Searching preference of the parasitoid, *Anisopteromalus calandrae* (Howard) for different stages of the host *Callosobruchus maculatus* in the laboratory. *Researches in Population Biology* **23**, 177–191.

Hérard, F.; Keller, M. A.; Lewis, W. J.; and Tumlinson, J. H. 1988. Beneficial arthropod behavior mediated by airborne semiochemicals. IV. Influence of host diet on host-oriented flight chamber responses of *Microplitus demolitor* Wilkinson. *Journal of Chemical Ecology* **14**, 1597–1606.

Heraty, J. M., and Barber, K. N. 1990. Biology of *Obeza floridana* (Ashmead) and *Pseudochalcura gibbosa* (Provancher) (Hymenoptera: Eucharitidae). *Proceedings of the Entomological Society of Washington* **92**, 248–258.

Heraty, J. M., and Darling, D. C. 1984. Comparative morphology of the planidial larvae of Eucharitidae and Perilampidae (Hymenoptera: Chalcidoidea). *Systematic Entomology* **9**, 309–328.

Herre, E. A. 1985. Sex ratio adjustment in fig wasps. *Science* **228**, 896–898.

Herre, E. A. 1987. Optimality, plasticity and selective regime in fig wasp sex ratios. *Nature* **329**, 627–629.

Herre, E. A. 1989. Coevolution of reproductive characteristics in 12 species of New World figs and their pollinator wasps. *Experientia* **45**, 637–647.

Herting, B. 1960. Biologie der wespaläarktischen Raupenfliegen (Diptera, Tachinidae). *Monographien zur Angewandte Entomologie* **16**, 1–202.

Hespenheide, H. A. 1979. Are there fewer parasitoids in the tropics? *American Naturalist* **113**, 766–769.

Hiehata, K.; Hirose, Y.; and Kimoto, M. 1976. The effect of host age on the parasitism by three species of *Trichogramma* (Hymenoptera: Trichogrammatidae), egg parasitoids of *Papilio xuthus* Linné (Lepidoptera: Papilionidae). *Japanese Journal of Applied Entomology and Zoology* **20**, 31–36.

Hill, C. C. 1926. *Platygaster hiemalis* Forbes, a parasite of the Hessian fly. *Journal of Agricultural Research* **32**, 261–275.

Hinton, H. E. 1955. Protective devices of endopterygote pupae. *Transactions of the Society for British Entomology* **12**, 49–92.

Hirose, Y.; Kimoto, H.; and Hichata, K. 1976. The effect of host aggregation on parasitism by *Trichogramma papilionis* Nagarkatti (Hym.: Trichogrammatidae), an egg parasitoid of *Papilio xuthis* L. (Lep.: Papilionidae). *Applied Entomology and Zoology* **11**, 116–125.

Hochberg, M. E., and Hawkins, B. A. 1992. Refuges as a predictor of parasitoid diversity. *Science* **255**, 973–976.

Hodgkin, J. 1990. Sex determination compared in *Drosophila* and *Caenorhabditis*. *Nature* **344**, 721–728.

Hoelscher, C. E., and Vinson, S. B. 1971. The sex ratio of a hymenopterous parasitoid, *Campoletis perdistinctus*, as affected by photoperiod, mating and temperature. *Annals of the Entomological Society of America* **64**, 1373–1376.

van der Hoeven, N., and Hemerik, L. 1990. Superparasitism as an ESS: To reject or not to reject, that is the question. *Journal of Theoretical Biology* **146**, 467–482.

Hoffmeister, T. 1992. Factors determining the structure and diversity of parasitoid complexes in tephritid fruit flies. *Oecologia* **89**, 288–297.

Hofsvang, T. 1988. Mechanisms of host discrimination and intraspecific competition in the aphid parasitoid *Ephedrus cerasicola*. *Entomologia Experimentalis et Applicata* **48**, 233–239.

Hogarth, W. L., and Diamond, P. 1984. Interspecific competition in larvae between entomophagous parasitoids. *American Naturalist* **124**, 552–560.

Hokkanen, H., and Pimentel, D. 1984. New approach for selecting biological control agents. *Canadian Entomologists* **116**, 1109–1121.

Hokkanen, H., and Pimentel, D. 1989. New associations in biological control: Theory and practice. *Canadian Entomologists* **121**, 829–840.

Hokyo, N.; Kiritani, K.; Nakusuji, F.; and Shiga, M. 1966. Comparative biology of the two scelionid egg parasites of *Nezara viridula* L. (Hemiptera: Pentatomidae). *Applied Entomology and Zoology* **1**, 94–102.

Holmes, H. B. 1972. Genetic evidence for fewer progeny and a higher percent males when *Nasonia vitripennis* oviposits in previously parasitized hosts. *Entomophaga* **17**, 79–88.

Holmes, J. C., and Bethel, W. M. 1972. Modification of intermediate host behaviour by parasites. *Zoological Journal of the Linnean Society* **51**, 123–149.

Holt, R. D. 1977. Predation, apparent competition and the structure of prey communities. *Theoretical Population Biology* **12**, 197–229.

Holt, R. D. 1984. Spatial heterogeneity, indirect interactions, and the coexistence of prey species. *American Naturalist* **124**, 377–406.

Holt, R. D., and Lawton, J. H. 1993. Apparent competition and enemy-free space in insect host-parasitoid communities. *American Naturalist* (in press).

Hopkins, M.J.G. 1984. The parasitoid complex associated with stem boring *Apion* (Col.: Curculionidae) feeding on *Rumex* species (Polygonaceae). *Entomologist's Monthly Magazine* **120**, 187–192.

Hopper, K. R., and King, E. G. 1984. Preference of *Microplitis croceipes* (Hymenoptera: Braconidae) for instars and species of *Heliothis* (Lepidoptera: Noctuidae). *Environmental Entomology* **13**, 1145–1150.

Hopper, K. R., and Woolson, E. A. 1991. Labeling a parasitic wasp, *Microplitus croceipes* (Hymenoptera, Braconidae) with trace elements for mark recapture studies. *Annals of the Entomological Society of America* **84**, 255–262.

Hoshiba, H.; Okada, I.; and Kusanagi, A. 1981. The diploid drone of *Apis cerana japonica* and its chromosomes. *Journal of Apicultural Research* **20**, 143–147.

Houston, A. I.; Krebs, J. R.; and Erichsen, J. T. 1980. Optimal prey choice and discrimination time in the great tit (*Parus major* L.). *Behavioural Ecology and Sociobiology* **6**, 169–175.

Houston, A. I.; Clark, C.; McNamara, J.; and Mangel, M. 1988. Dynamics models in behavioural and evolutionary ecology. *Nature* **332**, 29–34.

Houston, A. I.; McNamara, J. M.; and Godfray, H.C.J. 1991. The effect of variability on host feeding and reproductive success in parasitoids. *Bulletin of Mathematical Biology* **54**, 465–476.

Howard, L. O. 1897. A study in insect parasitism: A consideration of the parasites of the white-marked Tussock Moth, with an account of their habits and interrelations, and with description of new species. *United States Department of Agriculture Technical Series* **5**, 5–57.

Hubbard, S. F., and Cook, R. M. 1978. Optimal foraging by parasitoid wasps. *Journal of Animal Ecology* **47**, 593–604.

Hubbard, S. F.; Marris, G.; Reynolds, A.; and Rowe, G. W. 1987. Adaptive patterns in the avoidance of superparasitism by solitary parasitic wasps. *Journal of Animal Ecology* **56**, 387–401.

Huddleston, T., and Gauld, I. D. 1988. Parasitic wasps (Ichneumonoidea) in British light-traps. *Entomologist* **107**, 134–154.

Huger, A. M.; Skinner, S. W.; and Werren, J. H. 1985. Bacterial infections associated with the son-killer trait in the parasitoid wasp *Nasonia (=Mormoniella) vitripennis* (Hymenoptera: Pteromalidae). *Journal of Invertebrate Pathology* **46**, 272–280.

Hughes, R. N. 1979. Optimal diets under the energy maximization premise: The effects of recognition time and learning. *American Naturalist* **113**, 209–221.

Hulspas-Jordaan, P. M., and van Lenteren, J. C. 1978. The relationship between host-plant leaf structure and parasitization efficiency of the parasitic wasp *Encarsia formosa* Gahan (Hymenoptera: Aphelinidae). *Mededelingen van de Faculteit Landbouwwetenschappen Rijksuniversiteit Gent* **43**, 431–440.

Hung, A.C.F.; Vinson, S. B.; and Summerlin, J. W. 1974. Male sterility in the red imported fire ant, *Solenopsis invicta. Annals of the Entomological Society of America* **67**, 909–912.

Hunter, M. S. 1989. Sex allocation and egg distribution of an autoparasitoid, *Encarsia pergandiella* (Hymenoptera: Aphelinidae). *Ecological Entomology* **14**, 57–67.

Hunter, M. S. 1993. Sex allocation in a field population of an autoparasitoid. *Oecologia* (in press). **93**, 421–428.

Hunter, M. S.; Nur, U.; and Werren, J. H. 1993. Origin of males by genome loss in an autoparasitoid wasp. *Heredity* **70**, 162–171.

Hurlbutt, B. L. 1987. Sexual size dimorphism in parasitoid wasps. *Biological Journal of the Linnean Society* **30**, 63–89.

Hutchinson, G. E. 1951. Copepodology for the ornithologist. *Ecology* **32**, 571–577.

Ikawa, T., and Okabe, H. 1985. Regulation of egg number per host to maximise the reproductive success in the gregarious parasitoid *Apanteles glomeratus* L. (Hymenoptera: Braconidae). *Applied Entomology and Zoology* **20**, 331–339.

Ikawa, T., and Suzuki, Y. 1982. Ovipositional experience of the gregarious parasitoid, *Apanteles glomeratus* (Hymenoptera: Braconidae), influencing her discrimination of the host larvae, *Pieris rapae crucivora. Applied Entomology and Zoology* **17**, 119–126.

Ivanova-Kasas, O. M. 1972. Polyembryony in insects. In S. J. Counce and C. H. Waddington, eds., *Developmental Systems, Insects*, vol. 2 Academic Press, New York.

Ives, A. R. 1988. Covariance, coexistence and the population-dynamics of 2 competitors using a patchy resource. *Journal of Theoretical Biology* **133**, 345–361.

Ives, A. R. 1989. The optimal clutch size of insects when many females oviposit per patch. *American Naturalist* **133**, 671–687.

Ives, A. R. 1992. Continuous-time models of host-parasitoid interactions. *American Naturalist* **140**, 1–29.

Ives, W.G.H., and Muldrew, J. A. 1981. *Pristiphora erichsonii* (Hartig), Larch Sawfly (Hymenoptera: Tenthredinidae). In J. S. Kelleher and M. A. Hulme, eds., *Biological Control Programmes against Insects and Weeds in Canada, 1969–1980*, pp. 369–380. Commonwealth Agricultural Bureau, Slough, U.K.

Iwasa, Y. 1991. Asynchronous pupation of univoltine insects as evolutionary stable phenology. *Researches in Population Ecology* **33**, 213–227.

Iwasa, Y.; Higashi, M.; and Yamamura, N. 1981. Prey distribution as a factor determining the choice of optimal foraging strategy. *American Naturalist* **117**, 710–723.

Iwasa, Y.; Odendaal, F. J.; Murphy, D. D.; Ehrlich, P. R.; and Launer, A. E. 1983. Emergence patterns in butterflies: A hypothesis and a test. *Theoretical Population Biology* **23**, 363–379.

Iwasa, Y.; Suzuki, Y.; and Matsuda, H. 1984. Theory of oviposition strategy of parasitoids. I. Effect of mortality and limited egg number. *Theoretical Population Biology* **26**, 205–227.

Iwata, K. 1960. The comparative anatomy of the ovary of Hymenoptera. Part V. Ichneumonidae. *Acta Hymenopterologica* **1**, 115–169.

Iwata, K. 1966. The comparative anatomy of the ovary in Hymenoptera. Supplement on Ichneumonidae, *Coccygomimus luctuosa* Smith, *C. parnarae* Viereck and *C. pluto* Ashmead. *Acta Hymenopterologica* **2**, 133–135.

Jackson, C. G.; Cohen, A. C.; and Verdugo, C. L. 1988. Labeling *Anaphes ovijentatis* (Hymenoptera, Mymaridae), an egg parasite of *Lygus* spp. (Hemiptera, Miridae). *Annals of the Entomological Society of America* **81**, 919–922.

Jackson, D. J. 1958. Observations on the biology of *Caraphractus cinctus* Walker (Hymenoptera: Mymaridae), a parasite of the eggs of Dytiscidae I. Methods of rearing and numbers bred on different host eggs. *Transactions of the Royal Entomological Society, London* **110**, 533–566.

Jackson, D. J. 1964. Observations on the life-history of *Mestocharis bimacularis* (Dalman) (Hym. Eulophidae), a parasitoid of the eggs of Dytiscidae. *Opuscula Entomologica* **29**, 81–97.

Jackson, D. J. 1966. Observations on the biology of *Caraphractus cinctus* Walker (Hymenoptera: Mymaridae), a parasite of the eggs of Dytiscidae (Coleoptera) III. The adult life and sex ratio. *Transactions of the Royal Entomological Society, London* **118**, 23–49.

Jackson, D. J. 1969. Observations on the female reproductive organs and the poison apparatus of *Caraphractus cinctus* Walker (Hymenoptera: Mymaridae). *Zoological Journal of the Linnean Society* **48**, 59–81.

Jaenike, J. 1978. On optimal oviposition behaviour in phytophagous insects. *Theoretical Population Biology* **14**, 350–356.

Janssen, A. 1989. Optimal host selection by *Drosophila* parasitoids in the field. *Functional Ecology* **3**, 469–479.

Janzen, D. H. 1973. Sweep samples of tropical foliage insects: Effects of seasons, vegetation types, elevation, time of day, and insularity. *Ecology* **54**, 687–701.

Janzen, D. H. 1979. How to be a fig. *Annual Review of Ecology and Systematics* **10**, 13–51.

Janzen, D. H. 1981. The peak in North American ichneumonid species richness lies between 38° and 42°N. *Ecology* **62**, 532–537.

Janzen, D. H., and Pond C. M. 1975. A comparison, by sweep sampling, of the arthropod fauna of secondary vegetation in Michigan, England and Costa Rica. *Transactions of the Royal Entomological Society of London* **127**, 33–50.

Janzen, D. H.; Ataroff, M.; Farinas, M.; Reyes, S.; Rincon, N.; Soler, A.; Soriano, P.; and Vera, M. 1976. Changes in the arthropod community along an elevational transect in the Venezuelan Andes. *Biotropica* **8**, 193–203.

Jeffries, M. J., and Lawton, J. H. 1984. Enemy free space and the structure of ecological communities. *Biological Journal of the Linnean Society* **23**, 269–286.

Jenni, W. 1951. Beitrag zur Morphologie und Biologie der Cynipide *Pseudeucoila bo-*

chei Weld, eines Larvenparasiten von *Drosophila melanogaster* Meig. *Acta Zoologica, Stockholm* **32**, 177–254.

Jervis, M. A. 1979. Courtship, mating and "swarming" in *Aphelopus melaleucus* (Dalman) (Hymenoptera: Dryinidae). *Entomologists' Gazette* **30**, 191–193.

Jervis, M. A., and Kidd, N.A.C. 1986. Host-feeding strategies in hymenopteran parasitoids. *Biological Reviews* **61**, 395–434.

Johansson, A. S. 1950. Studies on the relation between *Apanteles glomeratus* L. (Hym.; Braconidae) and *Pieris brassicae* (Lepid.; Pieridae). *Norsk Entomologisk Tidsskrift* **8**, 145–186.

Johnson, B. 1959. Effect of parasitisation by *Aphidius plantensis* Brèthes on the developmental physiology of its host, *Aphis craccivora* Koch. *Entomologia Experimentalis et Applicata* **2**, 82–99.

Johnson, D. W. 1988. Eucharitidae (Hymenoptera: Chalcidoidea): Biology and potential for biological control. *Florida Entomologist* **71**, 528–537.

Johnson, J. B.; Miller, T. D.; Heraty, J. M.; and Merickel, F. W. 1986. Observations on the biology of two species of *Orasema* (Hymenoptera: Eucharitidae). *Proceedings of the Entomological Society of Washington* **88**, 542–549.

Johnson, N. F. 1992. Catalog of World species of Proctotrupoidea, exclusive of Platygasteridae. *Memoirs of the American Entomological Institute* **51**, 1–825.

Jones, D. 1987. Material from adult female *Chelonus* sp. directs expression of altered developmental program of host Lepidoptera. *Journal of Insect Physiology* **33**, 129–134.

Jones, D.; Jones, G.; and Hammock, B. D. 1981. Developmental and behavioural responses of larval *Trichoplusia ni* to parasitization by an imported braconid *Chelonus* sp. *Physiological Entomology* **6**, 387–394.

Jones, D. A. 1966. On the polymorphism of cyanogenesis in *Lotus corniculatus*. I. Selection by animals. *Canadian Journal of Cytology* **8**, 556–567.

Jones, E. P. 1937. The egg parasites of the cotton bollworm, *Heliothis armigera* (Hübn.) (*obsoleta* Fabr.) in Southern Rhodesia. *Report of the Mazoe Citrus Experimental Station* **1936**, 37–105.

Jones, R. E. 1987. Ants, parasitoids and the cabbage butterfly *Pieris rapae*. *Journal of Animal Ecology* **56**, 739–749.

Jones, R. L. 1981. Chemistry of semiochemicals involved in parasitoid-host and predator prey relationships. In D. A. Nordlund, R. L. Jones, and W. J. Lewis, eds., *Semiochemicals, Their Role in Pest Control*, pp. 239–250. John Wiley, New York.

Jones, R. L.; Lewis, W. J.; Bowman, M. C.; Beroza, M.; and Bierl, B. A. 1971. Host-seeking stimulant for parasite of corn earworm: Isolation, identification and synthesis. *Science* **173**, 842–843.

Jones, R. L.; Lewis, W. J.; Beroza, M.; Bierl, B. A.; and Sparks, A. N. 1973. Host-seeking stimulants (kairomones) for the egg parasite, *Trichogramma evanescens*. *Environmental Entomology* **2**, 593–596.

Jones, W. T. 1982. Sex ratio and host size in a parasitic wasp. *Behavioural Ecology and Sociobiology* **10**, 207–210

Joseph, K. J. 1958. Recherches sur les chalcidiens, *Blastophaga psenes* (L.) et *Philotrypesis caricae* (L.) du figuier (*Ficus carica* L.). *Annales des Sciences Naturelles Zoologie* **20**, 197–260.

Juliano, S. A. 1982. Influence of host age on host acceptability and suitability for a species of *Trichogramma* (Hymenoptera: Trichogrammatidae) attacking aquatic Diptera. *Canadian Entomologist* **114**, 713–740.

Kaiser, L.; Pham-Delegue, M. H.; Bakchine, E.; and Masson, C. 1989. Olfactory responses of *Trichogramma maidis* Pint. et Voeg.: Effect of chemical cues and behavioral plasticity. *Journal of Insect Behaviour* **2**, 701–712.

Kakehashi, N.; Suzuki, Y.; and Iwasa, Y. 1984. Niche overlap of parasitoids in host-parasitoid systems: Its consequences to single versus multiple introduction controversy in biological control. *Journal of Applied Ecology* **21**, 115–131.

Kanungo, K. 1955. Effect of superparasitism on sex ratio and mortality. *Current Science* **24**, 59–60.

Kareiva, P. 1987a. The ecology of invasions: Theory or anecdotes? (Book review.) *Ecology* **68**, 1556.

Kareiva, P. 1987b. Habitat fragmentation and the stability of predator-prey interactions. *Nature* **326**, 388–390.

Kareiva, P. 1990. Population dynamics in spatially complex environments: Theory and data. *Philosophical Transactions of the Royal Society, London B* **330**, 175–190.

Kareiva, P., and Odell, G. 1987. Swarms of predators exhibit preytaxis if individual predators use area restricted search. *American Naturalist* **130**, 233–270.

Karlin, S., and Lessard, S. 1986. *Theoretical Studies on Sex Ratio Evolution.* Princeton University Press, Princeton.

Karp, R. D. 1990. Cell-mediated-immunity in invertebrates. *Bioscience* **40**, 732–737.

Kasparayan, D. R. 1981. Ichneumonidae, subfamily Tryphoninae, tride Tryphonini. *Fauna of the USSR* **106**, 1–414 (English translation, original in Russian, 1973).

Kato, M. 1984. Mining pattern of the honeysuckle leaf-miner *Phytomyza lonicerae.* *Researches in Population Biology* **26**, 84–96.

Kato, M. 1985. The adaptive significance of leaf-mining pattern as an anti-parasitoid strategy: Theoretical study. *Researches in Population Biology* **27**, 265–275.

Kearns, C. W. 1934a. A hymenopterous parasite (*Cephalonomia gallicola* Ashm.) new to the cigarette beetle (*Lasioderma serricorne* Fab.). *Journal of Economic Entomology* **27**, 801–806.

Kearns, C. W. 1934b. Method of wing inheritance in *Cephalonomia gallicola* Ashm. (Bethylidae; Hymenoptera). *Annals of the Entomological Society of America* **27**, 533–541.

Kemner, N. A. 1926. Zur Kenntnis der Staphyliniden-Larven. II. Die Lebensweise und die parasitische Entwicklung der echten Aleochariden. *Entomolologisk Tidskrift* **47**, 133–170.

Kennedy, B. H. 1979. The effect of multilure on parasites of the European elm bark beetle, *Scolytus multistriatus.* *Bulletin of the Entomological Society of America* **25**, 116–118.

Kennett, C. E.; Huffaker, C. B.; and Finney, G. L. 1966. The role of an autoparasitic aphelinid, *Coccophagus utilus* Doutt, in the control of *Parlatoria oleae* (Colvée). *Hilgardia* **37**, 255–282.

Kerkut, G. A., and Gilbert, L. 1985. *Comprehensive Insect Physiology, Biochemistry and Pharmacology,* vols. 7 and 8. Pergamon Press, Oxford.

Kerr, W. E. 1974. Advances in cytology and genetics of bees. *Annual Review of Entomology* **19**, 253–268.

Khoo, B. K.; Forgash, A. J.; Respicio, N. C.; and Ramaswamy, S. B. 1985. Multiple progeny production by gypsy moth parasites *Brachymeria* spp. (Hymenoptera, Chalcididae), following exposure to diflubenzuron. *Environmental Entomology* **14**, 820–825.

King, B. H. 1987. Offspring sex ratios in parasitoid wasps. *Quarterly Review of Biology* **62**, 367–396.

King, B. H. 1988. Sex-ratio manipulation in response to host size by the parasitoid wasp *Spalangia cameroni*: A laboratory study. *Evolution* **42**, 1190–1198.

King, B. H. 1989a. Host-size dependent sex ratios among parasitoid wasps: Does host growth matter. *Oecologia* **78**, 420–426.

King, B. H. 1989b. A test of local mate competition theory with a solitary species of parasitoid wasp, *Spalangia cameroni*. *Oikos* **54**, 50–54.

King, B. H. 1990. Sex ratio manipulation by the parasitoid wasp *Spalangia cameroni* in response to host age: A test of the host-size model.. *Evolutionary Ecology* **4**, 149–156.

King, B. H., and Skinner, S. W. 1991a. Sex ratio in a new species of *Nasonia* with fully-winged males. *Evolution* **45**, 225–228.

King, B. H., and Skinner, S. W. 1991b. Proximal mechanisms of the sex ratio and clutch size responses of the wasp *Nasonia vitripennis* to parasitized hosts. *Animal Behaviour* **42**, 23–32.

King, P. E. 1961. A possible method of sex ratio determination in the parasitic Hymenoptera. *Nature* **189**, 330–331.

King, P. E. 1963. The rate of egg resorption in *Nasonia vitripennis* deprived of hosts. *Proceedings of the Royal Entomological Society, London A* **38**, 98–100.

King, P. E.; Askew, R. R.; and Sanger, C. 1969. The detection of parasitised hosts by males of *Nasonia vitripennis* (Walker) (Hymenoptera: Pteromalidae) and some possible implications. *Proceedings of the Royal Entomological Society of London A* **44**, 85–90.

Kirby, W. 1835. On the power and wisdom of God, as manifested in the creation of animals and in their history, habits and instincts. *Bridgewater Treatises* **7**, 243–244.

Kirby, W., and Spence, W. 1816. *An Introduction to Entomology or Elements of the Natural History of Insects.* Longman, Hurst, Rees, Orme and Brown, London.

Kirkpatrick, M., and Ryan, M. J. 1991. The evolution of mating preferences and the paradox of the lek. *Nature* **350**, 33–38

Kishi, Y. 1970. Difference in the sex ratio of the pine bark weevil parasite, *Dolichomitus* sp. (Hymenoptera: Ichneumonidae), emerging from different host species. *Applied Entomology and Zoology* **5**, 126–132.

Kitano, H. 1982. Effect of the venom of the gregarious parasitoid, *Apanteles glomeratus*, on its hemocytic encapsulation by the host, *Pieris*. *Journal of Invertebrate Pathology* **40**, 61–67.

Kitano, H. 1986. The role of *Apanteles glomeratus* venom in the defensive response of its host, *Pieris rapae crucivora*. *Journal of Insect Physiology* **32**, 369–375.

Kitano, H., and Natatsuji, N. 1978. Resistance of *Apanteles* eggs to the haemocytic encapsulation by their habitual host *Pieris*. *Journal of Insect Physiology* **24**, 261–271.

Kitano, H.; Wago, H.; and Arakawa, T. 1990. Possible role of teratocytes of the gregarious parasitoid, *Cotesia* (=*Apanteles*) *glomerata*, in the suppression of phenoloxidase activity in the larval host, *Pieris rapae crucivora*. *Archives of Insect Biochemistry and Physiology* **13**, 177–185.

Kjellberg, F., and Valdeyron, G., eds., 1984. *Minisymposium: Figs and Fig Insects.* CNRS, Montpellier, France.

Kjellberg, F.; Gouyon, P.-H.; Ibrahim, M.; Raymond, M.; and Valdeyron, G. 1987. The stability of the symbiosis between dioecious figs and their pollinators: A study of *Ficus carica* L. and *Blastophaga psenes* L. *Evolution* **41**, 693–704.

Klomp, H. 1981. Parasitic wasps as sleuthhounds: Response of an ichneumon wasp to the trail of its host. *Netherlands Journal of Zoology* **31**, 762–772.

Klomp, H., and Teerink, B. J. 1962. Host selection and number of eggs per oviposition in the egg parasite *Trichogramma embryophagum* Htg. *Nature* **195**, 1020–1021.

Klomp, H., and Teerink, B. J. 1967. The significance of oviposition rate in the egg parasite, *Trichogramma embryophagum* Htg. *Archives Neerlandaises de Zoologie* **17**, 350–375.

Klomp, H., and Teerink, B. J. 1978a. The elimination of supernumerary larvae of the gregarious egg parasitoid *Trichogramma embryophagum* (Hym.: Trichogrammatidae) in eggs of the host *Ephestia kuehniella* (Lep.: Pyralidae). *Entomophaga* **23**, 153–159.

Klomp, H., and Teerink, B. J. 1978b. The epithelium of the gut as a barrier against encapsulation by blood cells in three species of parasitoids of *Bupalus piniarius* (Lep.; Geometridae). *Netherlands Journal of Zoology* **28**, 132–139.

Klomp, H.; Teerink, B. J.; and Wei Chun Ma. 1980. Discrimination between parasitized and unparasitized hosts in the egg parasite *Trichogramma embryophagum* (Hym.: Trichogrammatidae): A matter of learning and forgetting. *Netherlands Journal of Zoology* **30**, 254–277.

Knowlton, N., and Parker, G. A. 1979. An evolutionarily stable strategy approach to indiscriminate spite. *Nature* **279**, 419–421.

Kochetova, N. I. 1972. [The effect of population density of *Trichogramma* females on the sex ratio of their progeny.] *Ekologiya* **3**, 84–86 (in Russian).

Kochetova, N. I. 1978. Factors determining the sex ratio in some entomophagous hymenoptera. *Entomological Review* **57**, 1–5.

Kogan, M., and Legner, E. F. 1970. A biosystematic revision of the genus *Muscidifurax* (Hymenoptera: Pteromalidae) with descriptions of four new species. *Canadian Entomologist* **102**, 1268–1290.

Koptur, S. 1985. Alternative defenses against herbivores in *Inga* (Fabaceae: Mimosoideae) over an elevational gradient. *Ecology* **66**, 1639–1650.

Kornhauser, S. J. 1919. The sexual characteristics of the membracid *Thelia bimaculata* (Fab.). I. External changes induced by *Aphelopus theliae* (Gahan). *Journal of Morphology* **32**, 531–635.

Kouamé, K. L., and Mackauer, M. 1991. Influence of aphid size, age and behaviour on host choice by the parasitoid wasp *Ephedrus californicus*: A test of host-size models. *Oecologia* **88**, 197–203.

Krebs, J. R., and Davies, N. B. 1987. *An Introduction to Behavioural Ecology.* Blackwell Scientific, Oxford.

Krebs, J. R., and Kacelnik, A. 1991. Decision-making. In J. R. Krebs, and N. B. Davies, eds., *Behavioural Ecology, an Evolutionary Approach*, 3d ed., pp. 105–136. Blackwell Scientific, Oxford.

Krebs, J. R., and McCleery, R. H. 1984. Optimization in behavioural ecology. In J. R. Krebs and N. B. Davies, eds., *Behavioural Ecology, an Evolutionary Approach*, 2d ed., pp. 91–121. Blackwell Scientific, Oxford.

Kuenzel, N. 1975. Population dynamics of protelean parasites (Hymenoptera: Aphelinidae) attacking a natural population of *Trialeurodes packardi* (Homoptera: Aley-rodidae) and new host records for two species. *Proceedings Entomological Society of Washington* **79**, 400–404.

Kunzel, J. G.; Gossniklaus-Buergin, C.; Karpells, S. T.; and Lazrein, B. 1990. Aryl-phorin of *Trichoplusia ni*: Characterization and parasite-induced precocious in-crease in titer. *Archives of Insect Biochemistry and Physiology* **13**, 117–126.

Kurosu, U. 1985. Male altruism and wing polymorphism in a parasitic wasp. *Journal of Ethology* **3**, 11–19.

Lack, D. 1947. The significance of clutch size. *Ibis* **89**, 309–352.

Laidlaw, H. H.; Gomes, F. P.; and Kerr, W. E. 1956. Estimation of the number of lethal alleles in a panmictic population of *Apis mellifera* L. *Genetics* **41**, 179–188.

Laing, D. R., and Caltagirone, L. E. 1969. Biology of *Habrobracon lineatellae* (Hyme-noptera, Braconidae). *Canadian Entomologist* **101**, 135–142.

Laing, J. 1937. Host-finding by insect parasites. I. Observations on the finding of hosts by *Alysia manducator, Mormoniella vitripennis* and *Trichogramma evanescens*. *Journal of Animal Ecology* **6**, 298–317.

Lanier, G. N.; Birch, M. C.; Schmitz, R. F.; and Furniss, M. M. 1972. Pheromones of *Ips pini* (Coleoptera; Scolytidae): Variations in response among three populations. *Canadian Entomologist* **104**, 1917–1923.

LaSalle, J., and Gauld, I. D. 1991. Parasitic Hymenoptera and the biodiversity crisis. *Redia* **74**, 315–334

Lathrop, F. H., and Newton, R. C. 1933. The biology of *Opius melleus* Gahan, a parasite of the blueberry maggot. *Journal of Agricultural Research* **46**, 143–160.

Latta, B. 1987. Adaptive and non-adative suicide in aphids (Scientific Correspon-dence). *Nature* **330**, 701.

Lawrence, P. O. 1981a. Host vibration: A cue to host location by the parasitoid *Bioste-res longicaudatus. Oecologia* **48**, 249–251.

Lawrence, P. O. 1981b. Interference competition and optimal host selection in the parasitic wasp *Biosteres longicaudatus. Annals of the Entomological Society of America* **74**, 540–544.

Lawrence, P. O. 1982. *Biosteres longicaudatus*: Developmental dependence on host (*Anastrepha suspensa*) physiology. *Experimental Parasitology* **53**, 396–405.

Lawrence, P. O. 1986. Host-parasite hormonal interactions: An overview. *Journal of Insect Physiology* **32**, 295–298.

Lawrence, P. O. 1990. The biochemical and physiological effects of insect hosts and the development and ecology of their insect parasitoids: An overview. *Archives of Insect Biochemistry and Physiology* **13**, 217–228.

Lawton, J. H. 1978. Host-plant influences on insect diversity: The effects of space and time. In L. A. Mound and N. Waloff eds., *Diversity of Insect Faunas,* pp. 105–125. Blackwell Scientific, Oxford.

Lawton, J. H. 1983. Plant architecture and the diversity of phytophagous insects. *An-nual Review of Entomology* **28**, 23–39.

Lawton, J. H. 1986. The effects of parasitoids on phytophagous insect communities. In J. K. Waage and D. Greathead, eds., *Insect Parasitoids,* pp. 265–287. Academic Press, London.

Lawton, J. H., and McNeill, S. 1979. Between the devil and the deep blue sea: On the problems of being a herbivore. In R. M. Anderson, B. D. Turner, and L. R. Taylor, eds., *Population Dynamics,* pp. 223–244. Blackwell Scientific, Oxford.

Lawton, J. H., and Schröder, D. 1977. Effects of plant type, size of geographical range and taxonomic isolation on the number of insect species associated with British plants. *Nature* **265**, 137–140.

Lawton, J. H., and Strong, D. R. 1981. Community patterns and competition in folivorous insects. *American Naturalist* **118**, 317–338.

Legner, E. F. 1969. Reproductive isolation and size variation in the *Muscidifurax* complex. *Annals of the Entomological Society of America* **62**, 382–385.

Legner, E. F. 1977. Temperature, humidity and depth of habitat influencing host destruction and fecundity of muscoid fly parasites. *Entomophaga* **22**, 199–206.

Legner, E. F. 1979. Prolonged culture and inbreeding effects on reproductive rates of two pteromalid parasites of muscoid flies. *Annals of the Entomological Society of America* **72**, 114–118.

Legner, E. F. 1985a. Effects of scheduled high temperature on male production in thelytokous *Muscidifurax uniraptor* (Hymenoptera: Pteromalidae). *Canadian Entomologist* **117**, 383–389.

Legner, E. F. 1985b. Natural and induced sex ratio changes in populations of thelytokous *Muscidifurax uniraptor* (Hymenoptera: Pteromalidae). *Annals of the Entomological Society of America* **78**, 398–402.

Legner, E. F. 1987a. Transfer of thelytoky to arrhenotokous *Muscidifurax raptor* Girault and Sanders (Hymenoptera: Pteromalidae). *Canadian Entomologist* **119**, 265–271.

Legner, E. F. 1987b. Inheritance of gregarious and solitary oviposition in *Muscidifurax raptorellus* Kogan and Legner (Hymenoptera: Pteromalidae). *Canadian Entomologist* **119**, 791–808.

Legner, E. F. 1988a. *Muscidifurax raptorellus* (Hymenoptera: Pteromalidae) females exhibit post mating behaviour typical of the male genome. *Annals of the Entomological Society of America* **81**, 524–527.

Legner, E. F. 1988b. Studies of four thelytokous Puerto Rican isolates of *Muscidifurax uniraptor* (Hymenoptera: Pteromalidae). *Entomophaga* **33**, 269–280.

Legner, E. F. 1988c. Quantification of heterotic behaviour in parasitic Hymenoptera. *Annals of the Entomological Society of America* **81**, 657–681.

Legner, E. F. 1988d. Hybridization in principal parasitoids of synanthropic Diptera: The genus *Muscidifurax* (Hymenoptera: Pteromalidae). *Hilgardia* **56**, 1–36.

Legner, E. F. 1989a. Wary genes and accretive inheritance in Hymenoptera. *Annals of the Entomological Society of America* **82**, 245–249.

Legner, E. F. 1989b. Paternal influences in males of *Muscidifurax raptorellus* (Hymenoptera: Pteromalidae). *Entomophaga* **34**, 307–320.

Legner, E. F. 1989c. Phenotypic expression of polygenes in *Muscidifurax raptorellus* (Hymenoptera: Pteromalidae), a synanthropic fly parasitoid. *Entomophaga* **34**, 523–530.

Legner, E. F. 1991. Estimations of number of active loci, dominance and heritability in polygenic inheritance of gregarious behavior in *Muscidifurax raptorellus* (Hymenoptera, Pteromalidae). *Entomophaga* **36**, 1–18.

Leigh, E. G.; Herre, E. A.; and Fischer, E. A. 1985. Sex allocation in animals. *Experientia* **41**, 1265–1276.

Leius, K. 1967. Influence of wild flowers on parasitism of tent caterpillar and codling moth. *Canadian Entomologist* **93**, 771–780.

van Lenteren, J. C. 1972. Contact chemoreceptors on the ovipositor of *Pseudeucoila bochei* Weld (Cynipidae). *Netherlands Journal of Zoology* **22**, 347–350.

van Lenteren, J. C. 1976. The development of host discrimination and the prevention of superparasitism in the parasite *Pseudeucoila bochei* Weld. *Netherlands Journal of Zoology* **26**, 1–83.

van Lenteren, J. C. 1981. Host discrimination by parasitoids. In D. A. Nordlund, R. L. Jones, and W. J. Lewis, eds., *Semiochemicals, Their Role in Pest Control,* pp. 153–180. John Wiley, New York.

van Lenteren, J. C. 1991. Encounters with parasitized hosts: To leave or not to leave a patch. *Netherlands Journal of Zoology* **41**, 144–157.

van Lenteren, J. C., and Bakker, K. 1975. Discrimination between parasitized and unparasitized hosts in the parasitic wasp *Pseudeucoila bochei*: A matter of learning. *Nature* **254**, 417–419.

van Lenteren, J. C., and Bakker, K. 1978. Behavioural aspects of the functional response of a parasite (*Pseudocoila bochei* Weld) to its host (*Drosophila melanogaster*). *Netherlands Journal of Zoology* **28**, 213–233.

van Lenteren, J. C.; Bakker, K.; and van Alphen, J.J.M. 1978. How to analyse host discrimination. *Ecological Entomology* **3**, 71–75.

Leon, J. A. 1985. Germination strategies. In P. J. Greenwood, P. H. Harvey, and M. Slatkin, eds., *Evolution. Essays in Honour of John Maynard Smith,* pp. 129–142. Cambridge University Press, Cambridge, U.K.

Lessells, C. M. 1985. Parasitoid foraging: Should parasitism be density-dependent? *Journal of Animal Ecology* **57**, 27–41.

Lewis, W. J. 1970. Life history and anatomy of *Microplitis croceipes* (Hymenoptera: Braconidae), a parasite of *Heliothis* spp. (Lepidoptera: Noctuidae). *Annals of the Entomological Society of America* **63**, 67–70.

Lewis, W. J., and Jones, R. L. 1971. Substance that stimulates host-seeking by *Microplitis croceipes* (Hymenoptera: Braconidae), a parasite of *Heliothis* species. *Annals of the Entomological Society of America* **64**, 557–558.

Lewis, W. J., and Redlinger, L. J. 1969. Suitability of eggs of the almond moth *Cadra cautella* of various ages for parasitism by *Trichogramma evanescens*. *Annals of the Entomological Society of America* **62**, 1482–1484.

Lewis, W. J., and Takasu, K. 1990. Use of learned odours by a parasitic wasp in accordance with host and food needs. *Nature* **348**, 635–636.

Lewis, W. J., and Tumlinson, J. H. 1988. Host detection by chemically mediated associative learning in a parasitic wasp. *Nature* **331**, 257–259.

Lewis, W. J., and Vinson, S. B. 1968. Immunological relationships between the parasite *Cardiochiles nigriceps* Viereck and certain *Heliothis* species. *Journal of Insect Physiology* **14**, 613–626.

Lewis, W. J.; Jones, R. L., and Redlinger, L. J. 1971a. Moth odour: A method of host-finding by *Trichogramma evanescens*. *Journal of Economic Entomology* **64**, 557–558.

Lewis, W. J.; Snow, J. W.; and Jones, R. L. 1971b. A pheromone trap for studying populations of *Cardiochiles nigriceps*, a parasite of *Heliothis virescens*. *Journal of Economic Entomology* **64**, 1417–1421.

Lewis, W. J.; Jones, R. L.; and Sparks, A. N. 1972. A host-seeking stimulant for the egg parasite *Trichogramma evanescens*: Its source and a demonstration of its laboratory and field activity. *Annals of the Entomological Society of America* **65**, 1087–1089.

Lewis, W. J.; Jones, R. L.; Nordlund, D. A.; and Sparks, A. N. 1975a. Kairomones and their use for management of entomophagous insects: II. Evaluation for increasing rates of parasitization by *Trichogramma* spp. in the field. *Journal of Chemical Ecology* **1**, 343–347.

Lewis, W. J.; Jones, R. L.; Nordlund, D. A.; and Gross, H. R. 1975b. Kairomones and their use for management of entomophagous insects: II. Mechanisms causing increase in rate of parasitization by *Trichogramma* spp. *Journal of Chemical Ecology* **1**, 349–360.

Lewis, W. J.; Nordlund, D. Q.; Gueldner, R. C.; Teel, P. D.; and Tumlinson, J. H. 1982. Kairomones and their use for management of entomophagous insects. XIII. Kairomonal activity for *Trichogramma* spp. of abdominal tips, feces, and a synthetic sex pheromone blend of *Heliothis zea* (Boddie) moths. *Journal of Chemical Ecology* **8**, 1323–1332.

Lewis, W. J.; Vet, L.E.M.; Tumlinson, J. H.; van Lenteren, J. C.; and Papaj, D. R. 1990. Variations in parasitoid foraging behavior: Essential element of a sound biological control theory. *Environmental Entomology* **19**, 1183–1193.

Lewontin, R. C. 1965. Selection for colonizing ability. In H. G. Baker and G. L. Stebbins, eds., *The Genetics of Colonizing Species*, pp. 79–94. Academic Press, New York.

Lindroth, C. H. 1971. Disappearance as a protective factor. A supposed case of Batesian mimicry among beetles (Coleoptera: Carabidae and Chrysomelidae). *Entomologica Scandinavica* **2**, 41–48.

Liu, S.-S. 1985. Development, adult size and fecundity of *Aphidius sonchi* reared in two instars of aphid host, *Hyperomyzus lactucae*. *Entomologia Experimentalis et Applicata* **37**, 41–48.

Liu, S.-S.; and Morton, R. 1986. Distribution of superparasitization in the aphid parasite *Aphidius sonchi*. *Entomologia Experimentalis et Applicata* **40**, 141–145.

Liu, S.-S.; Morton, R.; and Hughes, R. D. 1984. Oviposition preferences of a hymenopterous parasites for certain instars of its aphid host. *Entomologia Experimentalis et Applicata* **35**, 249–254.

Lloyd, D. C. 1940. Host selection by hymenopterous parasites of the moth *Plutella maculipennis* Curtis. *Proceedings of the Royal Society B* **128**, 451–484.

Loke, W. H., and Ashley, T. R. 1984. Behavioral and biological responses of *Cotesia marginiventris* to kairomones of the Fall Armyworm, *Spodoptera frugiperda*. *Journal of Chemical Ecology* **10**, 521–529.

Lubbock, J. 1862. On two aquatic Hymenoptera, one of which uses its wings in swimming. *Transactions of the Linnean Society, London* **24**, 135–142.

Luck, R. F. 1990. Evaluation of natural enemies for biological control: A behavioral approach. *Trends in Ecology and Evolution* **5**, 196–199.

Luck, R. F., and Podoler, H. 1985. Competitive exclusion of *Aphytis lingnanensis* by *A. melinus*: Potential role of host size. *Ecology* **66**, 904–913.

Luck, R. F.; Podoler, H.; and Kfir, R. 1982. Host selection and egg allocation behaviour by *Aphytis melinus* and *A. lignanensis*: Comparison of two facultatively gregarious parasitoids. *Ecological Entomology* **7**, 397–408.

Lyons, L. A. 1977. Parasitism of *Neodiprion sertifer* (Hymenoptera: Diprionidae) by *Exenterus* spp. (Hymenoptera: Ichneumonidae) in Ontario, 1962–1972, with notes on the parasites. *Canadian Entomologist* **109**, 555–564.

Lyttle, T. W.; Sandler, L. M.; Prout, T.; and Perkins, D. D., eds., 1991. The genetics and evolutionary biology of meiotic drive (symposium proceedings). *American Naturalist* **137**, 281–456.

McAllister, M. K., and Roitberg, B. D. 1987. Adaptive suicidal behaviour in pea aphids. *Nature* **328**, 797–799.

McAllister, M. K.; Roitberg, B. D.; and Weldon, K. L. 1990. Adaptive suicide in pea aphids: Decisions are cost sensitive. *Animal Behaviour* **40**, 167–175.

MacArthur, R. H., and Pianka, E. R. 1966. On the optimal use of a patchy environment. *American Naturalist* **100**, 603–609.

McAuslane, H. J.; Vinson, S. B.; and Williams, H. J. 1990a. Influence of host plant on mate location by the parasitoid *Campoletis sonorensis* (Hymenoptera: Ichneumonidae). *Environmental Entomology* **19**, 26–31.

McAuslane, H. J.; Vinson, S. B.; and Williams. 1990b. Effect of host diet on flight behaviour of the parasitoid *Campoletis sonorensis* (Hymenoptera: Ichneumonidae). *Journal of Entomological Science* **25**, 562–570.

McBrien, H., and Mackauer, M. 1990. Heterospecific larval competition and host discrimination in two species of aphid parasitoids: *Aphidius ervi* and *Aphidius smithi*. *Entomologia Experimentalis et Applicata* **56**, 145–153.

McBrien, H., and Mackauer, M. 1991. Decision to superparasitise based on larval survival: Competition between aphid parasitoids: *Aphidius ervi* and *Aphidius smithi*. *Entomologia Experimentalis et Applicata* **59**, 145–150.

McCall, P. J.; Turlings, T.C.J.; Lewis, W. J.; and Tumlinson, J. H. 1993. The role of plant volatiles in host location by the specialist parasitoid *Microplitus croceipes* Cresson (Braconidae: Hymenoptera). *Journal of Insect Behavior.* **6**, 625–639.

McColloch, J. W., and Yuasa, H. 1915. Further data on the life economy of the cinch bug egg parasite. *Journal of Economic Entomology* **8**, 248–261.

McCullogh, P., and Nelder, J. A. 1983. *Generalized Linear Models.* Chapman and Hall, London.

McGugan, B. M. 1955. Certain host-parasite relationships involving the spruce budworm. *Canadian Entomologist* **87**, 178–187.

Mackauer, M. 1967. Wirtsbindung und parallele Evolution parasitischer Hymenoptera. I. Allgemeines und Parasiten der Homompteran, 1. *Angewandte Parasitologie* **8**, 21–39.

Mackauer, M. 1973. Host selection and host suitability in *Aphidius smithi* (Hymenoptera: Aphidiidae). In A. D. Lowe, ed., *Perspectives in Aphid Biology,* pp. 20–29. Entomological Society of New Zealand, Christchurch.

Mackauer, M. 1976. An upper boundary for the sex ratio in a haplodiploid insect. *Canadian Entomologist* **108**, 1399–1402.

Mackauer, M. 1986. Growth and developmental interactions in some aphids and their hymenopteran parasites. *Journal of Insect Physiology* **32**, 275–280.

Mackauer, M. 1990. Host discrimination and larval competition in solitary endoparasitoids. In M. Mackauer, L. E. Ehler, and J. Roland, eds., *Critical Issues in Biological Control,* pp. 41–62. Intercept, Andover, U.K.

Mackensen, O. 1951. Viability and sex determination in the honey bee (*Apis mellifera*). *Genetics* **36**, 500–509.

McNair, J. N. 1982. Optimal giving-up times and the marginal value theorem. *American Naturalist* **119**, 511–529.

Macnair, M. R., and Parker, G. A. 1979. Modes of parent-offspring conflict. III. Intrabrood conflict. *Animal Behaviour* **27**, 1202–1209.

McNamara, J. M. 1982. Optimal patch use in a stochastic environment. *Theoretical Population Biology* **21**, 269–288.

McNamara, J. M., and Houston, A. I. 1986. The common currency for behavioral decisions. *American Naturalist* **127**, 358–378.

McNamara, J. M., and Houston, A. I. 1990. The value of fat reserves in terms of avoiding starvation. *Acta Biotheoretica* **38**, 37–61.

McNeil, J. N., and Rabb, R. L. 1973. Physical and physiological factors in diapause initiation of two hyperparasites of the tobacco hornworm, *Manduca sexta*. *Journal of Insect Physiology* **19**, 2107–2118.

Madden, J. L. 1968. Behavioural responses of parasites to the symbiotic funus associated with *Sirex noctilio* F. *Nature* **218**, 189–190.

Maindonald, J. H., and Markwick, N. P. 1986. The avoidance of superparasitism in four species of parasitic wasp—mathematical models and experimental results. *Researches in Population Ecology* **28**, 1–16.

Malo, F. 1961. Phoresy of *Xenufens*, a parasite of *Caligo eurilochus Journal of Economic Entomology* **54**, 465–466.

Malyshev, S. I. 1968. *Genesis of the Hymenoptera and the Phases of Their Evolution* Translated from the Russian by B. Haigh, O. W. Richards, and B. Uvarov. Methuen, London.

Mangel, M. 1987a. Modelling behavioral decisions of insects. In Y. Cohen, ed., *Lecture Notes in Biomathematics* **73**, pp. 1–18. Springer-Verlag, Berlin.

Mangel, M. 1987b. Oviposition site selection and clutch size in insects. *Journal of Mathematical Biology* **25**, 1–22.

Mangel, M. 1989a. Evolution of host selection in parasitoids: Does the state of the parasitoid matter? *American Naturalist* **133**, 688–705.

Mangel, M. 1989b. An evolutionary explanation of the motivation to oviposit. *Journal of Evolutionary Biology* **2**, 157–172.

Mangel, M., and Clark, C. W. 1986. Towards a unified foraging theory. *Ecology* **67**, 1127–1138.

Mangel, M., and Clark, C. W. 1988. *Dynamic Modeling in Behavioral Ecology.* Princeton University Press, Princeton.

Mangold, J. R. 1978. Attraction of *Euphasiopteryx ochracea, Corthrella* sp. and gryllids to broadcast songs of the southern mole cricket. *Florida Entomologist* **61**, 57–61.

Marchal, P. 1936. Recherches sur la biologie et de développement des Hyménoptères: Les Trichogrammes. *Annales des Épiphyties* **22**, 447–550.

Marchal, P. 1898. Le cycle evolutif de l'*Encyrtus fuscicollis. Societe Entomologie de France Bulletin* **1898**, 109–111.

Marris, G. C.; Hubbard, S. F.; and Scrimgeour, C. 1992. The discrimination of genetic similarity and its effect on the oviposition behaviour of solitary insect parasitoids. Manuscript.

Marshall, G.A.K. 1902. Five years' observations and experiments (1896–1901) on the bionomics of south African insects, chiefly directed to the investigation of mimicry and warning colours. *Transactions of the Royal Entomological Society of London* **1902**, 287–384.

Marston, N., and Ertle, L. R. 1969. Host age and parasitism by *Trichogramma minutum* (Hymenoptera: Trichogrammatidae). *Annals of the Entomological Society of America* **62**, 1476–1482.

Masner, L. 1959. A revision of the ecitophilous diapriid genus *Mimopria* Holmgren (Hymenoptera: Proctotrupoidea). *Insectes Sociaux* **6**, 361–367.

Masner, L. 1976. Notes on the ecitophilous diapriid genus *Mimopria* Holmgren (Hymenoptera: Proctotrupoidea: Diapriidae). *Canadian Entomologist* **108**, 123–126.

Masner, L. 1977. A new genus of ecitophilous diapriid wasps from Arizona (Hymenoptera: Proctotrupoidea: Diapriidae). *Canadian Entomologist* **109**, 33–36.

Mason, W.R.M. 1964. Regional colour patterns in the parasitic Hymenoptera. *Canadian Entomologist* **96**, 132–134.

Mason, W.R.M. 1967. Specialization in the egg structure of *Exentrus* (Hymenoptera: Ichneumonidae) in relation to distribution and abundance. *Canadian Entomologist* **99**, 375–384.

le Masurier, A. D. 1987. A comparative study of the relationship between host size and brood size in *Apanteles* spp. (Hymenoptera: Braconidae). *Ecological Entomology* **12**, 383–393.

le Masurier, A. D. 1991. Effect of host size on clutch size in *Cotesia glomerata*. *Journal of Animal Ecology* **60**, 107–118.

Matthews, R. W. 1974. Biology of Braconidae. *Annual Review of Entomology* **19**, 15–32.

Matthews, R. W.; Matthews, J. O.; and Crankshaw, O. 1979. Aggregation in male parasitic wasps of the genus *Megarhyssa*. I: Sexual discrimination, tergal stroking and description of associated anal structures and behaviour. *Florida Entomologist* **62**, 3–8.

Mauricio, R., and Bowers, M. D. 1990. Do caterpillars disperse their damage? Larval foraging behaviour of two specialist herbivores, *Euphydryas phaeton* (Nymphalidae) and *Pieris rapae* (Pieridae). *Ecological Entomology* **15**, 153–161.

Maw, M. G. 1960. Notes on the larch sawfly, *Pristiphora erichsonii* (Htg.) (Hymenoptera: Tenthredinidae), in Great Britain. *Entomologist's Gazette* **11**, 43–49.

May, R. M. 1978. Host-parasitoid systems in patchy environments: A phenomenological model. *Journal of Animal Ecology* **47**, 833–843.

May, R. M., and Hassell, M. P. 1981. The dynamics of multiparasitoid-host interactions. *American Naturalist* **124**, 552–560.

Mayer, K. 1934. Beiträge zur Sinnesphysiologie der Schlupfwespe *Nemeritis canescens* Grav. (Hym.: Ichneumonidae, Ophioninae). *Arbeiten über Physiologische und Angewandte Entomologie* **1**, 245–248.

Maynard Smith, J. 1978. *The Evolution of Sex*. Cambridge University Press, Cambridge, U.K.

Maynard Smith, J. 1984. Untitled appendix to a paper by P. H. Harvey: Intrademic group selection and the sex ratio. In R. M. Sibly and R. H. Smith, eds., *Behavioural Ecology* (British Ecological Society Symposium 25), pp. 72–73. Blackwell Scientific, Oxford.

Meelis, E. 1982. Egg distribution of insect parasitoids: A survey of models. *Acta Biotheoretica* **31**, 109–126.

Melander, A. L. and Brues, C. T. 1903. Guests and parasites of the burrowing bee, *Halictus*. *Biological Bulletin* **5**, 1–27.

Memmott, J., and Godfray, H.C.J. 1993. Parasitoid webs. In J. LaSalle and I. D. Gauld, eds., *Hymenoptera and Biodiversity*, pp. 217–234. CAB International Press, Wallingford, U.K.

Mertins, J. W. 1980. Life history and behaviour of *Laelius pedatus* a gregarious bethylid ectoparasitoid of *Anthrenus verbasci*. *Annals of the Entomological Society of America* **73**, 686–693.

Mertins, J. W. 1985. *Laelius utilis* (Hym.: Bethylidae), a parasitoid of *Anthrenus fuscus* (Col.: Dermestidae) in Iowa. *Entomophaga* **30**, 65–68.

Messenger, P. S., and van den Bosch, R. 1971. The adaptability of introduced biological control agents. In C. B. Huffaker, ed., *Biological Control*, pp. 68–92. Plenum, New York.

Mesterton-Gibbons, M. 1988. On the optimal compromise for a dispersing parasitoid. *Journal of Mathematical Biology* **26**, 375–385.

Micha, S. G.; Wellings, P. W.; and Morton, R. 1992. Time-related rejection of parasitized hosts in the aphid parasitoid, *Aphidius ervi. Entomologia Experimentalis et Applicata* **62**, 155–161.

Michener, C. D. 1969. Immature stages of a chalcidoid parasite tended by allodapine bees (Hymenoptera: Perilampidae and Anthophoridae). *Journal of the Kansas Entomological Society* **42**, 247–250.

Michod, R. E., and Levin, B. R. 1988. *The Evolution of Sex.* Sinauer, Sunderland, Massachusetts.

Mickel, C. E. 1924. An analysis of bimodal variation in size of the parasite *Dasymutilla bioculata* Cresson. *Entomological News* **35**, 236–242.

Mickel, C. E. 1973. John Ray: Indefatigable student of nature. *Annual Review of Entomology* **18**, 1–18.

Milligan, B. G. 1985. Evolutionary divergence and character displacement in two phenotypically variable competing species. *Evolution* **39**, 1207–1222

Mills, N. J. 1991. Searching strategies and attack rates of parasitoids of the ash bark beetle (*Leperisinus varius*) and its relevance to biological control. *Ecological Entomology* **16**, 461–470.

Mitchell, W. A. 1990. An optimal control theory of diet selection: The effects of resource depletion and exploitative competition. *Oikos* **58**, 16–24.

Mitchell, W. C., and Mau, R.F.L. 1971. Response of the female southern green stink bug and its parasite, *Trichopoda pennipes*, to male stink bug pheromones. *Journal of Economic Entomology* **64**, 856–859.

Mohamed, M. A., and Coppel, H. C. 1986. Sex-ratio regulation in *Brachymeria intermedia*, a pupal gypsy-moth parasitoid. *Canadian Journal of Zoology* **64**, 1412–1415.

Mollema, C. 1988. Genetical Aspects of Resistance in a Host-parasitoid Interaction. Ph.D., diss., University of Leiden.

Mollema, C. 1991. Heritability estimates of host selection behaviour by the *Drosophila* parasitoid *Asobara tabida. Netherlands Journal of Zoology* **41**, 174–183.

Monteith, L. G. 1956. Influence of host movement on selection of hosts by *Drino bohemica* Mesn. as determined in an olfactometer. *Canadian Entomologist* **88**, 583–586.

Monteith, L. G. 1960. Influence of plants other than the food plants of their host on host-finding by the tachinid parasites. *Canadian Entomologist* **92**, 641–652.

Monteith, L. G. 1963. Habituation and associative learning in *Drino bohemica* Mesn. *Canadian Entomologist* **95**, 418–426.

Moore, D. 1983. Hymenopterous parasitoids of stem-boring Diptera in perennial ryegrass (*Lolium perenne*) in Britain. *Bulletin of Entomological Research* **73**, 601–608.

Moran, N., and Hamilton, W. D. 1980. Low nutritive quality as defense against herbivores. *Journal of Theoretical Biology* **86**, 247–254.

Moratorio, M. S. 1977. Aspects of the biology of *Anagrus* spp. (Hymenoptera: Mymaridae), with special reference to host-parasitoid relationships. Ph.D., diss., University of London.

Morris, R. F. 1976. Influence of genetic changes and other variables on the encapsulation of parasites of *Hyphantria cunea*. *Canadian Entomologist* **108**, 673–684.

Morrison, G.; Auerbach, M.; and McCoy, E. D. 1979. Anomalous diversity of tropical parasitoids: A general phenomenon? *American Naturalist* **114**, 303–307.

Mountford, M. D. 1968. The significance of litter size. *Journal of Animal Ecology* **37**, 363–367.

Mudd, A., and Corbet, S. A. 1973. Mandibular gland secretion of larvae of the stored product pests *Anagasta kuehniella, Ephestia cautella, Plodia interpunctella* and *Ephestia elutella*. *Entomologia Experimentalis et Applicata* **16**, 291–292.

Mudd, A., and Corbet, S. A. 1982. Response of the ichneumonid parasite *Nemeritis canescens* to kairomones from the flour moth *Ephestia kuehniella*. *Journal of Chemical Ecology* **8**, 843–850.

Mudd, A.; Walters, J.H.H.; and Corbet, S. A. 1984. Relative kairomonal activities of 2-acylcyclohexane-1,3-diones in eliciting oviposition behaviour from the parasite *Nemeritis canescens* (Grav). *Journal of Chemical Ecology* **10**, 1597–1601.

Mueller, L. D. 1988. Density-dependent population growth and natural selection in food-limited environments. *American Naturalist* **132**, 786–809.

Mueller, T. F. 1983. The effect of plants on the host relations of a specialist parasitoid of *Heliothis* larvae. *Entomologia Experimentalis et Applicata* **34**, 78–84.

Muesebeck, C.F.W. 1922. *Zygobothria nidicola*, an important parasite of the brown-tail moth. *United States Department of Agriculture Bulletin* **1088**, 1–9.

Muldrew, J. A. 1953. The natural immunity of the larch sawfly (*Pristiphora erichsonii* Htg.) to the introduced parasite *Mesoleius thenthredinis* Morley in Manitoba and Saskatchewan. *Canadian Journal of Zoology* **31**, 313–332.

Müller, C. B., and Schmid-Hempel, R. 1992. To die for host or parasitoid. *Animal Behaviour* **44**, 177–179.

Murdoch, W. W. 1969. Switching in general predators: Experiments on predator specificity and stability of prey populations. *Ecological Monographs* **39**, 335–354.

Murdoch, W. W., and Oaten, A. 1975. Predation and population stability. *Theoretical Population Biology* **12**, 263–285.

Murdoch, W. W., and Stewart-Oaten, A. 1989. Aggregation by parasitoids and predators: Effects on equilibrium and stability. *American Naturalist* **134**, 288–310.

Murdoch, W. W.; Briggs, C. J.; Nisbet, R. M.; Gurney, W.S.C.; and Stewart-Oaten, A. 1992. Aggregation and stability in metapopulation models. *American Naturalist* **140**, 41–58.

Murray, M. G. 1985. Figs (*Ficus* spp.) and fig wasps (Chalcidoidea, Agaonidae): Hypotheses for an ancient symbiosis. *Biological Journal of the Linnean Society* **26**, 69–81.

Murray, M. G. 1987. The closed environment of the fig receptacle and its influence on male conflict in the old world fig wasp *Philotrypesis pilosa*. *Animal Behaviour* **35**, 488–506.

Murray, M. G. 1989. Environmental constraints on fighting in flightless male fig wasps. *Animal Behaviour* **38**, 186–193.

Murray, M. G. 1990. Comparative morphology and mate competition of flightless male fig wasps. *Animal Behaviour* **39**, 434–443.

Murray, M. G. and Gerrard, R. J. 1984. Conflict in the neighbourhood: Models where close relatives are in direct competition. *Journal of Theoretical Biology* **111**, 237–246.

Murray, M. G., and Gerrard, R. J. 1985. Putting the challenge into resource exploitation: A model of contest competition. *Journal of Theoretical Biology* **115**, 367–389.

Myers, J. G. 1928. Further notes on *Rhyssa* and *Ibalia*, parasitising *Sirex cyaneus* Fabr. *Bulletin of Entomological Research* **19**, 317–323.

Myers, J. H. 1981. Interactions between western tent caterpillars and wild rose—a test of some general plant herbivore hypotheses. *Journal of Animal Ecology* **50**, 11–25.

Nachtigall, W. 1974. *Insects in Flight*. Translated from the German. George Allen and Unwin, London.

Nadel, H. 1987. Male swarms discovered in Chalcidoidea (Hymenoptera: Encyrtidae, Pteromalidae). *Pan-Pacific Entomologist* **63**, 242–246.

Nadel, H., and van Alphen, J.J.M. 1987. The role of host and host plant odours in the attraction of a parasitoid, *Epidinocarsus lopezi*, to the habitat of its host, the cassava mealybug, *Phenacoccus manihoti*. *Entomologia Experimentalis et Applicata* **45**, 181–186.

Naito, T., and Suzuki, H. 1991. Sex determination in the sawfly, *Athalia rosae ruficornis* (Hymenoptera): Occurrence of triploid males. *Journal of Heredity* **82**, 101–104.

Nappi, A. J. 1975. Parasite encapsulation in insects. In K. Maramorosch and R. E. Shope, eds., *Invertebrate Immunity*, pp. 293–326. Academic Press, New York.

Nappi, A. J., and Carton, Y. D. 1986. Cellular immume responses and their genetic aspects in *Drosophila*. In M. Brehélin, ed., *Immunity in Invertebrates*, pp. 171–187. Springer-Verlag, Berlin.

Nealis, V. G. 1986. Responses to host kairomones and foraging behavior of the insect parasite *Cotesia rubecula* (Hymenoptera: Braconidae). *Canadian Journal of Zoology* **64**, 2393–2398.

Nealis, V. G. 1990. Factors affecting the rate of attack by *Cotesia rubecula* (Hymenoptera: Braconidae). *Ecological Entomology* **15**, 163–168.

Nechols, J. R., and Kikuchi, R. S. 1985. Host selection of the spherical mealybug (Homoptera: Pseudococcidae) by *Anagyrus indicus* (Hymenoptera: Encyrtidae): Influence of host stage on parasitoid oviposition, development, sex ratio and survival. *Environmental Entomology* **14**, 32–37.

Nell, H. W., and van Lenteren, J. C. 1982. Gastheerdiscriminatie bij *Pachycrepoides vindemiae*: Een voorbeeld van bezettingstype-concurrentie bij sluipwespen. *Vakblad voor Biologien* **62**, 2–6.

Nell, H. W.; Sevenster van der Lelie, L. A.; Woets, J.; and van Lenteren, J. C. 1976. The parasite-host relationship between *Encarsia formosa* (Hymenoptera: Aphelinidae) and *Trialeurodes vaporariorum* (Homoptera: Aleyrodidae). II. Selection of host stages for oviposition and feeding by the parasite. *Zeitschrift für Angewandte Entomologie* **81**, 372–376.

Neser, S. 1973. Biology and behaviour of *Euplectrus* near *laphygmae* Ferriere (Hymenoptera: Eulophidae). *Entomological Memoirs, Department of Agricultural and Technical Services, South Africa* **32**, 1–31.

Nettles, W. C., Jr.; Morrison, R. K.; Xie, Z.-N.; Ball, D.; Vinson, S. B.; and Shenkir, C. A. 1982. Synergistic action of potassium chloride and magnesium sulfate on parasitoid oviposition. *Science* **218**, 164–166.

Noda, T., and Hirose, Y. 1989. "Males second" strategy in the allocation of sexes by the parasitic wasp *Gryon japonicum. Oecologia* **81**, 145–148.

Noldus, L.P.J.J. 1989. Semiochemicals, foraging behaviour and quality of entomophagous insects for biological control. *Journal of Applied Entomology* **108**, 425–451.

Noldus, L.P.J.J., and van Lenteren, J. C. 1985. Kairomones for the egg parasite *Trichogramma evanescens* Westwood. Effect of volatile substances released by two of its hosts *Pieris brassicae* L and *Mamestra brassicae* L. *Journal of Chemical Ecology* **11**, 781–791.

Noldus, L.P.J.J.; Potting, R.P.J.; and Barendregt, H. E. 1991. Moth sex pheromone adsorption to a leaf surface: Bridge in time for chemical spies. *Physiological Entomology* **3**, 329–344.

Nordlund, D. A., and Lewis, W.J. 1976. Terminology of chemical releasing stimuli in intraspecific and interspecific interactions. *Journal of Chemical Ecology* **2**, 211–220.

Nordlund, D. A.; Lewis, W. J.; Todd, J. W.; and Chalfant, R. B. 1977. Kairomones and their use for management of entomophagous insects. VII. The involvement of various stimuli in the differential response of *Trichogramma pretiosum* Riley to two suitable hosts. *Journal of Chemical Ecology* **3**, 513–518.

Nordlund, D. A.; Jones, R. L.; and Lewis, W. J., eds., 1981. *Semiochemicals, Their Role in Pest Control,* pp. 51–78. John Wiley, New York.

Nordlund, D. A.; Lewis, W. J.; and Altieri, M. A. 1988. Influences of plant produced allelochemicals on the host and prey selection of entomophagous insects. In P. Barbosa and D. K. Letourneau, eds., *Novel Aspects of Insect-Plant Interactions,* pp. 65–90. John Wiley, New York.

Noyes, J. S. 1989. The diversity of Hymenoptera in the tropics with special reference to parasitoids in Sulawesi. *Ecological Entomology* **14**, 197–207.

Nunney, L. 1985. Female-biased sex ratios: Individual or group selection? *Evolution* **39**, 349–361.

Nunney, L., and Luck, R. F. 1988. Factors influencing the optimum sex ratio in structured populations. *Journal of Theoretical Biology* **33**, 1–30.

Nur, U.; Werren, J. H.; Eickbush, D. G.; Burke, W. D.; and Eickbush, T. H. 1988. A "selfish" B chromosome that enhances its transmission by eliminating the paternal genome. *Science* **240**, 512–514.

Nuttall, M. J. 1973. Pre-emergence fertilisation of *Megarhyssa nortoni nortoni* (Hymenoptera: Ichneumonidae). *New Zealand Entomology* **5**, 112–117.

Obara, M., and Kitano, H. 1974. Studies on the courtship behavior of *Apanteles glomeratus* 1. Experimental studies on releaser of wing vibrating behaviour in the male. *Kontyû* **42**, 208–214.

Obrycki, J. J. 1986. The influence of foliar pubescence on entomophagous species. In D. J. Boethel and R. D. Eikenbary eds., *Interactions of Plant Resistance and Parasitoids and Predators of Insects,* pp. 61–83. Ellis Horwood, Chichester, U.K.

Odebiyi, J. A., and Oatman, E. R. 1977. Biology of *Agathis unicolor* (Schrottky) and *Agathis gibbosa* (Say) (Hymenoptera: Braconidae), primary parasites of the potato tuberworm. *Hilgardia* **45**, 123–152.

Oldroyd, H. 1964. *Natural History of Flies.* Weidenfeld and Nicholson, London.

Olmi, M. 1994. The family Dryinidae. In P. Hansson and I. D. Gauld, eds., *The Hymenoptera of Costa Rica.* Oxford University Press, Oxford (in press).

Olson, D., and Pimentel, D. 1974. Evolution of resistance in a host population to attacking parasite. *Environmental Entomology* **3**, 621–624.

O'Neill, S. L.; Giordano, R.; Colbert, A.M.M.; Karr, T. L.; and Robertson, H.M. 1992. 16S rRNA phylogenetic analysis of the bacterial endosymbionts associated with cytoplasmic incompatibility in insects. *Proceedings of the National Academy of Sciences, U.S.A.* **89**, 2699–2702.

Onillon, J. C. 1990. The use of natural enemies for the biological control of whiteflies. In D. Gerling, ed., *Whiteflies: Their Bionomics, Pest Status and Management,* pp. 287–314. Intercept, Andover, Hampshire, U.K.

Opler, P. A. 1974. Oaks as evolutionary islands for leaf-mining insects. *American Science* **62**, 67–73.

Opp, S. B., and Luck, R. F. 1986. Effects of host size on selected fitness components of *Aphytis melinus* and *A. lingnanensis* (Hymenoptera: Aphelinidae). *Annals of the Entomological Society of America* **79**, 700–704.

Orgel, L. E., and Crick, F.H.C. 1980. Selfish DNA: The ultimate parasite. *Nature* **284**, 604–607.

Orzack, S. H. 1986. Sex-ratio control in a parasitic wasp, *Nasonia vitripennis* II. Experimental analysis of an optimal sex ratio model. *Evolution* **40**, 341–356.

Orzack, S. H. 1990. The comparative biology of second sex ratio evolution within a natural population of a parasitic wasp, *Nasonia vitripennis. Genetics* **124**, 385–396.

Orzack, S. H., and Parker, E. D. 1986. Sex ratio control in a parasitic wasp, *Nasonia vitripennis* I. Genetic variation in facultative sex ratio production. *Evolution* **40**, 331–340.

Orzack, S. H., and Parker, E. D. 1990. Genetic variation for sex ratio traits within a natural population of a parasitic wasp. *Genetics* **124**, 373–384.

Owen, D. F., and Owen, J. 1974. Species diversity in temperate and tropical Ichneumonidae. *Nature* **249**, 583–584.

Owen, R. E. 1983. Sex ratio adjustment in *Asobara persimilis* (Hymenoptera: Braconidae), a parasitoid of *Drosophila. Oecologia* **59**, 402–404.

Pacala, S., and Hassell, M. P. 1991. The persistence of host parasitoid associations in a patchy environment. II. Evaluation of field data. *American Naturalist* **138**, 584–605.

Pacala, S.; Hassell, M. P., and May, R. M. 1990. Host-parasitoid associations in patchy environments. *Nature* **344**, 150–153.

Pak, G. A. 1986. Behavioural variations among strains of *Trichogramma* spp. A review of the literature on host-age selection. *Journal of Applied Entomology* **101**, 55–64.

Pallewatta, P.K.T.N.S. 1986. Factors affecting progeny and sex allocation by the egg parasitoid *Trichogramma evanescens* Westwood. Ph.D. diss., University of London.

Pantel, J. 1910–1912. Recherches sur les Dipteres a larves entomobies. *Cellule* **26**, 27–216; **29**, 7–289.

Papaj, D. R. and Prokopy, R. J. 1989. Ecological and evolutionary aspects of learning in phytophagous insects. *Annual Review of Entomology* **34**, 315–350.

Papaj, D. R., and Vet, L.E.M. 1990. Odor learning and foraging success in the parasitoid, *Leptopilina heterotoma. Journal of Chemical Ecology* **16**, 3137–3150.

Parker, E. D., and Orzack, S. H. 1985. Genetic variation for the sex ratio in *Nasonia vitripennis. Genetics* **110**, 93–105.

Parker, G. A. 1970. The reproductive behaviour and the nature of sexual selection in *Scatophaga stercoraria* L. II. The fertilzation rate and the spatial and temporal relationships of each sex around the site of mating and oviposition. *Journal of Animal Ecology* **39**, 205–228.

Parker, G. A. 1974. Courtship persistence and female-guarding as male time-investment strategies. *Behaviour* **48**, 157–184.

Parker, G. A., and Courtney, S. P. 1983. Seasonal incidence: Adaptive variation in the timing of life history stages. *Journal of Theoretical Biology* **105**, 147–155.

Parker, G. A., and Courtney, S. P. 1984. Models of clutch size in insect oviposition. *Theoretical Population Biology* **26**, 27–48.

Parker, G. A., and Macnair, M. R. 1979. Models of parent-offspring conflict. IV. Suppression: Evolutionary retaliation by the parent. *Animal Behaviour* **27**, 1210–1235.

Parker, H. L. 1931. *Macrocentrus gifuensis* Ashmead, a polyembryonic braconid parasite of the European Corn Borer. *U.S. Department of Agriculture Technical Bulletin* **230**.

Parrish, D. S., and Davis, D. W. 1978. Inhibition of diapause in *Bathyplectes curculionis*, a parasite of the alfalfa weevil. *Annals of the Entomological Society of America* **71**, 103–107.

Partridge, L. 1983. Non-random mating and offspring fitness. In P. Bateson, ed., *Mate Choice,* pp. 227–255. Cambridge University Press, Cambridge, U.K.

Payne, J. A., and Wood, B. W. 1984. Rubidium as a marking agent for the hickory shuckworm (*Cydia caryari*) (Lepidoptera, Tortricidae). *Environmental Entomology* **13**, 1519–1521.

Pemberton, C. E., and Willard, H. F. 1918. A contribution to the biology of fruitfly parasites in Hawaii. *Journal of Agricultural Research* **15**, 419–465.

Petters, R. M.; Grosch, D. S.; and Olson, C. S. 1978. A flightless mutation in the wasp *Habrobracon juglandis*. *The Journal of Heredity* **69**, 113–116.

Picard, F., and Rabaud, E. 1914. Sur le parasitisme externe des Braconides. *Bulletin de la Société Entomologique de France* **1914**, 266–269.

Pickering, J. 1980. Larval competition and brood sex ratios in the gregarious parasitoid *Pachysomoides stupidus*. *Nature* **283**, 291–292.

Piek, T., ed., 1986. *Venoms of the Hymenoptera,* Academic Press, London.

Piek, T., and Owen, M. D. 1982. *Hymenoptera Venom Systems*. Academic Press, London.

Pierce, N. E., and Elgar, M. A. 1985. The influence of ants on host plant selection by *Jalmenus evagoras*, a mymecophilous lycaenid butterfly. *Behavioural Ecology and Sociobiology* **16**, 209–222.

Pierce, N. E., and Mead, P. S. 1981. Parasitoids as selective agents in the symbiosis between lycaenid butterfly larvae and ants. *Science* **211**, 1185–1187.

Pierce, W. D. 1910. On some phases of parasitism displayed by insect enemies of weevils. *Journal of Economic Entomology* **3**, 451–458.

Pierce, W. D., and Holloway, T. E. 1912. Notes on the biology of *Chelonus texanus* Cress. *Journal of Economic Entomology* **6**, 425–428.

Pimentel, D. 1963. Introducing parasites and predators to control native pests. *Canadian Entomologist* **95**, 785–792.

Pimentel, D. 1968. Population regulation and genetic feedback. *Science* **159**, 1432–1437.

Pimentel, D., and Al-Hafidh, R. 1965. Ecological control of a parasite population by genetic evolution in the parasite-host system. *Annals of the Entomological Society of America* **58**, 1–6.

Pimentel, D., and Stone, F. A. 1968. Evolution and population ecology of parasite-host systems. *Canadian Entomologist* **100**, 655–662.

Pimentel, D.; Nagel, W. P.; and Madden, J. L. 1963. Space-time structure of the environment and the survival of parasite-host systems. *American Naturalist* **97**, 141–167.

Pimentel, D.; Levin, S. A.; and Olson, D. A. 1978. Coevolution and the stability of exploiter-victim systems. *American Naturalist* **112**, 119–125

Pimm, S. L., and Lawton, J. H. 1980. Are food webs divided into compartments? *Journal of Animal Ecology* **49**, 879–898.

Polaszek, A. 1991. Egg parasitism in Aphelinidae (Hymenoptera: Chalcidoidea) with special reference to *Centrodora* and *Encarsia* species. *Bulletin of Entomological Research* **81**, 97–106.

Polgár, L.; Mackauer, M.; and Völkl, W. 1991. Diapause induction in two species of aphid parasitoids: The influence of aphid morph. *Journal of Insect Physiology* **37**, 699–702.

Pope, R. D., ed. 1977. Kloet and Hincks, *A Checklist of British Insects*, 2d ed. (completely revised), part 3, Coleoptera and Strepsiptera. *Handbooks for the Identification of British Insects* **11**(3), 1–105.

Poulin, R. 1992. Altered behaviour in parasitised bumblebees: Parasite manipulation or adaptive suicide? *Animal Behaviour* **44**, 174–177.

Powell, D. 1938. The biology of *Cephalonomia tarsalis* (Ash.), a vespoid wasp (Bethylidae: Hymenoptera) parasitic on saw-toothed grain beetle. *Annals of the Entomological Society of America* **31**, 44–49.

Powell, W.; and Zhang, Z. L. 1983. The reactions of two cereal aphid parasitoids, *Aphidius uzbekistanicus* and *A. ervi*, to host aphids and their food plants. *Physiological Entomology* **8**, 439–443.

Price, G. R. 1970. Selection and covariance. *Nature* **227**, 520–521.

Price, G. R. 1972. Extension of covariance selection mathematics. *Annals of Human Genetics* **35**, 485–490.

Price, P. W. 1970a. Trail odours: Recognition by insects parasitic in cocoons. *Science* **170**, 546–547.

Price, P. W. 1970b. Biology and host exploitation by *Pleolophus indistinctus* (Hymenoptera: Ichneumonidae). *Annals of the Entomological Society of America* **63**, 1502–1509.

Price, P. W. 1972a. Behaviour of the parasitoid *Pleolophus indistinctus*. *Annals of the Entomological Society of America* **63**, 1502–1509.

Price, P. W. 1972b. Parasitoids utilizing the same host: Adaptive nature of differences in size and form. *Ecology* **53**, 190–195.

Price, P. W. 1973a. Reproductive strategies in parasitoid wasps. *American Natualist* **107**, 684–693.

Price, P. W. 1973b. Parasitoid strategies and community organization. *Environmental Entomology* **2**, 263–626.

Price, P. W. 1974. Strategies for egg production. *Evolution* **28**, 76–84.

Price, P. W. 1975. Reproductive strategies of parasitoids. In P. W. Price, ed., *Evolutionary Strategies of Parasitoids*, pp. 87–111. Plenum, New York.

Price, P. W. 1980. *Evolutionary Biology of Parasites,* Princeton University Press, Princeton.

Price, P. W. 1981. Semiochemicals in evolutionary time. In D. A. Nordlund, R. L. Jones, and W. J. Lewis, eds., *Semiochemicals, Their Role in Pest Control,* pp. 251–279. John Wiley, New York.

Price, P. W. 1984. Communities of specialisis: Vacant niches in ecological and evolutionary time. In D. R. Strong, D. Simberloff, L. G. Abele, and A. B. Thistle, eds., *Ecological Communities: Conceptual Issues and the Evidence,* pp. 510–523. Princeton University Press, Princeton.

Price, P. W. 1990. Evaluating the role of natural enemies in latent and eruptive species: New approaches in life table construction. In A. D. Watt, S. R. Leather, M. D. Hunter, and N.A.C. Kidd, eds., *Population Dynamics of Forest Insects,* pp. 221–232. Intercept, Andover, U.K.

Price, P. W. 1991. Evolutionary theory of host and parasitoid interactions. *Biological Control* 1, 83–93.

Price, P. W., and Clancy, K. M. 1986. Interactions among three trophic levels: Gall size and parasitoid attack. *Ecology* 67, 1593–1600.

Price, P. W., and Pschorn-Walcher, H. 1988. Are galling insects better protected against parasitoids than exposed feeders? A test using tenthredinid sawflies. *Ecological Entomology* 13, 195–205.

Price, P. W.; Bouton, C. E.; Gross, P.; McPheron, B. A.; Thompson, J. N.; and Weis, A.E. 1980. Interactions among three trophic levels: Influence of plants on interactions between insect herbivores and natural enemies. *Annual Review of Ecology and Systematics* 11, 41–65

Price, P. W.; Fernandes, G. W.; and Waring, G. L. 1987. Adaptive nature of insect galls. *Environmental Entomology* 16, 15–24.

Price, P. W.; Cobb, N.; Craig, T. P.; Fernandes, G. W.; Itami, J. K.; Mopper, S.; and Preszler, R. W. 1990. Insect herbivore population dynamics on trees and shrubs: New approaches relevant to latent and eruptive species and life table development. In E. A. Bernays, ed., *Insect-Plant Interactions,* vol. 2, pp. 2–38. CRC Press, Boca Raton.

Prokopy, R. J. 1972. Evidence for a pheromone deterring repeated oviposition in apple maggot flies. *Environmental Entomology* 1, 326–332.

Prokopy, R. J., and Webster, R. P. 1978. Oviposition deterring pheromone of *Rhagoletis pomonella*-kairomone for its parasitoid *Opius lectus. Journal of Chemical Ecology* 4, 481–494.

Pschorn-Walcher, H. 1957. Probleme der Wirtswahl parasitischer Insekten. *Wanderversammlung Deutsche Entomologische Gesellschaft* 8, 79–85.

Pschorn-Walcher, H. 1965. Die Wirtsspezifität der parasitischen Hymenopteren in ökologisch-phylogenetischer Betrachtung (unter besonderer Berücksichtigung der Blattwespenparasiten). *Wanderversammlung Deutsche Entomologische Gesellschaft* 10, 55–63.

Pschorn-Walcher, H., and Altenhofer, E. 1989. The parasitoid community of leafmining sawflies (Fenusini and Heterarthrini): A comparative analysis. *Zoologischer Anzeiger* 222, 37–57.

Pschorn-Walcher, H., and Zwölfer, H. 1968. Konkurrenzerscheinungen in Parasitenkomplexen als Problem der biologischen Schädlingsbekampfung. *Anzeiger für Schädlingskunde* 41, 71–76.

Purrington, F. F., and Uleman, J. S. 1972. Brood size of the parasitic wasp *Hyssopus thymus* (Hymenoptera: Eulophidae): Functional correlation with the mass of a cryptic host. *Annals of the Entomological Society of America* **65**, 280–281.

Putters, F. A., and van den Assem, J. 1985. Precise sex ratio in a parasitic wasp: The result of counting eggs. *Behavioural Ecology and Sociobiology* **17**, 265–270.

Puttler, B. 1967. Biology of *Hyposoter exiguae* (Hymenoptera: Ichneumonidae), a parasite of lepidopterous larvae. *Annals of the Entomological Society of America* **54**, 25–30.

Puttler, B. 1967. Interrelationships of *Hypera postica* (Coleoptera: Curculionidae) and *Bathyplectes curculionis* (Hymenoptera: Ichneumonidae) in the eastern United States with particular reference to encapsulation of the parasite eggs by the weevil larvae. *Annals of the Entomological Society of America* **60**, 1031–1038.

Puttler, B. 1974. *Hypera postica* and *Bathyplectes curculionis*: Encapsulation of parasite eggs by host larvae in Missouri and Arkansas. *Environmental Entomology* **3**, 881–882.

Puttler, B., and van den Bosch, R. 1959. Partial immunity of *Laphygma exigua* (Huebner) to the parasite *Hyposoter exiguae* (Vireck). *Journal of Economic Entomology* **52**, 327–329.

Quicke, D.L.J. 1984. Evidence for the function of white-tipped ovipositor sheaths in Braconinae (Hymenoptera: Braconidae). *Proceedings and Transactions of the British Entomological and Natural History Society* **17**, 71–79.

Quicke, D.L.J. 1986. Preliminary notes on homeochromatic associations within and between afrotropical Braconinae (Hym., Braconidae) and Lamiinae (Col., Cerambycidae). *Entomologist's Monthly Magazine* **122**, 97–109.

Quicke, D.L.J. 1988. Host relationships in the Braconinae (Hymenoptera: Braconidae)—how little we know! *Entomological Society of Queensland News Bulletin* **16**, 85–92.

Quicke, D.L.J. 1992. Nocturnal Australasian Braconinae (Hym.; Braconidae). *Entomologist's Monthly Magazine* **128**, 33–37.

Quicke, D.L.J.; Ingram, S. N.; Proctor, J.; and Huddleston, T. 1992. Batesian and Mullerian mimicry between species with connected life histories with a new example involving braconid wasp parasites of *Phoracantha* beetles. *Journal of Natural History* **26** 1013–1034.

Rabaud, E. 1922. Note sur la camportement de *Rielia manticida* Kieff.; Proctotrupide parasite des ootheques de mantes. *Bulletin de la Société Zoologique de France* **47**, 10–15.

Rabb, R. L., and Bradley, J. R. 1970. Marking host eggs by *Telenomus sphingis*. *Annals of the Entomological Society of America* **63**, 1053–1056.

Rajendram, G. F., and Hagen, K. S. 1974. *Trichogramma* oviposition into artificial substrates. *Environmental Entomology* **3**, 399–401.

Rathcke, B. J., and Price, P. W. 1976. Anomalous diversity of tropical ichneumonid parasitoids: A predation hypothesis. *American Naturalist* **110**, 889–902.

Ratzeburg, J.T.C. 1844. *Die Ichneumonen der Forestinsekten in forstlicher und entomologischer Beziehung,* Berlin.

Raw, A. 1968. The behaviour of *Leopoldius coronatus* (Rond.) (Dipt. Conopidae) towards its hymenopterous hosts. *Entomologist's Monthly Magazine* **104**, 54.

Read, D. P.; Feeny, P. P.; and Root, R. B. 1970. Habitat selection by the aphid parasite *Diaeretiella rapae* and its hyperparasitoid, *Charips brassicae*. *Canadian Entomologist* **102**, 1567–1578.

Reichstein, T.; von Euw, J.; Parsons, J. A.; and Rothschild, M. 1968. Heart poisons in the monarch butterfly. *Science* **161**, 861–866.

Rejmanek, M., and Stary P. 1979. Connectance in real biotic communities and critical values for stability of modal ecosystems. *Nature* **280**, 311–313.

Reuter, O. M. 1913. *Lebensgewohnheiten und Instinkte der Insekten,* Friedlander, Berlin.

Rhoades, D. F., and Cates, R. H. 1976. Toward a general theory of plant antiherbivore chemistry. In J. W. Wallace and R. L. Mansell, eds., *Biochemical Interactions between Insects and Plants,* pp. 168–213. Plenum, New York.

Rice, R. E. 1968. Observations on host selection by *Tomicobia tibialis* Ashmead (Hymenoptera: Pteromalidae). *Contribution of the Boyce Thompson Institute* **24**, 53–56.

Rice, R. E. 1969. Response of some predators and parasites of *Ips confusus* (Lec.) (Coleoptera: Scolytidae) to olfactory attractants. *Contribution of the Boyce Thompson Institute* **24**, 189–194.

Richards, O. W. 1939. The British Bethylidae (*s.l,*) (Hymenoptera). *Transactions of the Royal Entomological Society,* **89**, 299–305.

Richards, O. W. 1940. The biology of the small white butterfly (*Pieris rapae*), with special reference to the factors controlling its abundance. *Journal of Animal Ecology* **9**, 243–288.

Richards, O. W., and Waloff, N. 1948. The hosts of four British Tachinidae (Diptera). *Entomologist's Monthly Magazine* **84**, 127.

Richardson, P. M.; Holmes, W. P.; and Saul, G. B. 1987. The effect of tetracycline on nonreciprocal cross incompatibility in *Mormoniella (=Nasonia) vitripennis. Journal of Invertebrate Pathology* **50**, 176–183.

Richerson, J. V., and Borden, J. H. 1972. Host finding by heat perception in *Coeloides brunneri* (Hymenoptera: Braconidae). *Canadian Entomologist* **104**, 1877–1881.

Richerson, J. V., and DeLoach, C. J. 1972. Some aspects of host selection by *Perilitus coccinellae. Annals of the Entomological Society of America* **65**, 834–839.

Ridley, M. 1983. *The Explanation of Organic Diversity,* Oxford University Press, Oxford.

Ridley, M. 1988a. Why not to use species in comparative tests. *Journal of Theoretical Biology* **136**, 361–364.

Ridley, M. 1988b. Mating frequency and fecundity in insects. *Biological Reviews* **63**, 509–549.

Ridley, M. 1993. A sib competitive relation between clutch size and mating frequency in parasitic Hymenoptera. *American Naturalist* (in press).

Ridsdill Smith, T. J. 1970. The behaviour of *Hemithynnus hyalinatus* (Hymenoptera: Tiphiidae), with notes on some other Thynninae. *Journal of the Australian Entomological Society* **9**, 196–208.

Rilett, R. O. 1949. The biology of *Cephalonomia waterstoni* Gahan. *Canadian Journal of Research, Series D* **27**, 93–111.

Rizki, R. M., and Rizki, T. M. 1984a. The cellular defense system of *Drosophila melanogaster,* In R. C. King and H. Akai, eds., *Insect Ultrastructure,* vol. 2, pp. 579–605. Plenum, New York.

Rizki, R. M., and Rizki, T. M. 1984b. Selective destruction of a host blood cell type by a parasitoid wasp. *Proceedings of the National Academy of Sciences, USA* **81**, 6154–6158.

Rizki, R. M., and Rizki, T. M. 1990. Parasitoid virus-like particles destroy *Drosophila* cellular immunity. *Proceedings of the National Academy of Sciences, USA* **87**, 8388–8392.

Robacker, D. C., and Hendry, L. B. 1977. Neral and geranial: Components of the sex pheromone of the parasitic wasp *Itoplectis conquisitor*. *Journal of Chemical Ecology* **3**, 563–577.

Robacker, D. C.; Weaver, K. M.; and Hendry, L. B. 1976. Sexual communication and associative learning in the parasitic wasp *Itoplectis conquisitor*. *Journal of Chemical Ecology* **2**, 39–48.

Roeske, C. N.; Seiber, J. N.; Brower, L.; and Moffitt, C. M. 1976. Milkweed cardenolides and their comparative processing by monarch butterflies (*Danaus plexippus* L.). *Recent Advances in Phytochemistry* **10**, 93–167.

Rogers, D. J. 1972. Random search and insect population models. *Journal of Animal Ecology* **41**, 369–383.

Rogers, D. J. 1975. A model for avoidance of superparasitism by solitary insect parasitoids. *Journal of Animal Ecology* **44**, 623–638.

Roitberg, B. D. 1990. Optimistic and pessimistic fruit flies: Measuring the cost of estimation errors. *Behaviour* **114**, 65–82.

Roitberg, B. D., and Lalonde, R. G. 1991. Host marking enhances parasitism risk for a fruit-infesting fly *Rhagoletis basiola*. *Oikos* **61**, 389–393.

Roitberg, B. D., and Mangel, M. 1988. On the evolutionary ecology of marking pheromones. *Evolutionary Ecology* **2**, 289–315.

Roitberg, B. D., and Myers, J. H. 1978. Adaption of alarm pheromone responses of the pea aphid *Acyrthosiphon pisum* (Harris). *Canadian Journal of Zoology* **56**, 103–108.

Roitberg, B. D., and Prokopy, R. J. 1987. Insects that mark host plants. *BioScience* **37**, 400–406.

Roitberg, B. D.; Mangel, M.; Lalonde, R.G.; Roitberg, C.A.; van Alphen, J.J.M.; and Vet, L. 1992. Seasonal dynamic shifts in patch exploitation by parasitic wasps. *Behavioural Ecology* **3**, 156–165.

Rojas-Rousse, D.; Eslami, J.; and Periquet, G. 1988. Reproductive strategy of *Dinarmus vagabundus* Timb. (Hym.; Pteromalidae): Real sex ratio, sequence of emitting diploid and haploid eggs and effects of inbreeding on progeny. *Journal of Applied Entomology* **106**, 276–285.

Roland, J. 1986. Parasitism of winter moth in British Columbia during build-up of its parasitoid *Cyzenis albicans* attack rate on oak *v* apple. *Journal of Animal Ecology* **55**, 215–234.

Roland, J.; Evans, W. G.; and Myers, J. H. 1989. Manipulation of oviposition patterns of the parasitoid *Cyzenis albicans* (Tachinidae) in the field using plant extracts. *Journal of Insect Behaviour* **2**, 487–503.

Romstöck-Völkl, M. 1990. Host refuges and spatial patterns of parasitism in an endophytic host-parasitoid system. *Ecological Entomology* **15**, 321–331.

Rosen, D., ed. 1981. *The Role of Hyperparasitism in Biological Control: A Symposium*. University of California Press, Berkeley.

Rosenheim, J. A., and Rosen, D. 1991. Foraging and oviposition decisions in the parasitoid *Aphytis lingnanensis*: Distinguishing the influences of egg load and experience. *Journal of Animal Ecology* **60**, 873–894.

Ross, K. G., and Fletcher, D.J.C. 1985. Genetic origin of male diploidy in the fire ant *Solenopsis invicta*, and its evolutionary significance. *Evolution* **39**, 888–903.

Rotary, N., and Gerling, D. 1973. The influence of some external factors upon the sex ratio of *Bracon hebetor* Say (Hymenoptera: Braconidae). *Environmental Entomology* **2**, 134–138.

Roth, J. P.; King, E. G.; and Thompson, A. C. 1978. Host location behaviour by the tachinid, *Lixiopaga diatraeae*. *Environmental Entomology* **7**, 794–798.

Rotheram, S. M. 1967. Immune surface of eggs of a parasitic insect. *Nature* **214**, 700.

Rotheray, G. E., and Barbosa, P. 1984. Host related factors affecting oviposition behavior in *Brachymeria intermedia*. *Entomologia Experimentalis et Applicata* **35**, 141–145.

Rothschild, G.H.L. 1966. Notes on two hymenopterous egg parasites of Delphacidae (Hom.). *Entomologist's Monthly Magazine* **102**, 5–9.

Ryan, R. B., and Rudinsky, J. A. 1962. Biology and habits of the Douglas fir beetle parasite, *Coeloides brunneri* Viereck (Hymenoptera: Braconidae) in Western Oregon. *Canadian Entomologist* **94**, 748–763.

Ryan, S. L., and Saul, G. B. 1968. Post-fertilization effect of incompatibility factors in *Mormoniella*. *Molecular and General Genetics* **103**, 24–36.

Ryan, S. L.; Saul, G. B.; and Conner, G. W. 1985. Aberrant segregation of *R*-locus genes in male progeny from incompatible crosses in *Mormoniella*. *Journal of Heredity* **76**, 21–26.

Sabelis, M. W., and de Jong, M.C. M. 1988. Should all plants recruit bodyguards? Conditions for a polymorphic ESS of synomone production in plants. *Oikos* **53**, 247–252.

Sahad, K. A. 1982. Biology and morphology of *Gonatocerus* sp. (Hymenoptera, Mymaridae), an egg parasitoid of the Green Rice Leafhopper, *Nephottetix cincticeps* Uhler (Homoptera, Deltocephalidae). I. Biology. *Kontyû* **50**, 467–476.

Sahad, K. A. 1984. Biology of *Anagrus optabilis* (Perkins) (Hymenoptera, Mymaridae), an egg parasitoid of delphacid planthoppers. *Esakia* **22**, 129–144.

Salt, G. 1931. Parasites of wheat-stem sawfly, *Cephus pygmaeus* Linnaeus, in England. *Bulletin of Entomological Research* **22**, 479–545.

Salt, G. 1932. Superparasitism by *Collyria calcitrator* Grav. *Bulletin of Entomological Research* **23**, 211–216.

Salt, G. 1934. Experimental studies in insect parasitism. II. Superparasitism. *Proceedings of the Royal Society of London B* **114**, 455–476.

Salt, G. 1935. Experimental studies in insect parasitism. III. Host selection. *Proceedings of the Royal Society of London B*, **117**, 413–435.

Salt, G. 1936. Experimental studies in insect parasitism. IV. The effect of superparasitism on populations of *Trichogramma evanescens*. *Journal of Experimental Biology* **13**, 363–375.

Salt, G. 1937a. Experimental studies in insect parasitism. V. The sense used by *Trichogramma* to distinguish between parasitized and unparasitized hosts. *Proceedings of the Royal Society of London B* **122**, 57–75.

Salt, G. 1937b. The egg-parasite of *Sialis lutaria*: A study of the influence of the host upon a dimorphic parasite. *Parasitology* **29**, 539–553.

Salt, G. 1938. Experimental studies in insect parasitism. VI. Host suitability. *Bulletin of Entomological Research* **29**, 223–246.

Salt, G. 1939. Further notes on *Trichogramma semblidis*. *Parasitology* **30**, 511–522.

Salt, G. 1940. Experimental studies in insect parasitism. VII. The effects of different hosts on the parasite *Trichogramma evanescens* Westw. (Hym.; Chalcidoidea). *Proceedings of the Royal Entomological Society of London A* **15**, 81–124.

Salt, G. 1941. The effects of hosts upon their insect parasites. *Biological Reviews* **16**, 239–264.

Salt, G. 1952. Trimorphism in the ichneumonid parasite *Gelis corruptor. Quarterly Journal of Microscopical Science* **93**, 453–475.

Salt, G. 1956. Experimental studies in insect parasitism. IX. The reactions of a stick insect to an alien parasite. *Proceedings of the Royal Society of London B*, **146**, 93–108.

Salt, G. 1958. Parasite behaviour and the control of insect pests. *Endeavour* **17**, 145–148.

Salt, G. 1961. Competition among insect parasitoids. Mechanisms in biological competition, *Symposium of the Society for Experimental Biology* **15**, 96–119.

Salt, G. 1963. The defence reactions of insects to metazoan parasites. *Parasitology* **53**, 527–642.

Salt, G. 1965. Experimental studies in insect parasitism. XIII. The haemocytic reaction of a caterpillar to eggs of its habitual parasite. *Proceedings of the Royal Society of London B* **162**, 303–318.

Salt, G. 1968. The resistance of insect parasitoids to the defence reactions of their hosts. *Biological Review* **43**, 200–232.

Salt, G. 1970. *The Cellular Defence Reactions of Insects,* Cambridge University Press, Cambridge, U.K.

Salt, G. 1971. Teratocytes as a means of resistance to cellular defence reactions. *Nature* **232**, 639.

Salt, G., and van den Bosch, R. 1967. The defence reactions of three species of *Hypera* (Coleoptera: Curculionidae) to an ichneumon wasp. *Journal of Invertebrate Pathology* **9**, 164–177.

Sandlan, K. 1979a. Host feeding and its effects on the physiology and behaviour of the ichneumonid parasite *Coccygomimus turionellae. Physiological Entomology* **4**, 383–392.

Sandlan, K. 1979b. Sex ratio regulation in *Coccygomimus turionellae* Linnaeus (Hymenoptera: Ichneumonidae) and its ecological implications. *Ecological Entomology* **4**, 365–378.

Sato, H. 1990. Parasitoid complexes of lepidopteran leafminers on oaks (*Quercus dentata* and *Quercus mongolica*) in Hokkaido, Japan. *Ecological Research* **5**, 1–8.

Sato, Y.; Tanaka, T.; Imafuku, M.; and Hidaka, T. 1983. How does diurnal *Apanteles kariyai* parasitise and egress from a nocturnal host larva? *Kontyû* **51**, 128–139.

Saunders, D. S. 1965. Larval diapause of maternal origin: Induction of diapause in *Nasonia vitripennis* (Walk.) (Hymenoptera: Pteromalidae). *Journal of Experimental Biology* **12**, 569–581.

Saunders, D. S. 1966. Larval diapause of maternal origin. II. The effect of photoperiod and temperature in *Nasonia vitripennis. Journal of Insect Physiology* **12**, 569–581.

Saunders, D. S. 1982. *Insect Clocks,* Pergamon Press, Oxford.

Schieferdecker, H. 1969. Der Gregärparasitismus von *Trichogramma* (Hymenoptera, Trichogrammatidae). *Beiträge zur Entomologie* **19**, 507–521.

Schmeider, R. G. 1933. The polymorphic forms of *Melittobia chalybii* Ashmead and the determining factors involved in their production. *Biological Bulletin* **65**, 338–354.

Schmeider, R. G. 1938. The sex ratio in *Melittobia chalybii* Ashmead, gametogenesis and cleavage in females and in haploid males. *Biological Bulletin* **74**, 256–266.

Schmeider, R. G., and Whiting, P. W. 1947. Reproductive economy in the chalcidoid wasp *Melittobia. Genetics* **32**, 29–37.

Schmid-Hempel, R., and Müller, C. B. 1991. Do parasitised bumblebees forage for their colony? *Animal Behaviour* **41**, 910–912.

Schmid-Hempel, P., and Schmid-Hempel, R. 1990. Endoparasitic larvae of conopid flies alter pollination behavior of bumblebees. *Naturwissenschaften* **77**, 450–452.

Schmid-Hempel, R., and Schmid-Hempel, P. 1991. Endoparasitic flies, pollen-collection by bumblebees and a potential host-parasitic conflict. *Oecologia* **87**, 227–232.

Schmid-Hempel, P.; Müller, C.; Schmid-Hempel, R.; and Shykoff, J. A. 1990. Frequency and ecological correlates of parasitism by conopid flies (Conopidae, Diptera) in a population of bumblebees. *Insectes Socieux* **37**, 14–30.

Schmidt, J. M., and Smith, J.J.B. 1985a. Host volume measurement by the parasitoid wasp *Trichogramma minutum*: The roles of curvature and surface area. *Entomologia Experimentalis et Applicata* **39**, 213–221.

Schmidt, J. M., and Smith, J.J.B. 1985b. The mechanism by which the parasitoid *Trichogramma minutum* responds to host clusters *Entomologia Experimentalis et Applicata* **39**, 287–294.

Schmidt, J. M., and Smith, J.J.B. 1986. Correlations between body angles and substrate curvature in the parasitoid wasp *Trichogramma minutum*: A possible mechanism of host radius measurement. *Journal of Experimental Biology* **125**, 271–285.

Schmidt, J. M., and Smith, J.J.B. 1987a. Measurement of host curvature by the parasitoid wasp *Trichogramma minutum*, and its effect on host examination and progeny allocation. *Journal of Experimental Biology* **129**, 151–164.

Schmidt, J. M., and Smith, J.J.B. 1987b. The effect of host pacing on the clutch size and parasitization rate of *Trichogramma minutum* Riley. *Entomologia Experimentalis et Applicata* **43**, 125–132.

Schmidt, J. M., and Smith, J.J.B. 1987c. The measurement of exposed host volume by the parasitoid wasp *Trichogramma minutum* and the effects of wasp size. *Canadian Journal of Zoology* **65**, 2837–2845.

Schmidt, J. M., and Smith, J.J.B. 1987d. The external sensory morphology of the legs and hairplate system of female *Trichogramma minutum*. *Proceedings of the Royal Society of London B* **232**, 323–366.

Schmidt, J. M., and Smith, J.J.B. 1987e. Short interval time measurement by a parasitoid wasp. *Science* **237**, 903–905.

Schmidt, J. M., and Smith, J.J.B. 1989. Host examination walk and oviposition site selection of *Trichogramma minutum*: Studies on spherical hosts. *Journal of Insect Behavior* **2**, 143–171.

Schmidt, O., and Schuchmann-Feddersen, I. 1989. The role of virus-like particles in parasitoid-host-interactions of insects. *Subcell Biochemistry* **15**, 91–102.

Schmidt, O.; Andersson, K.; Will, A.; and Schuchmann-Feddersen, I. 1990. Viruslike particle proteins from a hymenopteran endoparasitoid are reated to a protein component of the immune system in the lepidopteran host. *Archives of Insect Biochemistry and Physiology* **13**, 107–115.

Schneider, F. 1950. Die Entwicklung des Syrphidenparasiten *Diplazon fissorius*, Grav. (Hym.; Ichneumonidae) in uni-, oligo- und polyvoltinen Wirten und sein Verhalten bei parasitärer Aktivierung der Diapauselarven durch *Diplazon pectoratius* Grav. *Mitteilungen der Schweizerischen Entomologischen Gesellschaft* **23**, 155–194.

Schneider, F. 1951. Einige physiologische Beziehungen zwischen Syrphidenlarven und ihren Parasiten. *Zeitschrift für Angewandte Entomologie* **33**, 150–162.

Schneiderman, H. A., and Horwitz, J. 1958. The induction and termination of facultative diapause in the chalcid wasps *Mormoniella vitripennis* (Walker) and *Tritneptis klugii* (Ratzeburg). *Journal of Experimental Biology* **35**, 520–551.

Schröder, D. 1974. A study of the interactions between the internal parasites of *Rhyacionia buoliana* (Lepidoptera: Olethreutidae). *Entomophaga* **19**, 145–171.

Schuster, M. F. 1965. Studies on the biology of *Dusmetia sangwani* (Hymenoptera: Encyrtidae). *Annals of the Entomological Society of America* **58**, 272–275.

Seger, J. 1983. Partial bivoltinism may cause sex-ratio biases that favour eusociality. *Nature* **301**, 59–62.

Sequeira, R., and Mackauer, M. 1992a. Covariance of adult size and developmental time in the parasitoid wasp *Aphidius ervi* in relation to the size of its host, *Acyrthosiphon pisum*. *Evolutionary Ecology* **6**, 34–44.

Sequeira, R., and Mackauer, M. 1992b. Nutritional ecology of an insect host-parasitoid association: The pea aphid-*Aphidius ervi* system. *Ecology* **73**, 183–189.

Settle, W. H., and Wilson, L.T. 1990. Invasion by the variegated leafhopper and biotic interactions: Parasitism, competition and apparent competition. *Ecology* **71**, 1461–1470.

Shahjahan, M. 1974. *Erigeron* flowers as a food and attractive odor source for *Peristenus pseudopallipes*, a braconid parasitoid of the tarnished plant bug. *Environmental Entomology* **3**, 69–72.

Shapiro, A. M. 1976. Beau geste? *American Naturalist* **110**, 900–902.

Shaw, M. R. 1981a. Delayed inhibition of host development by the nonparalyzing venoms of parasitic wasps. *Journal of Invertebrate Pathology* **37**, 215–221.

Shaw, M. R. 1981b. Parasitic control: Section A. General information. In J. Feltwell, *Large White Butterfly: The Biology, Biochemistry and Physiology of* Pieris brassicae (Linnaeus), pp. 401–407. W. Junk, The Hague.

Shaw, M. R. 1983. On[e] evolution of endoparasitism: The biology of Rogadinae (Braconidae). *Contribution of the American Entomological Institute* **20**, 307–328.

Shaw, M. R., and Askew, R. R. 1976. Parasites. In J. Heath, ed., *The Moths and Butterflies of Great Britain and Ireland. 1, Micropterigidae-Heliozelidae*, pp. 24–56. Blackwell Scientific, Oxford.

Shaw, M. R., and Huddleston, T. 1991. Classification and biology of braconid wasps. *Handbooks for the Identification of British Insects*, vol. 7, part 11. Royal Entomological Society of London, London.

Shaw, S. R. 1988. Euphorine phylogeny: The evolution of diversity in host-utilization by parasitoid wasps (Hymenoptera: Braconidae). *Ecological Entomology* **13**, 323–335.

Sheehan, W., and Hawkins, B. A. 1991. Attack strategy as an indicator of host range in metopiine and pimpline Ichneumonidae (Hymenoptera). *Ecological Entomology* **16**, 129–131.

Sheehan, W., and Shelton, A. M. 1989. The role of experience in plant foraging by the aphid parasitoid *Diaretiella rapae* (Hymenoptera: Aphidiidae). *Journal of Insect Behaviour* **2**, 743–759.

Shelford, R.W.C. 1902. Observations on some mimetic insects and spiders from Borneo and Singapore. *Proceedings of the Zoological Society of London* **1902**, 230–284.

Shiga, M., and Nakanishi, A. 1968. Variation in the sex ratio of *Gregopimpla himala-yensis* Cameron (Hymenoptera: Ichneumonidae) parasitic on *Malacosoma neus-tria testacea* Molschulsky (Lepidoptera: Lasiocampidae) with considerations on the mechanism. *Kontyû* **36**, 369–376.

Silvestri, F. 1906. Contribuzioni alla conoscenza biologica degli Imenotteri parassiti. Biologia del *Litomastix truncatellus* (Dalm.) (2° nota preliminare). *Bollettino del Laboratoria di Zoologia Generale e Agraria della R. Scuola Superiore d'Agricoltura in Portici* **6**, 3–51.

Silvestri, F. 1909. Consideration of the existing condition of agricultural entomology in the United States of North America, and suggestions which can be gained from it for the benefit of Italian agriculture. *Hawaiian Forester and Agriculturist* **6**, 287–336.

Simbolotti, G.; Putters, F. A.; and van den Assem, J. 1987. Rates of attack and control of the offspring sex ratio in the parasitic wasp *Lariophagus distinguendus* in an environment where host quality varies. *Behaviour* **100**, 1–32.

Simmonds, F. J. 1947. Improvement of the sex ratio of a parasite by selection. *Canadian Entomology* **79**, 41–44.

Simmonds, F. J. 1953. Observations on the biology and mass-breeding of *Spalangia drosophilae* Ashm. (Hym.; Spalangidae), a parasite of the frit fly, *Oscinella frit* (L.). *Bulletin of Entomological Research* **44**, 773–778.

Singh, R., and Sinha, T. B. 1980. Bionomics of *Trioxys* (*Binodoxys*) Subba Rao and Sharma, an aphidid parasitoid of *Aphis craccivora* Koch. VII. Sex ratio of the parasitoid in field populations. *Entomon* **4**, 269–275.

Skaife, S. H. 1921. A tachinid parasite of the honey bee. *South African Journal of Science* **17**, 196–200.

Skellam, J. G. 1951. Random dispersal in theoretical populations. *Biometrika* **38**, 196–218.

Skinner, S. W. 1982. Maternally-inherited sex ratio in the parasitoid wasp *Nasonia vitripennis*. *Science* **215**, 1133–1134.

Skinner, S. W. 1985. Son-killer: A third extrachromosomal factor affecting the sex ratio in the parasitoid wasp, *Nasonia* (*=Mormoniella) vitripennis*. *Genetics* **109**, 745–759.

Skinner, S. W. 1986. Paternal transmission of an extrachromosomal factor in a wasp: Evolutionary implications. *Heredity* **59**, 47–53.

Skinner, S. W. 1985. Clutch size as an optimal foraging problem for insects. *Behavioural Ecology and Sociobiology* **17**, 231–238.

Skinner, S. W., and Werren, J. H. 1980. The genetics of sex determination in *Nasonia vitripennis*. *Genetics* **94**, 98–106.

Slansky, F. 1986. Nutritional ecology of endoparasitic insects and their hosts: An overview. *Journal of Insect Physiology* **32**, 255–261.

Slansky, F., and Scriber, J. M. 1985. Food consumption and utlization. In G. A. Kerkut and L. I. Gilbert, eds., *Comprehensive Insect Physiology, Biochemistry, Physiology and Pharmacology*, vol. 4, pp. 87–163. Pergamon Press, Oxford.

Slatkin, M. 1980. Ecological character displacement. *Ecology* **61**, 163–177.

Smiley, J. T. 1978. Plant chemistry and the evolution of host specificity: New evidence from *Heliconius* and *Passiflora*. *Science* **201**, 745–747.

Smilowitz, Z. 1974. Relationships between the parasitoid *Hyposoter exiguae* (Vireck) and cabbage looper, *Trichoplusia ni* (Hübner): Evidence for endocrine involvement in successful parasitism. *Annals of the Entomological Society of America* **67**, 317–320.

Smith, A.D.M., and Maelzer, D. A. 1986. Aggregation of parasitoids and density-independence of parasitism in field populations of the wasp *Aphytis melinus* and its host, the red scale, *Aonidiella aurantii. Ecological Entomology* **11**, 425–434.

Smith, C. C., and Fretwell, S. D. 1974. The optimal balance between size and number of offspring. *American Naturalist* **108**, 499–506.

Smith, H. D. 1932. *Phaeogenes nigridens* Wesmael, an important ichneumonid parasite of the pupa of the European corn borer. *United States Department of Agriculture Technical Bulletin* **331**, 1–45.

Smith, H. S. 1916. An attempt to redefine the host relationships exhibited by entomophagous insects. *Journal of Economic Entomology* **9**, 477–486.

Smith, H. S. 1941. Status of biological control of scale insects. *Californian Citrusgrower* **26**, 75–77.

Smith, K.G.V., ed. 1976. Kloet and Hincks, *A Checklist of British Insects,* 2d ed. (completely revised), part 5, Diptera and Siphonaptera. *Handbooks for the Identification of British Insects* **11(5)**, 1–139.

Smith, M. A., and Cornell, H. V. 1979. Hopkins host-selection in *Nasonia vitripennis* and its implications for sympatric speciation. *Animal Behaviour* **27**, 365–370.

Smith, O. J. 1952. Biology and behavior of *Microctonus vittatae* Muesebeck (Braconidae), with descriptions of its immature stages. *University of California Publications in Entomology* **9**, 315–343.

Smith, R. H., and Lessells, C. M. 1985. Oviposition, ovicide and larval competition in granivorous insects. In R. M. Sibly and R. H. Smith, eds., *Behavioural Ecology* (British Ecological Society Symposium 25), pp. 423–448. Blackwell Scientific, Oxford.

Smith, R. H., and Shaw, M. R. 1980. Haplodiploid sex ratios and the mutation rate. *Nature* **287**, 728–729.

Smith, S. G., and Wallace, D. R. 1971. Allelic sex determination in a lower hymenopteran, *Neodiprion nigroscutum* Midd. *Canadian Journal of Genetics and Cytology* **13**, 617–621.

Smith Trail, D. R. 1980. Behavioral interactions between parasites and host: Host suicide and the evolution of complex life cycles. *American Naturalist* **116**, 77–91.

Snell, G. D. 1935. The determination of sex in *Habrobracon. Proceedings of the National Academy of Sciences, USA* **21**, 446–453.

Sokolowski, M. B. 1980. Foraging strategies of *Drosophila melanogaster*: A chromosomal analysis. *Behavioural Genetics* **10**, 291–302.

Sokolowski, M. B., and Turlings, T.C.J. 1987. *Drosophila* parasitoid-host interactions: Vibrotaxis and ovipositor searching from the host's perspectives. *Canadian Journal of Zoology* **65**, 461–464.

Sokolowski, M. B.; Bauer, S. J.; Wai Ping, V.; Rodriguez, L.; Wong, J. L.; and Kent, C. 1986. Ecological genetics and behaviour of *Drosophila melanogaster* larvae in nature. *Animal Behaviour* **34**, 403–408.

Soper, R. S.; Shewell, G. E.; and Tyrrell, D. 1976. *Colcondamyia auditrix* nov. sp. (Diptera: Sarcophagidae), a parasite which is attracted by the mating song of its host, *Okanagana rimosa* (Homoptera: Cicadidae). *Canadian Entomologist* **108**, 61–68.

Southwood, T.R.E. 1957. Observations on swarming in the Braconidae (Hymenoptera) and Coniopterygidae (Neuroptera). *Proceedings of the Royal Entomological Society of London A* **32**, 80–82.

Southwood, T.R.E. 1961. The number of species of insect associated with various trees. *Journal of Animal Ecology* **30**, 1–8.

Southwood, T.R.E. 1977. Habitat, the templet for ecological strategies? *Journal of Animal Ecology* **46**, 337–365.

Southwood, T.R.E. 1988. Tactics, strategies and templets. *Oikos* **52**, 3–18.

Spencer, H. 1926. Biology of parasites and hyperparasites of aphids. *Annals of the Entomological Society of America* **19**, 119–153.

Spiers, D. C.; Sherratt, T. N.; and Hubbard, S. F. 1991. Parasitoid diets: Does superparasitism pay? *Trends in Ecology and Evolution* **6**, 22–25.

Spradbery, J. P. 1969. The biology of *Pseudorhyssa sternata* Merrill (Hym.; Ichneumonidae), a cleptoparasite of siricid wood wasps. *Bulletin of Entomological Research* **59**, 291–297.

Spradbery, J. P. 1970a. The biology of *Ibalia drewseni* Borries, a parasite of siricid woodwasps. *Proceedings of the Entomological Society London B* **45**, 104–113.

Spradbery, J. P. 1970b. Host finding by *Rhyssa persuasoria*, an ichneumonid parasite of siricid woodwasps. *Animal Behaviour* **18**, 103–114.

Stamp, N. E. 1981a. Effect of group size on parasitism in a natural population of the Baltimore Checkerspot *Euphydryas phaeton. Oecologia* **49**, 201–206.

Stamp, N. E. 1981b. Behavior of parasitized aposematic caterpillars: Advantages to the parasitoid or the host? *American Naturalist* **118**, 715–725.

Stamp, N. E. 1982. Searching behaviour of parasitoids for web-making caterpillars: A test of optimal searching theory. *Journal of Animal Ecology* **51**, 387–396.

Stary, P. 1966. *Aphid Parasites of Czechoslovakia. A Review of the Czechoslovak Aphidiidae (Hymenoptera)*. W. Junk, The Hague.

Stary, P. 1970. *Biology of Aphid Parasites*. W. Junk, The Hague.

Steffan, J. R. 1961. Comportement de *Lasiochalcidia igiliensis* Ms.; Chalcidide parasite de Fourmilions. *Compte rendu de l'Académie des Sciences* **253**, 2401–2403.

Stephens, D. W., and Krebs, J. R. 1986. *Foraging Theory*. Princeton University Press, Princeton.

Sternlicht, M. 1973. Parasitic wasps attracted by the sex pheromone of their coccid host. *Entomophaga* **18**, 339–342.

Stewart-Oaten, A. 1977. Optimal foraging in patches: A case for stochasticity. *Theoretical Population Biology* **12**, 263–285.

Stiling, P. D. 1987. The frequency of density-dependence in host-parasitoid systems. *Ecology* **68**, 884–865.

Stiling, P. D., and Strong, D. R. 1982. Egg density and the intensity of parasitism in *Prokelesia marginata* (Homoptera: Delphacidae). *Ecology* **63**, 1630–1635.

Stoltz, D. B. 1981. A putative baculovirus in the ichneumonid parasitoid *Mesoleius tenthredinis. Canadian Journal of Microbiology* **27**, 116–122.

Stoltz, D. B. 1986. Interactions between parasitoid-derived products and host insects: An overview. *Journal of Insect Physiology* **32**, 347–350.

Stoltz, D. B. 1990. Evidence for chromosomal transmission of polydnavirus DNA. *Journal of General Virology* **71**, 1051–1056.

Stoltz, D. B., and Guzo, D. 1986. Apparent haemocytic transformations associated with parasitoid-induced inhibition of immunity in *Malacosoma disstria* larvae. *Journal of Insect Physiology* **32**, 377–388.

Stoltz, D. B., and Vinson, S. B. 1979. Viruses and parasitism in insects. *Advances in Virus Research* **24**, 125–171.

Stoltz, D. B.; Krell, P. J.; Summers, M. D.; and Vinson, S. B. 1984. Polydnaviridae—a proposed family of insect viruses with segmented, double-stranded, circular DNA genomes. *Intervirology* **21**, 1–4.

Stoltz, D. B.; Krell, P. J.; Cook, D.; MacKinnon, E. A.; and Lucarotti, C. J. 1988. An unusual virus from the parasitic wasp *Cotesia melanoscella. Virology* **162**, 311–320.

Stouthamer, R. 1990. Evidence for microbe-mediated parthenogenesis in Hymenoptera. *Proceedings and Abstracts Vth International Colloquium on Invertebrate Pathology and Microbial Control, Adelaide, Australia,* pp. 417–421.

Stouthamer, R. 1991. Effectiveness of several antibiotics in reverting thelytoky to arrhenotoky in *Trichogramma. Les Colloques de l'INRA* **56**, 119–122.

Stouthamer, R., and Luck, R. F. 1993. Influence of microbe-associated parthenogenesis on the fecundity of *Trichogramma deion* and *T. pretiosum* (Hymenoptera, Trichogrammatidae). *Entomologia Experimentalis et Applicata.* **67**, 183–192.

Stouthamer, R., and Werren, J. H. 1993. Microbes associated with parthenogenesis in wasps of the genus *Trichogramma. Journal of Invertebrate Pathology* **61**, 6–9.

Stouthamer, R.; Luck, R. F.; and Hamilton, W. D. 1990a. Antibiotics cause parthenogenetic *Trichogramma* (Hymenoptera/Trichogrammatidae) to revert to sex. *Proceedings of the National Academy of Sciences, USA* **87**, 2424–2427

Stouthamer, R.; Pinto, J. D.; Platner, G. R.; and Luck, R. F. 1990b. Taxonomic status of thelytokous forms of *Trichogramma* (Hymenoptera: Trichogrammatidae). *Annals of the Entomological Society of America* **83**, 475–481.

Stouthamer, R.; Breeuwer, J.A.J.; Luck, R. F.; and Werren, J. H. 1993. Molecular identification of microorganisms associated with parthenogenesis. *Nature* **361**, 66–68.

Strand, M. R. 1986. The physiological interactions of parasitoids with their hosts and their influence on reproductive strategies. In J. K. Waage and D. Greathead, eds., *Insect Parasitoids,* pp. 97–136. Academic Press, London.

Strand, M. R. 1988. Variable sex ratio strategy of *Telenomus heliothidis* (Hymenoptera: Scelionidae): Adaptation to host and conspecific density. *Oecologia* **77**, 219–224.

Strand, M. R. 1989a. Development of the polyembryonic parasitoid *Copidosoma floridanum* in *Trichoplusia ni. Entomologia Experimentalis et Applicata* **50**, 37–46.

Strand, M. R. 1989b. Oviposition behaviour and progeny allocation of the polyembryonic wasp *Copidosoma floridanum* (Hymenoptera: Encyrtidae). *Journal of Insect Behaviour* **2**, 355–369.

Strand, M. R. 1989c. Clutch size, sex ratio and mating by the polyembryonic encyrtid *Copidosoma floridanum* (Hymenoptera: Encyrtidae). *Florida Entomologist* **72**, 32–42.

Strand, M. R. 1993. Progeny and sex allocation by *Copidosoma floridanum* (Hymenoptera: Encyrtidae): Implications for brood composition in the field. Manuscript.

Strand, M. R., and Dover, B. A. 1991. Developmental disruption of *Pseudoplusia in-*

cludens and *Heliothis virescens* larvae by the calyx fluid and venom of *Microplitus demolitor*. *Archives of Insect Biochemistry and Physiology* **18**, 131–145.

Strand, M. R., and Godfray, H.C.J. 1989. Superparasitism and ovicide in parasitic Hymenoptera: A case study of the ectoparasitoid *Bracon hebetor. Behavioural Ecology and Sociobiology* **24**, 421–432.

Strand, M. R., and Noda, T. 1991. Alterations in the haemocytes of *Pseudoplusia includens* after parasitism by *Microplitus demolitor. Journal of Insect Physiology* **37**, 839–850.

Strand, M. R., and Vinson, S. B. 1982a. Behavioural response of the parasitoid, *Cardiochiles nigriceps* to a kairomone. *Entomologia Experimentalis et Applicata* **31**, 308–315.

Strand, M. R., and Vinson, S. B. 1982b. Source and characterization of an egg recognition kairomone of *Telenomus heliothidis*, a parasitoid of *Heliothis virescens. Physiological Entomology* **7**, 83–90.

Strand, M. R., and Vinson, S. B. 1983a. Factors affecting host recognition and acceptance in the egg parasitoid *Telenomus heliothidis* (Hymenoptera: Scelionidae). *Environmental Entomology* **12**, 1114–1119.

Strand, M. R., and Vinson, S. B. 1983b. Analysis of an egg recognition kairomone of *Telenomus heliothidis* (Hymenoptera: Scelionidae), isolation and function. *Journal of Chemical Ecology* **9**, 423–432.

Strand, M. R., and Vinson, S. B. 1984. Facultative hyperparasitism by the egg parasitoid *Trichogramma pretiosum* (Hymenoptera, Trichogrammatidae). *Annals of the Entomological Society of America* **77**, 679–686.

Strand, M. R., and Vinson, S. B. 1985. *In vitro* culture of *Trichogramma pretiosum* on an artificial medium. *Entomologia Experimentalis et Applicata* **39**, 203–209.

Strand, M. R., and Wong, E. A. 1991. The growth and role of *Microplitus demolitor* teratocytes in parasitism of *Pseudoplusia includens. Journal of Insect Physiology* **37**, 503–515.

Strand, M. R.; Quarles, J.M.; Meola, S. M.; and Vinson, S. B. 1985. Cultivation of teratocytes of the egg parasitoid *Telenomus heliothidis* (Hymenoptera: Scelionidae). *In Vitro Cellular and Developmental Biology* **21**, 361–367.

Strand, M. R.; Meola, S. M.; and Vinson, S. B. 1986. Correlating pathological symptoms in *Heliothis virescens* eggs with development of the parasitoid *Telenomus heliothidis. Journal of Insect Physiology* **32**, 389–402.

Strand, M. R.; Williams, H. J.; Vinson, S. B.; and Mudd, A. 1989. Kairomonal activities of 2-acylcyclohexane-1,3-diones produced by *Ephestia kuehniella* Zeller in eliciting searching behavior by the parasitoid *Bracon hebetor* (Say). *Journal of Chemical Ecology* **15**, 1491–1500.

Strand, M. R.; Johnson, J. A.; and Dover, B. A. 1990. Ecdysteroid and juvenile hormone esterase profiles of *Trichoplusua ni* parasitized by the polyembryonic wasp *Copidosoma floridanum. Archives of Insect Biochemistry and Physiology* **13**, 41–51.

Strand, M. R.; Goodman, W. G.; and Baehrecke, E. H. 1991a. The juvenile hormone titer of *Trichoplusia ni* and its potential role in embryogenesis of the polyembryonic wasp *Copidosoma floridanum. Insect Biochemistry* **21**, 205–214.

Strand, M. R.; Baehrecke, E. H.; and Wong, E.A. 1991b. The role of host endocrine factors in the development of polyembryonic parasitoids. *Biological Control* **1**, 144–152.

Streams, F. A. 1971. Encapsulation of insect parasites in superparasitized hosts. *Entomologia Experimentalis et Applicata* **14**, 484–490.

Streams, F. A., and Greenberg, L. 1969. Inhibition of the defence reaction of *Drosophila melanogaster* parasitized simultaneously by the wasps *Pseudeucolia bochei* and *P. mellipes. Journal of Invertebrate Pathology* **13**, 371–377.

Strickland, E. H. 1923. Biological notes on parasites of prairie cutworms. *Bulletin of the Department of Agriculture, Canada* **26**, 1–40.

van Strien-van Liempt, W.T.F.H., and van Alphen, J.J.M. 1981. The absence of interspecific host discrimination in *Asobara tabida* Nees and *Leptopilina heterotoma* (Thomson) coexisting larval parasitoids of *Drosophila* species. *Netherlands Journal of Zoology* **31**, 701–712.

van Strien-van Liempt, W.T.F.H., and Hofker, C. D. 1985. Host selection in two coexisting parasitoids of *Drosophila, Asobara tabida* Nees von Esenbeck and *Leptopilina heterotoma* (Thomson). *Netherlands Journal of Zoology* **35**, 693–706.

Strong, D. R. 1974. Nonasymptotic species richness models and the insects of British trees. *Proceedings of the National Academy of Sciences, USA* **71**, 2766–2769.

Strong, D. R. 1979. Biogeographic dynamics of insect plant communities. *Annual Review of Entomology* **24**, 89–119.

Strong, D. R. 1989. Density independence in space and inconsistent temporal relationships for host mortality caused by a fairyfly parasitoid. *Journal of Animal Ecology* **58**, 1065–1076.

Strong, D. R.; Lawton, J. H.; and Southwood, T.R.E. 1984. *Insects on Plants.* Blackwell Scientific, Oxford.

Styer, E. L.; Hamm, J. J.; and Nordlund, D. A. 1987. A new virus associated with the parasitoid *Cotesia marginiventris* (Hymenoptera: Braconidae): Replication in noctuid host larvae. *Journal of Invertebrate Pathology* **50**, 302–309.

Subba Rao, B. R., and Sharma, A. K. 1962. Studies on the biology of *Trioxys indicus* Subba Rao and Sharma 1958, a parasite of *Aphis gossypii* Glover. *Proceedings of the National Institute of Science India B* **28**, 164–182.

Sugimoto, T. 1977. Ecological studies on the relationship between the ranunculus leaf mining fly *Phytomyza ranunculi* Schrank (Diptera: Agromyzidae) and its parasite, *Kratochviliana* sp. (Hymenoptera: Eulophidae) from the viewpoint of spatial structure. I. Analysis of searching and attacking behaviours of the parasite. *Applied Entomology and Zoology* **12**, 87–103.

Sugimoto, T., and Tsujimoto, S. 1988. Stopping rules of host search by the parasitoid *Chrysocharis pentheus* (Hymenoptera: Eulophidae), in host patches. *Researches in Population Biology* **30**, 123–133.

Sugimoto, T.; Uenishi, M.; and Machida, F. 1986. Foraging for patchily-distributed leaf-miners by the braconid *Dapsilarthra rufiventris* (Hymenoptera: Braconidae) I. Discrimination of previously searched leaflets. *Applied Entomology and Zoology* **21**, 500–508.

Sugimoto, T.; Murakami, H.; and Yamazaki, R. 1987. Foraging for patchily-distributed leaf-miners by the braconid *Dapsilarthra rufiventris* (Hymenoptera: Braconidae) II. Stopping rules for host search. *Journal of Ethology* **5**, 95–103.

Sugimoto, T.; Shimono, Y.; Hata, Y.; Nakai, A.; and Yahara, M. 1988a. Foraging for patchily-distributed leaf-miners by the braconid *Dapsilarthra rufiventris* (Hymenoptera: Braconidae) III. Visual and acoustic cues to a close range patch-location. *Applied Entomology and Zoology* **23**, 113–121.

Sugimoto, T.; Ichikawa, T.; Mitomi, M.; and Sakuratani, Y. 1988b. Foraging for patchily-distributed leaf-miners by the braconid *Dapsilarthra rufiventris* (Hymenoptera: Braconidae) IV. Analysis of sounds emitted by a feeding host. *Applied Entomology and Zoology* **23**, 209–211.

Sugimoto, T.; Kawado, K.; and Tadera, K. 1988c. Responses to the host mine by two parasitic wasps, *Dapsilarthra rufiventris* (Hymenoptera: Braconidae) and *Chrysocharis pentheus* (Hymenloptera: Eulophidae). *Journal of Ethology* **6**, 55–58.

Sugimoto, T.; Minkenberg, O.P.J.M.; Takabayashi, J.; Dicke, M.; and van Lenteren, J. V. 1990. Foraging for patchily-distributed leaf-miners by the parasitic wasp, *Dacnusa sibirica*. *Researches in Population Biology* **32**, 381–389.

Sullivan, D. J. 1987. Insect hyperparasitism. *Annual Review of Entomology*, **32**, 49–70.

Summy, K. R.; Gilstrap, F. E.; and Hart, W. G. 1985. *Aleurocanthus woglumi* (Hom.; Aleyrodidae) and *Encarsia opulenta* (Hym.; Enycyrtidae): Density dependent relationship between adult parasite aggregation and mortality of the host. *Entomophaga* **30**, 1107–1112.

Sunose, T. 1978. Studies on extended diapause in *Hasegawia sasacola* Monzen (Diptera, Cecidomyiidae) and its parasites. *Kontyû* **46**, 400–415.

Sutherland, W. J. 1983. Aggregation and the "ideal free" distribution. *Journal of Animal Ecology* **52**, 821–828.

Suzuki, Y., and Hiehata, K. 1985. Mating systems and sex ratios in the egg parasitoids, *Trichogramma dendrolimi* and *T. papilionis* (Hymenoptera: Trichogrammatidae). *Animal Behaviour* **33**, 1223–1227.

Suzuki, Y., and Iwasa, Y. 1980. A sex ratio theory of gregarious parasitoids. *Researches in Population Biology* **22**, 366–382.

Suzuki, Y.; Tsuji, H.; and Sasakawa, M. 1984. Sex allocation and effects of superparasitism on secondary sex ratios in the gregarious parasitoid, *Trichogramma chilonis* (Hymenoptera: Trichogrammatidae). *Animal Behaviour* **32**, 478–484.

Syme, P. D. 1977. Observations on the longevity and fecundity of *Orgilus obscurator* (Hymenoptera: Braconidae) and the effects of certain food on longevity. *Canadian Entomologist* **109**, 995–1000.

Tagawa, J. 1977. Localization and histology of the female sex pheromone producing gland in the parasitic wasp *Apanteles glomeratus*. *Journal of Insect Physiology* **23**, 49–56.

Tagawa, J. 1987. Post mating changes in the oviposition tactics of the parasitic wasp *Apanteles glomeratus* L. (Hymenoptera, braconidae). *Applied Entomology and Zoology* **22**, 537–542.

Tagawa, J., and Kitano H. 1981. Mating behaviour of the braconid wasp *Apanteles glomeratus* L. (Hymenoptera: Braconidae) in the field. *Applied Entomology and Zoology* **16**, 345–350.

Takabayashi, J., and Takahashi, S. 1985. Host selection behavior of *Anicetus benificus* Ishii et Yasumatsu (Hymenoptera: Encyrtidae). 3. Presence of ovipositional stimulants in the scale wax of the genus *Ceroplastes*. *Applied Entomology and Zoology* **20**, 173–178.

Takagi, M. 1985. The reproductive strategy of the gregarious parasitoid, *Pteromalus puparum* (Hymenoptera: Pteromalidae). 1. Optimal number of eggs in a single host. *Oecologia* **68**, 1–6.

Takagi, M. 1986. The reproductive strategy of the gregarious parasitoid, *Pteromalus puparum* (Hymenoptera: Pteromalidae). 2. Host size discrimination and regulation of the number and sex ratio of progeny in a single host. *Oecologia* **70**, 321–325.

Takagi, M. 1987. The reproductive strategy of the gregarious parasitoid, *Pteromalus puparum* (Hymenoptera: Pteromalidae). 3. Superparasitism in a field population. *Oecologia* **71**, 321–324.

Takahashi, F. 1963. Changes in some ecological characters of the almond moth caused by the selective action of an ichneumon wasp in their interacting system. *Researches in Population Ecology* **5**, 117–129.

Takasu, K., and Hirose, Y. 1988. Host discrimination in the parasitoid *Ooencyrtus nezarae*: The role of the egg stalk as an external marker. *Entomologia Experimentalis et Applicata* **47**, 45–48.

Takasu, K., and Hirose, Y. 1991. The parasitoid *Ooencyrtus nezarae* (Hymenoptera: Encyrtidae) prefers hosts parasitised by conspecifics over unparasitized hosts. *Oecologia* **87**, 319–323.

Tanaka, T. 1987. Effect of the venom of the endoparasitoid *Apanteles kariyai* Watanabe, on the cellular defence reaction of the host, *Pseudaletra seprata* Walker. *Journal of Insect Physiology* **33**, 413–420.

Tanaka, T., and Wago, H. 1990. Ultrastructural and functional maturation of teratocytes of *Apanteles karizai*. *Archives of Insect Biochemistry and Physiology* **13**, 187–197.

Tauber, C. A., and Tauber, M. J. 1989. In D. Otte and J. Endler, eds., *Speciation and Its Consequences,* pp. 307–344. Sinauer, Sunderland, Massachusetts.

Tauber, M. J.; Tauber, C. A.; Nechols, J. R.; and Obrycki, J. J. 1983. Seasonal activity of parasitoids: Control by external, internal and genetic factors. In V. K. Brown and I. Hodek, eds., *Diapause and Life Cycle Strategies in Insects*, pp. 87–108. Dr W. Junk, The Hague.

Tauber, M. J.; Tauber, C. A.; and Masaki, S. 1986. *Seasonal Adaptations of Insects*. Oxford University Press, New York.

Taylor, P. D. 1981. Intra-sex and inter-sex sibling interactions as sex ratio determinants. *Nature* **291**, 64–66.

Taylor, P. D. 1988. Inclusive fitness models with two sexes. *Theoretical Population Biology* **34**, 145–168.

Taylor, P. D., and Bulmer, M. G. 1980. Local mate competition and the sex ratio. *Journal of Theoretical Biology* **86**, 409–419.

Taylor, P. D., and Sauer, A. 1980. The selective advantage of sex-ratio homeostasis. *American Naturalist* **116**, 305–310.

Taylor, T.H.C. 1937. *The Biological Control of an Insect in Fiji, an Account of the Coconut Leaf-mining Beetle and Its Parasite Complex*. Commonwealth Institute of Entomology, London.

Tepedino, V. J. 1988. Host discrimination in *Monodontomerus obsoletus* Fabricius (Hymenoptera: Torymidae), a parasite of the alphalpha leafcutting bee *Megachile rotundata* (Fabricius) (Hymenoptera: Megachilidae). *Journal of the New York Entomological Society* **96**, 113–118.

Thomas, J. A., and Elmes, G. W. 1993. Specialised searching and the aggressive use by a parasitoid whose host—the butterfly *Maculinea rebeli*—inhabits ant nests. *Animal Behaviour*. **45**, 593–602.

Thompson, J. N. 1986. Oviposition behaviour and searching efficiency in a natural population of a braconid parasitoid. *Journal of Animal Ecology* **55**, 351–360.

Thompson, S. N. 1983. Metabolic and physiological effects of metazoan endoparasitoids on their host species. *Comparative Biochemistry and Physiology* **74B**, 183–211.

Thompson, S. N. 1986a. The metabolism of insect parasites (parasitoids): An overview. *Journal of Insect Physiology* **32**, 421–423.

Thompson, S. N. 1986b. Nutrition and in vitro culture of insect parasitoids. *Annual Review of Entomology* **31**, 197–219.

Thompson, S. N., and Barlow, J. S. 1974. The fatty acid composition of parasitic Hymenoptera and its possible biological significance. *Annals of the Entomological Society of America* **67**, 627–632.

Thornhill, R., and Alcock, J. 1983. *The Evolution of Insect Mating Systems*. Harvard University Press, Cambridge, Massachusetts.

Thorpe, W. H. 1933. Notes on the natural control of *Coleophora laricella*, the larch case bearer. Biology and morphology of *Angitia nana*. *Bulletin of Entomological Research* **24**, 273–277.

Thorpe, W. H., and Caudle, H. B. 1938. A study of the olfactory responses of insect parasites to the food plant of their host. *Parasitology* **30**, 523–528.

Thorpe, W. H., and Jones, F.G.W. 1937. Olfactory conditioning in a parasitic insect and its relation to the problem of host selection. *Proceedings of the Royal Entomological Society B* **124**, 56–81.

Thurston, R., and Fox, P. M. 1972. Inhibition by nicotine of emergence of *Apanteles congregatus* from its host, the tobacco hornworm. *Annals of the Entomological Society of America* **65**, 547–550.

Timberlake, P. H. 1910. Observations on the early stages of two aphidiine parasites of aphids. *Psyche* **17**, 125–130.

Timberlake, P. H. 1912. Experimental parasitism, a study of the biology of *Limnerium validum* (Cresson). *United States Bureau of Entomology Technical Series* **19**, 71–92.

Timberlake, P. H. 1916. Note on an interesting case of two generations of a parasite reared from the same individual host. *Canadian Entomologist* **48**, 89–91.

Toft, C. A. 1984. Activity budgets in two species of bee flies (*Lordotus*: Bombyliidae, Diptera): A comparison of species and sexes. *Behavioural Ecology and Sociobiology* **14**, 287–296.

Toft, C. A. 1989a. Population structure and mating system of a desert bee fly (*Lordotus pulchrissimus*) (Diptera: Bombyliidae). 1. Male demography and interactions. *Oikos* **54**, 345–358.

Toft, C. A. 1989b. Population structure and mating system of a desert bee fly (*Lordotus pulchrissimus*) (Diptera: Bombyliidae). 2. Female demography, copulations and characteristics of swarm sites. *Oikos* **54**, 359–369.

de Toledo, A. A. 1942. On the bionomics of *Aphidius matricariae* Hal.; a braconid parasite of *Myzus persicae*. *Parasitology* **34**, 141–151.

Tomlinson, I. 1987. Adaptive and non-adative suicide in aphids (Scientific Correspondence). *Nature* **330**, 701.

Tothill, J. D.; Taylor, T.H.C.; and Paine, R. W. 1930. *The Coconut Moth in Fiji*. Imperial Bureau of Entomology, London.

Townes, H. 1939. Protective odors among the Ichneumonidae (Hymenoptera). *Bulletin of the Brooklyn Entomological Society* **34**, 29–30.

Townes, H., and Townes, M. 1960. Ichneumon-flies of America north of America: 2, Subfamilies Ephialtinae, Xoridinae, Acaenitinae. *Bulletin of the United States National Museum* **216**, 1–676.

Townsend, C.H.T. 1908. A record of results from rearings and dissections of Tachinidae. *United States Bureau of Entomology Technical Series 12*, **6**, 95–118.

Townsend, C.H.T. 1934–1939. *Manual of Myiology,* Itaquaquecetuba, São Paolo, Brazil.

Tremblay, E. 1966. Ricerche sugli Imenotteri parassiti. III. Osservationi sulla competizione intraspecifica negli Aphidiinae (Hymenoptera: Braconidae). *Bollettino del Laboratorio di Entomologia Agraria "Filippo Silvestri"* **24**, 209–225.

Tripp, H. A. 1961. The biology of a hyperparasite, *Euceros frigidus* Cress. (Ichneumonidae) and description of the planidial stage. *Canadian Entomologist* **93**, 40–58.

Trivers, R. L. 1974. Parent-offspring conflict. *American Zoologist* **14**, 249–264.

Trivers, R. L., and Willard, D. E. 1973. Natural selection of parental ability to vary the sex ratio of offspring. *Science* **179**, 90–92.

Tscharntke, T. 1992. Coexistence, tritrophic interactions and density dependence in a species-rich parasitoid complex. *Journal of Animal Ecology* **61**, 59–68.

Turlings, T.C.J.; and Tumlinson, J. H. 1991. Do parasitoids use herbivore-induced plant chemical defenses to locate hosts? *Florida Entomologist* **74**, 42–50.

Turlings, T.C.J., and Tumlinson, J. H. 1992. Systemic release of chemical signals by herbivore-injured corn. *Proceedings of the National Academy of Sciences, USA.* **89**, 8399–8402.

Turlings, T.C.J.; van Batenburg, F.D.H.; and van Strien-van Liempt, W.T.F.H. 1985. Why is there no interspecific host discimination in the two coexisting larval parasitoids of *Drosophila* species, *Leptopilina heterotoma* (Thomson) and *Asobara tabida* (Nees). *Oecologia* **67**, 352–359.

Turlings, T.C.J.; Tumlinson, J. H.; Lewis, W. J.; and Vet, L.E.M. 1989. Beneficial arthropod behaviour mediated by airborne semiochemicals. VIII. Learning of host-related odors induced by a brief contact experience with host by-products in *Cotesia marginiventris* (Cresson), a generalist larval parasitoid. *Journal of Insect Behaviour* **2**, 217–225.

Turlings, T.C.J.; Scheepmaker, J.W.A.; Vet, L.E.M.; Tumlinson, J. H.; Lewis, W. J.; and Vet, L.E.M. 1990a. How contact foraging experiences affect the preferences for host-related odors in the larval parasitoid *Cotesia marginiventris* (Cresson) (Hymenoptera: Braconidae). *Journal of Chemical Ecology* **16**, 1577–1589.

Turlings, T.C.J.; Tumlinson, J. H.; and Lewis, W. J. 1990b. Exploitation of herbivore-induced plant odours by host seeking parasitic wasps. *Science* **250**, 1251–1253.

Turlings, T.C.J.; Tumlinson, J. H.; Eller, F. J.; and Lewis, W. J. 1991a. Larval-damaged plants: Source of volatile synonomes that guide the parasitoid *Cotesia marginiventris* to the microhabitat of its host. *Entomologia Experimentalis et Applicata* **58**, 75–82.

Turlings, T.C.J.; Tumlinson, J. H.; Heath, R. H.; Proveaux, A. T.; and Doolittle, R. E. 1991b. Isolation and identification of allelochemicals that attract the larval parasitoid, *Cotesia marginiventris* (Cresson), to the microhabitat of one of its hosts. *Journal of Chemical Ecology* **17**, 2235–2251.

Turlings, T.C.J.; Wäckers, F. L.; Vet, L.E.M.; Lewis, W. J.; and Tumlinson, J. H. 1992. Learning of host-loction cues by hymenopterous parasitoids. In A. C. Lewis and D. R. Papaj, eds., *Insect Learning: Ecological and Evolutionary Perspectives,* pp. 51–78. Chapman and Hall, New York.

Turnbull, A. L., and Chant, D. A. 1961. The practice and theory of biological control of insects in Canada. *Canadian Journal of Zoology* **39**, 697–753.

Turnock, W. J. 1973. Factors influencing the fall emergence of *Bessa harveyi* (Tachinidae: Diptera). *Canadian Entomologist* **105**, 399–409.

Ulenberg, S. A. 1985. The systematics of the fig wasp parasites of the genus *Apocrypta* Coqueral. *Verhandelingen der Koninklijke Nederlandse Akademie van Wetenschappen. Afd. Natuurkunde* Second series, part 83.

Ullyet, G. C. 1936. Host selection by *Microplectron fuscipennis* (Chalcidoidea, Hymenoptera). *Proceedings of the Royal Society of London B* **120**, 253–291.

Ullyet, G. C. 1953. Biomathematics and insect population problems. *Entomological Society of South Africa Memoirs* **2**, 1–89.

Unruh, T. R.; Gonzalez, D.; and Gordh, G. 1984. Electrophoretic studies on parasitic Hymenoptera and implications for biological control. *Proceedings of the XVIIth International Congress of Entomology*, p. 705.

Utida, A. 1957. Population fluctuation: An experimental and theoretical approach. *Cold Spring Harbor Symposium in Quantitative Biology* **22**, 139–152.

Uyenoyama, M. K., and Bengtsson, B. O. 1981. Towards a genetic theory for the evolution of the sex ratio II. Haplodiploid and diploid models with sibling and parental control of the brood sex ratio and brood size. *Theoretical Population Biology* **20**, 57–79.

Uyenoyama, M. K., and Bengtsson, B. O. 1982. Towards a genetic theory for the evolution of the sex ratio III. Parental and sibling control of brood investment under partial sib-mating. *Theoretical Population Biology* **22**, 43–68.

Valladares, G., and Lawton, J. H. 1991. Host-plant selection in the holly leaf-miner: Does mother know best? *Journal of Animal Ecology* **60**, 227–240.

Varley, G. C. 1937. Description of the eggs and larvae of four species of chalcidoid hymenoptera parasitic on the knapweed gall-fly. *Proceedings of the Royal Entomological Society of London B* **6**, 122–130.

Varley, G. C.; Butler, B. A.; and Butler, C. G. 1933. The acceleration of development in insects by parasitism. *Parasitology* **25**, 263–268.

Velthuis, H.H.W.; Velthuis-Kluppell, F. M.; and Bossink, G.A.H. 1965. Some aspects of the biology and population dynamics of *Nasonia vitripennis* (Walker) (Hymenoptera: Pteromalidae). *Entomologia Experimentalis et Applicata* **8**, 205–227.

Venkatraman, T. V.; and Chacko, M. J. 1961. Some factors influencing the efficiency of *Goniozus marasmi* Kurian, a parasite of the Maize and Jowar Leaf Roller. *Proceedings of the Indian Academy of Science B*, **6**, 275–283.

Verai, E. J. 1942. On the bionomics of *Aphidius matricariae* Hal., a braconid parasite of *Myzus persicae*. *Parasitology* **34**, 141–151.

Verner, J. 1965. Selection for the sex ratio. *American Naturalist* **99**, 419–422.

Vet, L.E.M. 1983. Host-habitat location through olfactory cues by *Leptopilina clavipes* (Hartig) (Hym: Eucoilidae), a parasitoid of fungivorous *Drosophila*: The influence of conditioning. *Netherlands Journal of Zoology* **33**, 225–248.

Vet, L.E.M. 1985a. Olfactory microhabitat location in some eucoilid and alysiine species (Hymenoptera), larval parasitoids of Diptera. *Netherlands Journal of Zoology* **35**, 486–496.

Vet, L.E.M. 1985b. Response to kairomones by some alysiine and eucoilid parasitoid species (Hymenoptera), larval parasitoids of Diptera. *Netherlands Journal of Zoology* **35**, 486–496.

Vet, L.E.M., and van Alphen, J.J.M. 1985. A comparative functional approach to the host detection behaviour of parasitic wasps. I. A qualitative study on Eucoilidae and Alysiinae. *Oikos* **44**, 478–486.

Vet, L.E.M., and Bakker, K. 1985. A comparative functional approach to the host detection behaviour of parasitic wasps. II. A quantitative study on eight eucoilid species. *Oikos* **44**, 487–498.

Vet, L.E.M., and Dicke, M. 1992. Ecology of infochemical use by natural enemies in a tritrophic context. *Annual Review of Entomology* **37**, 141–172.

Vet, L.E.M., and Groenewold, A. W. 1990. Semiochemicals and learning in parasitoids. *Journal of Chemical Ecology* **16**, 3119–3135.

Vet, L.E.M., and van der Hoeven, R. 1984. Comparison of the behavioural responses of two *Leptopilina* species (Hymenoptera: Eucoilidae), living in different microhabitats, to kairomone of their host (Drosophilidae). *Netherlands Journal of Zoology* **34**, 220–227.

Vet, L.E.M., and van Opzeeland, K. 1984. The influence of conditioning on olfactory microhabitat and host location in *Asobara tabida* (Nees) and *A. rufescens* (Foerster) (Braconidae, Alysiinae), larval parasitoids of *Drosophila. Oecologia* **63**, 171–177.

Vet, L.E.M., and van Opzeeland, K. 1985. Olfactory microhabitat selection in *Leptopilina heterotoma* (Thompson) (Hym.: Eucoilidae), a parasitoid of Drosophilidae. *Netherlands Journal of Zoology* **35**, 497–504.

Vet, L.E.M., and Schoonman, G. 1988. The influence of previous foraging experience on microhabitat acceptance in *Leptopilina heterotoma. Journal of Insect Behaviour* **1**, 387–392.

Vet, L.E.M.; van Lenteren, J. C.; Heymans, M.; and Meelis, E. 1983. An airflow olfactometer for measuring olfactory responses of hymenopterous parasitoids and other small insects. *Physiological Entomology* **8**, 97–106.

Vet, L.E.M.; Janse, C. J.; van Achterberg, C.; and van Alphen, J.J.M. 1984a. Microhabitat location and niche segregation in two sibling species of drosophilid parasitoids: *Asobara tabida* (Nees) and *A. rufescens* (Foerster) (Braconidae: Alysiinae). *Oecologia* **61**, 182–188.

Vet, L.E.M.; Meyer, M.; Bakker, K.; and van Alphen, J.J.M. 1984b. Intra- and interspecific host discrimination in *Asobara* (Hymenoptera) larval endoparasitoids of Drosophilidae: Comparison between closely related and less closely related species. *Animal Behaviour* **32**, 871–874.

Vet, L.E.M.; Lewis, W. J.; Papaj, D. R.; and van Lenteren, J. C. 1990. A variable response model for parasitoid foraging behavior. *Journal of Insect Behaviour* **3**, 471–491.

Vet, L.E.M.; Wäckers, F. L.; and Dicke, M. 1991. How to hunt for hiding hosts: The reliability-detectability problem in foraging parasitoids. *Netherlands Journal of Zoology* **41**, 202–213.

Viggiani, G. 1984. Bionomics of the Aphelinidae. *Annual Review of Entomology* **29**, 257–276.

Viggiani, G., and Battaglia, D. 1983. Courtship and mating behaviour in a few Aphelinidae. *Bollettino del Laboratorio di Entomologia Agraria "Filippo Silvestri"* **40**, 89–96.

Viggiani, G., and Mazzone, P. 1978. Morfologia, biologia e utilizazzione di *Prospaltella lahorensis* How. (Hym. Aphelinidae), parassita esotico introdotto in Italia per la lotta biologica al *Dialeurodes citri* (Asm.). *Bollettino del Laboratorio di Entomologia Agraria "Filippo Silvestri"* **35**, 99–161.

Viktorov, G. A. 1968. [The influence of the population density upon the sex ratio in *Trissolcus grandis* Thoms. (Hymenoptera, Sceliondae).] *Zoologicheskii Zhurnal* **47**, 1035–1039 [in Russian].

Viktorov, G. A., and Kochetova, N. I. 1971. [The significance of population density in the regulation of the sex ratio of *Trissolcus volgensis* (Hymenoptera, Scelionidae).] *Zoologicheskii Zhurnal* **50**, 1753–1755 [in Russian].

Viktorov, G. A., and Kochetova, N. I. 1973a. [The role of trace pheromones in regulating the sex ratio in *Trissolcus grandis* (Hymenoptera, Scelionidae)]. *Zhurnal Obshchei Biologii* **34**, 559–562 [in Russian].

Viktorov, G. A., and Kochetova, N. I. 1973b. [On the regulation of the sex ratio in *Dahlbominus fuscipennis* Zett. (Hymenoptera, Eulophidae).] *Éntomologicheskoe Obozrenie* **52**, 651–657 [in Russian].

Vinson, S. B. 1968. Source of a substance in *Heliothis viresecens* that elicits a searching response in its habitual parasite, *Cardiochiles nigriceps*. *Annals of the Entomological Society of America* **58**, 869–871.

Vinson, S. B. 1972a. Competition and host discrimination between two species of tobacco budworm parasitoids. *Annals of the Entomological Society of America* **65**, 229–236.

Vinson, S. B. 1972b. Factors involved in successful attack on *Heliothis virescens* by the parasitoid *Cardiochiles nigriceps*. *Journal of Invertebrate Pathology* **20**, 118–123.

Vinson, S. B. 1976. Host selection by insect parasitoids. *Annual Review of Entomology* **21**, 109–133.

Vinson, S. B. 1981. Habitat location. In D. A. Nordlund, R. L. Jones and W. J. Lewis, eds., *Semiochemicals, Their Role in Pest Control,* pp. 51–78. John Wiley, New York.

Vinson, S. B. 1984. How parasitoids locate their hosts: A case of insect espionage. In T. Lewis, *Insect Communication,* pp. 325–348. Academic Press, London.

Vinson, S. B. 1985. The behaviour of parasitoids. In G. A. Kerkut and L. I. Gilbert, eds., *Comprehensive Insect Physiology, Biochemistry and Pharmacology,* pp. 417–469. Pergamon Press, New York.

Vinson, S. B. 1990a. How parasitoids deal with the immune system of their host: An overview. *Archives of Insect Biochemistry and Physiology* **13**, 3–27.

Vinson, S. B. 1990b. Physiological interactions between the host genus *Heliothis* and its guild of parasitoids. *Archives of Insect Biochemistry and Physiology* **13**, 63–81.

Vinson, S. B., and Iwantsch, G. F. 1980a. Host suitability for insect parasitoids. *Annual Review of Entomology* **25**, 397–419.

Vinson, S. B., and Iwantsch, G. F. 1980b. Host regulation by insect parasitoids. *Quarterly Review of Biology* **55**, 143–165.

Vinson, S. B., and Lewis, W. J. 1965. A method of host selection by *Cardiochiles nigriceps*. *Journal of Economic Entomology* **58**, 869–871.

Vinson, S. B., and Scott, J. R. 1974. Particles containing DNA associated with the oocyte of an insect parasitoid. *Journal of Invertebrate Pathology* **25**, 375–378.

Vinson, S. B.; Jones, R. L.; Sonnet, P. E.; Bierl, B. A.; and Beroza, M. 1975. Isolation, identification and synthesis of host-seeking stimulants for *Cardiochiles nigriceps*, a parasitoid of the tobacco budworm. *Entomologia Experimentalis et Applicata* **18**, 443–450.

440 • REFERENCES

Vinson, S. B.; Henson, R.D.; and Barfield, C. S. 1976. Ovipositional behaviour of *Bracon mellitor* Say (Hymenoptera: Braconidae), a parasitoid of the boll weevil (*Anthonomus grandis* Boh.). I.Isolation and identification of a synthetic releaser of oviposition probing. *Journal of Chemical Ecology* **2**, 157–164.

Vinson, S. B.; Barfield, C. S.; and Henson, R. D. 1977. Oviposition behaviour of *Bracon mellitor*, a parasitoid of the boll weevil (*Anthonomus grandis*). II. Associative learning. *Physiological Entomology* **2**, 157–164.

Vinson, S. B.; Harlan, D. P.; and Hart, W. G. 1978. Response of the parasitoid *Microterys flavus* to brown soft scale and its honeydew. *Environmental Entomology* **7**, 874–878.

Vinson, S. B.; Edson, K. M.; and Stoltz, D. B. 1979. Effect of a virus associated with the reproductive system of the parasitoid wasp, *Campoletis sonorensis*, on host weight gain. *Journal of Invertebrate Pathology* **34**, 133–137.

Visser, M. E. 1991. Prey selection by predators depleting a patch: An ESS model. *Netherlands Journal of Zoology* **41**, 63–80.

Visser, M. E. 1993. Adaptive self- and conspecific superparasitism in the solitary parasitoid *Leptopilina heterotoma*. *Behavioural Ecology* **4**, 22–28.

Visser, M. E., and Driessen, G. 1991. Indirect mutual interference in parasitoids. *Netherlands Journal of Zoology* **41**, 214–227.

Visser, M. E., and Sjerps, M. J. 1991. Optimal diet in depletable patches: A comparison of two papers. *Oikos* **62**, 80–82.

Visser, M. E.; van Alphen, J.J.M.; and Nell, H. W. 1990. Adaptive superparasitism and patch time allocation in solitary parasitoids: The influence of the number of parasitoids depleting a patch. *Behaviour* **114**, 21–36.

Visser, M. E.; van Alphen, J.J.M.; and Hemerik, L. 1992a. Adaptive superparasitism and patch time allocation in solitary parasitoids: An ESS model. *Journal of Animal Ecology* **61**, 93–101.

Visser, M. E.; van Alphen, J.J.M.; and Nell, H. W. 1992b. Adaptive superparasitism and patch time allocation in solitary parasitoids: The influence of pre-patch experience. *Behavioural Ecology and Sociobiology* **31**, 163–171.

Visser, M. E.; Luyckx, B.; Nell, H. W.; and Boskamp, G.J.F. 1992c. Adaptive superparasitism in solitary parasitoids: Marking of parasitised hosts in relation to the payoff from superparasitism. *Ecological Entomology* **17**, 76–82.

Vlug, H. J. 1993. *Catalogue of the Platygastridae of the World (Hymenopteroum Catalogue, nova editio, pars 19)*. SPB Academic Publishing, The Hague (in press).

Völkl, W., and Mackauer, M. 1990. Age-specific host discrimination by the aphid parasitoid *Ephedrus californicus* Baker (Hyemnoptera: Aphidiidae). *Canadian Entomologist* **122**, 349–361.

Volterra, V. 1926. Variations and fluctuations of the numbers of individuals in animal species living together. Reprinted in 1931 in R. N. Chapman, ed., *Animal Ecology*. McGraw-Hill, New York.

Voukassovitch, P. 1927. Observations biologiques sur le *Macrocentrus abdominalis* Fab.; braconide parasite. *Compte Rendu des Séances de la Société de Biologie* **96**, 379–381.

Waage, J. K. 1978. Arrestment responses of a parasitoid, *Nemeritis canescens*, to a contact chemical produced by its host, *Plodia interpunctella*. *Physiological Entomology* **3**, 135–146.

Waage, J. K. 1979. Foraging for patchily-distributed hosts by the parasitoid, *Nemeritis canescens. Journal of Animal Ecology* **48**, 353–371.

Waage, J. K. 1982a. Sib-mating and sex ratio strategies in scelionid wasps. *Ecological Entomology* **7**, 103–112.

Waage, J. K. 1982b. Sex ratio and population dynamics of natural enemies—some possible interactions. *Annals of Applied Biology* **101**, 159–164.

Waage, J. K. 1983. Aggregation in field parasitoid populations: Foraging time allocation by a population of *Diadegma* (Hymenoptera, Ichneumonidae). *Ecological Entomology* **8**, 447–453.

Waage, J. K. 1986. Family planning in parasitoids: Adaptive patterns of progeny and sex allocation. In J. K. Waage and D. Greathead, eds., *Insect Parasitoids*. pp. 63–95. Academic Press, London.

Waage, J. K. 1990. Ecological theory and the selection of biological control agents. In M. Mackauer, L. E. Ehler, and J. Roland, eds., *Critical Issues in Biological Control*, pp. 135–157. Intercept, Andover, Hants., U.K.

Waage, J. K., and Godfray, H.C.J. 1985. Reproductive strategies and population ecology of insect parasitoids. In R. M. Sibly and R. H. Smith, eds., *Behavioural Ecology* (British Ecological Society Symposium 25), pp. 449–470. Blackwell Scientific, Oxford.

Waage, J. K., and Greathead, D., eds. 1986. *Insect Parasitoids*. Academic Press, London.

Waage, J. K., and Greathead, D. 1988. Biological control: Challenges and opportunities. *Philosophical Transactions of the Royal Society of London B* **318**, 111–126.

Waage, J. K., and Lane, J. A. 1984. The reproductive strategy of a parasitic wasp. II. Sex allocation and local mate competition in *Trichogramma evanescens. Journal of Animal Ecology* **53**, 417–426.

Waage, J. K., and Ng, S.M. 1984. The reproductive strategy of a parasitic wasp. I. Optimal progeny allocation in *Trichogramma evanescens. Journal of Animal Ecology* **53**, 401–415.

Wäckers, F. L., and Lewis, W. J. 1993. Olfactory and visual learning and their interaction in host site location by *Microplitus croceipes. Biological control* (in press).

Wago, H., and Kitano, H. 1985. Effects of the venom from *Apanteles glomeratus* on the hemocytes and hemolymph of *Pieris rapae crucivora. Applied Entomology and Zoology* **20**, 103–110.

Wajnberg, E.; Prévost, G.; and Boulétreau, M. 1985. Genetic and epigenetic variation in *Drosophila* larvae suitable to a hymenopterous endoparasitoid. *Entomophaga* **30**, 187–190.

Wajnberg, E.; Pizzol, J.; and Babault, M. 1989. Genetic variation in progeny allocation in *Trichogramma maidis. Entomologia Experimentalis et Applicata* **53**, 177–187.

Wajnberg, E.; Boulétreau, M.; Prévost, G.; and Fouillet, P. 1990. Developmental relationships between *Drosophila* larvae and their endoparasitoid *Leptopilina* (Hymenoptera: Cynipidae) as affected by crowding. *Archives of Insect Biochemistry and Physiology* **13**, 239–245.

Walde, S. J., and Murdoch, W. W. 1988. Spatial density dependence in insect parasitoids. *Annual Review of Entomology* **33**, 441–466.

Walker, I. 1959. Die Abwehrreaktion des Wirtes *Drosophila melanogaster* gegen die zoophage Cynipidae *Pseudeucoila bochei* Weld. *Revue Suisse de Zoologie* **66**, 569–632.

Walker, I. 1962. *Drosophila* und *Pseudeucoila*. II. Selektionsversuche zur Steigerung der Resistenz des Parasitens gegen die Abwehrreaktion des Wirtes. *Revue Suisse de Zoologie* **69**, 209–227.

Walker, I. 1967. Effect of population density on the viability and fecundity in *Nasonia vitripennis* Walker [Hymenoptera, Pteromalidae]. *Ecology* **48**, 294–301.

Waloff, N., and Jervis, M. A. 1987. Communities of parasitoids associated with leaf-hoppers and planthoppers in Europe. *Advances in Ecological Research* **17**, 281–376.

Walter, G. H. 1983a. "Divergent male ontogenies" in Aphelinidae (Hymenoptera: Chalcidoidea): A simplified classification and a suggested evolutionary sequence. *Biological Journal of the Linnean Society* **19**, 63–82.

Walter, G. H. 1983b. Differences in host relationships between male and female heter-onomous hyperparasitoids (Aphelinidae: Chalcidoidea): A review of host loca-tion, oviposition and pre-imaginal physiology and morphology. *Journal of the Entomological Society of South Africa* **46**, 261–282.

Walter, G. H. 1986. Suitability of a diphagous parasitoid, *Coccophagus bartletti* An-necke and Insley (Hymenoptera, Aphelinidae), for sex ratio studies: Ovipositional and host-feeding behaviour. *Journal of the Entomological Society of South Africa* **49**, 141–152.

Wardle, A. R. 1990. Learning of host microhabitat colour by *Exeristes roborator* (F.) (Hymenoptera: Ichneumonidae). *Animal Behaviour* **39**, 914–923.

Wardle, A. R., and Borden, J. H. 1985. Age-dependent associative learning by *Exeristes roboratur* (F.) (Hymenoptera: Ichneumonidae). *Canadian Entomologist* **117**, 605–616.

Wardle, A. R., and Borden, J. H. 1990. Learning of host microhabitat form by *Exeristes roborator* (F.) (Hymenoptera: Ichneumonidae). *Journal of Insect Behaviour* **3**, 251–263.

Washburn, J. O., and Cornell, H. V. 1979. Chalcid parasitoid attack on a gall wasp population (*Acraspis hirta* [Hymenoptera: Cynipidae]) on *Quercus prinus* (Faga-ceae). *Canadian Entomologist* **111**, 391–400.

Webb, B. A., and Summers, M. D. 1990. Venom and viral expression products of the endoparasitic wasp *Campoletis sonorensis* share epitopes and related sequences. *Proceedings of the National Academy of Sciences, USA* **87**, 4961–4965.

Weis, A. E. 1982a. Resource utilization patterns in a community of gall-attacking para-sitoids. *Environmental Entomology* **11**, 809–815.

Weis, A. E. 1982b. Use of a symbiotic fungus by the gall maker *Asteromyia carbonifera* to inhibit attack by the parasitoid *Torymus capite*. *Ecology* **63**, 1602–1605.

Weis, A. E. 1983. Patterns of parasitism by *Torymus capite* on hosts distributed in small patches. *Journal of Animal Ecology* **52**, 867–878.

Weis, A. E., and Abrahamson, W. G. 1985. Potential selective pressures by parasitoids on a plant-herbivore interaction. *Ecology* **66**, 1261–1269.

Weis, A. E.; Price, P. W.; and Lynch, M. 1983. Selective pressures on the clutch size in the gall maker *Asteromyia carbonifera*. *Ecology* **64**, 688–695.

Weis, A. E.; Abrahamson, W. G., and McCrea, K. D. 1985. Host gall size and oviposi-tion success by the parasitoid *Eurytoma gigantea*. *Ecological Entomology* **10**, 341–348.

Weis, A. E.; McCrea, K. D.; and Abrahamson, W. G. 1989. Can there be an escalating arms race without coevolution? Implications from a host-parasitoid simulation. *Evolutionary Ecology* **3**, 361–370.

Wellings, P. W. 1991. Host location and oviposition on animals. In W. J. Bailey and J. Ridsdill-Smith, eds., *Reproductive Behaviour of Insects,* pp. 75–107. Chapman and Hall, Melbourne.

Wellings, P. W.; Morton, R.; and Hart P. J. 1986. Primary sex-ratio and differential progeny survivorship in solitary haplo-diploid parasitoids. *Ecological Entomology* **11**, 341–348.

Wells, A. 1992. The first parasitic Trichoptera. *Ecological Entomology* **17**, 299–302.

van Welzen, C.R L., and Waage, J. K. 1987. Adaptive responses to local mate competition by the parasitoid *Telenomus remus. Behavioural Ecology and Sociobiology* **21**, 359–367.

Werren, J. H. 1980. Sex ratio adaptations to local mate competition in a parasitic wasp. *Science* **208**, 1157–1159.

Werren, J. H. 1983. Sex ratio evolution under local mate competition in a parasitic wasp. *Evolution* **37**, 116–124.

Werren, J. H. 1984a. A model for sex ratio selection in parasitic wasps: Local mate competition and host quality effects. *Netherlands Journal of Zoology* **34**, 81–96.

Werren, J. H. 1984b. Brood size and sex ratio regulation in the parasitic wasp *Nasonia vitripennis* (Walker) (Hymenoptera: Pteromalidae). *Netherlands Journal of Zoology* **34**, 123–143.

Werren, J. H. 1987a. Labile sex ratios in wasps and bees. *Bioscience* **37**, 498–506.

Werren, J. H. 1987b. The coevolution of autosomal and cytoplasmic sex ratio factors. *Journal of Theoretical Biology* **124**, 317–334.

Werren, J. H. 1991. The paternal-sex-ratio chromosome of *Nasonia. American Naturalist* **137**, 392–402.

Werren, J. H. 1993. The evolution of inbreeding in haplodiploid organisms. In N. Thornhill, ed., *The Natural History of Inbreeding and Outbreeding: Theoretical and Empirical Perspectives*, pp. 42–59. University of Chicago Press, Chicago.

Werren, J. H., and van den Assem, J. 1986. Experimental analysis of a paternally inherited extrachromosomal factor. *Genetics* **114**, 217–233.

Werren, J. H., and Beukeboom, L. 1993. Population genetics of a parasitic chromosome. I. Theoretical analysis of *psr* in subdivided populations. *American Naturalist* (in press).

Werren, J. H., and Charnov, E. L. 1978. Facultative sex ratios and population dynamics. *Nature* **272**, 349–350.

Werren, J. H., and Simbolotti, G. 1989. Combined effects of host quality and local mate competition on sex allocation in *Lariophagus distinguendus. Evolutionary Ecology* **3**, 203–213.

Werren, J. H.; Skinner, S. W.; and Charnov, E. L. 1981. Paternal inheritance of a daughterless sex ratio factor. *Nature* **293**, 467–468.

Werren, J. H.; Skinner, S. W.; and Huger, A. M. 1986. Male-killing bacteria in a parasitic wasp. *Science* **231**, 990–992.

Werren, J. H.; Nur, U.; and Eickbush, D. 1987. An extrachromosomal factor causing loss of paternal chromosomes. *Nature* **327**, 75–76.

Werren, J. H.; Nur, U.; and Wu, C.-I. 1988. Selfish genetic elements. *Trends in Ecology and Evolution* **3**, 297–302.

Weseloh, R. M. 1976a. Behavioural responses of the parasite, *Apanteles melanoscelus*, to gypsy moth silk. *Environmental Entomology* **5**, 1128–1132.

Weseloh, R. M. 1976b. Behavior of forest insect parasitoids. In J. F. Anderson and H.K. Kaya, eds., *Perspectives in Forest Entomology*, pp. 99–110. Academic Press, New York.

Weseloh, R. M. 1976c. Dufour's gland: Source of sex pheromone in a hymenopterous parasitoid. *Science* **193**, 695–697.

Weseloh, R. M. 1977. Effects on behavior of *Apanteles melanoscelus* females caused by modifications in extraction, storage, and presentation of gypsy moth silk kairomones. *Journal of Chemical Ecology* **3**, 723–735.

Weseloh, R. M. 1981. Host location by parasitoids. In D. A. Nordlund, R. L. Jones and W. J. Lewis, eds., *Semiochemicals, Their Role in Pest Control*, pp. 79–96. John Wiley, New York.

West, C. 1985. Factors underlying the late seasonal appearance of the lepidopterous leaf-mining guild on oak. *Ecological Entomology* **10**, 111–120.

Wharton, R. A. 1988. Classification of the braconid subfamily Opiinae (Hymenoptera). *Canadian Entomologist* **120**, 333–360.

White, H. C., and Grant, B. 1977. Olfactory cues as a factor in frequency-dependent mate selection in *Mormoniella vitripennis*. *Evolution* **31**, 829–835.

Whitfield, J. B. 1990. Parasitoids, polydnaviruses and endosymbiosis. *Parasitology Today* **6**, 381–384.

Whitham, T. G. 1980. The theory of habitat selection: Examined and extended using *Pemphigus* aphids. *American Naturalist* **115**, 449–466.

Whiting, A. R. 1961. Genetics of *Habrobracon*. *Advances in Genetics* **10**, 333–406.

Whiting, A. R. 1967. The biology of the parasitic wasp *Mormoniella vitripennis*. *Quarterly Review of Biology* **42**, 333–406.

Whiting, P. W. 1939. Sex determination and reproductive economy in *Habrobracon*. *Genetics* **24**, 110–111.

Whiting, P. W. 1943. Multiple alleles in complementary sex determinatio of *Habrobracon*. *Genetics* **28**, 365–382.

Whiting, P. W.; Greb, R. J.; and Speicher, B. R. 1934. A new type of sex intergrade. *Biological Bulletin* **66**, 152–165.

Whittaker, P. L. 1984. The insect fauna of mistletoe *(Phoradendron tomentosum* Loranthaceae) in Southern Texas. *Southwestern Naturalist* **29**, 435–444.

Wiackowski, S. K. 1962. Studies on the biology and ecology of *Aphidius smithi* Sharma and Subba Rao (Hymenoptera, Braconidae), a parasite of the pea aphid, *Acyrthosiphon pisum* (Harr.) (Homoptera, Aphididae). *Polskie Pismo Entomologiczne* **32**, 253–310.

Wiklund, C., and Fagerström, T. 1977. Why do males emerge before females? A hypothesis to explain the incidence of protandry in butterflies. *Oecologia* **31**, 153–158.

Wilkes, A. 1947 The effects of selective breeding on the laboratory propagation of insect parasites. *Proceedings of the Royal Society of London B* **134**, 227–245.

Wilkes, A. 1963. Environmental causes of variation in the sex ratio of an arrhenotokous insect, *Dahlbominus fuliginosus*. *Canadian Entomologist* **95**, 20–47.

Willers, D.; Lehmann-Danzinger, H.; and Führer, E. 1982. Antibacterial and antimycotic effect of a newly discovered secretion from larvae of an endoparasitic insect, *Pimpla turionellae*. *Archives of Microbiology* **133**, 225–231.

Williams, G. C. 1966. Natural selection, the costs of reproduction and a refinement of Lack's principle. *American Naturalist* **100**, 687–690.

Williams, G. C. 1975. *Sex and Evolution*. Princeton University Press, Princeton.

Williams, G. C. 1979. The question of adaptive variation in sex ratio in out-crossed vertebrates. *Proceedings of the Royal Society of London B* **205**, 567–580.

Williams, H. J.; Elzen, G. W.; and Vinson, S. B. 1988. Parasitoid-host-plant interactions emphasizing cotton (*Gossypium*). In P. Barbosa and D. Letourneau, eds., *Novel Aspects of Insect-Plant Allelochemicals and Host Specificity,* pp. 171–200. John Wiley, New York.

Williams, J. R. 1972. The biology of *Physcus seminotus* Silv., and *P. subflavus* Annecke and Insley (Aphelinidae), parasites of the sugar-cane scale insect *Aulacaspis tegalensis* (Zhnt.) (Diaspidae). *Bulletin of Entomological Research* **61**, 463–484.

Williams, J. R. 1977. Some features of sex-linked hyperparasitism in Aphelinidae (Hymenoptera). *Entomophaga* **22**, 345–350.

Williams, T. 1991. Host selection and sex ratio in a heteronomous hyperparasitoid. *Ecological Entomology* **16** 377–386.

Wilson, D. S., and Colwell, R. K. 1981. Evolution of sex ratio in structured demes. *Evolution* **35**, 882–897.

Wilson, E. O. 1971. *The Insect Societies*. Belknap Press of Harvard University Press, Cambridge, Massachusetts.

Wilson, F. 1961. Adult reproductive behaviour in *Asolcus basalis* (Hymenoptera: Scelionidae). *Australian Journal of Zoology* **9**, 739–751.

Wilson, K., and Lessells, C. M. 1993. Evolution of clutch size in insects. I. A review of static optimality models. *Journal of Evolutionary Biology* (in press).

Wing, M. W. 1951. A new genus and species of myrmecophilous Diapriidae with taxonomic and biological notes on related forms (Hymenoptera). *Transactions of the Royal Entomological Society of London* **102**, 195–210.

Wong, H. R. 1974. The identification and origin of the strains of the larch sawfly, *Pristiphora erichsonii* (Hymenoptera: Tenthredinidae), in North America. *Canadian Entomologist* **106**, 1121–1131.

Worthley, H. N. 1924. The biology of *Trichopoda pennipes* Fab. (Diptera, Tachinidae), a parasite of the common squash bug. *Psyche* **31**, 7–16.

Woycke, J. 1965. Genetic proof of the origin of drones from fertilised eggs of the honeybee. *Journal of Apicultural Research* **4**, 7–11.

Woycke, J. 1976. Population genetic studies on sex alleles in the honeybee using the example of the Kangaroo Island bee sanctuary. *Journal of Apicultural Research* **15**, 105–123.

Woycke, J. 1979. Sex determination in *Apis cerana indica*. *Journal of Apicultural Research* **18**, 122–127.

Wylie, H. G. 1965. Discrimination between parasitized and unparasitized housefly pupae by females of *Nasonia vitripennis* (Walk.) (Hym.: Pteromalidae). *Canadian Entomologist* **97**, 279–286.

Wylie, H. G. 1966. Some mechanisms that affect the sex ratio of *Nasonia vitripennis* (Walk.) (Hymenoptera: Pteromalidae) reared from superparasitized housefly pupae. *Canadian Entomologist* **98**, 645–653.

Wylie, H. G. 1967. Some effects of host size on *Nasonia vitripennis* and *Muscidifurax raptor* (Hymenoptera: Pteromalidae). *Canadian Entomologist* **99**, 742–748.

Wylie, H. G. 1970. Oviposition restraint of *Nasonia vitripennis* (Walk.) (Hym.: Pteromalidae) on hosts parasitized by another hymenopterous species. *Canadian Entomologist* **102**, 886–894.

Wylie, H. G. 1973. Control of egg fertilization by *Nasonia vitripennis* (Hymenoptera: Pteromalidae) when laying on parasitized housefly pupae. *Canadian Entomologist* **105**, 709–718.

Wylie, H. G. 1976. Interference among females of *Nasonia vitripennis* (Hymenoptera, Pteromalidae) and its effect on sex ratio of the progeny. *Canadian Entomology* **108**, 655–661.

Wylie, H. G. 1979. Sex ratio variability of *Muscidifurax zaraptor* (Hymenoptera, Pteromalidae). *Canadian Entomology* **111**, 105–109.

Yamada, Y. 1988. Optimal use of patches by parasitoids with a limited fecundity. *Researches on Population Ecology* **30**, 235–250.

Yasumatsu, K. 1937. Some observations on *Thalessa citraria* (Olivier) and *Thalessa superbiens* Morley. *Akitu* **1**, 33–43.

Yeargan, K. V., and Braman, S. K. 1986. Life history of the parasite *Diolcogaster facetosa* (Weed) (Hymenoptera: Braconidae) and its behavioral adaptation to the defensive response of a lepidopteran host. *Annals of the Entomological Society of America* **79**, 1029–1033.

Yeargan, K. V., and Braman, S. K. 1989. Life-history of the hyperparasitoid *Mesochorus discitergus* (Hymenoptera, Ichneumonidae) and tactics used to overcome the defensive behavior of the green cloverworm (Lepidoptera, Noctuidae). *Annals of the Entomological Society of America* **82**, 393–398.

Yokoyama, S., and Nei, M. 1979. Population dynamics of sex-determining alleles in honey bees and self-incompatibility alleles in plants. *Genetics* **91**, 609–626.

Young, S. Y., and Yearian, W. C. 1990. Transmission of nuclear polyhedrosis virus by the parasitoid *Microplitis croceipes* (Hymenoptera: Braconidae) to *Heliothis virescens* (Lepidoptera: Noctuidae) on soybean. *Environmental Entomology* **19**, 251–256.

Zareh, N.; Westoby, M.; and Pimentel, D. 1980. Evolution in a laboratory host-parasitoid system and its effect on population kinetics. *Canadian Entomologist* **112**, 1049–1060.

Zaslavsky, V. A., and Umarova, T. Y. 1981. Photoperiodic and temperature control of diapause in *Trichogramma evanescens* Westw. (Hymenoptera, Trichogrammatidae). *Entomological Review* **60**, 1–12.

Zchori-Fein, E., Roush, R. T.; and Hunter, M. S. 1992. Female production induced by antibiotic treatment in *Encarsia formosa*. *Experientia* **48**, 102–105.

Zinna, G. 1961. Ricerche sugli insetti entomofagi. II. Specializzazione entomoparassitica negli Aphelinidae: Studio morfologico, etologico e fisiologico del *Coccophagus bivittatus* Compere, nuova parassita del *Coccus hesperidum*. L. per l'Italia. *Bollettino del Laboratorio di Entomologia Agraria "Filippo Silvestri"* **19**, 301–358.

Zinna, G. 1962. Ricerche sugli insetti entomofagi. III. Specializzazione entomoparassitica negli Aphelinidae: Interdipendenze biocenotiche tra due specie associate. Studio morfologico, etologico e fisiologico del *Coccophagoides similis* (Masi) e

Azotus matritensis Mercet. *Bollettino del Laboratorio di Entomologia Agraria "Filippo Silvestri"* **20**, 73–182.

Zwölfer, H. 1971. The structure and effect of parasite complexes attacking phytophagous host insects. In P. J. den Boer and G. R. Gradwell, eds., *Dynamics of Populations: Proceedings of the Advanced Study Institute on "Dynamics of Numbers in Populations" (Oosterbeck, 1970)*, pp. 405–418. Centre for Agricultural Publishing and Documentation, Wageningen, The Netherlands.

Zwölfer, H. 1975. Speciation and niche diversification in phytophagous insects. *Verhandlungen der Deutschen Zoologischen Gesellschaft* **67**, 394–401.

Zwölfer, H. 1979. Strategies and counterstrategies in insect population systems competing for space and food in flower heads and plant galls. *Fortschritte der Zoologie* **25**, 331–353.

Zwölfer, H., and Kraus, M. 1957. Biocoenotic studies on the parasites of two fir- and two oak-tortricids. *Entomophaga* **2**, 173–196.

Author Index

Index to senior authors and to junior authors in papers with two authors.

Subject Index

adelphoparasitoids. *See* heteronomous aphelinids

adult size, effect on fitness, 109–114

aggregation, 61–65, 79–82

Allee effect: evolution of gregariousness, 123–125; evolution of thelytoky, 218

allelic-diversity sex determination, 153

allelochemicals, 30

allomone, 30

antennation, 49, 144, 198

antimicrobial secretions, 242

ants: defensive morphs, 38, 291; mimicry of, 294, 300–301; parasitism of, 38, 41; parasitism of guests, 34, 291

apparency, 332–333

apparent competition, 350–352

aptery, 283–284, 292–298

aquatic parasitoids, 15–16

area-restricted searching, 50

arrestant chemicals, 30

Arsenophonous. See bacteria

asymmetric composition responses, 104

asymmetric density responses, 104

attractant chemicals, 30

autoparasitoids, *see* heteronomous aphelinids

B chromosomes, 213, 221

bacteria: as incompatibility microorganisms (*Wolbachia*), 217; as sex ratio distorters (*Arsenophonous*), 216–217; causing thelytoky (*Wolbachia*), 218–221

balanced mortality hypothesis, 309–314

behavioral ecology research programme, 4–5; criticisms of, 180–182

behavioral mechanisms: clutch size, 119–121; host acceptance, 94–96; host location, 59–61, 62, 65–74; sex ratio, 184–189, 198–199; superparasitism, 143–146

bet-hedging, 74

carniveroid, 6

castration, 251

cellular defense response. *See* encapsulation

character displacement, 343

cleptoparasitism, 9, 33

clutch size: change after mating, 222–224; correlation with duration of oviposition, 141; and encapsulation, 239–240; evolution in host, 77; experimental studies, 106–119; genetic variation in, 222–224; and sex ratio, 167, 178–182; in solitary parasitoids, 125–126; theory, 99–106

co-cladogenesis, 355–356

coevolution: ecological, 349–350; host defenses and parasitoid countermeasures, 244–248; spatial distribution, 77

coexistence of parasitoid species, 339–346

Coleoptera, as parasitoids, 20–21, 41

community ecology, 321–363

comparative studies: clutch size, 124; host acceptance, 48–50; life histories, 319–321; mating frequency, 277–278; mating strategies, 271–272; sex ratio, 169–170

competition: apparent, 350–352; between gregarious larvae, 104–106, 190, 257–258; for mates, 261, 267–271 (*see also* local mate competition and fighting); between parasitoid species, 339–346; between solitary larvae, 122–125, 255–258; after superparasitism, 135–137

competitive exclusion, 342–343

complementary sex determination, 153–156

conditional sex expression, 192

constrained sex allocation: causes of, 280–282; importance of, 206–207; pheromone production, 276; predicted sex ratio, 204–206; selection to remove constraint, 206–207

cost of reproduction: in clutch size theory, 101–103; experimental studies, 114–116

counter-balanced competition, 340

countermeasures. *See* parasitoid countermeasures (against host defenses)

courtship. *See* mating

crystal cell, 232

CsV (*Campoletis sonorensis* virus), 242–243

cultural inheritance, 43

cuticle: in host defense, 288; in parasitoid defense, 298

cyclical parthenogenesis, 23–24, 297

cytology: use of, in identifying sex, 203

Taxonomic Index

The species of hosts and parasitoids mentioned in the book are listed in this index together with their family and order affiliations. The abbreviations used for orders are Hym.: Hymenoptera; Col.: Coleoptera; Hem.: Hemiptera (i.e., Homoptera plus Heteroptera); Lep.: Lepidoptera; Dip.: Diptera; Neu.: Neuroptera. I have attempted to use modern nomenclature, but in some cases this has not been possible, for example, when a correct determination would require examination of voucher material. Where the name I use differs from that in the original publication, the old name is given in brackets after an equal sign.